Traveling Wave Antennas

CARLTON H. WALTER

Professor Emeritus of Electrical Engineering

The Ohio State University

QP PENINSULA PUBLISHING

WESTPORT, CONNECTICUT USA

Traveling Wave Antennas

Copyright © by Carlton H. Walter

Reprint Edition published by:
Peninsula Publishing
Westport, Connecticut, USA

E-mail: sales@PeninsulaPublishing.com
Telephone: 203-292-5621
Website: http://www.PeninsulaPublishing.com

Library of Congress Catalog Number 89-064104

ISBN-10: 0-932146-51-1

ISBN-13: 978-0932146-51-9

Printed in the United States of America

To my wife
SALLY
and
my son
CARL

PREFACE

The primary objective of this book is to present under one cover much of the material on traveling wave antennas that previously had been found only in reports, journals and handbooks. *Traveling Wave Antennas* was first published in 1965 by McGraw Hill and was followed several years later by a Russian edition. In 1970 Dover republished the book with a few minor corrections. Now in 1990, *Traveling Wave Antennas* is being republished again, this time by Peninsula Publishing, by virtue of renewed interest in "skin" antennas for radar, communications, navigation and countermeasures systems in high performance military aircraft.

A traveling wave antenna implies a continuous structure that is long in terms of wavelength. It is frequently an entirely different structure from a conventional stub, dipole, reflector, or array of such elements. However, since almost any radiating source of finite size and many arrays can be represented by traveling waves, the material in this book is more general than the title may imply. The emphasis is on continuous and quasi-continuous sources. Although conventional arrays of slot or dipole elements can be treated as traveling wave sources, such antennas are quite well known and details are not given here.

The general subjects covered in the book are analysis, synthesis, design, excitation, and applications of traveling wave antennas. The first chapter reviews antenna fundamentals and defines the terms that are used in the text. Chapter 2 presents analytical methods for determining the fields of an antenna when the source distribution is known, and Chapter 3 treats the inverse problem of determining a source distribution that will produce a specified field.

The problem of designing a physical structure to produce a specified source distribution is covered in Chapter 4, and design data for a number of sources are given in Chapters 5 and 6 and also in chapter 8, where practical traveling wave antennas are described. The problem of traveling wave excitation is treated in Chapter 7.

In skimming the book, one may get the impression that the treatment is theoretical and at a high level because of the large number of equations, some of which look rather complicated. The material has been taught at The Ohio State University at the first-year graduate level following undergratuate fields and antennas courses and the first graduate fields course. Although the results of some of the more advanced mathematical techniques such as saddle point integration and the Wiener-Hopf method are used, these techniques are perhaps best covered in advanced fields courses. Details of such methods are not given here; little more than differential and integral calculus, complex variables, and vector analysis are needed to handle most of the material in this book. With discretion on the part of the instructor and with elaboration of Chapter 1, *Traveling Wave Antennas* can be used as a first text on antennas at either the advanced undergraduate level or first-year graduate level.

The book was not written primarily as a textbook. On the contrary, it started out to be a handbook on traveling wave antennas, but the inclusion of analytical methods for determining design data and the opportunity to teach the material which produced pressure for more details led to the present content which should be useful to researchers, development engineers, and students.

It is impossible to give proper credit to all who helped in the preparation of this book. I tried to acknowledge correctly the works of the many engineers that make up this book. Unfortunately I could acknowledge only those works familiar to me, so inevitablility there were omissions and I apologize for this.

I am grateful for the Air Force contracts out of Wright Field that sponsored The Ohio State University work described in this book, and the understanding and encouragement of the Antenna Laboratory Directors at that time, C. A. Levis and R. A. Fouty, are deeply appreciated.

Finally, I am especially thankful to Dr. T. E. Tice, who encouraged me to start this book; to my EE870 students, who provided many interesting discussions and helped to correct mistakes; and to Martha, who typed both the rough draft and final manuscript.

Carlton H. Walter

CONTENTS

Contents

Contents

CHAPTER 1
INTRODUCTION AND DEFINITIONS

1-1. INTRODUCTION

Antennas are studied from the point of view that a traveling wave source is a fundamental radiating element. The currents and fields (the source distribution) of any continuous antenna structure may be expressed in terms of traveling waves just as a wave form in circuit theory is expressed as a series of sine waves by Fourier analysis. In this respect the material that is presented here may be considered as "traveling wave" analysis.

The primary aim in presenting the material that follows is to provide the necessary theoretical and experimental background for the analysis, synthesis, and design of any radiating system that can be represented by one or more traveling waves.

The emphasis in the text is on continuous sources (such as long wire antennas, polyrod antennas, and aperture antennas) and quasi-continuous sources (such as arrays of closely spaced nonresonant slots or holes in waveguides). A conventional array of resonant slots or dipoles with half-wavelength or greater spacing often can be satisfactorily represented by a continuous source. Likewise, a continuous source can be approximated by an array.[1] Certain aspects of arrays of discrete elements are introduced at var-

---

[1] Illustrations of the representation of continuous sources by arrays are given in Sec. 2-18. The representation of an array by a continuous source is illustrated in Sec. 3-2G.

ious places in the text. However, conventional array theory is well covered in the literature[1] and is not given comprehensive treatment here.

1-2.   DEFINITIONS AND ANTENNA FUNDAMENTALS

Some antenna fundamentals as well as definitions of some of the principal terms used in the following chapters will now be presented. Other terms will be defined as they are used in the text. Where terms are not defined, the reader is referred to IEEE (formerly IRE) standards. Throughout the text the rationalized mks system of units will be used.

Antenna

An antenna as used in electromagnetic radiation may be defined as the transition region between free space and a guiding device or transmission line. The guiding device transports electromagnetic energy to or from the antenna. In the former case we have a "transmitting" antenna, and in the latter a "receiving" antenna. An antenna can and does take on many sizes and shapes. It may be a conductor such as a piece of wire, it may be a rod of dielectric, it may be a slot, or it may be an open-ended waveguide, to mention a few. A list of practical antennas would be quite large.

Radiation Pattern

The antenna property with which we shall be most concerned is the radiation pattern. The power that an observer receives from an antenna by means of a test probe is generally a function of the observer's position with respect to the antenna. A graph of the received power at a constant radius is called the *power pattern* of the antenna. It is the spatial distribution of radiated energy. The spatial variation of the electric (or magnetic) field, on the other hand, is called a *field pattern*.

Radiation Intensity

The power per unit area in the field of an antenna is known as the *power density*. Multiplying the power density by the square of the radial distance from the antenna gives the power per unit solid angle, which is called *radiation intensity*.

Directivity

The ratio of the radiation intensity in a particular direction to the average radiation intensity of the antenna is the directivity of the antenna in

[1] J. D. Kraus, "Antennas," chaps. 4 and 11, McGraw-Hill, New York, 1950; S. Silver, "Microwave Antenna Theory and Design," chap. 9, McGraw-Hill, New York, 1949; E. C. Jordan, "Electromagnetic Waves and Radiating Systems," chap. 12, Prentice-Hall, Englewood Cliffs, N.J., 1950; H. Jasik (ed.), "Antenna Engineering Handbook," chaps. 5 and 9, McGraw-Hill, New York, 1961.

that direction. Unless stated otherwise, the direction is taken to be that of maximum radiation. In terms of the radiation intensity $U(\theta,\phi)$, the directivity D is given by

$$D = \frac{4\pi U_{max}(\theta,\phi)}{\int U(\theta,\phi)\,d\Omega} \tag{1-1}$$

where $d\Omega$ ($= \sin\theta\,d\theta\,d\phi$) is an element of solid angle.

Gain

The usual figure of merit of an antenna is its *gain*. Antenna gain is defined as the ratio of the maximum radiation intensity to the maximum radiation intensity of some reference antenna with the same power input. The reference antenna is generally understood to be an isotropic source, in which case we have the *absolute gain* of the antenna. If some other reference such as a dipole is used, it should be clearly stated.

Antenna gain includes the efficiency of the antenna and is the product of the directivity and the efficiency. While directivity is frequently computed, gain is usually measured by comparison with a reference antenna such as a dipole or horn. This gives relative gain. If the absolute gain of the reference antenna is known, as it usually is, then the absolute gain of the test antenna can be readily determined.

Polarization

Polarization is simply the orientation of the electric field of an antenna. The principal polarizations are horizontal and vertical taken with reference to the Earth's surface.

Many antennas are designed to radiate (or receive) either vertical or horizontal polarization (or some orientation of electric field between the horizontal and the vertical). These are called *linearly polarized* antennas. If the antenna responds to two orthogonal field components with some time phase between the field components, the antenna is said to be *elliptically polarized*. For the special case where the magnitudes of the two orthogonal components are equal and the phase angle is ± 90 deg, the antenna is *circularly polarized*. The polarization is said to be *right-circular polarization* if the electric field is rotating in the clockwise direction when the electromagnetic wave is observed from the antenna as it travels away from the antenna. Conversely, counterclockwise rotation for a receding wave corresponds to *left-circular polarization*.[1] Polarization generally is a function of direction from the antenna, but careful design can usually produce the desired polarization characteristic.

---

[1] IRE Standards on Radio Wave Propagation (definition of terms), p. 2, 1942, Supplement to *Proc. IRE*, **30**(7), part III.

Impedance

Impedance is the ratio of voltage to current at a pair of terminals or the ratio of the appropriate components of electric and magnetic fields at a point. One can represent an antenna by its Thévenin equivalent circuit as illustrated in Fig. 1-1b or the dual in Fig. 1-1c. The impedance Z_a is the impedance at the antenna terminals with no load attached and with all generators replaced by impedances equal to the internal impedances of the generators. The voltage V is the voltage across the antenna terminals with no load attached. In general the impedance Z_a will be complex and it will include antenna loss. For a lossless antenna the real part of Z_a is the radiation resistance of the antenna. Antenna impedance, per se, will be of little concern in this text, since most practical traveling wave antennas can be readily matched to conventional transmission lines. In fact in many cases the traveling wave antenna structure is a simple perturbation of conventional waveguide.

Antenna Aperture

Some antennas such as horns and reflectors have well-defined physical apertures. Any antenna, however, when used as a receiving antenna will collect a certain amount of energy from an incident electromagnetic wave. Thus the antenna can be thought of as having an aperture. If W is the power (watts) delivered to a load impedance at the antenna terminals and P_i is the power density (watts per square meter) incident upon the antenna, the *effec-*

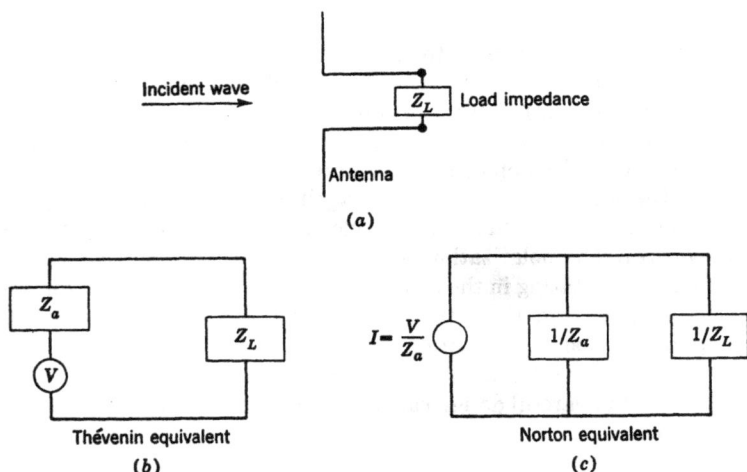

Figure 1-1. Antenna and equivalent circuits.

tive aperture of the antenna is defined as

$$A_e = \frac{W}{P_i} \tag{1-2}$$

The effective aperture depends upon the load impedance and the polarization and direction of the incident wave. For the general case where polarizations and impedances are not matched, Tai[1] has evaluated Eq. (1-2) as

$$A_e = pq \frac{\lambda^2 D(\theta,\phi)}{4\pi} \tag{1-3}$$

where $D(\theta,\phi)$ = directivity of the antenna in the direction of the incident field

q = impedance mismatching factor
p = polarization mismatching factor
λ = free space wavelength

The quantity q is found to be

$$q = \frac{4R_a R_L}{(R_a + R_L)^2 + (X_a + X_L)^2} \tag{1-4}$$

where $Z_L = R_L + jX_L$ = load impedance
$Z_a = R_a + jX_a$ = antenna impedance defined at the load terminals

The quantity p is given by

$$p = \frac{|\mathbf{h} \cdot \mathbf{E}^i|^2}{|\mathbf{h}|^2 |\mathbf{E}^i|^2} \tag{1-5}$$

where $\mathbf{h} = h_\theta \theta + h_\phi \phi$ = vector effective height of the antenna[2]

$\mathbf{E}^i = E_\theta \theta + E_\phi \phi$ = incident electric field

and boldface denotes vector quantities. Tai[3] has shown that for a randomly polarized incident field the average value of p is 0.5.

The vector height $\mathbf{h}$ is related to the far field of the antenna by

$$\mathbf{E}(r,\theta,\phi,t) = \frac{-j\omega\mu_0 I_0 \mathbf{h}(\theta,\phi)}{4\pi r} e^{j(\omega t - kr)} \tag{1-6}$$

where $\mathbf{E}$ is the far field of the antenna with current I_0 flowing into the antenna terminals. The usual spherical coordinates (r,θ,ϕ) are used, μ_0 is the permeability of free space, ω is the angular frequency of the field which is taken to vary sinusoidally with time, and k is the free-space phase constant $(2\pi/\lambda)$.

[1] C. T. Tai, On the Definition of the Effective Aperture of Antennas, *IRE Trans. Antennas Propagation*, **AP-9**(2): 224, March, 1961.

[2] G. Sinclair, The Transmission and Reception of Elliptically Polarized Waves, *Proc. IRE*, **38**:148–151, February, 1950.

[3] C. T. Tai, *loc. cit.*

The vector height is quite useful in that it is defined so that the open circuit voltage V at the antenna terminals is simply

$$V = \mathbf{E}^i \cdot \mathbf{h} \tag{1-7}$$

For the special case where the polarization of the antenna is matched to that of the incident wave ($p = 1$), the load impedance is matched to the antenna ($q = 1$), and the incoming wave is in the direction of maximum directivity of the antenna, Eq. (1-3) reduces to

$$A_e = \frac{\lambda^2 D}{4\pi} \tag{1-8}$$

Equation (1-8) is the usual expression for effective aperture,[1] but it should be noted that it is the optimum case.

The effective aperture is in many cases of the same order as the physical area (physical aperture), particularly for large-aperture antennas. For pyramidal horns the effective aperture is approximately 50 to 80 per cent of the physical aperture. For parabolic-reflector antennas the effective aperture is generally 50 to 65 per cent of the physical aperture. In large dipole-reflector arrays the effective aperture can be equal to or even somewhat greater than the physical area of the array.

Although effective aperture and physical aperture are the most widely used aperture concepts in antenna work, two other apertures, *scattering aperture* and *loss aperture*, are also used.[2] The *scattering aperture* is defined as the ratio of the power reradiated from an antenna to the power density of the incident wave. The *loss aperture* is defined as the ratio of the power lost in the antenna to the power density of the incident wave. All three apertures (effective, scattering, and loss) can be added together to give a total aperture called the *collecting aperture*.[2]

Fundamental Source of Radiation

The fundamental source of radiation will be taken to be a time-varying current. This does not rule out radiation from a field distribution. We shall merely replace by equivalent currents any fields from which radiation is considered to take place. The tangential magnetic field at a surface of radiation (the aperture of a horn, for example) is related to an equivalent electric surface current $\mathbf{J}$ by the relation

$$\mathbf{J} = \mathbf{n} \times \mathbf{H} \tag{1-9}$$

where $\mathbf{H}$ is the magnetic field and $\mathbf{n}$ is a unit vector (dimensionless) normal to the surface and pointing into the region in which radiation takes place.

[1] IRE Standards on Antennas, 1948.
[2] J. D. Kraus, "Antennas," chap. 3, McGraw-Hill, New York, 1950.

If **H** is expressed in amperes per meter, then **J** also will be expressed in amperes per meter. In a similar manner the tangential electric field may be related to a surface current by the relation

$$\mathbf{K} = \mathbf{E} \times \mathbf{n} \tag{1-10}$$

where **E** is the electric field and **K** is magnetic surface current. The quantities **E** and **K** will have dimensions of volts per meter. We do not know of any actual magnetic currents (flow of magnetic charge), but the magnetic current defined by Eq. (1-10) is quite useful. The use of equivalent currents is discussed in more detail under Huygens' Principle in Chap. 2.

There are perhaps some instances in which it would be convenient to represent a radiating system by means of a volume current distribution. However, in the material that follows we shall work in terms of surface currents and line currents.

Line Sources

A line source of electric current is defined as an electric conductor of infinitesimal cross section carrying current **I**. Similarly a line of magnetic current is defined as a magnetic conductor of infinitesimal cross section carrying magnetic current **M**. The fields of line sources are relatively easy to determine, and since a continuous current may be represented as a sum of infinitesimal line elements, the line current concept is very useful. The details of determining the fields of line current sources are discussed in Chap. 2.

Point Sources

If the volume occupied by a time-varying current is infinitesimally small, we say that we have a *point source*. The concept of point source is quite useful in array theory and is extended to include any antenna, regardless of size and complexity, where the observation point is far enough from the antenna that the antenna fields appear to emanate from a point.

Reciprocity

The principle of reciprocity states that the response of a system is unchanged when source and measurer are interchanged. As applied to antennas this means that the transmitting and receiving patterns of an antenna are the same.

Reciprocity may be demonstrated by referring to Fig. 1-2. Antenna 1 represents the antenna we wish to analyze or synthesize. Antenna 2 represents a small test antenna that may be used as a receiving antenna when No. 1 is transmitting or as a transmitting antenna when No. 1 is receiving. The two antennas and the region through which coupling takes place constitute a two-port network as represented in Fig. 1-2b. If an equivalent two-port network can be made up of linear, passive, bilateral elements, then in

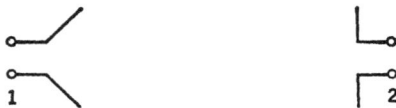

(a) Two-antenna system

Figure 1-2. Antenna system as a two-port network.

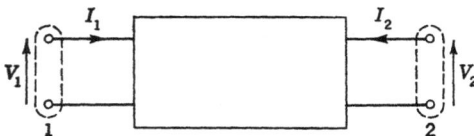

(b) Equivalent two-port network

terms of the usual circuit notation one can readily show that

$$\frac{V_1}{I_2}\bigg|_{I_1=0} = Z_{12} \tag{1-11}$$

$$\frac{V_2}{I_1}\bigg|_{I_2=0} = Z_{21} \tag{1-12}$$

$$Z_{12} = Z_{21} \tag{1-13}$$

where V and I are voltage and current, respectively, and Z_{12} and Z_{21} are the mutual or transfer impedances of the system. If the antennas are far apart, the self-impedances

$$Z_{11} = \frac{V_1}{I_1}\bigg|_{I_2=0} \tag{1-14}$$

and

$$Z_{22} = \frac{V_2}{I_2}\bigg|_{I_1=0} \tag{1-15}$$

become the conventional antenna impedances and the mutual impedance becomes the conventional antenna pattern. Thus, Eq. (1-13) is a statement of reciprocity. The interpretation of an antenna system in terms of circuit theory concepts is a very powerful tool and is discussed in Chap. 3, where use is made of Fourier and Laplace transform theory.

Reciprocity also can be established in terms of field quantities by starting with Maxwell's equations.[1] Manipulation of Maxwell's curl equations for two sets of fields ($\mathbf{E}_1,\mathbf{H}_1$ and $\mathbf{E}_2,\mathbf{H}_2$) and corresponding sources ($\mathbf{J}_1,\mathbf{K}_1$ and $\mathbf{J}_2,\mathbf{K}_2$) gives

$$-\nabla \cdot (\mathbf{E}_1 \times \mathbf{H}_2 - \mathbf{E}_2 \times \mathbf{H}_1) = \mathbf{E}_1 \cdot \mathbf{J}_2 + \mathbf{H}_2 \cdot \mathbf{K}_1$$
$$- \mathbf{E}_2 \cdot \mathbf{J}_1 - \mathbf{H}_1 \cdot \mathbf{K}_2 \tag{1-16}$$

where $\mathbf{J}_1,\mathbf{K}_1$ and $\mathbf{J}_2,\mathbf{K}_2$ are source currents and can be taken to be equivalent

[1] See, for example, R. F. Harrington, "Time-Harmonic Electromagnetic Fields," p. 116, McGraw-Hill, New York, 1961.

currents on surfaces surrounding antennas 1 and 2, respectively (more on this in the discussion of Huygens' principle in Chap. 2). Fields E_1 and H_1 are produced by antenna 1 (sources J_1 and K_1), and fields E_2 and H_2 are produced by J_2 and K_2. At any point where sources do not exist, Eq. (1-16) reduces to

$$\nabla \cdot (E_1 \times H_2 - E_2 \times H_1) = 0 \qquad (1\text{-}17)$$

which is known as the Lorentz reciprocity theorem. Integrating Eq. (1-17) throughout a source-free region and applying the divergence theorem give the integral form of the Lorentz reciprocity theorem:

$$\int (E_1 \times H_2 - E_2 \times H_1) \cdot ds = 0 \qquad (1\text{-}18)$$

If the source regions are included, integration of Eq. (1-16) yields

$$-\int_S (E_1 \times H_2 - E_2 \times H_1) \cdot ds$$
$$= \int (E_1 \cdot J_2 - H_1 \cdot K_2 - E_2 \cdot J_1 + H_2 \cdot K_1)\, dv \qquad (1\text{-}19)$$

Letting surface S become a sphere of infinite extent while the sources are confined to a volume of finite extent, the surface integral vanishes and Eq. (1-19) reduces to

$$\int_{\text{Vol. source 2}} (E_1 \cdot J_2 - H_1 \cdot K_2)\, dv$$
$$= \int_{\text{Vol. source 1}} (E_2 \cdot J_1 - H_2 \cdot K_1)\, dv \qquad (1\text{-}20)$$

This is a very useful form of the reciprocity theorem. Although the notation is that for the two-antenna system in Fig. 1-2, the equation applies to any two sets of sources of finite extent. The integrals in Eq. (1-20) have been given the name *reaction* by Rumsey.[1] By definition the reaction of field a on source b is

$$<a,b> = \int (E_a \cdot J_b - H_a \cdot K_b)\, dv \qquad (1\text{-}21)$$

In reaction notation reciprocity is expressed by

$$<a,b> = <b,a> \qquad (1\text{-}22)$$

that is, the reaction of field a on source b is equal to the reaction of field b on source a.

If one considers the volumes covered by the integrals in Eq. (1-20) to include just the antenna terminal regions as indicated by the dotted lines in

[1] V. H. Rumsey, The Reaction Concept in Electromagnetic Theory, *Phys. Rev.*, ser. 2, **94**(6):1483–1491, June 15, 1954.

Fig. 1-2*b*, then Eq. (1-20) reduces to

$$V_1^{oc}I_2 = V_2^{oc}I_1 \qquad\qquad (1\text{-}23)$$

or

$$\frac{V_1^{oc}}{I_1} = \frac{V_2^{oc}}{I_2} \qquad\qquad (1\text{-}24)$$

The quantity V^{oc} means open circuit voltage; that is, I_1 flowing into the terminals of antenna 1 produces voltage V_1^{oc} across the open terminals of antenna 2. It can be seen that Eq. (1-24) is the same as Eq. (1-13).

Traveling Wave

A *traveling wave* is a disturbance (electromagnetic in this case) that propagates with a definite phase velocity v. That is, in order for a point on the disturbance to maintain a constant phase, it must travel with velocity v. Perhaps the simplest example is an unattenuated, monochromatic plane wave in an unbounded, homogeneous medium. If the wave has amplitude A, angular frequency ω ($\omega = 2\pi \times$ frequency f), and is traveling in the positive y direction, it may be represented by

$$F(y) = A e^{j(\omega t - \beta y)} \qquad\qquad (1\text{-}25)$$

Unless stated otherwise, the time convention $e^{j\omega t}$ will be implied. Thus the plane wave may be expressed simply as

$$F(y) = A e^{-j\beta y} \qquad\qquad (1\text{-}26)$$

The quantity β is called the phase constant. It is given in terms of the wavelength (distance between two consecutive equal phase points) as

$$\beta = \frac{2\pi}{\text{wavelength}} \qquad\qquad (1\text{-}27)$$

The phase velocity of the wave is given by

$$v = \frac{\omega}{\beta} \qquad\qquad (1\text{-}28)$$

In the case of a wave traveling in free space the phase velocity v is equal to c, the velocity of light in free space (approximately 3×10^8 m/sec). The free-space wavelength is denoted λ, and the free-space phase constant is

$$k = \frac{2\pi}{\lambda} \qquad\qquad (1\text{-}29)$$

The representation of a plane wave in free space traveling at an arbitrary angle with respect to a rectangular coordinate system as shown in Fig. 1-3 is

$$F(x,y,z) = A e^{-jk(x \sin \eta \cos \xi + y \sin \eta \sin \xi + z \cos \eta)} \qquad\qquad (1\text{-}30)$$

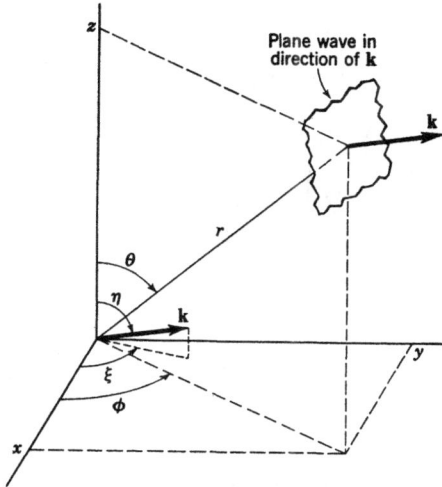

Figure 1-3.   Plane wave.

For a plane wave propagating in the y direction and polarized in the z direction the electric field is

$$\mathbf{E} = \mathbf{z}E_0 e^{-jky} \tag{1-31}$$

The corresponding magnetic field has only an x component which is equal to $E_0 e^{-jky}/120\pi$, where 120π is the intrinsic impedance of free space.

A traveling wave must satisfy Maxwell's equations and thus the wave equation. This is discussed in detail in Chap. 2, where it is found that the wave equation for a wave propagating in free space is given by

$$\nabla^2 \mathbf{E} + k^2 \mathbf{E} = 0 \tag{1-32}$$

The plane wave given by Eq. (1-30) satisfies Eq. (1-32). A cylindrical wave as illustrated in Fig. 1-4 also satisfies Eq. (1-32). A solution of Eq. (1-32) in cylindrical coordinates involves Hankel functions where the Hankel function $H_0^{(2)}(k\rho)$ corresponds to an outward traveling wave. The electric field for the wave in Fig. 1-4 is given by

$$\mathbf{E} = \mathbf{z}E_0 H_0^{(2)}(k\rho) \tag{1-33}$$

and the magnetic field is

$$\mathbf{H} = \frac{\boldsymbol{\phi} E_0}{j\omega\mu} \frac{d}{d\rho} H_0^{(2)}(k\rho) \tag{1-34}$$

For large $k\rho$ the asymptotic behavior of $H_0^{(2)}$ gives

$$H_0^{(2)}(k\rho) \approx \sqrt{\frac{2}{\pi k\rho}}\, e^{-j(k\rho - \pi/4)} \tag{1-35}$$

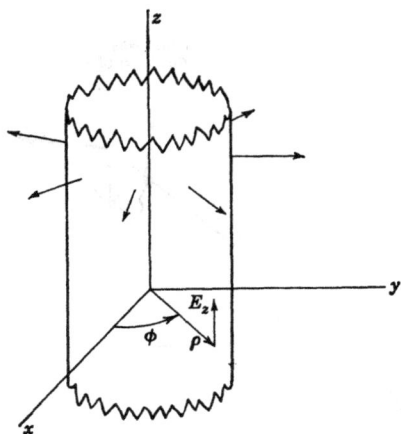

Figure 1-4.   Cylindrical wave.

Thus in the limit as $\rho \rightarrow \infty$,

$$\frac{E_z}{-H_\phi} = 120\pi \tag{1-36}$$

That is, the relationship between **E** and **H** for a cylindrical wave at large ρ approaches that for the plane wave case. The electric and magnetic fields in Eqs. (1-33) and (1-34) would be produced by an infinite line of electric current along the z axis. A line of magnetic current would produce the same field configuration with **E** and **H** and μ and ϵ interchanged. More will be said of the duality of electric and magnetic current sources in Chap. 2.

A spherical traveling wave, as illustrated in Fig. 1-5, is also a solution of the wave equation. The radial variation is of the form e^{-jkr}/r. This would be produced by an isotropic radiator. If we attempt, however, to define an isotropic radiator of electromagnetic energy, we run into an ambiguity with

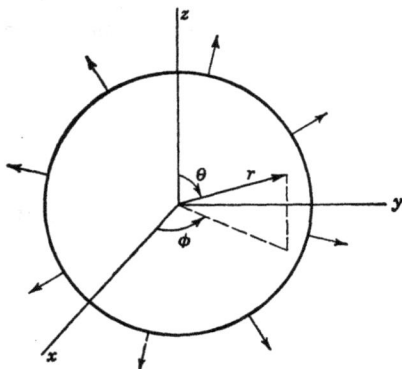

Figure 1-5.   Spherical wave.

the field vectors at $\theta = 0$ and π. Strictly speaking, the isotropic spherical wave is a scalar wave (a spherical wave of sound, for example). However, we frequently refer to an isotropic source in antenna work, especially as a reference for specifying gain.

Traveling Wave Antenna

A *traveling wave antenna* is defined as an antenna for which the fields and currents that produce the antenna pattern may be represented by one or more traveling waves, usually in the same direction. Examples of early traveling wave antennas are the long wire and rhombic antennas.

If the structure supporting a traveling wave is properly terminated, the reflected wave will be quite small and essentially a traveling wave distribution will exist. A standing wave antenna, however, can be considered as a traveling wave antenna with waves traveling in opposite directions. The pattern of a half-wave dipole has been obtained in this manner.[1]

Other examples of traveling wave antennas are the polyrod, long slots in waveguides, helical antennas, and various surface wave antennas. Aperture antennas such as horns and parabolic reflectors can be treated as traveling wave antennas. If the main beam of radiation is normal to the aperture, as it usually is, the wave travels over the aperture with infinite phase velocity.

To many engineers a traveling wave antenna is synonymous with radiation from a continuous source. Actually many arrays of discrete elements can be satisfactorily approximated by a traveling wave current or field distribution. Examples are Yagi-Uda arrays and arrays of slots and holes in waveguides. If the elements are closely spaced ($<\lambda/2$), the structure may be referred to as a quasi-continuous source.

Nonuniform Waves

A wave that represents a source of electromagnetic radiation generally is not so simple in form as the free-space waves described above. Consider a line of electric current I along the z axis. In general I will be a complex quantity having amplitude and phase and may be expressed as

$$I(z) = |I(z)|e^{j\psi(z)} \tag{1-37}$$

where time variation $e^{j\omega t}$ is assumed. A complex current as represented by Eq. (1-37) can be interpreted as a wave with nonuniform amplitude and phase velocity. Such a wave may be represented as a sum of waves having uniform phase velocities. That is, the wave may be represented as

$$I(z) = \sum_n A_n e^{-j\beta_n z} \tag{1-38}$$

where the individual waves have a constant amplitude A_n and a constant

[1] Kraus, *op. cit.*, chap. 5.

phase velocity ω/β_n. In general, A_n will be complex. The wave $I(z)$ also may be represented in terms of damped waves as

$$I(z) = \sum_n B_n e^{-\gamma_n z} \qquad (1\text{-}39)$$

where γ_n is a complex propagation constant. The propagation constant γ is given by

$$\gamma = \alpha + j\beta \qquad (1\text{-}40)$$

The quantity α is the attenuation constant, which is expressed in nepers per meter where an attenuation of 1 neper/m reduces the traveling wave by a factor of $1/e$ over a distance of 1 m. More will be said on the subject of representing nonuniform waves in Chap. 2.

The propagation constant γ is a convenient quantity with which to work. It is readily determined theoretically and experimentally for many traveling wave structures. Data in the form of equations or graphs giving α and β as functions of antenna geometry are found in Chaps. 5 and 6. These data provide the basis for antenna design as described in Chap. 4. A particular value of γ may be associated with a pole location on the complex γ plane. This is quite useful in pattern synthesis and is discussed further in Chap. 3.

Fast and Slow Waves

As pointed out above, a traveling wave with angular frequency ω and phase constant β has a phase velocity $v = \omega/\beta$. A traveling wave may be classed as a fast or slow wave depending on whether v is greater than or less than the velocity c of a plane wave in free space. Thus a wave having $c/v < 1$ is called a *fast wave* and a wave having $c/v > 1$ is called a *slow wave*.

Leaky Wave

A *leaky wave* is a traveling wave that continuously loses energy owing to radiation. Generally the fields decay along the structure in the direction of propagation and build up in the direction away from the structure for such a wave, but there are exceptions to the decay along the structure in cases such as tapered amplitude sources and also active antennas where amplification is distributed along the radiating structure (see Prob. 6-9). Most leaky wave antennas are fast wave structures. A number of practical leaky wave structures are described in Chap. 5.

Surface Wave

Barlow and Brown[1] define a surface wave simply as the wave propagating along an interface between two different media without radiation. It is distinguished from the leaky wave by the fact that there is not a continuous

---

[1] H. M. Barlow and J. Brown, "Radio Surface Waves," chap. 1, Oxford, Fair Lawn, N.J., 1962.

radiation of energy but rather the surface wave is bound to the surface and radiation takes place only at curvatures, nonuniformities, and discontinuities. Radiation is taken to mean energy converted from the surface wave field to some other form.

The chief characteristics of a surface wave are that its phase velocity is less than that of the surrounding medium, usually free space, and for a plane uniform structure the fields decay exponentially away from the structure. Power flow is parallel to the structure except when losses are involved, in which case there is some power flow into the structure. A surface wave is sometimes referred to as a *trapped wave*.

The classifications of leaky wave and surface wave are usually applied to uniform structures. Nonuniformities in the structure can cause one type of wave to behave like the other. For example, a curved or modulated surface wave structure may have energy leakage and thus behave like a leaky wave structure.

Continuous Source versus Array

By a *continuous source* we mean a radiating structure which at the macroscopic level has no discontinuities or interruptions except at the ends of the radiating portion of the structure. Thus the real or equivalent currents that produce the radiation field, mathematically speaking, can be said to be continuous, single valued, and finite over the interval corresponding to the extent of the source. A polyrod antenna is an example of a continuous source. Figure 1-6*a* illustrates the sort of amplitude that one might find for an

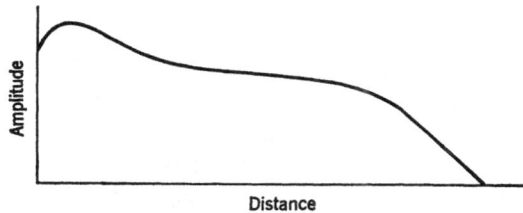

(*a*) Continuous source

Figure 1-6.   Comparing a continuous source with an array.

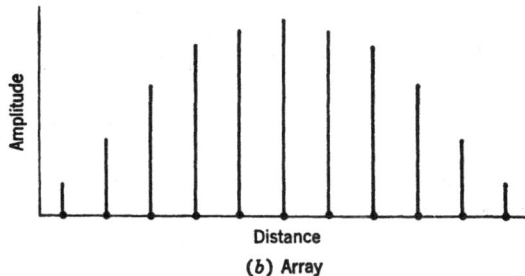

(*b*) Array

$$E(\theta) \sim \int_0^L A(l)e^{j\psi(l)}\, e^{jkl\cos\theta}\, dl$$

(a) Continuous source

$$E(\theta) \sim \sum_{n=0}^N A_n e^{j\psi_n}\, e^{jnkd\cos\theta}$$

(b) Uniformly spaced array

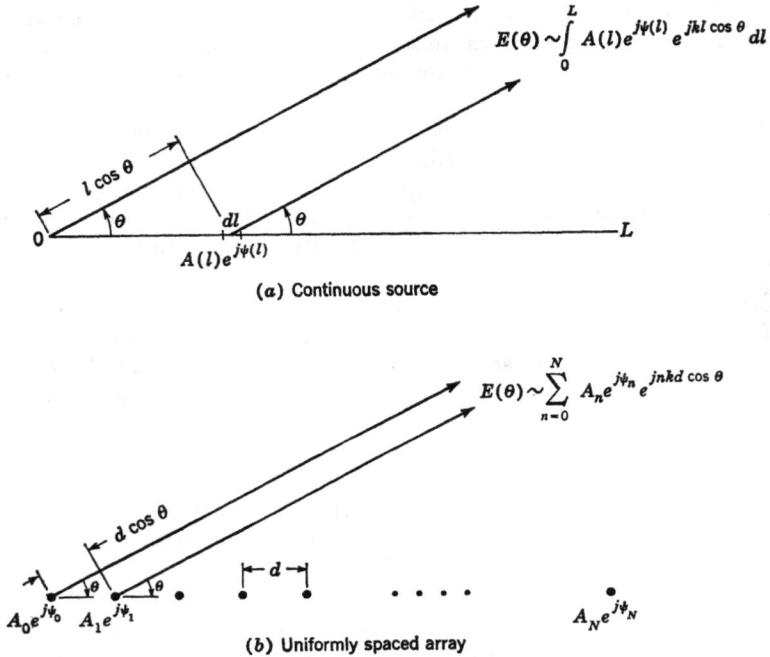

Figure 1-7.   Comparison of radiation from a continuous source and an array.

equivalent current along the polyrod. As we shall see in Chap. 2, the field of such a continuous source is in the form of an integral. This is illustrated in Fig. 1-7a, where the integral sums up the contributions from the elements of the source.

An *array*, on the other hand, is a group of radiating elements spaced some distance apart and with the current on each element having a particular amplitude and phase. An example of an amplitude distribution along an array is shown in Fig. 1-6b. The field of the array is obtained by superimposing the fields of the individual elements. This leads to a series representation for the field as illustrated in Fig. 1-7b. This is in contrast to the integral representation for the continuous source. Other than their physical differences this is the main distinction between a continuous source and an array, and for small element spacings (one-half wavelength or less) even this distinction tends to disappear, as the series and the integral are approximately equal for array-element amplitudes equal to the continuous distribution at corresponding points along the two sources. Thus array theory can be applied to continuous sources and traveling wave theory (for continuous sources) can be applied to arrays. In Chap. 5 we shall see that some structures that are taken to be continuous sources are actually arrays of closely spaced radiating elements. Such an array may be called a quasi-continuous source.

Endfire and Broadside Sources

Two terms that are frequently used in antenna work are *endfire* and *broadside*. An *endfire source* is a source of linear extent for which maximum radiation is along the linear axis. A dielectric rod (polyrod) is an example of an endfire source. Principal radiation is along the axis of the rod. An endfire source generally has a phase velocity less than the velocity of light in free space; that is, $c/v > 1$.

A *broadside source*, on the other hand, is one for which maximum radiation is normal to the source, which is usually a line or a planar aperture. Examples of broadside planar apertures are horn antennas and parabolic reflectors. A broadside source is characterized by an infinite phase velocity over the aperture; that is, $c/v = 0$.

The preceding definitions and discussion will suffice to introduce the subject of traveling wave antennas. Further definitions are required, of course, but it is deemed more appropriate to present them as required throughout the text of the material that follows.

ADDITIONAL REFERENCES

1. J. D. Kraus, "Antennas," chaps. 1 and 2, McGraw-Hill, New York, 1950.
2. S. Silver, "Microwave Antenna Theory and Design," chaps. 1–3, McGraw-Hill, New York, 1949.
3. H. Jasik (ed.), "Antenna Engineering Handbook," chaps. 1, 2, and 16, McGraw-Hill, New York, 1961.
4. M. I. Skolnik, "Introduction to Radar Systems," chap. 7, McGraw-Hill, New York, 1962.

CHAPTER 2

FIELDS OF
A TRAVELING
WAVE SOURCE

2-1. INTRODUCTION

Application of Huygens' principle enables one to represent a source by an equivalent distribution of electric and magnetic currents. The calculation of the fields due to current distributions is discussed, and the special cases of the line source, the rectangular source, and the circular source are considered in some detail. The total field is considered to be made up of three zones: the far field, the Fresnel region, and the near field. These regions are defined, and mathematical techniques appropriate to the different regions are presented.

In many practical cases the evaluation of the integral equation for the fields of a source is extremely difficult. A series approximation of the integral may be used in such cases. This is interpreted physically as replacing a continuous source by an array of discrete radiators.

The method of stationary phase is utilized in determining the direction of maximum radiation of a source, and some consideration is given to the problem of finding the radiated fields of a source on a finite ground plane.

2-2. HUYGENS' PRINCIPLE

Nearly three centuries ago Huygens[1] proposed as an explanation for the bending of light at the edge of an obstacle the rule that each point on a wave front may be regarded as a new source of waves. The adaptation of this

[1] Huygens, "Traité de la lumière," 1690.

(a) Actual system

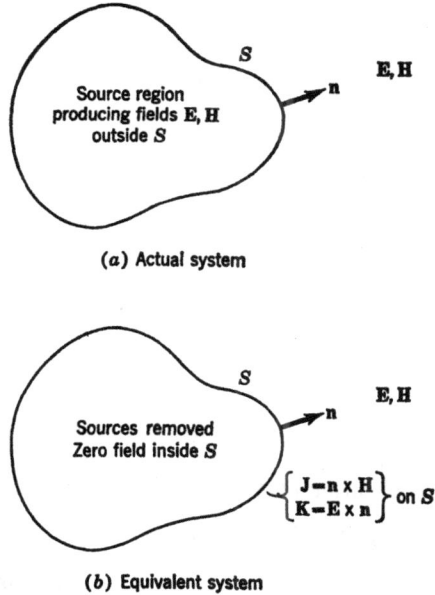

Figure 2-1.   Illustration   of   Huygens'
principle

(b) Equivalent system

principle to electric and magnetic fields is usually referred to as the equivalence principle.[1]

Let the source or sources be contained in a finite volume enclosed by the surface S in a homogeneous isotropic medium, usually free space. According to Huygens' principle we may remove the source and replace it by a current sheet coinciding with surface S. The magnitude and phase of the currents must be such as to produce the same field distribution outside S as existed with the original source. This current distribution is called an *equivalent* current distribution, and the fields outside S may be evaluated in terms of these equivalent currents on S.

The equivalence principle follows from the uniqueness theorem for electromagnetic fields. The uniqueness theorem states[2] that the field in a region is uniquely determined by the sources within the region plus the tangential components of electric field E over a surface S enclosing the region or the tangential components of magnetic field H over S or tangential E over part of S and tangential H over the remainder of S.

Let E,H be the field outside the surface S enclosing the source region in Fig. 2-1. Furthermore, let the field E,H be the original field due to the sources within S. Let the field within S be zero (sources removed). To support this

[1] S. A. Schelkunoff, *Bell System Tech. J.*, January, 1936, pp. 92–112.
[2] For proof of the uniqueness theorem see, for example, R. F. Harrington, "Time-Harmonic Electromagnetic Fields," p. 100, McGraw-Hill, New York, 1961.

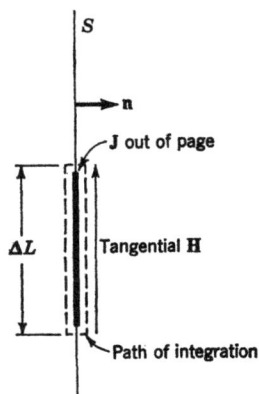

Figure 2-2. Boundary conditions at surface S.

total field (zero within S and $\mathbf{E},\mathbf{H}$ outside S) there must exist surface current densities $\mathbf{J}$ and $\mathbf{K}$ on S to account for the discontinuity in $\mathbf{E}$ and $\mathbf{H}$ at S. Application of Ampère's law at the surface S as illustrated in Fig. 2-2 gives

$$\oint \mathbf{H} \cdot \mathbf{dl} = I = J \, \Delta l \qquad (2\text{-}1)$$

Equation (2-1) tells us that the discontinuity in the tangential magnetic field is equal to the surface electric-current density $\mathbf{J}$. The relationship in vector form is given as

$$\mathbf{J} = \mathbf{n} \times \mathbf{H} \qquad (2\text{-}2)$$

where the unit vector $\mathbf{n}$ points into the source-free region. A similar application of Faraday's law ($\oint \mathbf{E} \cdot \mathbf{dl} = -K \, \Delta l$) gives

$$\mathbf{K} = \mathbf{E} \times \mathbf{n} \qquad (2\text{-}3)$$

The discontinuity in the tangential electric field is equal to a fictitious surface magnetic-current $\mathbf{K}$. The positive sense for $\mathbf{K}$ is chosen to preserve symmetry in Maxwell's equations.

From the uniqueness theorem, the fields of the surface current densities $\mathbf{J}$ and $\mathbf{K}$ will be the postulated fields, that is, the original field $\mathbf{E},\mathbf{H}$ exterior to S and zero interior to S. Thus the original fields exterior to S are given in terms of the tangential components of the fields on the surface S, which is a statement of Huygens' principle. An element of the surface S is called a Huygens source.

Actually it is not necessary to specify zero field inside S. Any other field $\mathbf{E}_i,\mathbf{H}_i$ could have been selected, in which case the surface current densities $\mathbf{J}$ and $\mathbf{K}$ become

$$\mathbf{J} = \mathbf{n} \times (\mathbf{H} - \mathbf{H}_i) \qquad (2\text{-}4)$$
$$\mathbf{K} = (\mathbf{E} - \mathbf{E}_i) \times \mathbf{n} \qquad (2\text{-}5)$$

This Huygens system of equivalent currents radiates into the same region outside S as the original system that produced fields $\mathbf{E},\mathbf{H}$. Also the interior region of S must be the same in all respects as the system that produced fields $\mathbf{E}_i,\mathbf{H}_i$.

In many practical situations a portion of the surface S will consist of an electric conductor. Only tangential $\mathbf{H}$ will exist over this portion, thus only electric current $\mathbf{J}$ will exist on S in this region, and it will be the same as the actual electric current on the conductor. One may think of S as having real electric currents over the conducting portion and equivalent currents $\mathbf{J}$ and $\mathbf{K}$ over aperture portions. However, there is no distinction between the fields of a real current and an equivalent current. Equivalent current $\mathbf{J}$ is treated as if it were real electric current, and likewise $\mathbf{K}$ is treated as if it were real magnetic current.

2-3. SOLUTION OF THE FIELD EQUATIONS IN TERMS OF THE SOURCES

Maxwell's equations may be written to include magnetic current density $\mathbf{m}$ (volts per square meter) and magnetic charge density ρ_m (magnetic charge per cubic meter). For a nondissipative medium and harmonic time variation of the form $e^{j\omega t}$ we have

$$\nabla \times \mathbf{E} = -\mathbf{m} - j\omega\mu\mathbf{H} \tag{2-6}$$

$$\nabla \times \mathbf{H} = \mathbf{i} + j\omega\epsilon\mathbf{E} \tag{2-7}$$

$$\nabla \cdot \mathbf{D} = \rho \tag{2-8}$$

$$\nabla \cdot \mathbf{B} = \rho_m \tag{2-9}$$

where $\mathbf{m}$ and $\mathbf{i}$ are magnetic and electric volume current densities, respectively. For simplicity the time dependence $e^{j\omega t}$ of all fields, currents, and charges will not be shown. The equations of continuity may be written

$$\nabla \cdot \mathbf{i} = -j\omega\rho \tag{2-10}$$

$$\nabla \cdot \mathbf{m} = -j\omega\rho_m \tag{2-11}$$

Equations for $\mathbf{E}$ and $\mathbf{H}$ in terms of the sources $\mathbf{i}$ and $\mathbf{m}$ can be obtained by taking the curl of Eqs. (2-6) and (2-7) and performing an integration based on a vector Green's theorem.[1] However, a more convenient method utilizes potential functions.

If we consider first that only electric currents and charges exist, the electric and magnetic fields are given by

$$\mathbf{E}^e = -\nabla\phi - j\omega\mathbf{A} \tag{2-12}$$

$$\mathbf{H}^e = \frac{1}{\mu}\nabla \times \mathbf{A} \tag{2-13}$$

[1] S. Silver, "Microwave Antenna Theory and Design," p. 80, McGraw-Hill, New York, 1949.

where

$$\phi = \frac{1}{4\pi\epsilon} \iiint \frac{\rho e^{-jkr'}}{r'} \, dv \qquad (2\text{-}14)$$

$$\mathbf{A} = \frac{\mu}{4\pi} \iiint \frac{\mathbf{i} e^{-jkr'}}{r'} \, dv \qquad (2\text{-}15)$$

where r' is the distance from a point in the source region to the point of observation. The quantities ϕ and $\mathbf{A}$ are the usual scalar electric and vector magnetic potentials, respectively.[1] The potentials ϕ and $\mathbf{A}$ are solutions of the nonhomogeneous wave equations

$$\nabla^2 \phi + \omega^2 \mu\epsilon \, \phi = -\frac{\rho}{\epsilon} \qquad (2\text{-}16)$$

$$\nabla^2 \mathbf{A} + \omega^2 \mu\epsilon \, \mathbf{A} = -\mu\mathbf{i} \qquad (2\text{-}17)$$

It can be shown by use of the continuity equation that

$$\phi = \frac{j}{\omega\mu\epsilon} \nabla \cdot \mathbf{A} \qquad (2\text{-}18)$$

thus Eq. (2-12) may be written

$$\mathbf{E}^e = \frac{-j}{\omega\mu\epsilon} \nabla(\nabla \cdot \mathbf{A}) - j\omega\mathbf{A} \qquad (2\text{-}19)$$

In a similar manner one may introduce a vector electric potential

$$\mathbf{F} = \frac{\epsilon}{4\pi} \iiint \frac{\mathbf{m} e^{-jkr'}}{r'} \, dv \qquad (2\text{-}20)$$

and the fields due to magnetic currents will be given by

$$\mathbf{E}^m = -\frac{1}{\epsilon} \nabla \times \mathbf{F} \qquad (2\text{-}21)$$

$$\mathbf{H}^m = \frac{-j}{\omega\mu\epsilon} \nabla(\nabla \cdot \mathbf{F}) - j\omega\mathbf{F} \qquad (2\text{-}22)$$

The total fields are given by

$$\mathbf{E} = \mathbf{E}^e + \mathbf{E}^m = \frac{-j}{\omega\mu\epsilon} \nabla(\nabla \cdot \mathbf{A}) - j\omega\mathbf{A} - \frac{1}{\epsilon} \nabla \times \mathbf{F} \qquad (2\text{-}23)$$

$$\mathbf{H} = \mathbf{H}^e + \mathbf{H}^m = \frac{1}{\mu} \nabla \times \mathbf{A} - \frac{j}{\omega\mu\epsilon} \nabla(\nabla \cdot \mathbf{F}) - j\omega\mathbf{F} \qquad (2\text{-}24)$$

By means of the vector identity

$$\nabla \times \nabla \times \mathbf{A} = \nabla(\nabla \cdot \mathbf{A}) - \nabla^2 \mathbf{A} \qquad (2\text{-}25)$$

[1] J. A. Stratton, "Electromagnetic Theory," p. 23, McGraw-Hill, New York, 1941.

and the wave equation

$$\nabla^2 \mathbf{A} + \omega^2 \mu \epsilon \mathbf{A} = 0 \tag{2-26}$$

applied to both $\mathbf{A}$ and $\mathbf{F}$ we can rewrite Eqs. (2-23) and (2-24) as

$$\mathbf{E} = -\frac{1}{\epsilon} \nabla \times \mathbf{F} - \frac{j}{\omega \mu \epsilon} \nabla \times \nabla \times \mathbf{A} \tag{2-27}$$

$$\mathbf{H} = \frac{1}{\mu} \nabla \times \mathbf{A} - \frac{j}{\omega \mu \epsilon} \nabla \times \nabla \times \mathbf{F} \tag{2-28}$$

Since we need concern ourselves only with surface currents in the application of Huygens' principle, the potentials $\mathbf{A}$ and $\mathbf{F}$ may be written

$$\mathbf{A} = \frac{\mu}{4\pi} \iint \mathbf{J} \frac{e^{-jkr'}}{r'} \, ds \tag{2-29}$$

$$\mathbf{F} = \frac{\epsilon}{4\pi} \iint \mathbf{K} \frac{e^{-jkr'}}{r'} \, ds \tag{2-30}$$

The fields of a current source also can be expressed in terms of a free-space dyadic Green's function[1] G, where G is a solution of the wave equation

$$\nabla^2 G(x,y,z \mid x',y',z') + k^2 G(x,y,z \mid x',y',z') \\ = -(\mathbf{xx} + \mathbf{yy} + \mathbf{zz}) \delta(x - x') \delta(y - y') \delta(z - z') \tag{2-31}$$

where δ is the unit impulse function and $(\mathbf{xx} + \mathbf{yy} + \mathbf{zz})$ is the unit dyadic. The boldface coordinates are unit vectors. The unprimed coordinates are for the point of observation, and the primed coordinates are for the source. Equation (2-31) has the same form as Eq. (2-17) except the source $\mu\mathbf{i}$ is replaced by a unit vector source. By inspection of Eq. (2-15) we can write the free-space Green's function as

$$G(x,y,z \mid x',y',z') = (\mathbf{xx} + \mathbf{yy} + \mathbf{zz}) \frac{e^{-jkr'}}{4\pi r'} \tag{2-32}$$

Each component of the unit vector source gives rise to a vector field. In this case the vector potential $\mathbf{A}$ can be expressed in terms of the free-space Green's function as

$$\mathbf{A}(x,y,z) = \mu \iiint G(x,y,z \mid x',y',z') \cdot \mathbf{i} \, dx' \, dy' \, dz' \tag{2-33}$$

The physical interpretation of Green's function is that it is the response of a system to a unit point source. In the case just considered Green's function gave the vector magnetic potential field due to a unit electric-current element. In general Green's function can be chosen to give the vector

[1] R. E. Collin, "Field Theory of Guided Waves," chap. 2, McGraw-Hill, New York, 1960.

magnetic potential, the electric field, or the magnetic field of a unit electric-current element or the vector electric potential, the electric field, or the magnetic field of a unit magnetic-current element. Once Green's function is known for a particular set of boundary conditions, the vector potential (or electric or magnetic field, depending on Green's function) can be obtained for any source distribution by integrating the product of Green's function and the source function over the volume of the source [as in Eq. (2-33)].

Use of Green's functions permits simplified notation, but in the text that follows we shall use the conventional vector potential functions.

2-4.   EVALUATION OF THE FIELDS OF THE SOURCES

The procedure for determining the fields outside a Huygens surface S is first to evaluate the electric and magnetic surface current densities on S in terms of the tangential magnetic and electric fields. It is assumed that there will be a surface S for which tangential **H** and **E** will be either known or readily determined. The vector potentials are obtained by inserting the expressions for the currents into Eqs. (2-29) and (2-30). Equations (2-27) and (2-28) can be used to evaluate the fields in the region outside S.

For an antenna in the form of an aperture (or apertures) in a conducting surface (a slot antenna, for example) the surface S is usually taken as the surface of the aperture and the conducting surface. Since it is a consequence of Huygens' principle that an equivalent current distribution can produce zero field within S, a sheet of electric conductor may be introduced just inside and infinitesimally close to S. This shorts out contributions from the electric currents **J** on S, and the same radiated field as before is obtained in terms of magnetic currents $\mathbf{K} = \mathbf{E} \times \mathbf{n}$ which exist only over the aperture region where **E** is the electric field that existed in the aperture of the original structure. It must be kept in mind, however, that the radiated fields are those of the distribution **K** in the presence of the conducting surface just within S.

In a similar manner one may introduce a magnetic conducting sheet just within S and find the radiated fields in terms of an electric-current distribution only. In general one may use a magnetic-current sheet **K** over part of S, lining that part with a perfect conductor of electricity, and an electric-current sheet **J** over the remainder of S, lining that part with a perfect conductor of magnetism.

If the conducting surface of the radiating device is an infinite plane, image theory enables us to find the radiated field in terms of a magnetic-current sheet 2**K** in place of the aperture and radiating into half space. This is often a good approximation for a slot antenna on a large ground plane. On the other hand if the antenna is a strip of conductor (such as a wire), the radiated field is found entirely from the electric current **J** on the surface of the conductor. For a polyrod antenna one could use a magnetic-current distribution **K** over the surface of the polyrod and the appropriate **J** and **K** on the feed, although in practice it would be more convenient and usually sufficiently accurate to represent the polyrod by a line source radiator.

Good results for endfire antennas such as the polyrod can be obtained by taking for the Huygens surface the transverse plane at the end of the source.[1] This plane, extending to infinity, encloses the source just as effectively as any other closed surface. For simplicity the fields on the plane usually are taken to be the transverse fields of the dominant mode of the structure. To get good agreement between theory and measurements in this case, it is sometimes necessary to superimpose on the radiation from the Huygens surface a component due to direct radiation from the feed.

Usually the conducting surface on which a traveling wave antenna is mounted is curved and of finite extent. Frequently the curved surface may be approximated by a plane surface. The currents on the rear surface of the structure are generally very small compared with those in the vicinity of the aperture. Thus a good approximation will result in most cases if the surface S over which integration is performed is replaced by a surface S' existing over the aperture and the aperture side of the conducting surface. Calculations are further simplified if S' is a plane surface.

The computation of the radiated field of a traveling wave antenna of arbitrary shape is, in general, extremely difficult. Fortunately in many antenna applications simple geometries are used. The line, the rectangle, and the circle are the most common. We shall simplify the problem further by assuming that the antenna is surrounded by free space.

In some instances a great deal of labor can be saved by making use of the dual nature of electric and magnetic sources and fields. This duality can be observed by comparing Eqs. (2-13) and (2-19) with (2-21) and (2-22), respectively. The appropriate interchange of quantities in going from a system of electric sources to a system of magnetic sources, or vice versa, is given in Table 2-1. Thus if one had computed the fields of an electric-current source **J**, the fields of a magnetic-current source of the same form as **J** would be obtained by substituting **K** for **J**, $\mathbf{H}^m$ for $\mathbf{E}^e$, $-\mathbf{E}^m$ for $\mathbf{H}^e$ and interchanging μ and ϵ.

2-5. THE LINE SOURCE

The line source will be taken as a cylinder of current of vanishingly small radius. Nevertheless, in practice, current distributions with transverse dimensions on the order of a half wavelength may be adequately represented by this idealized line source.

Because of the duality that exists between the fields of electric currents and the fields of magnetic currents, we need to solve only for the fields of

[1] J. Brown and J. O. Spector, The Radiating Properties of Endfire Aerials, *Proc. Inst. Elec. Engrs.*, part B, January, 1957, pp. 27–34.

Table 2-1.   Dual   Relations   in   Electric   and
Magnetic Systems

Electric system	Magnetic system
ρ	ρ_m
i	m
J	K
I	M
ϕ	Magnetic scalar potential
	$\left(\dfrac{1}{4\pi\mu} \displaystyle\iiint \rho_m \dfrac{e^{-jkr'}}{r'}\, dv\right)$
$\mathbf{E}^e$	$\mathbf{H}^m$
$\mathbf{H}^e$	$-\mathbf{E}^m$
μ	ϵ
ϵ	μ
σ_e	σ_m
Impedance	Admittance
Admittance	Impedance

one type of source. The fields of the other type are obtained immediately
by interchanging the dual quantities given in Table 2-1.

Let us consider the fields of an electric-current line source. A very
general line source may be represented by

$$I(z',t) = |I(z',t)|e^{j\psi(z',t)} \tag{2-34}$$

where both amplitude $|I(z',t)|$ and phase $\psi(z',t)$ vary with time t and dis-
tance z'. We shall restrict our discussion to a monochromatic source of
angular frequency ω. Thus, Eq. (2-34) may be written as

$$I(z') = |I(z')|e^{j\psi(z')} \tag{2-35}$$

where time variation is understood to be of the form $e^{j\omega t}$.

Let the source and coordinate system be as shown in Fig. 2-3. Primed
coordinates will be used at the source, and unprimed coordinates will be
used at the point of observation P. The vector potential is given by

$$A_s = \frac{\mu}{4\pi} \int_{z_1}^{z_2} I(z') \frac{e^{-jkr'}}{r'}\, dz' \tag{2-36}$$

The resulting magnetic and electric fields are

$$\mathbf{H} = \frac{1}{\mu}\, \nabla \times \mathbf{A} \tag{2-37}$$

$$\mathbf{E} = -\frac{j}{\omega\mu\epsilon}\, \nabla \times \nabla \times \mathbf{A} = -\frac{j}{\omega\epsilon}\, \nabla \times \mathbf{H} \tag{2-38}$$

Because of symmetry there are no variations in the ϕ direction; therefore, without loss of generality the point P can be located in the yz plane

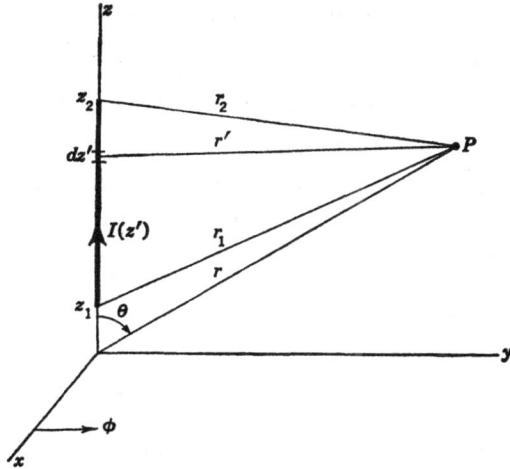

Figure 2-3. Electric-current line source.

(ϕ = 90 deg). The following relations hold:

$$r = (y^2 + z^2)^{1/2} \qquad (2\text{-}39)$$
$$r_1 = [y^2 + (z - z_1)^2]^{1/2} \qquad (2\text{-}40)$$
$$r' = [y^2 + (z - z')^2]^{1/2} \qquad (2\text{-}41)$$
$$r_2 = [y^2 + (z - z_2)^2]^{1/2} \qquad (2\text{-}42)$$

In cylindrical coordinates (ρ,ϕ,z) in which ρ is the radial distance from the z axis, Eq. (2-37) gives

$$H_\phi = -\frac{1}{\mu}\frac{\partial A_z}{\partial \rho} \qquad (2\text{-}43)$$

However, we have restricted our analysis to the yz plane where $\rho = y$; therefore

$$H_\phi = -\frac{1}{\mu}\frac{\partial A_z}{\partial y} \qquad (2\text{-}44)$$

Applying Eq. (2-44) to Eq. (2-36) and interchanging the order of differentiation and integration give

$$H_\phi = -\frac{1}{4\pi}\int_{z_1}^{z_2}\frac{\partial}{\partial y}\left[\frac{I(z')e^{-jkr'}}{r'}\right]dz' \qquad (2\text{-}45)$$

Performing the differentiation,

$$H_\phi = \frac{1}{4\pi}\int_{z_1}^{z_2}\left[\frac{jkyI(z')e^{-jkr'}}{r'^2} + \frac{yI(z')e^{-jkr'}}{r'^3}\right]dz' \qquad (2\text{-}46)$$

No exact solution exists for Eq. (2-46) for an arbitrary source. Numerical integration methods can be used to obtain approximate results; however, these methods will not be discussed here.

There is an exact solution to Eq. (2-46) for the special case

$$I(z') = I_0 e^{-jkz'} \tag{2-47}$$

where I_0 is a constant and k is the phase constant of free space. This is the current distribution of what is called an ordinary endfire source. Such a source tends to radiate along the axis, i.e., off the end of the source, hence the name endfire. The term "ordinary" is applied to the special case where the phase velocity of the wave is equal to the velocity of light in free space.

For this case Eq. (2-46) reduces to

$$H_\phi = \frac{I_0}{4\pi} \int_{z_1}^{z_2} \left[\frac{jkye^{-jk(z'+r')}}{r'^2} + \frac{ye^{-jk(z'+r')}}{r'^3} \right] dz' \tag{2-48}$$

The integrand in Eq. (2-48) is a perfect differential. It can be shown that

$$\frac{d}{dz'} \left[\frac{e^{-jk(z'+r')}}{r'(r'+z'-z)} \right] = \frac{-jke^{-jk(z'+r')}}{r'^2} - \frac{e^{-jk(z'+r')}}{r'^3} \tag{2-49}$$

Performing the integration gives

$$H_\phi = \frac{-I_0 y}{4\pi} \left[\frac{e^{-jk(z_2+r_2)}}{r_2(r_2+z_2-z)} - \frac{e^{-jk(z_1+r_1)}}{r_1(r_1+z_1-z)} \right] \tag{2-50}$$

Making use of the relations $y^2 = r_1{}^2 - (z-z_1)^2$ and $y^2 = r_2{}^2 - (z-z_2)^2$ gives

$$H_\phi = -\frac{I_0}{4\pi y} \left[\left(1 + \frac{z-z_2}{r_2} \right) e^{-jk(z_2+r_2)} - \left(1 + \frac{z-z_1}{r_1} \right) e^{-jk(z_1+r_1)} \right] \tag{2-51}$$

In terms of spherical coordinates, Eq. (2-51) becomes

$$H_\phi = -\frac{I_0}{4\pi r \sin\theta} \left[\left(1 + \frac{r\cos\theta - z_2}{r_2} \right) e^{-jk(z_2+r_2)} \right.$$
$$\left. - \left(1 + \frac{r\cos\theta - z_1}{r_1} \right) e^{-jk(z_1+r_1)} \right] \tag{2-52}$$

where

$$r_2 = (r^2 + z_2{}^2 - 2rz_2 \cos\theta)^{1/2}$$
$$r_1 = (r^2 + z_1{}^2 - 2rz_1 \cos\theta)^{1/2}$$

The electric field can be determined from the relation $\mathbf{E} = (1/j\omega\epsilon)\nabla \times \mathbf{H}$. Thus

$$E_r = \frac{1}{j\omega\epsilon r \sin\theta} \frac{\theta}{\partial\theta} (H_\phi \sin\theta) \tag{2-53}$$

$$E_\theta = \frac{-1}{j\omega\epsilon r} \frac{\partial}{\partial r} (rH_\phi) \tag{2-54}$$

The resulting electric field components of the ordinary endfire electric-current source are

$$
\begin{aligned}
E_r = \frac{I_0}{j\omega\epsilon 4\pi r} &\left\{ \left[\frac{z_2(r\cos\theta - z_2)}{r_2{}^3} + \frac{jkz_2(r\cos\theta - z_2)}{r_2{}^2} \right. \right. \\
&+ \left. \frac{(1 + jkz_2)}{r_2} \right] e^{-jk(z_2+r_2)} - \left[\frac{z_1(r\cos\theta - z_1)}{r_1{}^3} + \frac{jkz_1(r\cos\theta - z_1)}{r_1{}^2} \right. \\
&+ \left. \left. \frac{(1 + jkz_1)}{r_1} \right] e^{-jk(z_1+r_1)} \right\} \quad (2\text{-}55)
\end{aligned}
$$

$$
\begin{aligned}
E_\theta = \frac{-I_0}{j\omega\epsilon 4\pi r \sin\theta} &\left(\left\{ \left[\frac{(r\cos\theta - z_2)}{r_2{}^3} + \frac{jk(r\cos\theta - z_2)}{r_2{}^2} + \frac{jk}{r_2} \right] \right. \right. \\
&\times (r - z_2\cos\theta) - \left. \frac{\cos\theta}{r_2} \right\} e^{-jk(z_2+r_2)} - \left\{ \left[\frac{(r\cos\theta - z_1)}{r_1{}^3} \right. \right. \\
&+ \left. \frac{jk(r\cos\theta - z_1)}{r_1{}^2} + \frac{jk}{r_1} \right] (r - z_1\cos\theta) \\
&- \left. \left. \frac{\cos\theta}{r_1} \right\} e^{-jk(z_1+r_1)} \right) \quad (2\text{-}56)
\end{aligned}
$$

It will be shown later (Sec. 2-9) that these exact expressions for an endfire source can be used to obtain approximate fields for an arbitrary source.

2-6. TRAVELING WAVE REPRESENTATION OF AN ARBITRARY LINE SOURCE

An arbitrary electric-current line source along the z axis may be expressed as

$$
I(z') = |I(z')|e^{j\psi(z')} \quad (2\text{-}57)
$$

where time convention $e^{j\omega t}$ is assumed. One may also express the complex current $I(z')$ in terms of its real and imaginary parts as

$$
I(z') = \operatorname{Re} I(z') + j \operatorname{Im} I(z') \quad (2\text{-}58)
$$

The quantity $I(z')$ may be represented as a discrete spectrum of unattenuated traveling waves by Fourier series analysis, as a continuous spectrum of unattenuated traveling waves by Fourier transform analysis, or as a continuous spectrum of attenuated traveling waves by Laplace transform analysis.

Recall from circuit theory that a periodic time function $f(t)$ of period T may be represented by the exponential Fourier series as

$$
f(t) = \sum_{n=-\infty}^{\infty} c_n e^{jn\omega t} \quad (2\text{-}59)
$$

where the coefficients c_n are given by

$$
c_n = \frac{1}{T} \int_0^T f(t) e^{-jn\omega t} \, dt \quad (2\text{-}60)
$$

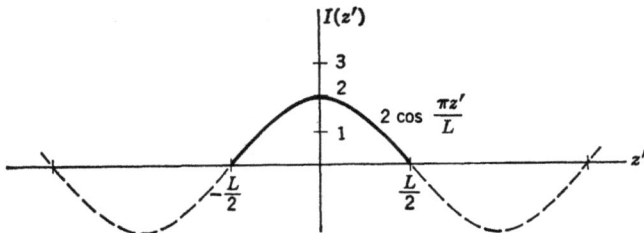

Figure 2-4. Showing $I(z')$ as a periodic function.

Equations (2-59) and (2-60) apply when $f(t)$ represents a finite periodic function with at most a finite number of maxima and minima and a finite number of discontinuities in every finite interval. These are known as the *Dirichlet conditions* and must be satisfied in order for the Fourier series to be a valid representation of $f(t)$.

In an analogous manner the current $I(z')$ may be represented as

$$I(z') = \sum_{n=-\infty}^{\infty} I_n e^{-j\beta_n z'} \tag{2-61}$$

where the coefficient I_n of the unattenuated traveling wave $e^{-j\beta_n z'}$ is given by

$$I_n = \frac{1}{p} \int_{z_1}^{z_1+p} I(z') e^{j\beta_n z'} \, dz' \tag{2-62}$$

For convenience $I(z')$ is made periodic with period p a multiple of source length L. This does not affect the fields of the source $I(z')$ because in the determination of A and then H or E we shall utilize only the section of $I(z')$ that exists over the actual source length L. This is illustrated in Fig. 2-4 for the source function

$$I(z') = 2 \cos \frac{\pi z'}{L} \tag{2-63}$$

The traveling wave representation of $I(z')$ in Eq. (2-63) is

$$I(z') = e^{j\pi z'/L} + e^{-j\pi z'/L} \tag{2-64}$$

The potential A for this case is

$$A_z = \frac{\mu}{4\pi} \left[\int_{-L/2}^{L/2} \frac{e^{-j(kr' + \pi z'/L)}}{r'} \, dz' + \int_{-L/2}^{L/2} \frac{e^{-j(kr' - \pi z'/L)}}{r'} \, dz' \right] \tag{2-65}$$

In general the potential A has a term corresponding to each uniform wave in the series representation for the current, and the electric and magnetic fields will consist of a superposition of the fields due to the individual uniform waves that make up the source.

In the example just given the period p is equal to twice the source length. This is a convenient choice in general, and it gives

$$\beta_n = \frac{n\pi}{L} = n\beta \tag{2-66}$$

Thus Eqs. (2-61) and (2-62) may be written

$$I(z') = \sum_{n=-\infty}^{\infty} I_n e^{-j(n\pi z'/L)} \tag{2-67}$$

$$I_n = \frac{1}{2L} \int_{z_1}^{z_1+2L} I(z') e^{j(n\pi z'/L)} \, dz' \tag{2-68}$$

In comparing Eqs. (2-67) and (2-68) with (2-59) and (2-60) it is seen that time t corresponds to distance z' and angular frequency ω corresponds to negative phase constant $-\beta$. Furthermore

$$\omega = \frac{2\pi}{T} \tag{2-69}$$

where T is the period of the time function $f(t)$ and

$$\beta = \frac{2\pi}{2L} \tag{2-70}$$

where $2L$ is the period that has been chosen for the source function $I(z')$. The quantity $2L$ can be interpreted as the wavelength of the fundamental wave of the source. As we shall observe later, β corresponds to a particular direction in space for the radiated field of the source.

Fourier series, then, enables us to represent an arbitrary line current as a series of traveling waves. The I_n and β_n constitute a discrete spectrum of uniform traveling waves. In general the coefficients I_n will be complex.

On the other hand we can consider $I(z')$ as a nonperiodic function and make use of Fourier transform theory. From circuit theory the Fourier transform pair may be expressed as

$$f(t) = \int_{-\infty}^{\infty} g(\omega) e^{j\omega t} \, d\omega \tag{2-71}$$

and

$$g(\omega) = \frac{1}{2\pi} \int_{-\infty}^{\infty} f(t) e^{-j\omega t} \, dt \tag{2-72}$$

The analogous equations for a current source $I(z')$ are

$$I(z') = \int_{-\infty}^{\infty} F(\beta) e^{-j\beta z'} \, d\beta \tag{2-73}$$

and

$$F(\beta) = \frac{1}{2\pi} \int_{-\infty}^{\infty} I(z') e^{j\beta z'} \, dz' \tag{2-74}$$

In this case $F(\beta)$ constitutes a continuous spectrum of uniform traveling waves. We shall observe later that $F(\beta)$ is proportional to the far-field pattern of the source $I(z')$. In general both $I(z')$ and $F(\beta)$ are complex.

One also may make use of Laplace transform theory for which the transform equations are

$$I(z') = \frac{1}{2\pi j} \int_{c-j\infty}^{c+j\infty} F(\gamma)e^{-\gamma z'}\, d\gamma \tag{2-75}$$

and

$$F(\gamma) = \int_0^\infty I(z')e^{\gamma z'}\, dz' \tag{2-76}$$

This is the usual one-sided Laplace transform pair expressed in terms of the variables z' and γ. The quantity γ is the complex propagation constant defined in Chap. 1, and $F(\gamma)$ constitutes a continuous spectrum of waves whose amplitudes vary exponentially with z'. Equations (2-73) to (2-76) will be considered in more detail in Chap. 3.

2-7.   FAR FIELDS OF A LINE SOURCE

The analysis of an arbitrary source is usually simplified by expressing the source as a series of traveling waves. It is assumed that superposition applies and thus the fields of an arbitrary source can be obtained by summing the fields of the individual currents that make up the source. Unfortunately the only traveling wave source for which we have an exact solution is the ordinary endfire wave ($\beta = k$). Thus traveling wave analysis generally does not give the exact fields of an arbitrary source, but it does work quite well with the usual procedure for dividing the total field of a source into three zones which are commonly referred to as the near field, the Fresnel region, and the far field. The boundaries of these zones are determined by the nature of the approximations in the integral of Eq. (2-46).

The distance r' from an element of the line source to the point of observation is given by Eq. (2-41). Expressing r' in spherical coordinates gives

$$r' = (r^2 - 2rz' \cos\theta + z'^2)^{1/2} \tag{2-77}$$

Equation (2-77) can be expanded, using the binomial theorem, as

$$r' \approx r - z' \cos\theta + \frac{z'^2 \sin^2\theta}{2r} + \frac{z'^3 \cos\theta \sin^2\theta}{2r^2} + \cdots \tag{2-78}$$

where terms above third order in z' have been neglected. The various regions of the total field can be defined by means of Eq. (2-78).

The far-field region is that region in which r is large enough that only the first-order terms in Eq. (2-78) need be considered. The source appears to be a point radiator. The length of the source is significant only in the phase term $e^{-jkr'}$ in which the first two terms of Eq. (2-78) are used. Geometrically this means that the rays r' and r in Fig. 2-3 are parallel. This is strictly true for an arbitrary point P at infinity only, but the region in which this is a satisfactory approximation defines the far-field region. Furthermore, we may neglect the second term in Eq. (2-46) and replace r'^2 in the denominator of the first term by r^2. In neglecting the $1/r'^3$ term we are assuming that

$$r \gg \frac{\lambda}{2\pi} \tag{2-79}$$

This inequality certainly holds true in the far field, and it also will be used in the Fresnel region approximation.

With these usual far-field simplifications and the use of the relation $y = r \sin \theta$, Eq. (2-46) becomes

$$H_\phi = \frac{jk \sin \theta e^{-jkr}}{4\pi r} \int_{z_1}^{z_2} I(z') e^{jkz' \cos \theta} \, dz' \tag{2-80}$$

The integral in Eq. (2-80) is frequently called the *diffraction* integral. It can be evaluated exactly for many practical amplitude and phase distributions.

The far electric field can be obtained from Eqs. (2-7) and (2-80), which give

$$E_\theta = -\frac{1}{j\omega\epsilon r} \frac{\partial}{\partial r} (rH_\phi) = \frac{k}{\omega\epsilon} H_\phi \tag{2-81}$$

The quantity $k/\omega\epsilon$ is the intrinsic impedance of the medium (120π ohms for free space).

In the far field the only significant field components are those that are transverse to the direction of propagation, and it can be shown that Eq. (2-19) reduces to

$$\mathbf{E} = -j\omega\mathbf{A} \tag{2-82}$$

which in the present case may be written

$$E_\theta = -j\omega A_\theta = j\omega \sin \theta A_z \tag{2-83}$$

Equations (2-81) and (2-83) are identical. Therefore, in far-field calculations of an electric-current source it is convenient first to find the electric field from Eq. (2-82) and then determine the magnetic field from the electric field and the impedance of the medium.

The far field of either an electric or magnetic line source of current is an outward traveling spherical wave. The radial distance is large enough that the wave appears to be a plane wave to the observer. The field contains one electric field component and one magnetic field component. The two fields are orthogonal and related by the intrinsic impedance (or admittance) of the medium. The far-field components have a radial variation of the form $1/r$, and the field distribution with respect to θ and ϕ is the same over all spheres centered at the origin; that is, the fields vary with distance in magnitude but not in form. In practice the far-field region is taken to be $r \gtrsim 2L^2/\lambda$ for a source of length L centered at the origin. The physical significance of this distance is that a radiating point source located $2L^2/\lambda$ or more from a receiving source of length L would produce a spherical wave front such that over source length L the phase would vary no more than $22\frac{1}{2}$ deg. For most practical purposes this is an adequate approximation of an incident plane wave. Thus if the receiving source were rotated, its terminal voltage as a function of θ and ϕ would be proportional to the far field of the antenna.

If $k \cos \theta$ is replaced by β in Eq. (2-80) and the limits on the integral changed to $-\infty$ and ∞, we have the Fourier integral of Eq. (2-74) except for a proportionality factor. Thus the far-field or "diffraction" integral is one of a Fourier transform pair. The other integral of the pair [Eq. (2-73)] gives the source function in terms of the far field. This is the integral that is of principal interest in antenna synthesis and will be discussed in detail in Chap. 3.

As an example of the far field of a line source let us consider an electric-current source of the form

$$I(z') = I_0 e^{-\gamma_0 z'} = I_0 e^{-(\alpha_0 + j\beta_0)z'} \tag{2-84}$$

with coordinates as given in Fig. 2-3. This is the form of a wave that exists along a traveling wave structure of uniform cross section. Structures that support waves of this type are described in Chap. 5.

The vector potential for the source in Eq. (2-84) is given by

$$A_z = \frac{\mu I_0}{4\pi} \int_{z_1}^{z_2} \frac{e^{-\gamma_0 z' - jkr'}}{r'} \, dz' \tag{2-85}$$

Utilizing the far-field approximations and taking the θ component of **A** give

$$A_\theta = \frac{-\mu I_0 \sin \theta e^{-jkr}}{4\pi r} \int_{z_1}^{z_2} e^{(jk \cos \theta - \gamma_0)z'} \, dz' \tag{2-86}$$

The electric field is given by Eq. (2-82) as

$$E_\theta = \frac{j\omega\mu I_0 \sin \theta e^{-jkr}}{4\pi r} \int_{z_1}^{z_2} e^{(jk \cos \theta - \gamma_0)z'} \, dz' \tag{2-87}$$

Performing the integration gives

$$E_\theta = \frac{j\omega\mu I_0 \sin \theta e^{-jkr}}{4\pi r} \frac{e^{(jk \cos \theta - \gamma_0)z_2} - e^{(jk \cos \theta - \gamma_0)z_1}}{jk \cos \theta - \gamma_0} \tag{2-88}$$

Replacing z_2 by $z_1 + L$, where L is the length of the source, gives

$$E_\theta = \frac{j\omega\mu I_0 \sin \theta e^{-jkr} e^{(jk \cos \theta - \gamma_0)z_1}}{4\pi r} \frac{e^{(jk \cos \theta - \gamma_0)L} - 1}{jk \cos \theta - \gamma_0} \tag{2-89}$$

Equation (2-89) gives the far field of a traveling wave source of electric current having constant phase velocity and exponential variation in amplitude. For the special case where the amplitude is uniform ($\alpha_0 = 0$) Eq. (2-89) reduces to

$$E_\theta = \frac{j\omega\mu I_0 \sin \theta e^{-jkr} e^{j(k \cos \theta - \beta_0)z_1}}{4\pi r} \frac{e^{j(k \cos \theta - \beta_0)L} - 1}{j(k \cos \theta - \beta_0)} \tag{2-90}$$

Equation (2-90) can be expressed as

$$E_\theta = E_0 \frac{\sin X_0}{X_0} \tag{2-91}$$

where

$$E_0 = \frac{j\omega\mu I_0 L \sin \theta e^{-jkr} e^{j(k \cos \theta - \beta_0)(z_1+L/2)}}{4\pi r} \tag{2-92}$$

and

$$X_0 = \frac{(k \cos \theta - \beta_0)L}{2} \tag{2-93}$$

The phase term $e^{j(k \cos \theta - \beta_0)(z_1+L/2)}$ in Eq. (2-92) appears because of the displacement of the source with respect to the origin of the coordinate system. This term would disappear if the origin were at the center of the source (that is, $z_1 = -L/2$).

The $(\sin X)/X$ function is well known, and some graphs of this function are included in the Appendix. Equation (2-91) in conjunction with the Fourier series expansion in Eq. (2-67) eliminates the integration of an arbitrary source function $I(z')$ in the integral of Eq. (2-80). If the arbitrary source $I(z')$ can be represented by Eq. (2-67), then the far field is given as

$$E_\theta = \sum_{n=-\infty}^{\infty} E_n \frac{\sin X_n}{X_n} \tag{2-94}$$

where E_n and X_n are given by Eqs. (2-92) and (2-93) with I_n and β_n substituted for I_0 and β_0, respectively. The I_n and β_n may be determined from Eqs. (2-68) and (2-66), respectively.

It is important to remember that both I_n and E_n are in general complex. It may be necessary in many cases to use numerical or graphical techniques for determining I_n; however, the problem of determining I_n is not so formidable as it may at first appear, since it usually requires relatively few terms to describe $I(z')$ adequately. There will be instances, of course, where it will be more convenient to use the diffraction integral of Eq. (2-80) rather than Eq. (2-94) to determine the far field. Equation (2-94) is a good choice whenever $I(z')$ can be readily represented by a few terms in Eq. (2-67).

2-8. FRESNEL REGION OF A LINE SOURCE

At distances less than $2L^2/\lambda$ we enter what is called the Fresnel region of the source. In this region the second-order term is included in the expansion of r' [see Eq. (2-78)] for the phase term $e^{-jkr'}$. However, it is still sufficiently accurate to drop the second term in Eq. (2-46) and replace r'^2 in the denominator of the first term by r^2. Thus the Fresnel zone magnetic field of an electric-current source is obtained from Eq. (2-46) as

$$H_\phi = \frac{jk \sin \theta e^{-jkr}}{4\pi r} \int_{z_1}^{z_2} I(z') e^{j\left(kz' \cos \theta - \frac{kz'^2 \sin^2 \theta}{2r}\right)} dz' \qquad (2\text{-}95)$$

In general, numerical methods are required to evaluate Eq. (2-95). For a uniform current distribution Eq. (2-95) reduces to

$$H_\phi = \frac{jkI_0 \sin \theta e^{-jkr}}{4\pi r} \int_{z_1}^{z_2} e^{j\left[(k \cos \theta - \beta_0)z' - \frac{kz'^2 \sin^2 \theta}{2r}\right]} dz' \qquad (2\text{-}96)$$

The exponent in Eq. (2-96) is a quadratic of the form

$$az' - b^2 z'^2 = \left(\frac{a}{2b}\right)^2 - \left(bz' - \frac{a}{2b}\right)^2 \qquad (2\text{-}97)$$

where $a = k \cos \theta - \beta_0$ and $b^2 = (k \sin^2 \theta)/2r$. If we make a change of variable,

$$\sqrt{\frac{\pi}{2}}\, u = bz' - \frac{a}{2b} \qquad (2\text{-}98)$$

Eq. (2-96) becomes

$$H_\phi = \frac{jkI_0 \sin \theta e^{-jkr}}{4\pi r b \sqrt{2/\pi}} e^{j(a/2b)^2} \int_{\sqrt{2/\pi}(bz_1 - a/2b)}^{\sqrt{2/\pi}(bz_1 - a/2b)} e^{-j(\pi/2)u^2} du \qquad (2\text{-}99)$$

The integral in Eq. (2-99) is the standard form of the Fresnel integral. The integral is evaluated as

$$\int_{u_1}^{u_2} e^{-j(\pi/2)u^2} du = C(u_2) - C(u_1) - j[S(u_2) - S(u_1)] \qquad (2\text{-}100)$$

where

$$C(u) = \int_0^u \cos \frac{\pi u^2}{2} du \qquad (2\text{-}101)$$

and

$$S(u) = \int_0^u \sin \frac{\pi u^2}{2} du \qquad (2\text{-}102)$$

Tables of $C(u)$ and $S(u)$ are available in the literature, or one may use a graph of $C(u)$ and $S(u)$ known as Cornu's spiral.[1]

The electric field in the Fresnel region of an electric-current source can be obtained from Eqs. (2-7) and (2-95). Again when all terms with radial variation other than $1/r$ are neglected as in the far-field case, the only significant component is E_θ, and it is related to H_ϕ by the intrinsic impedance of the medium as in the far field. Furthermore, Eq. (2-83) may be used in the Fresnel region to obtain the electric field directly from the vector potential **A**, where **A** has only a z component given by

$$A_z = \frac{\mu e^{-jkr}}{4\pi r} \int_{z_1}^{z_2} I(z') e^{j\left(kz'\cos\theta - \frac{kz'^2\sin^2\theta}{2r}\right)} dz' \qquad (2\text{-}103)$$

Equation (2-96) gives the Fresnel field of a uniform current source. Fortunately it has been possible to evaluate this integral for a source with constant amplitude and arbitrary phase constant. Thus if an arbitrary source $I(z')$ can be expressed as a Fourier series by Eq. (2-67), then the electric or magnetic fields can be found as a series of Fresnel integrals. The electric field of $I(z')$ would be

$$E_\theta = \frac{j\omega\mu \sin\theta e^{-jkr}}{4\pi r b \sqrt{2/\pi}} \sum_n I_n e^{j(a_n/2b)^2}$$
$$\times \{C(u_2) - C(u_1) - j[S(u_2) - S(u_1)]\}_n \qquad (2\text{-}104)$$

where

$$a_n = k\cos\theta - \beta_n \qquad (2\text{-}105)$$
$$b^2 = \frac{k\sin^2\theta}{2r} \qquad (2\text{-}106)$$
$$u_1 = \sqrt{\frac{2}{\pi}}\left(bz_1 - \frac{a_n}{2b}\right) \qquad (2\text{-}107)$$
$$u_2 = \sqrt{\frac{2}{\pi}}\left(bz_2 - \frac{a_n}{2b}\right) \qquad (2\text{-}108)$$

and I_n and β_n are determined from Eqs. (2-68) and (2-66), respectively.

In most respects the far-field relationships carry over into the Fresnel region. This is a result of the way in which the regions are defined, the only difference being the additional phase term for the Fresnel region in the approximation for $e^{-jkr'}$. Perhaps the most important effect of this term is that the θ and ϕ variations of the Fresnel fields *do* change with radial distance r as contrasted to the far field, in which the fields do not change form with r.

The point where one stops using the Fresnel approximations depends on how accurately one needs to know the field at that point. A practical lower

[1] See, for example, A. van Wijngaarden and W. L. Scheen, "Tables of Fresnel Integrals," Ak. van Wetenschappen, first section, Vol. 19, No. 4, North Holland Publishing Company, Amsterdam, 1949, or F. A. Jenkins and H. E. White, "Fundamentals of Optics," 3d ed., pp. 364 and 368, McGraw-Hill, New York, 1957.

limit for the Fresnel region may be taken as the distance r for which the z'^3 term in the binomial expansion of r' [see Eq. (2-78)] produces a phase error of not more than $\pi/8$ rad. This is analogous to the far-field case in which the z'^2 term produced a phase error of less than $\pi/8$ rad for $r > 2L^2/\lambda$. Equating the z'^3 term in the expansion of r' to $\lambda/16$ gives a lower limit of $0.62\sqrt{L^3/\lambda}$ for r. Thus for a source of length L the Fresnel region may be taken to be

$$0.62\sqrt{\frac{L^3}{\lambda}} \lesssim r \lesssim \frac{2L^2}{\lambda} \qquad (2\text{-}109)$$

The Fresnel region does not have rigid boundaries. The transitions to the far and near fields are gradual. While $2L^2/\lambda$ is commonly used as the boundary between the far field and the Fresnel region, the distance L^2/λ is frequently used with satisfactory results. Although we are considering at present the line source, the zones we are defining apply in general to a planar source with L the maximum dimension. Furthermore, if both the measured and the measuring antennas have apertures of maximum extent L and l, respectively, the sum $L + l$ should be used in place of L.

The lower limit, $0.62\sqrt{L^3/\lambda}$, given in Eq. (2-109) is an easily remembered criterion for defining the boundary between the Fresnel region and the near field. It gives a minimum radial distance from the center of the source if one were to measure a Fresnel polar pattern comparable to the usual far-field pattern. However, Fresnel patterns are often determined along a surface parallel to the source as illustrated in Fig. 2-5. If a second antenna of length l is placed at this surface, the Fresnel boundary must be modified by using $L + l$ in place of L.

Criteria other than Eq. (2-109) have been developed for defining the boundary between the Fresnel region and the near field. Some of these are

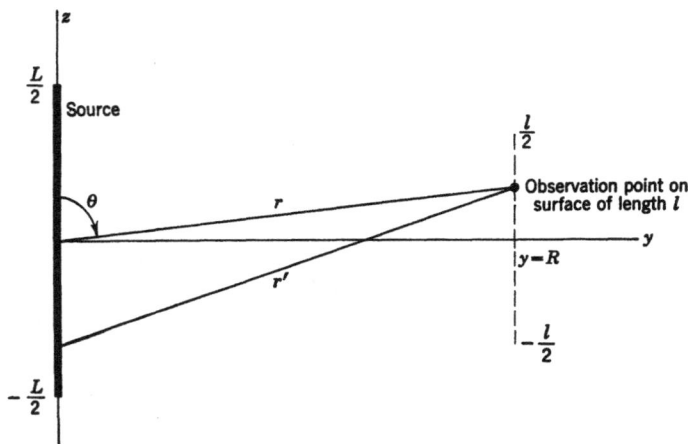

Figure 2-5.   Geometry for determining Fresnel pattern over a plane surface.

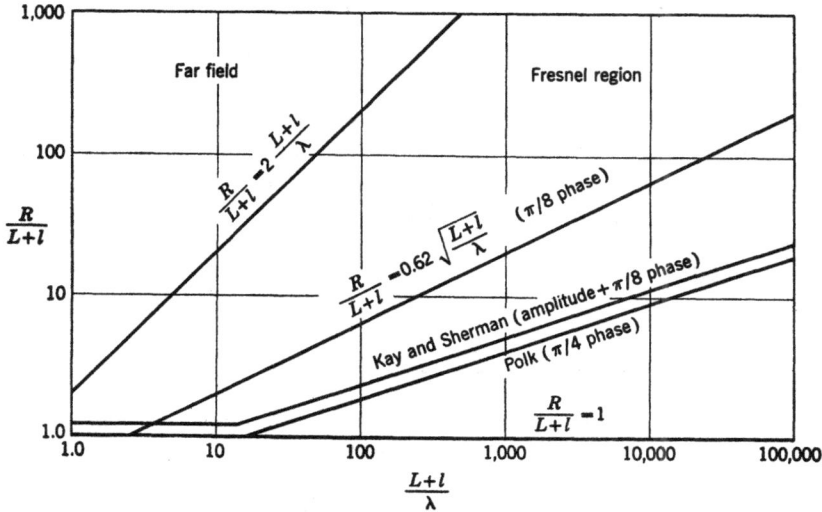

Figure 2-6.   Comparison of Fresnel zone criteria.

compared in Fig. 2-6. For large sources a criterion of one aperture dimension is the least restrictive of those illustrated. Next comes Polk's criterion,[1] which is based on a $\pi/4$ phase error and gives $0.397 \sqrt[3]{(L + l)^4/\lambda}$ as the boundary between the Fresnel zone and the near field. .

A criterion that includes both amplitude and phase effects was proposed by Kay[2] and also used by Sherman.[3] This criterion requires that the error in using the first two terms of the binomial expansion for r' in the exponent of $e^{-jkr'}/r'$ not exceed $\lambda/16$. This results in the inequality

$$\frac{\lambda}{16} > \left\{ \left[y^2 + \left(\frac{L + l}{2} \right)^2 \right]^{1/2} - y - \frac{(L + l)^2}{8y} \right\} \tag{2-110}$$

In addition it is required that the percentage difference between the true range r' and the axial distance R for a source centered at the origin in Fig. 2-5 be no more than the percentage difference between the in-phase component of a vector with phase $\pi/8$ (the maximum phase error) and the magnitude of that vector. This gives an amplitude criterion

$$R \gtrsim 0.923 \left[y^2 + \left(\frac{L + l}{2} \right)^2 \right]^{1/2} \tag{2-111}$$

[1] C. Polk, Optical Fresnel-zone Gain of a Rectangular Aperture, *IRE Trans. Antennas Propagation*, **AP-4**:65–69, January, 1956; also "The Fresnel Region of Large Aperture Antennas," Ph.D. Dissertation, University of Pennsylvania, Philadelphia, 1956.

[2] A. F. Kay, Near Field Gain of Aperture Antennas, *IRE Trans. Antennas Propagation*, **AP-8**:586–593, November, 1960.

[3] J. W. Sherman, III, Properties of Focused Apertures in the Fresnel Region, *IRE Trans. Antennas Propagation*, **AP-10**:399–408, July, 1962.

which reduces to

$$\frac{R}{L+l} \gtrsim 1.19 \tag{2-112}$$

The results of Kay and Sherman in Fig. 2-6 show that the amplitude term dominates for apertures smaller than about 15 wavelengths (λ).

The expansion used in determining the Fresnel region over a surface parallel to the source [see Eq. (2-164)] can produce serious error if the region of observation is large. The inequality following Eq. (2-164) should be observed.

2-9. NEAR–FIELD REGION OF A LINE SOURCE

At distances so small that the contributions at the point of observation from the elements of the source vary significantly in both amplitude and phase with distance r, the Fresnel region approximations are no longer valid in Eq. (2-46). This region constitutes the near field and extends from the Fresnel region to the boundary of the source.

For an arbitrary amplitude and phase along the line source a satisfactory approximation to the near field can be obtained by using a series representation for Eq. (2-46). This can be accomplished by breaking the line source into segments of length ΔL_n as shown in Fig. 2-7 and summing the contributions from the various segments. If we make each segment an ordinary endfire source, then the exact fields of each segment are given by Eqs. (2-52), (2-55), and (2-56), where $z_1 = z_n$, $z_2 = z_n + \Delta L_n$, and $I_0 = I_n e^{j\psi_n}$. The quantities I_n and ψ_n are the appropriate amplitude and phase at the beginning of the nth segment. The form of the approximation is illustrated in Fig. 2-8. The sawtooth nature of the phase distribution is due to the linear phase delay along each endfire segment. The accuracy of the approximation depends on the

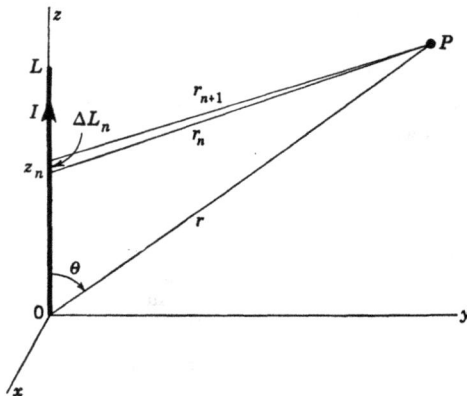

Figure 2-7. Quantizing a line source into endfire segments.

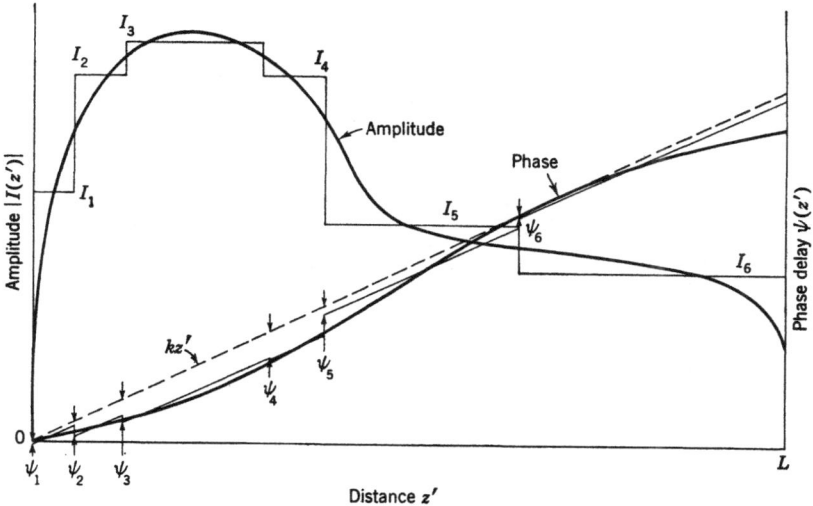

Figure 2-8.   Approximating a line source with ordinary endfire segments. ψ is negative (phase delay) for a wave in the positive z direction.

number of segments that are used. The lengths of the segments need not be the same. As illustrated in Fig. 2-8 the segment length may be much larger in a region of slowly varying amplitude and phase than in a region of rapid variation. This method may be applied to the Fresnel region and the near field, and gives the radial component as well as transverse field components.

Another series representation that is useful in the near field and Fresnel region is obtained by breaking up or quantizing the line source into segments so short that the distance to the point of observation while in the near field of the total source is in the far field of each segment. The far fields of all the segments are then summed at the point of observation with the proper amplitudes and phases determined by the original source distribution just as before. In this case it is usually most convenient to assume that each segment is a source of constant amplitude I_n and constant phase velocity with phase constant β_n. The choice of β_n may be made by picking a value that appears to give the best sawtooth approximation to the actual phase $\psi(z')$. We are not restricted to the free-space phase constant k as in the previous case. The quantity β_n may even be made zero, in which case the segment would be a broadside radiator.

To illustrate this method, a source of length L is broken into four segments of equal lengths as shown in Fig. 2-9. The electric field of the total source along the line $y = $ constant is given approximately as

$$E_z = \frac{-j\omega\mu \sin^2 \theta}{4\pi} \int_{-L/2}^{L/2} I(z') \frac{e^{-jkr'}}{r'} dz' \qquad (2\text{-}113)$$

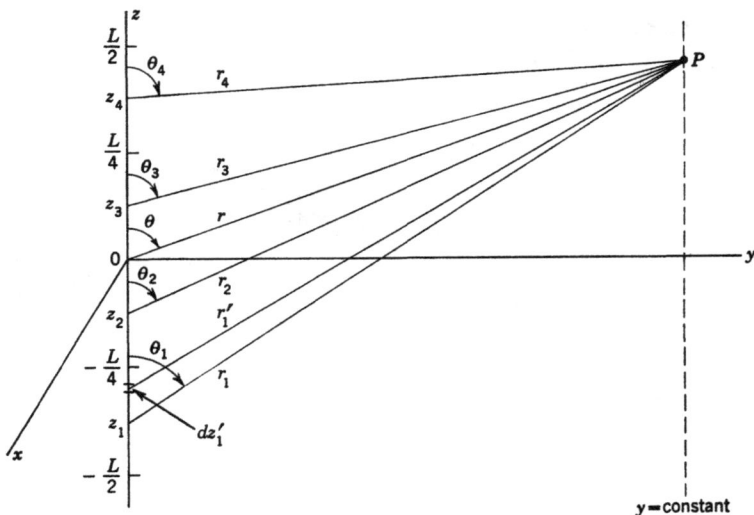

Figure 2-9. Illustrating the application of the far-field analysis to the near field.

Equation (2-113) may also be written

$$E_z = \frac{-j\omega\mu \sin^2\theta}{4\pi} \left[\int_{-L/2}^{-L/4} I(z') \frac{e^{-jkr'}}{r'} \, dz' + \int_{-L/4}^{0} I(z') \frac{e^{-jkr'}}{r'} \, dz' \right.$$
$$\left. + \int_{0}^{L/4} I(z') \frac{e^{-jkr'}}{r'} \, dz' + \int_{L/4}^{L/2} I(z') \frac{e^{-jkr'}}{r'} \, dz' \right] \quad (2\text{-}114)$$

If we consider that each term of Eq. (2-114) represents a source of length $L/4$, then the far-field region of each source is given by

$$r_n \gtrless \frac{2}{\lambda} \left(\frac{L}{4} \right)^2 \quad (2\text{-}115)$$

where r_n is measured from the center of the segment to the point of observation. Thus the far-field boundary is reduced by the square of the number of elements in the total source. This reduction in range for which the far-field approximation is valid may be preserved for the complete source if Eq. (2-114) is written as

$$E_z = \frac{-j\omega\mu}{4\pi} \left[\sin^2\theta_1 \int_{-L/8}^{L/8} I_1(z_1') \frac{e^{-jkr_1'}}{r_1'} \, dz_1' \right.$$
$$+ \sin^2\theta_2 \int_{-L/8}^{L/8} I_2(z_2') \frac{e^{-jkr_2'}}{r_2'} \, dz_2' + \sin^2\theta_3 \int_{-L/8}^{L/8} I_3(z_3') \frac{e^{-jkr_3'}}{r_3'} \, dz_3'$$
$$\left. + \sin^2\theta_4 \int_{-L/8}^{L/8} I_4(z_4') \frac{e^{-jkr_4'}}{r_4'} \, dz_4' \right] \quad (2\text{-}116)$$

where z_n' is the source coordinate for the nth segment and r_n' is the distance from an element dz'_n on the nth segment to the point of observation (see Fig. 2-9).

Each term in Eq. (2-116) can be viewed as a separate source whose far-field region is given by Eq. (2-115). If all sources are of the same form, each term can be readily determined from a standard source centered at the origin of its coordinate system. For convenience let each segment have a constant amplitude I_n and phase constant β_n. The amplitude I_n may be chosen to give the step approximation to $|I(z')|$ as illustrated in Fig. 2-10, and the phase constant β_n may be chosen to give the best sawtooth approximation to the phase.

With this simplification Eq. (2-116) becomes

$$E_z = \frac{-j\omega\mu}{4\pi}\left[I_1 e^{j\psi_1} \sin^2\theta_1 \int_{-L/8}^{L/8} \frac{e^{-j(kr_1'+\beta_1 z_1')}}{r_1'}\,dz_1'\right.$$
$$+ I_2 e^{j\psi_2} \sin^2\theta_2 \int_{-L/8}^{L/8} \frac{e^{-j(kr_2'+\beta_2 z_2')}}{r_2'}\,dz_2'$$
$$+ I_3 e^{j\psi_3} \sin^2\theta_3 \int_{-L/8}^{L/8} \frac{e^{-j(kr_3'+\beta_3 z_3')}}{r_3'}\,dz_3'$$
$$\left. + I_4 e^{j\psi_4} \sin^2\theta_4 \int_{-L/8}^{L/8} \frac{e^{-j(kr_4'+\beta_4 z_4')}}{r_4'}\,dz_4'\right] \quad (2\text{-}117)$$

Because of the assumption of constant amplitude and phase velocity for the segments, the integrations in Eq. (2-117) are readily performed and give

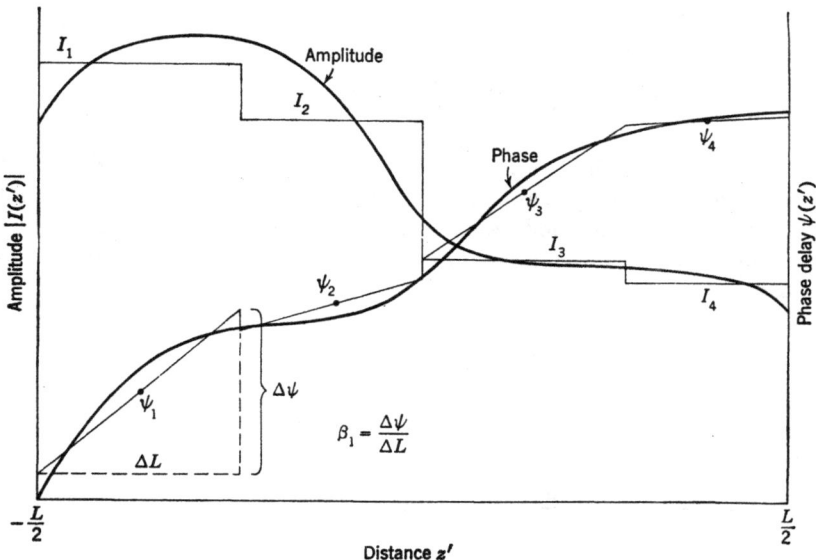

Figure 2-10. Representing arbitrary $I(z')$ by four uniform sources of equal lengths.

the $(\sin X)/X$ function obtained in Eq. (2-91). Equation (2-117) may be written

$$E_z = \frac{-j\omega\mu L}{16\pi} \sum_{n=1}^{4} \frac{I_n e^{j(\psi_n - kr_n)}}{r_n} \sin^2 \theta_n \frac{\sin X_n}{X_n} \qquad (2\text{-}118)$$

where

$$X_n = \left[\frac{k(z - z_n)}{r_n} - \beta_n\right]\frac{L}{8} \qquad (2\text{-}119)$$

$$r_n = \sqrt{(z - z_n)^2 + y^2} \qquad (2\text{-}120)$$

The quantity I_n is the amplitude of the nth segment, and ψ_n is the phase at the center of the nth segment as illustrated in Fig. 2-10. The distances z_n and r_n are illustrated in Fig. 2-9.

Equation (2-118) may be generalized to a source of length L divided into N segments. In this case the minimum distance for which the result is a valid representation of the source fields is

$$y = \frac{2L^2}{\lambda N^2} \qquad (2\text{-}121)$$

and the electric field is given as

$$E_z = \frac{-j\omega\mu L}{4N\pi} \sum_{n=1}^{N} \frac{I_n e^{j(\psi_n - kr_n)}}{r_n} \sin^2 \theta_n \frac{\sin X_n}{X_n} \qquad (2\text{-}122)$$

where X_n now becomes

$$X_n = \left[\frac{k(z - z_n)}{r_n} - \beta_n\right]\frac{L}{2N} \qquad (2\text{-}123)$$

and β_n is chosen to give the best sawtooth approximation to $\psi(z')$.

The y component of electric field can be found in the same manner as the z component, thus giving the total near field of the source.

The above method may be extended to other than uniform segments to obtain better accuracy in approximating $I(z')$. This method is easier to apply than the previous method based on the ordinary endfire segments, but in general it may require a greater number of terms for regions quite near the source.

2-10.   THE RECTANGULAR SOURCE

The rectangular source will be taken to be a plane rectangular sheet of current with coordinate system as shown in Fig. 2-11. The vector potentials are given by

$$\mathbf{A} = \frac{\mu}{4\pi} \int_0^L \int_{-W/2}^{W/2} \frac{\mathbf{J}(x',z')e^{-jkr'}}{r'} dx' \, dz' \qquad (2\text{-}124)$$

$$\mathbf{F} = \frac{\epsilon}{4\pi} \int_0^L \int_{-W/2}^{W/2} \frac{\mathbf{K}(x',z')e^{-jkr'}}{r'} dx' \, dz' \qquad (2\text{-}125)$$

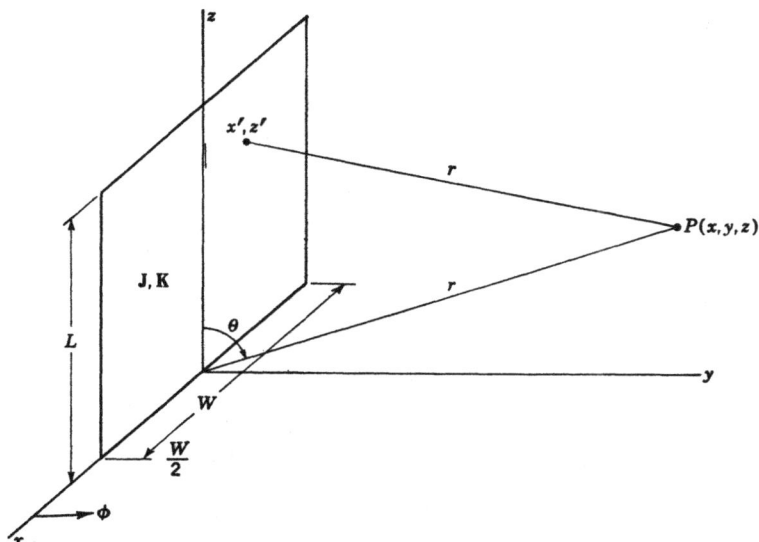

Figure 2-11.   Rectangular source.

where primed coordinates are used at the source and unprimed coordinates are used at the point of observation. The quantities $\mathbf{J}(x',z')$ and $\mathbf{K}(x',z')$ are in general complex. The distance r' can be expressed in rectangular coordinates as

$$r' = [(x - x')^2 + y^2 + (z - z')^2]^{1/2} \tag{2-126}$$

Equation (2-126) can be converted to spherical coordinates by using the relations

$$x = r \sin \theta \cos \phi \tag{2-127}$$
$$y = r \sin \theta \sin \phi \tag{2-128}$$
$$z = r \cos \theta \tag{2-129}$$

If the source consists of both electric and magnetic currents, the resultant field is a superposition of the fields of the individual currents. As in the case of the line source we shall utilize the dual nature of electric- and magnetic-current sources and consider in detail only an electric-current source.

The magnetic field of an electric-current source is obtained by taking the curl of Eq. (2-124). Interchanging the order of integration and differentiation gives

$$\mathbf{H} = \frac{1}{4\pi} \int_0^L \int_{-W/2}^{W/2} \nabla \times \left[\mathbf{J}(x',z') \frac{e^{-jkr'}}{r'} \right] dx'\, dz' \tag{2-130}$$

where the differentiation is with respect to the unprimed coordinates. Using the vector identity

$$\nabla \times \mathbf{J}f = \nabla f \times \mathbf{J} + f(\nabla \times \mathbf{J}) \tag{2-131}$$

reduces Eq. (2-130) to

$$\mathbf{H} = \frac{1}{4\pi} \int_0^L \int_{-W/2}^{W/2} \frac{e^{-jkr'}}{r'} \left(jk + \frac{1}{r'} \right) (\mathbf{J} \times \mathbf{r}') \, dx' \, dz' \tag{2-132}$$

where $\mathbf{r}'$ is a unit vector along r'.

There is no exact solution to Eq. (2-132) for an arbitrary $\mathbf{J}(x',z')$. The same approximate methods that were used for the line source may be applied to the rectangular aperture, including the division of the total field into near field, Fresnel region, and far field.

2-11. FAR–FIELD REGION OF A RECTANGULAR SOURCE

As in the case of the line source, the far field of a rectangular source is the least troublesome of the three regions. The far field is taken to be that region for which $r \gtrless 2D^2/\lambda$, where D is the largest dimension of the aperture. With the usual far-field simplifications, Eq. (2-132) gives

$$H_x = \frac{-jk \sin \theta \sin \phi e^{-jkr}}{4\pi r} \int_0^L \int_{-W/2}^{W/2} J_z(x',z') e^{jk(x' \sin \theta \cos \phi + z' \cos \theta)} \, dx' \, dz' \tag{2-133}$$

$$H_y = \frac{-jke^{-jkr}}{4\pi r} \int_0^L \int_{-W/2}^{W/2} [J_x(x',z') \cos \theta - J_z(x',z') \sin \theta \cos \phi] \\ \times e^{jk(x' \sin \theta \cos \phi + z' \cos \theta)} \, dx' \, dz' \tag{2-134}$$

$$H_z = \frac{jk \sin \theta \sin \phi e^{-jkr}}{4\pi r} \int_0^L \int_{-W/2}^{W/2} J_x(x',z') e^{jk(x' \sin \theta \cos \phi + z' \cos \theta)} \, dx' \, dz' \tag{2-135}$$

For any vector $\mathbf{U}$ the transformation from rectangular to spherical coordinates can be obtained by means of the relations

$$U_r = U_x \sin \theta \cos \phi + U_y \sin \theta \sin \phi + U_z \cos \theta \tag{2-136}$$
$$U_\theta = U_x \cos \theta \cos \phi + U_y \cos \theta \sin \phi - U_z \sin \theta \tag{2-137}$$
$$U_\phi = -U_x \sin \phi + U_y \cos \phi \tag{2-138}$$

Thus the far magnetic field in spherical coordinates is

$$H_\theta = \frac{-jk \sin \phi e^{-jkr}}{4\pi r} \int_0^L \int_{-W/2}^{W/2} J_x(x',z') e^{jk(x' \sin \theta \cos \phi + z' \cos \theta)} \, dx' \, dz' \tag{2-139}$$

$$H_\phi = \frac{jke^{-jkr}}{4\pi r} \int_0^L \int_{-W/2}^{W/2} [\sin \theta J_z(x',z') - \cos \theta \cos \phi J_x(x',z')] \\ \times e^{jk(x' \sin \theta \cos \phi + z' \cos \theta)} \, dx' \, dz' \tag{2-140}$$

The far electric field can be found from Eqs. (2-139) and (2-140) and the intrinsic impedance of free space. Only transverse components are significant, and they are

$$E_\theta = \frac{\omega\mu}{k} H_\phi \tag{2-141}$$

and

$$E_\phi = -\frac{\omega\mu}{k} H_\theta \tag{2-142}$$

The far-field components may also be found from $\mathbf{E} = -j\omega\mathbf{A}$ by taking only transverse components. Thus

$$E_\theta = -j\omega A_\theta \tag{2-143}$$

and

$$E_\phi = -j\omega A_\phi \tag{2-144}$$

Utilizing Eq. (2-137) we obtain from Eq. (2-143)

$$E_\theta = -j\omega A_x \cos\theta\cos\phi + j\omega A_z \sin\theta \tag{2-145}$$

Using the far-field approximations and substituting A_x and A_z from Eq. (2-124) into (2-145) give

$$E_\theta = \frac{j\omega\mu e^{-jkr}}{4\pi r} \int_0^L \int_{-W/2}^{W/2} [\sin\theta J_z(x',z') - \cos\theta\cos\phi J_x(x',z')]$$
$$\times e^{jk(x'\sin\theta\cos\phi + z'\cos\theta)}\, dx'\, dz' \tag{2-146}$$

Similarly

$$E_\phi = \frac{j\omega\mu\sin\phi\, e^{-jkr}}{4\pi r} \int_0^L \int_{-W/2}^{W/2} J_z(x',z') e^{jk(x'\sin\theta\cos\phi + z'\cos\theta)}\, dx'\, dz' \tag{2-147}$$

Equations (2-146) and (2-147) are identical with Eqs. (2-141) and (2-142), respectively.

2-12. QUASI LINE SOURCE

For a source of infinitesimal width ΔW and $\mathbf{J} = \mathbf{z}J_z(z')$, where $J_z\,\Delta W = I(z')$, Eq. (2-140) reduces to Eq. (2-80) for a line source of electric current. In most practical problems no serious error is introduced if Eq. (2-80) is used for ΔW as large as $\lambda/2$. A narrow or quasi line source may also consist of transverse current elements; that is, $\mathbf{J} = \mathbf{x}J_x(z')$. The quantity $J_x\,\Delta W$ is termed the current moment of the source. If ΔW does not exceed $\lambda/2$ for this case, Eqs. (2-146) and (2-147) may be written as

$$E_\theta = \frac{-j\omega\mu\,\Delta W\,\cos\theta\,\cos\phi e^{-jkr}}{4\pi r}\int_0^L J_z(z')e^{jkz'\cos\theta}\,dz' \tag{2-148}$$

$$E_\phi = \frac{j\omega\mu\,\Delta W\,\sin\phi e^{-jkr}}{4\pi r}\int_0^L J_z(z')e^{jkz'\cos\theta}\,dz' \tag{2-149}$$

2-13. PATTERN MULTIPLICATION

The current distribution $\mathbf{J}$ for the rectangular aperture in Fig. 2-11 is usually a function of both x' and z'. In many cases it is possible to express $\mathbf{J}$ as

$$\mathbf{J}(x',z') = \mathbf{x}J_x(x')J_z(z') + \mathbf{z}J_z(x')J_z(z') \tag{2-150}$$

When this is the case, Eqs. (2-146) and (2-147) become

$$E_\theta = \frac{j\omega\mu\,\sin\theta e^{-jkr}}{4\pi r}\int_{-W/2}^{W/2} J_z(x')e^{jkx'\sin\theta\cos\phi}\,dx'$$

$$\times\int_0^L J_z(z')e^{jkz'\cos\theta}\,dz' - \frac{j\omega\mu\,\cos\theta\,\cos\phi e^{-jkr}}{4\pi r}$$

$$\times\int_{-W/2}^{W/2} J_z(x')e^{jkx'\sin\theta\cos\phi}\,dx'\int_0^L J_z(z')e^{jkz'\cos\theta}\,dz' \tag{2-151}$$

$$E_\phi = \frac{j\omega\mu\,\sin\phi e^{-jkr}}{4\pi r}\int_{-W/2}^{W/2} J_z(x')e^{jkx'\sin\theta\cos\phi}\,dx'$$

$$\times\int_0^L J_z(z')e^{jkz'\cos\theta}\,dz' \tag{2-152}$$

Each integral is of the form of the far field of a line current. That is, each integral itself is a far-field pattern. Thus the total pattern may be expressed in terms of product patterns when the source distribution is separable [Eq. (2-150)]. This is called pattern multiplication. Most practical rectangular sources have separable distributions.

If the distributions in x' and z' are other than uniform traveling waves, a Fourier series representation may be used (see Sec. 2-6). The result for a separable distribution that has period $2W$ in x' and $2L$ in z' is

$$\mathbf{J}(x',z') = \mathbf{x}\sum_l I_l e^{-j(l\pi/W)x'}\sum_m I_m e^{-j(m\pi/L)z'}$$

$$+ \mathbf{z}\sum_n I_n e^{-j(n\pi/W)x'}\sum_p I_p e^{-j(p\pi/L)z'} \tag{2-153}$$

where

$$I_l = \frac{1}{2W}\int_{-W}^{W} J_z(x')e^{j(l\pi/W)x'}\,dx' \tag{2-154}$$

$$I_m = \frac{1}{2L}\int_{-L}^{L} J_z(z')e^{j(m\pi/L)z'}\,dz' \tag{2-155}$$

$$I_n = \frac{1}{2W}\int_{-W}^{W} J_z(x')e^{j(n\pi/W)x'}\,dx' \tag{2-156}$$

$$I_p = \frac{1}{2L}\int_{-L}^{L} J_z(z')e^{j(p\pi/L)z'}\,dz' \tag{2-157}$$

The electric-field components in the far field are found to be

$$E_\theta = \frac{j\omega\mu WL \sin\theta e^{-jkr}}{16\pi r} \sum_n I_n \frac{\sin X_n}{X_n} \sum_p I_p e^{j\left(k\cos\theta - \frac{p\pi}{L}\right)\frac{L}{2}}$$

$$\times \frac{\sin X_p}{X_p} - \frac{j\omega\mu WL \cos\theta\cos\phi e^{-jkr}}{16\pi r} \sum_l I_l \frac{\sin X_l}{X_l}$$

$$\times \sum_m I_m e^{j\left(k\cos\theta - \frac{m\pi}{L}\right)\frac{L}{2}} \frac{\sin X_m}{X_m} \quad (2\text{-}158)$$

$$E_\phi = \frac{j\omega\mu WL \sin\phi e^{-jkr}}{16\pi r} \sum_l I_l \frac{\sin X_l}{X_l} \sum_m I_m e^{j\left(k\cos\theta - \frac{m\pi}{L}\right)\frac{L}{2}} \frac{\sin X_m}{X_m} \quad (2\text{-}159)$$

where

$$X_l = \left(k\sin\theta\cos\phi - \frac{l\pi}{W}\right)\frac{W}{2} \quad (2\text{-}160)$$

$$X_m = \left(k\cos\theta - \frac{m\pi}{L}\right)\frac{L}{2} \quad (2\text{-}161)$$

$$X_n = \left(k\sin\theta\cos\phi - \frac{n\pi}{W}\right)\frac{W}{2} \quad (2\text{-}162)$$

$$X_p = \left(k\cos\theta - \frac{p\pi}{L}\right)\frac{L}{2} \quad (2\text{-}163)$$

2-14. FRESNEL AND NEAR-FIELD REGIONS OF A RECTANGULAR SOURCE

For distances less than $2D^2/\lambda$ down to distances on the order of $0.62\sqrt{D^3/\lambda}$, where D is the diagonal of the source, the Fresnel approximations described in Sec. 2-8 are usually satisfactory. For the phase term of $e^{-jkr'}/r'$ the distance r' may be expanded as

$$r' \approx y + \frac{(x-x')^2 + (z-z')^2}{2y} \quad (2\text{-}164)$$

If $y > [(x-x')^2 + (z-z')^2]/2y$, the denominator term of $e^{-jkr'}/r'$ may be approximated by y. With these approximations Eq. (2-132) reduces to

$$H_x = \frac{-jke^{-jky}}{4\pi y} \int_0^L \int_{-W/2}^{W/2} J_z(x',z') e^{-jk\frac{(x-x')^2+(z-z')^2}{2y}} dx'\, dz' \quad (2\text{-}165)$$

$$H_y = \frac{-jke^{-jky}}{4\pi y^2} \int_0^L \int_{-W/2}^{W/2} [(z-z')J_x(x',z') - (x-x')J_z(x',z')]$$

$$\times e^{-jk\frac{(x-x')^2+(z-z')^2}{2y}} dx'\, dz' \quad (2\text{-}166)$$

$$H_z = \frac{jke^{-jky}}{4\pi y} \int_0^L \int_{-W/2}^{W/2} J_x(x',z') e^{-jk\frac{(x-x')^2+(z-z')^2}{2y}} dx'\, dz' \quad (2\text{-}167)$$

where H_y will be henceforth neglected because of the $1/y^2$ variation.

If $\mathbf{J}$ is separable, Eqs. (2-165) and (2-167) may be written

$$H_x = \frac{-jke^{-jky}}{4\pi y} \int_{-W/2}^{W/2} J_z(x')e^{-jk\frac{(x-x')^2}{2y}} dx' \int_{0}^{L} J_z(z')e^{-jk\frac{(z-z')^2}{2y}} dz'$$

(2-168)

and

$$H_z = \frac{jke^{-jky}}{4\pi y} \int_{-W/2}^{W/2} J_x(x')e^{-jk\frac{(x-x')^2}{2y}} dx' \int_{0}^{L} J_z(z')e^{-jk\frac{(z-z')^2}{2y}} dz'$$

(2-169)

For a uniform source of the form

$$J_z(x',z') = |J_z|e^{-j\beta_{zx}x'}e^{-j\beta_{zz}z'}$$

(2-170)

and

$$J_z(x',z') = |J_z|e^{-j\beta_{zx}x'}e^{-j\beta_{zz}z'}$$

(2-171)

Eqs. (2-168) and (2-169) become

$$H_x = \frac{-jk|J_z|e^{-jky}}{4\pi y} \int_{-W/2}^{W/2} e^{-j\left[\beta_{zx}x' + \frac{k(x-x')^2}{2y}\right]} dx'$$

$$\times \int_{0}^{L} e^{-j\left[\beta_{zz}z' + \frac{k(z-z')^2}{2y}\right]} dz' \quad (2\text{-}172)$$

and

$$H_z = \frac{jk|J_z|e^{-jky}}{4\pi y} \int_{-W/2}^{W/2} e^{-j\left[\beta_{zx}x' + \frac{k(x-x')^2}{2y}\right]} dx'$$

$$\times \int_{0}^{L} e^{-j\left[\beta_{zz}z' + \frac{k(z-z')^2}{2y}\right]} dz' \quad (2\text{-}173)$$

Equations (2-172) and (2-173) can be put into the form of standard Fresnel integrals. The exponents in the integrals of Eqs. (2-172) and (2-173) are quadratics of the form

$$\frac{kx^2}{2y} + \left(\beta_x - \frac{kx}{y}\right)x' + \frac{kx'^2}{2y} = a_x + b_x x' + c_x^2 x'^2$$

(2-174)

and

$$\frac{kz^2}{2y} + \left(\beta_z - \frac{kz}{y}\right)z' + \frac{kz'^2}{2y} = a_z + b_z z' + c_z^2 z'^2$$

(2-175)

Following the procedure outlined in Sec. 2-8, Eqs. (2-172) and (2-173) reduce to

$$H_x = \frac{-jk|J_z|e^{-j[ky+a_x+a_z-(b_x/2c_x)^2-(b_z/2c_z)^2]}}{8yc_xc_z} \{C(u_{x2}) - C(u_{x1})$$

$$- j[S(u_{x2}) - S(u_{x1})]\}\{C(u_{z2}) - C(u_{z1}) - j[S(u_{z2}) - S(u_{z1})]\} \quad (2\text{-}176)$$

and

$$H_z = -\frac{|J_z|}{|J|}H$$

(2-177)

where a_x, a_z, b_x, b_z, $c_x{}^2$, and $c_z{}^2$ are defined by Eqs. (2-174) and (2-175) and

$$u_{x2} = \sqrt{\frac{2}{\pi}}\left(c_x\frac{W}{2} + \frac{b_x}{2c_x}\right) \tag{2-178}$$

$$u_{x1} = -\sqrt{\frac{2}{\pi}}\left(c_x\frac{W}{2} - \frac{b_x}{2c_x}\right) \tag{2-179}$$

$$u_{z2} = \sqrt{\frac{2}{\pi}}\left(c_z L + \frac{b_z}{2c_z}\right) \tag{2-180}$$

$$u_{z1} = \sqrt{\frac{2}{\pi}}\frac{b_z}{2c_z} \tag{2-181}$$

The preceding results can be used to find the Fresnel fields of a non-uniform source. Assume that the source can be represented by Fourier series as

$$J_x(x',z') = J_x(x')J_x(z') = \sum_l I_l e^{-j\beta_{xl}x'} \sum_m I_m e^{-j\beta_{xm}z'} \tag{2-182}$$

and

$$J_z(x',z') = J_z(x')J_z(z') = \sum_n I_n e^{-j\beta_{zn}x'} \sum_p I_p e^{-j\beta_{zp}z'} \tag{2-183}$$

In this case the components of the magnetic field are

$$H_x = \frac{-jke^{-j(ky+a_x+a_z)}}{8yc_xc_z}\sum_n I_n e^{j(b_{zn}/2c_x)^2}\{C(u_{x2}) - C(u_{x1})$$

$$- j[S(u_{x2}) - S(u_{x1})]\}_n \sum_p I_p e^{j(b_{zp}/2c_z)^2}\{C(u_{z2}) - C(u_{z1})$$

$$- j[S(u_{z2}) - S(u_{z1})]\}_p \quad (2\text{-}184)$$

$$H_z = \frac{jke^{-j(ky+a_x+a_z)}}{8yc_xc_z}\sum_l I_l e^{j(b_{xl}/2c_x)^2}\{C(u_{x2}) - C(u_{x1})$$

$$- j[S(u_{x2}) - S(u_{x1})]\}_l \sum_m I_m e^{j(b_{xm}/2c_z)^2}\{C(u_{z2}) - C(u_{z1})$$

$$- j[S(u_{z2}) - S(u_{z1})]\}_m \quad (2\text{-}185)$$

The electric-field components can be found from the magnetic-field components and the impedance of the medium as

$$E_x = \frac{-\omega\mu}{k}H_z \tag{2-186}$$

and

$$E_z = \frac{\omega\mu}{k}H_x \tag{2-187}$$

Equations (2-184) to (2-187) are convenient for determining the Fresnel field on the surface $y = $ const when the rectangular source is located in the

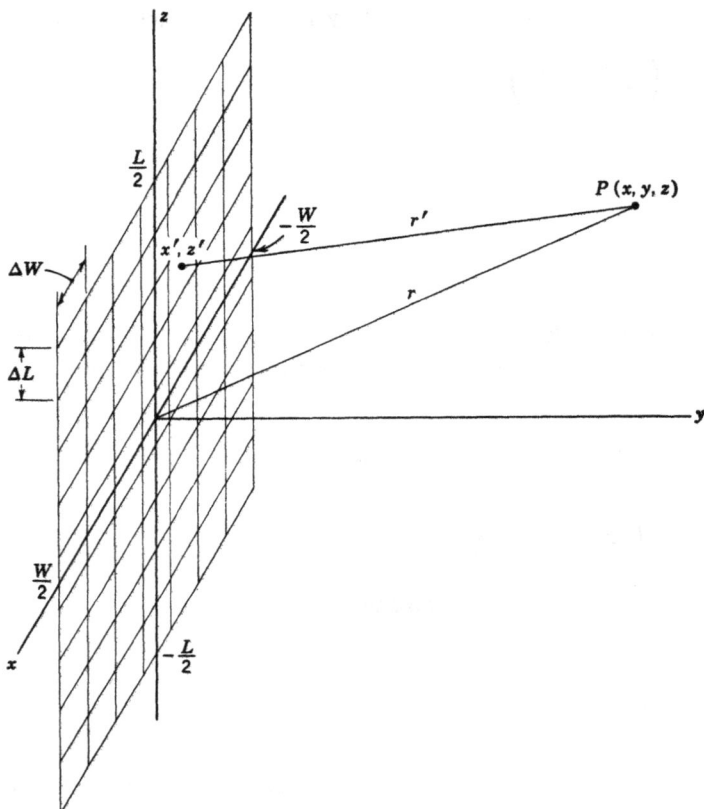

Figure 2-12.   Quantizing a rectangular source.

$y = 0$ plane. It must be kept in mind, however, that the use of these equations is restricted to the Fresnel zone defined in Sec. 2-8.

Approximate fields in the near-field region as well as the Fresnel region of a rectangular source can be found by breaking the source into small segments as in the case of the line source. Thus a source may be quantized as in Fig. 2-12. If ΔW is $\lambda/2$ or less and ΔL is short enough that the step amplitude and sawtooth phase approximations of Fig. 2-8 are adequate, the total field of the z component of electric current in the rectangular source can be found by summing the contributions of ordinary endfire sources of current

$$I_z(z') = \Delta W |J_z| e^{-jkz'} \tag{2-188}$$

where $|J_z|$ is constant. The fields of the ordinary endfire source are given by Eqs. (2-52), (2-55), and (2-56).

An x component of current in the rectangular source may be treated in a similar manner. The total fields of an ordinary endfire current

$$I_z(x') = \Delta L |J_x| e^{-jkx'} \tag{2-189}$$

may be found by following the procedure for $I(z')$ in Sec. 2-5. In this case ΔL usually should be no larger than $\lambda/2$.

Let us consider this method in more detail. Assume that the rectangular source in Fig. 2-12 has only a z component of surface current $\mathbf{J}$ and we desire to find the field on a surface $y = $ const. Let all the elements of the total source be of the same size so that the fields of each element are of the same form and are given by Eqs. (2-52), (2-55), and (2-56). It is only necessary to translate coordinates x' and z' to find the fields of each element at the point of observation in terms of the fields of a standard element centered at the origin. The component E_z for the element at the origin, the 0,0th element, is given by

$$
\begin{aligned}
E_{z,00} = \frac{\Delta W \, |J_{z,00}| e^{j\psi_{00}}}{j\omega\epsilon 4\pi r} & \left(\left\{ \left[\frac{(r\cos\theta - \Delta L/2)}{r_2{}^3} + \frac{jk(r\cos\theta - \Delta L/2)}{r_2{}^2} \right. \right. \right. \\
& \left. + \frac{jk}{r_2} \right] \left(r - \frac{\Delta L}{2}\cos\theta \right) - \frac{\cos\theta}{r_2} \right\} e^{-jk(r_2+\Delta L/2)} \\
& - \left\{ \left[\frac{(r\cos\theta + \Delta L/2)}{r_1{}^3} + \frac{jk(r\cos\theta + \Delta L/2)}{r_1{}^2} + \frac{jk}{r_1} \right] \right. \\
& \left. \left. \times \left(r + \frac{\Delta L}{2}\cos\theta \right) - \frac{\cos\theta}{r_1} \right\} e^{-jk(r_1-\Delta L/2)} \right)
\end{aligned}
\tag{2-190}
$$

The z component due to the m,nth element is

$$
\begin{aligned}
E_{z,mn} = \frac{\Delta W |J_{z,mn}| e^{j\psi_{mn}}}{j\omega\epsilon 4\pi r_{mn}} & \left(\left\{ \left[\frac{(r_{mn}\cos\theta_{mn} - \Delta L/2)}{r_{2,mn}^3} \right. \right. \right. \\
& \left. + \frac{jk(r_{mn}\cos\theta_{mn} - \Delta L/2)}{r_{2,mn}^2} + \frac{jk}{r_{2,mn}} \right] \left(r_{mn} - \frac{\Delta L}{2}\cos\theta_{mn} \right) \\
& \left. - \frac{\cos\theta_{mn}}{r_{2,mn}} \right\} e^{-jk(r_{2,mn}+\Delta L/2)} - \left\{ \left[\frac{(r_{mn}\cos\theta_{mn} + \Delta L/2)}{r_{1,mn}^3} \right. \right. \\
& \left. + \frac{jk(r_{mn}\cos\theta_{mn} + \Delta L/2)}{r_{1,mn}^2} + \frac{jk}{r_{1,mn}} \right] \\
& \left. \left. \times \left(r_{mn} + \frac{\Delta L}{2}\cos\theta_{mn} \right) - \frac{\cos\theta_{mn}}{r_{1,mn}} \right\} e^{-jk(r_{1,mn}-\Delta L/2)} \right)
\end{aligned}
\tag{2-191}
$$

Figure 2-13 helps to explain the notation. The quantities $|J_{z,mn}|$ and ψ_{mn} are the amplitude and phase, respectively, at the point x_m, z_n. The other quantities are

$$r = (x^2 + y^2 + z^2)^{1/2} \tag{2-192}$$
$$r_{mn} = [(x - x_m)^2 + y^2 + (z - z_n)^2]^{1/2} \tag{2-193}$$
$$\cos\theta_{mn} = \frac{z - z_n}{r_{mn}} \tag{2-194}$$

$$r_{2,mn} = \left[(x - x_m)^2 + y^2 + \left(z - z_n - \frac{\Delta L}{2} \right)^2 \right]^{\frac{1}{2}} \tag{2-195}$$

$$r_{1,mn} = \left[(x - x_m)^2 + y^2 + \left(z - z_n + \frac{\Delta L}{2} \right)^2 \right]^{\frac{1}{2}} \tag{2-196}$$

$$x_m = m\, \Delta W \tag{2-197}$$

$$z_n = n\, \Delta L \tag{2-198}$$

where m and n are integers. The z component of electric field for the total source is given by

$$E_z = \sum_{m=-(M-1)/2}^{(M-1)/2} \sum_{n=-(N-1)/2}^{(N-1)/2} E_{z,mn} \tag{2-199}$$

where

$$M = \frac{W}{\Delta W} \tag{2-200}$$

$$N = \frac{L}{\Delta L} \tag{2-201}$$

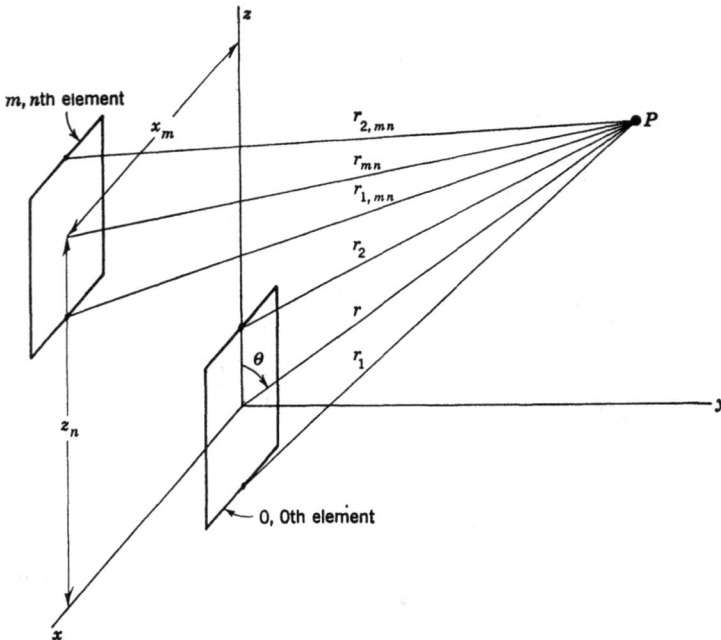

Figure. 2-13.   Coordinates for the m,nth element.

and ΔW and ΔL are chosen so that M and N are odd integers. Furthermore, ΔW usually should be no larger than $\lambda/2$ and ΔL should be small enough so that the step amplitude and sawtooth phase adequately represent the actual current distribution.

As in the line source case, one may choose the segment size so that the point of observation is in the far field of each segment although in the Fresnel region or near field of the total source. Following the procedure in Sec. 2-9, the total source of Fig. 2-12 can be broken into M times N constant amplitude and constant phase velocity sources. The z component of electric field in this case is given as

$$E_z = \frac{j\omega\mu WL}{16MN\pi} \sum_{m=-(M-1)/2}^{(M-1)/2} \sum_{n=-(N-1)/2}^{(N-1)/2} |J_{z,mn}| \frac{e^{j(\psi_{mn}-kr_{mn})}}{r_{mn}} \sin^2\theta_{mn}$$
$$\times \frac{\sin X_m}{X_m} \frac{\sin X_n}{X_n} \quad (2\text{-}202)$$

where

$$X_m = \left[\frac{k(x - x_m)}{r_{mn}} - \beta_{xm}\right]\frac{W}{2M} \quad (2\text{-}203)$$

and

$$X_n = \left[\frac{k(z - z_n)}{r_{mn}} - \beta_{zn}\right]\frac{L}{2N} \quad (2\text{-}204)$$

The phase constants β_{xm} and β_{zn} are chosen to give the best approximation to $\psi(x',z')$. Equation (2-202) should be useful for

$$y \gtrsim \frac{2[(\Delta W)^2 + (\Delta L)^2]}{\lambda} \quad (2\text{-}205)$$

Generally Eq. (2-202) will be easier to apply than Eq. (2-199), although for the same number of terms, Eq. (2-199) should be more accurate than Eq. (2-202).

Although the Fresnel and near-field calculations are much more difficult than those for the far field, they are generally necessary for distances less than $2D^2/\lambda$ from the source. Figure 2-14 shows a comparison of fields evaluated by the Fresnel and far-field integrals for a 4-in.-square source at a wavelength of 1.18 in. The two computations are in substantial agreement in the region of $2D^2/\lambda$ but show increasing disagreement as the point of observation approaches the source. Figure 2-15 shows that the Fresnel computation is in reasonably good agreement with measurements to within a distance equal to the maximum dimension of the source for this case.

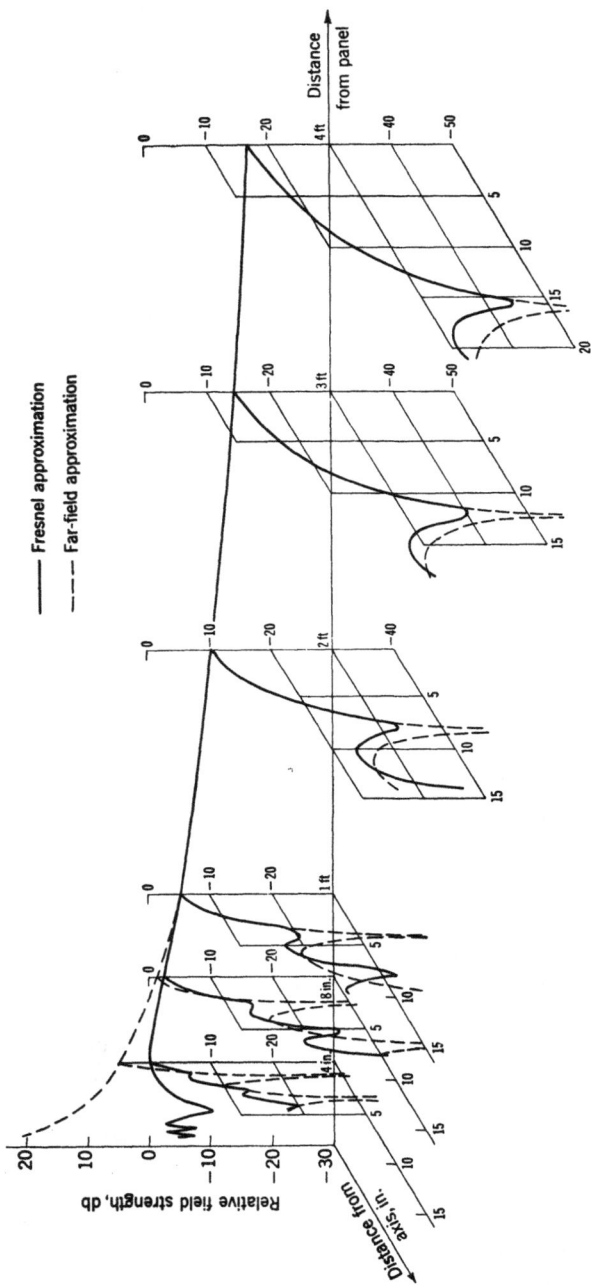

Figure. 2-14. Comparison of Fresnel and far-field computations for a 4-in.-square source at a wavelength of 1.18 in.

56

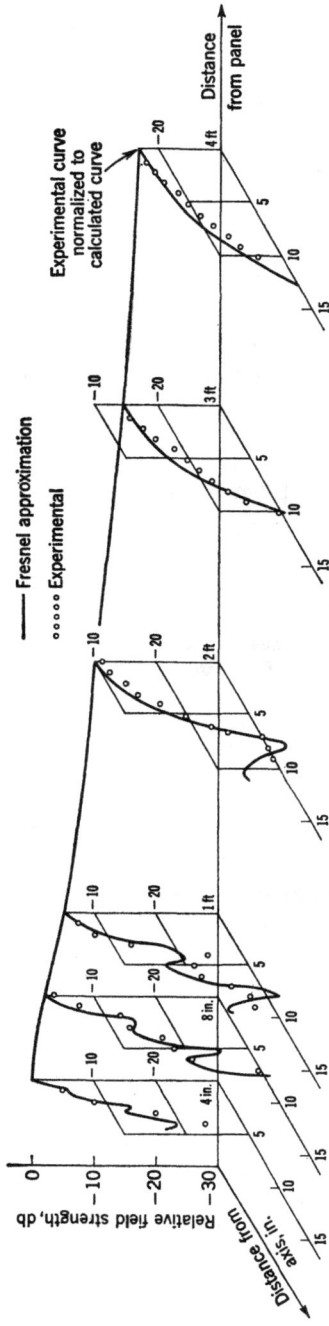

Figure 2-15. Comparison of Fresnel computations and measurements for a 4-in.-square source at a wavelength of 1.18 in.

57

2-15. THE CIRCULAR SOURCE

The circular source will be taken to be a plane circular sheet of current with coordinate system as shown in Fig. 2-16. The vector potentials are given by

$$\mathbf{A} = \frac{\mu}{4\pi} \int_0^{2\pi} \int_0^a \mathbf{J}(\rho',\phi') \frac{e^{-jkr'}}{r'} \rho' \, d\rho' \, d\phi' \tag{2-206}$$

$$\mathbf{F} = \frac{\epsilon}{4\pi} \int_0^{2\pi} \int_0^a \mathbf{K}(\rho',\phi') \frac{e^{-jkr'}}{r'} \rho' \, d\rho' \, d\phi' \tag{2-207}$$

The development of the expressions for the fields of a circular source parallels that of the rectangular source although details will differ because of the difference in geometry. Again we may utilize the dual nature of electric and magnetic sources and determine only the fields of an electric-current source. The expression for the magnetic field of an electric-current source is given by

$$\mathbf{H} = \frac{1}{4\pi} \int_0^{2\pi} \int_0^a \mathbf{\nabla} \times \mathbf{J} \frac{e^{-jkr'}}{r'} \rho' \, d\rho' \, d\phi' \tag{2-208}$$

where differentiation is with respect to the coordinates (r,θ,ϕ) at the point of observation P. By means of the vector identity

$$\mathbf{\nabla} \times \mathbf{J}f = \nabla f \times \mathbf{J} + f(\mathbf{\nabla} \times \mathbf{J}) \tag{2-209}$$

Eq. (2-208) reduces to

$$\mathbf{H} = \frac{1}{4\pi} \int_0^{2\pi} \int_0^a \frac{e^{-jkr'}}{r'} \left(jk + \frac{1}{r'} \right) (\mathbf{J} \times \mathbf{r}') \rho' \, d\rho' \, d\phi' \tag{2-210}$$

where $\mathbf{r}'$ is a unit vector along r'. The same Fresnel and far-field approximations that were used for the rectangular source may be applied to the circular source.

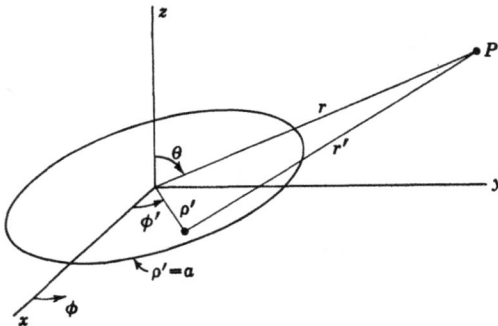

Figure 2-16.   Circular source.

2-16. FAR–FIELD REGION OF A CIRCULAR SOURCE

The far-field region is taken to be that region for which $r \gtrsim 8a^2/\lambda$. The distance r' can be expressed as

$$r' = [r^2 + \rho'^2 - 2r\rho' \sin \theta \cos (\phi - \phi')]^{\frac{1}{2}} \qquad (2\text{-}211)$$

Expanding Eq. (2-211) gives

$$r' \approx r - \rho' \sin \theta \cos (\phi - \phi') + \frac{\rho'^2}{2r} [1 - \sin^2 \theta \cos^2 (\phi - \phi')] \qquad (2\text{-}212)$$

For the far field we need only the first two terms of Eq. (2-212) in the exponent of e^{-jkr}/r' and only the first term in the denominator. At $r = 8a^2/\lambda$ the third term in Eq. (2-212) gives a maximum phase error of $\pi/8$ rad. Using these far-field approximations and neglecting $1/r'$ with respect to jk in Eq. (2-210) give

$$\mathbf{H} = \frac{jke^{-jkr}}{4\pi r} \int_0^{2\pi} \int_0^a [\mathbf{J}(\rho',\phi') \times \mathbf{r'}]e^{jk\rho' \sin \theta \cos (\phi-\phi')}\rho' \, d\rho' \, d\phi' \qquad (2\text{-}213)$$

The integral in Eq. (2-213), although of much simpler form than the integral in Eq. (2-208), may be difficult to evaluate for an arbitrary current distribution. The integration is quite straightforward, however, for a current distribution that is constant in amplitude and phase. Consider the case $\mathbf{J} = \mathbf{x}J_x$ where J_x is constant. Substituting into Eq. (2-213) and performing the cross product operation give

$$H_\theta = \frac{-jk \sin \phi J_x e^{-jkr}}{4\pi r} \int_0^{2\pi} \int_0^a e^{jk\rho' \sin \theta \cos (\phi-\phi')}\rho' \, d\rho' \, d\phi' \qquad (2\text{-}214)$$

$$H_\phi = \frac{-jk \cos \theta \cos \phi J_x e^{-jkr}}{4\pi r} \int_0^{2\pi} \int_0^a e^{jk\rho' \sin \theta \cos (\phi-\phi')}\rho' \, d\rho' \, d\phi' \qquad (2\text{-}215)$$

Integrating with respect to ϕ' gives

$$\int_0^{2\pi} e^{jk\rho' \sin \theta \cos (\phi-\phi')} \, d\phi' = 2\pi J_0(k\rho' \sin \theta) \qquad (2\text{-}216)$$

where J_0 is the Bessel function of the first kind of order zero. The quantity J with numerical subscripts will denote Bessel functions, whereas $\mathbf{J}$ with no subscripts or J with letter subscripts will denote surface current density. Integrating with respect to ρ' gives

$$2\pi \int_0^a J_0(k\rho' \sin \theta)\rho' \, d\rho' = \frac{2\pi a^2 J_1(ka \sin \theta)}{ka \sin \theta} \qquad (2\text{-}217)$$

The function $J_1(U)/U$ for the circular source can be compared with the function $(\sin X)/X$ encountered in the study of uniformly illuminated line sources and rectangular sources. Substituting the results of Eq. (2-217) into (2-214) and (2-215) gives

$$H_\theta = \frac{-ja \sin \phi J_x J_1(ka \sin \theta)e^{-jkr}}{2r \sin \theta} \tag{2-218}$$

$$H_\phi = \frac{-ja \cos \theta \cos \phi J_x J_1(ka \sin \theta)e^{-jkr}}{2r \sin \theta} \tag{2-219}$$

Equations (2-218) and (2-219) give the far magnetic field for a circular source with uniform electric-current density J_x. The electric field can be found from the intrinsic impedance of the medium as

$$E_\theta = \frac{k}{\omega\epsilon} H_\phi \tag{2-220}$$

$$E_\phi = \frac{-k}{\omega\epsilon} H_\theta \tag{2-221}$$

The far field of certain amplitude functions other than the uniform distribution can be readily evaluated for the circular aperture. For example, an amplitude taper of the form $[1 - (\rho'/a)^2]^p$, where p is a positive integer, has been used a great deal for parabolic reflector antennas.[1] Integrating with respect to ρ' gives

$$2\pi \int_0^a \left[1 - \left(\frac{\rho'}{a}\right)^2\right]^p J_0(k\rho' \sin \theta)\rho' \, d\rho' = \frac{\pi a^2 2^p p! J_{p+1}(ka \sin \theta)}{(ka \sin \theta)^{p+1}} \tag{2-222}$$

Thus a current distribution $J_x(\rho') = |J_x|\left[1 - \left(\frac{\rho'}{a}\right)^2\right]^p$ would produce a far magnetic field with components

$$H_\theta = \frac{-jk \sin \phi a^2 2^{p-2} p! |J_x| J_{p+1}(ka \sin \theta)e^{-jkr}}{r(ka \sin \theta)^{p+1}} \tag{2-223}$$

$$H_\phi = \frac{-jk \cos \theta \cos \phi a^2 2^{p-2} p! |J_x| J_{p+1}(ka \sin \theta)e^{-jkr}}{r(ka \sin \theta)^{p+1}} \tag{2-224}$$

In general when the aperture distribution is independent of ϕ', the far magnetic field is given by

$$\mathbf{H} = \frac{jke^{-jkr}}{2r} \int_0^a [\mathbf{J}(\rho') \times \mathbf{r}'] J_0(k\rho' \sin \theta)\rho' \, d\rho' \tag{2-225}$$

Solutions of Eq. (2-225) and the more general case given by Eq. (2-213) may be extremely difficult for an arbitrary aperture distribution. Numerical integration techniques or a series representation obtained by breaking the

[1] R. C. Spencer, Paraboloid Diffraction Patterns from the Standpoint of Physical Optics, *MIT Radiation Lab. Rept.* T-7, October 21, 1942; S. Silver, "Microwave Antenna Theory and Design," chap. 6, McGraw-Hill, New York, 1949.

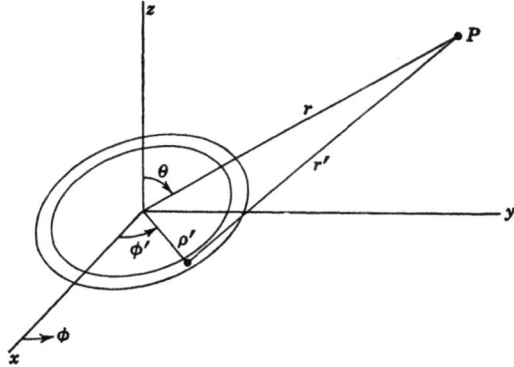

Figure 2-17. Ring source.

aperture into small segments usually give satisfactory results. The latter method will be discussed in some detail in Sec. 2-18.

A special form of circular source is illustrated in Fig. 2-17. This is simply a ring source which in practice may be a complete circle or some portion of a circle. Examples are circular slots, circular arrays of slots, stubs or dipoles, and the semicircular aperture of a two-dimensional Luneberg lens. Development of the fields of such a source is left as a problem at the end of the chapter.

2-17. FRESNEL AND NEAR–FIELD REGIONS OF A CIRCULAR SOURCE

For $r \gtrless 0.62 \sqrt{8a^3/\lambda}$, the Fresnel region approximations are usually satisfactory. Expanding r' in cylindrical coordinates gives

$$r' \approx z + \frac{\rho^2}{2z} - \frac{\rho\rho' \cos (\phi - \phi')}{z} + \frac{\rho'^2}{2z} \tag{2-226}$$

where terms in $1/z^2$ and above have been neglected. If we consider the field in a plane parallel to the xy plane due to a circular source of electric current in the xy plane as shown in Fig. 2-16 and restrict ourselves to $\rho < z$ and $z > 2a$, then Eq. (2-210) may be expressed as

$$\mathbf{H} = \frac{jke^{-jk(z+\rho^2/2z)}}{4\pi z} \int_0^{2\pi} \int_0^a [\mathbf{J}(\rho',\phi') \times \mathbf{r}']$$
$$\times e^{jk[(\rho\rho'/z) \cos (\phi-\phi')-\rho'^2/2z]} \rho' \, d\rho' \, d\phi' \tag{2-227}$$

If we consider the case where the electric-current source is constant in amplitude and phase, Eq. (2-227) becomes

$$\mathbf{H} = \frac{jke^{-jk(z+\rho^2/2z)}}{4\pi z} (\mathbf{J} \times \mathbf{r}')(C - jS) \tag{2-228}$$

where

$$C = \int_0^{2\pi} \int_0^a e^{jk(\rho\rho'/z)\cos(\phi-\phi')} \cos\left(\frac{k\rho'^2}{2z}\right) \rho'\, d\rho'\, d\phi' \tag{2-229}$$

$$S = \int_0^{2\pi} \int_0^a e^{jk(\rho\rho'/z)\cos(\phi-\phi')} \sin\left(\frac{k\rho'^2}{2z}\right) \rho'\, d\rho'\, d\phi' \tag{2-230}$$

Integrating with respect to ϕ' gives

$$C = 2\pi \int_0^a J_0\left(\frac{k\rho\rho'}{z}\right) \cos\left(\frac{k\rho'^2}{2z}\right) \rho'\, d\rho' \tag{2-231}$$

$$S = 2\pi \int_0^a J_0\left(\frac{k\rho\rho'}{z}\right) \sin\left(\frac{k\rho'^2}{2z}\right) \rho'\, d\rho' \tag{2-232}$$

Following a development due to Lommel and described by Rayleigh[1] the functions C and S can be written as

$$C = \frac{2\pi z}{k}\left[\cos\left(\frac{ka^2}{2z}\right) U_1 + \sin\left(\frac{ka^2}{2z}\right) U_2\right] \tag{2-233}$$

$$S = \frac{2\pi z}{k}\left[\sin\left(\frac{ka^2}{2z}\right) U_1 - \cos\left(\frac{ka^2}{2z}\right) U_2\right] \tag{2-234}$$

where

$$U_1 = \frac{a}{\rho} J_1\left(\frac{ka\rho}{z}\right) - \left(\frac{a}{\rho}\right)^3 J_3\left(\frac{ka\rho}{z}\right) + \left(\frac{a}{\rho}\right)^5 J_5\left(\frac{ka\rho}{z}\right) - \cdots \tag{2-235}$$

$$U_2 = \left(\frac{a}{\rho}\right)^2 J_2\left(\frac{ka\rho}{z}\right) - \left(\frac{a}{\rho}\right)^4 J_4\left(\frac{ka\rho}{z}\right) + \cdots \tag{2-236}$$

Equations (2-233) through (2-236) are convenient for $\rho > a$. For $\rho < a$ it is more convenient to use

$$C = \frac{2\pi z}{k}\left[\sin\left(\frac{k\rho^2}{2z}\right) + \sin\left(\frac{ka^2}{2z}\right) V_0 - \cos\left(\frac{ka^2}{2z}\right) V_1\right] \tag{2-237}$$

$$S = \frac{2\pi z}{k}\left[\cos\left(\frac{k\rho^2}{2z}\right) - \cos\left(\frac{ka^2}{2z}\right) V_0 - \sin\left(\frac{ka^2}{2z}\right) V_1\right] \tag{2-238}$$

where

$$V_0 = J_0\left(\frac{ka\rho}{z}\right) - \left(\frac{\rho}{a}\right)^2 J_2\left(\frac{ka\rho}{z}\right) + \left(\frac{\rho}{a}\right)^4 J_4\left(\frac{ka\rho}{z}\right) - \cdots \tag{2-239}$$

$$V_1 = \left(\frac{\rho}{a}\right) J_1\left(\frac{ka\rho}{z}\right) - \left(\frac{\rho}{a}\right)^3 J_3\left(\frac{ka\rho}{z}\right) + \cdots \tag{2-240}$$

For a circular source of radius a with an arbitrary current distribution **J**, Eq. (2-227) would apply for $r \gtrsim 0.62\sqrt{8a^3/\lambda}$. For $r < 0.62\sqrt{8a^3/\lambda}$, which constitutes the near field, one should use Eq. (2-210) without approxi-

---

[1] Lord Rayleigh, On Pin-hole Photography, *Phil. Mag.*, ser. V, pp. 87–99, 1891.

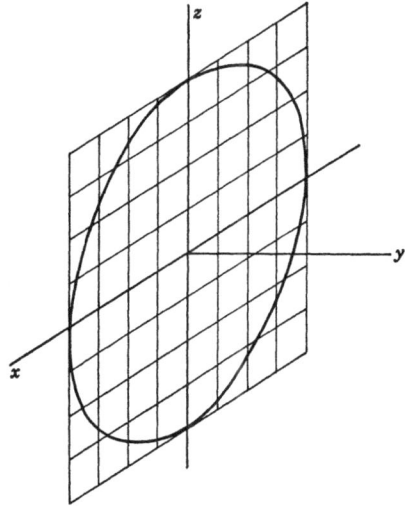

Figure 2-18.   Quantizing a circular source.

mation. In practice it would be necessary to resort to numerical integration techniques in evaluating either of these equations. In many cases it will be more convenient to break the surface of the source into small segments and apply the near-field techniques described in Sec. 2-14 for a rectangular source. If a quantized square is placed over the circular source as shown in Fig. 2-18, the results of Sec. 2-14 may be used directly. Zero amplitude is given to each segment which is located outside the circle. This technique may be used for a plane source of any shape.

2-18.   APPROXIMATE FAR–FIELD COMPUTATIONS

By now it should be quite obvious that the computation of the fields of an arbitrary current source may be very involved. Fortunately in practice most computations deal with far-field patterns because the point of observation is either in the far field of the total source or in the near-field or Fresnel region of the total source yet in the far field of each segment of a quantized source. The far-field approximations greatly simplify the integral expression for the field of a source. However, in many cases we are unable to get an exact solution of the far-field integral. Thus not only is the integral expression for the field an approximate one but the evaluation of the integral also may involve approximations. In spite of what appears to be an excessive number of approximations, experience has shown that satisfactory accuracy can be obtained.

Let us consider a line, or one-dimensional, source as illustrated in Fig. 2-19. The far-field pattern can be obtained by integrating the contribu-

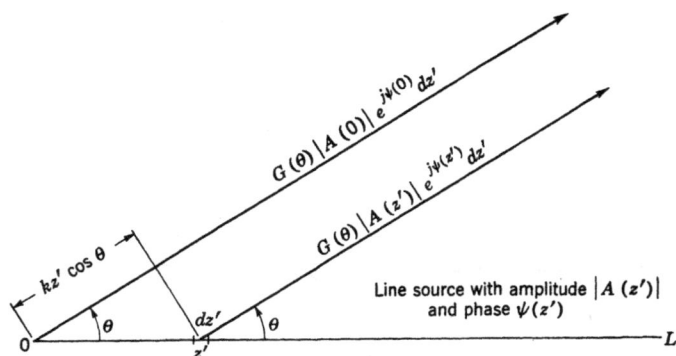

Figure 2-19.   Line source of current.

tions due to dz' from 0 to L as shown in Fig. 2-19 or by using Eq. (2-80). The relative far-field pattern is given by

$$E(\theta) = G(\theta) \int_0^L |A(z')| e^{j[\psi(z')+kz'\cos\theta]} \, dz' \tag{2-241}$$

where $G(\theta)$ is the pattern of an element dz' of the source. In general $|A(z')|$ and $\psi(z')$ may be functions such that an exact solution of Eq. (2-241) is not known. A standard procedure would be to employ numerical integration such as the trapezoidal rule or Simpson's rule.[1] A practical method that is very nearly the same as applying the trapezoidal rule is to represent the continuous source by an array of discrete elements which have the amplitude and phase of the continuous distribution at the points where the elements are located. An element of the approximating array will have the same radiation characteristics as the differential element dz', and the elements of the array need not be equally spaced. Experience with dipole arrays and continuous distributions indicates that an array with an element every $\lambda/2$ would, in most cases, give essentially the same pattern as a continuous distribution with the same amplitude and phase at corresponding points. For $\lambda/2$ spacing the integral in Eq. (2-241) is approximated by the series

$$\int_0^L |A(z')| e^{j[\psi(z')+kz'\cos\theta]} \, dz' \approx \sum_{m=0}^{2L/\lambda} A_m e^{j(\psi_m + m\pi\cos\theta)} \tag{2-242}$$

where A_m and ψ_m are the amplitude and phase, respectively, at distances $m\lambda/2$ along the source. Although the physical picture is that of an array

[1] W. C. Johnson, "Mathematical and Physical Principles of Engineering Analysis," chap. V, McGraw-Hill, New York, 1944.

of elements representing a continuous distribution, Eq. (2-242) is of the form that would be obtained by applying the trapezoidal rule except for the end terms of the series, which are off by a factor of 2.

Figures 2-20 to 2-23 illustrate the error that is involved in the array approximation. Figure 2-20 is the case of uniform amplitude and phase, i.e., broadside radiator. For this case the integral in Eq. (2-241) can be solved exactly and we obtain the familiar function $(\sin X)/X$, where $X = (\pi L/\lambda) \cos \theta$. The array representation for this case gives the function $(\sin NY)/(N \sin Y)$ where $Y = (\pi S/\lambda) \cos \theta$, $N = L/S$, and S is the distance between elements. Comparison of these patterns shows that very good

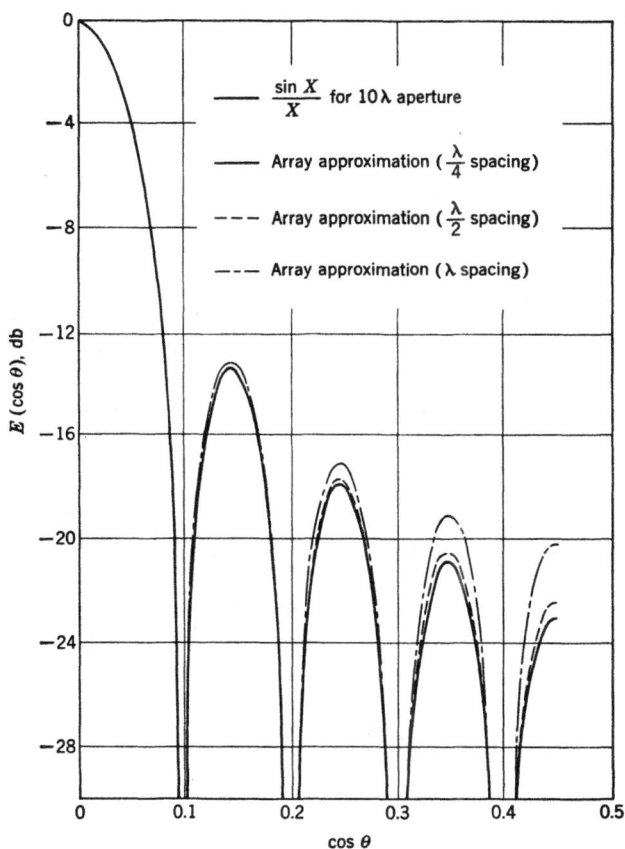

Figure 2-20. Computed far-field patterns comparing a continuous source with several arrays for $L = 10\lambda$, $A(z') = A_m = 1$, and $\psi(z') = \psi_m = 0$.

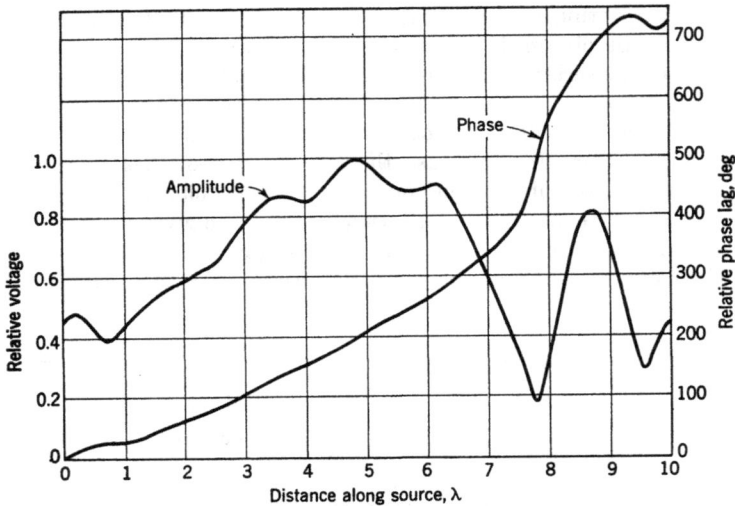

Figure 2-21. Example of nonuniform amplitude and phase distributions.

agreement is obtained when the elements are spaced $\lambda/2$ or less and reasonably good agreement is obtained for 1-λ spacing. Although this is the special case of a traveling wave source with infinite phase velocity, i.e., broadside radiation, the results are indicative of the accuracy of the array representation for any practical phase velocity.

The second case considered is that of the nonuniform amplitude and phase shown in Fig. 2-21. This source distribution has been obtained by the iterative method in Chap. 3 for a csc θ far-field pattern. Computed patterns for this case are shown in Fig. 2-22. The results indicate that, even for a source distribution that has rather large variations in amplitude and phase, the $\lambda/2$ separation gives a reasonably good approximation to the actual pattern and the $1 = \lambda$ spacing gives a fair approximation. The best agreement is in the region of the main beam.

The array approximation also may be used to represent surface and volume source distributions. The physical interpretation again is that a surface or volume array of discrete sources replaces the continuous distribution where the continuous distribution and the discrete elements have the same amplitude and phase at corresponding points. For a rectangular source of length L and width W and in the xz plane, the relative far field is

$$E(\theta,\phi) = G(\theta,\phi) \int_0^L \int_0^W |A(x',z')| e^{j\psi(x',z')} e^{jk(x' \sin \theta \cos \phi + z' \cos \theta)} \, dx' \, dz'$$

$$(2\text{-}243)$$

By the array approximation, Eq. (2-243) may be represented as

$$E(\theta,\phi) \approx G(\theta,\phi) \sum_{n=0}^{L/\Delta z'} \sum_{m=0}^{W/\Delta x'} A_{mn}e^{j\psi_{mn}}e^{jkm\,\Delta x'\,\sin\,\theta\,\cos\,\phi}e^{jkn\,\Delta z'\,\cos\,\theta} \qquad (2\text{-}244)$$

where $\Delta x'$ and $\Delta z'$ are the separations of the elements along the x' and z' coordinates, respectively. Equation (2-244) can be obtained from conventional array theory, which is well covered in the literature,[1] or it can be obtained from Eq. (2-202) by introducing the far-field approximations in the expansion of r_{mn} and noting that for $\Delta x'$ and $\Delta z'$ small ($\leqslant \lambda/2$), $(\sin X_n)/X_n$ and $(\sin X_m)/X_m$ are both approximately unity. For the special case where $\Delta x'$ and $\Delta z'$ are both equal to $\lambda/2$, Eq. (2-244) reduces to

$$E(\theta,\phi) \approx G(\theta,\phi) \sum_{n=0}^{2L/\lambda} \sum_{m=0}^{2W/\lambda} A_{mn}e^{j\psi_{mn}}e^{jm\pi\,\sin\,\theta\,\cos\,\phi}e^{jn\pi\,\cos\,\theta} \qquad (2\text{-}245)$$

An example of the two-dimensional array approximation for a circular source with nonuniform amplitude and phase is illustrated in Fig. 2-23.

[1] J. D. Kraus, "Antennas," chap. 4, McGraw-Hill, New York, 1950.

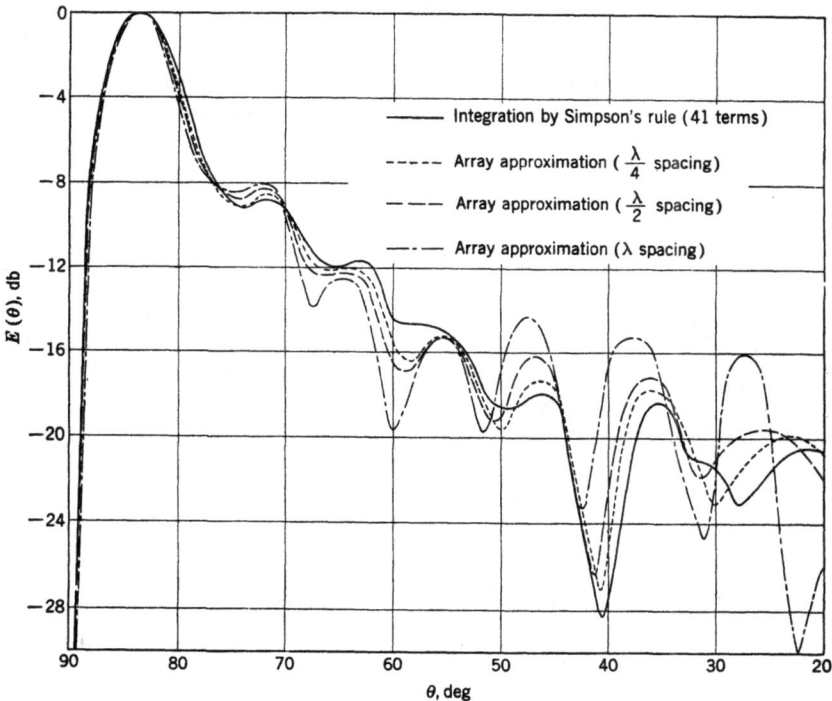

Figure 2-22. Computed far-field patterns for the source distribution given in Fig. 2-21.

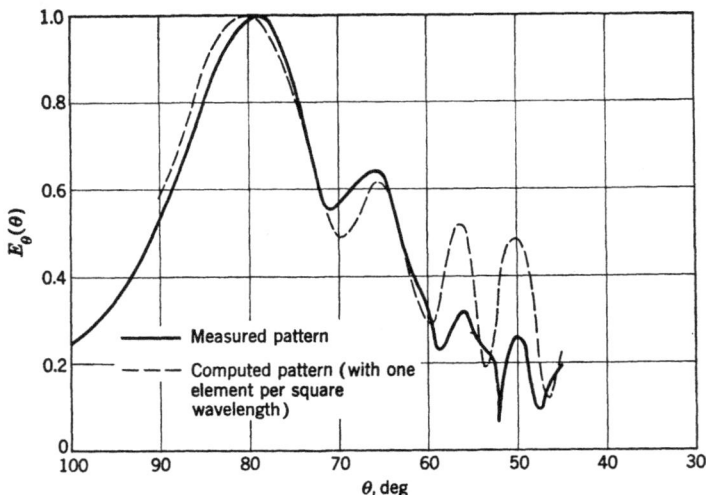

Figure 2-23. Comparison of measured and computed patterns for a 15-λ-diameter lens with nonuniform amplitude and phase (see Sec. 8-8).

Equation (2-244) may be applied to a plane source of any shape lying in the xz plane. The rectangle of length L and width W may be fitted over the source, and all elements lying outside the contour of the source will have zero magnitude.

2-19.   THE ANGLE OF MAXIMUM RADIATION BY THE PRINCIPLE OF STATIONARY PHASE

A problem of considerable practical importance is that of determining the angle of maximum radiation of a structure on which the current distribution is known. Consider a line source distribution of the form $|A(z')|e^{j\psi(z')}$. The relative far-field pattern of such a source is given by Eq. (2-241). The principle of stationary phase[1] states that for an integral of the form

$$P(\cos \theta) = \int_{z_1}^{z_2} U(z', \cos \theta)e^{jV(z', \cos \theta)} \, dz' \tag{2-246}$$

where U varies slowly with z', there is general cancellation of positive and negative portions of the integral except for ranges of stationary phase, i.e., where $\partial V/\partial z' = 0$. In other words, the main contribution of the integral is for those values of $\cos \theta$ for which $V(z', \cos \theta)$ has stationary values.

[1] G. N. Watson, "Theory of Bessel Functions," 2d ed., sec. 8.2, Cambridge, New York, 1945.

As an example let us consider a line source of length L in free space with uniform amplitude of unity and a constant phase velocity v. The integral in the far-field expression would be of the form

$$P(\cos \theta) = \int_0^L e^{-jkz'\left(\frac{c}{v}-\cos \theta\right)} dz' \tag{2-247}$$

where c is the velocity of light in free space. By the method of stationary phase the beam maximum should occur for the value of $\cos \theta$ for which

$$\frac{\partial}{\partial z'}\left[\left(\frac{c}{v} - \cos \theta\right) kz'\right] = 0 \tag{2-248}$$

Thus

$$\frac{c}{v} = \cos \theta_m \tag{2-249}$$

where θ_m is the angle of maximum radiation. This can be verified by performing the integration in Eq. (2-247) and finding the location of the maximum value of the resultant expression. For this case the principle of stationary phase gives the exact location of the maximum. This is not true in general because of the modifying effect of the amplitude function in Eq. (2-246). However, the method usually gives the location of the maximum with satisfactory accuracy.

The method of stationary phase can be extended to a two-dimensional source distribution. Consider a rectangular source in the xz plane. The relative far field for a source of amplitude $|A(x',z')|$ and phase $\psi(x',z')$ is given in Eq. (2-243) and can be rewritten as

$$E(\theta,\phi) = G(\theta,\phi) \int_{x_1}^{x_2} \int_{z_1}^{z_2} |A(x',z')| e^{j\xi(x',z',\theta,\phi)} dz' dx' \tag{2-250}$$

where $\xi = \psi(x',z') + kx' \sin \theta \cos \phi + kz' \cos \theta$. The main contribution of the integral in Eq. (2-250) is in the region where

$$\frac{\partial \xi}{\partial x'} = \frac{\partial \xi}{\partial z'} = 0 \tag{2-251}$$

2-20. EFFECT OF A FINITE GROUND PLANE ON THE FAR FIELD OF A SOURCE

For an antenna in or on a finite ground plane and having substantial radiation in the plane of the ground plane, the extent of the ground plane may have significant effect on the radiation pattern of the antenna owing to electric current on the ground plane. By Huygens' principle the antenna and ground plane may be replaced by a current sheet consisting of the actual electric currents on the antenna and ground plane and the equivalent electric and magnetic currents over any antenna apertures that exist on the structure. Let us assume that the only significant currents are those on the antenna side of the ground plane and denote this surface as S'. One approach to the problem is to short out the electric currents and work in terms of the magnetic currents in the presence of the electrical conducting surface of shape S' (see Sec. 2-4). The boundary value problem may be quite formidable for this approach. The usual procedure is to utilize both the electric and magnetic currents over S'.

The classical method for determining the electric surface current over S' is to assume that the electric-current distribution is approximately the same as the current distribution which would exist on that portion of an infinite ground plane which corresponds to the finite ground plane. Thus the problem reduces to one of finding the fields of a source or sources in the presence of an infinite ground plane.

To illustrate these points consider the antenna shown in Fig. 2-24. Assuming that we know the fields in the aperture WL or can obtain a reasonable approximation to them, we start by considering the aperture to be in an infinite ground plane. If $\mathbf{K}_a$ is the equivalent magnetic-current distribution over the aperture WL, by image theory we may remove the ground plane and find the fields of magnetic-current source $2\mathbf{K}_a$. Solving for the tangential magnetic field in the plane of the ground plane (xz plane in this case) gives the equivalent electric current $\mathbf{J}_g$ flowing in the plane of the ground plane. Thus the source whose fields we are to determine is the surface S' with magnetic current $\mathbf{K}_a$ and electric current $\mathbf{J}_a$ over WL and electric current $\mathbf{J}_g$ over the remainder of S'. If there is no ground plane, the

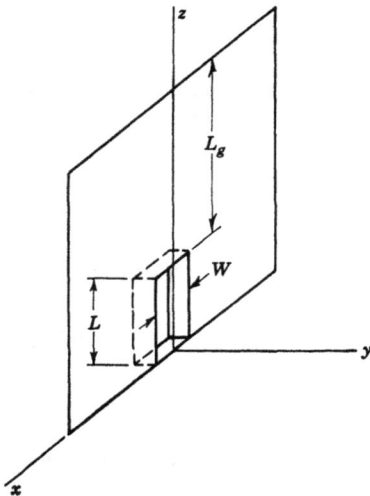

Figure 2-24. Slot backed by cavity in finite ground plane.

source is taken to be $\mathbf{K}_a$ and $\mathbf{J}_a$. The practical application of this method has been demonstrated by Elliott.[1]

Although the method just described is straightforward, solving for the ground plane currents and the radiation from the composite current sheet may get quite involved. This method can be simplified somewhat usually without serious loss in accuracy. The simplification arises from assuming that the electric current that is set up on the ground plane travels with a uniform amplitude and a constant phase velocity equal to the velocity of light in the medium surrounding the structure (free space in this case). If there is an aperture, as in Fig. 2-24, the ground plane current is adjusted in amplitude and phase to be a continuation of the equivalent electric current in the aperture. Since the resultant radiation from an aperture in a finite ground plane is a superposition of the fields of the electric- and magnetic-current sources, it is necessary to know the relative amplitudes and phases of these sources. If the aperture were in an infinite ground plane, the equivalent magnetic current in the aperture would produce the same far fields as the equivalent electric current in the aperture plus the actual electric current on the ground plane. For a finite ground plane the electric- and magnetic-current sources will have somewhat different far-field patterns. If the ground plane is large, it would appear to be a good approximation to adjust the amplitude and phase of the electric-current source relative to the magnetic-current source so that both sources would produce the same amplitude and phase at the maximum value of their respective far-field patterns as in the case of the infinite ground plane. Actually the adjustment of the two sources is not very critical. Good results are obtained in practice by letting the sources be in-phase at the origin and with amplitudes adjusted so that their pattern maxima are equal.

As an example of this simplified method for determining the effect of finite ground plane, let us assume that the aperture WL in Fig. 2-24 has fields in the form of a traveling wave propagating in the z direction with phase velocity v. If $v \lessapprox c$ and $\mathbf{K}_a = \mathbf{x}K_a$, where c is the velocity of light in free space, there will be substantial radiation in the direction of the z axis. Furthermore, if W is less than $\lambda/2$ or if the aperture distribution is separable, we can find the pattern in the yz plane by considering the one-dimensional problem of a line source lying on the z axis. The line source can be taken to be the type illustrated in Fig. 5-27b, in which case we have a length L of transverse magnetic-current elements with phase velocity v in the z direction and a line of longitudinal electric current with phase velocity v from 0 to L and a constant amplitude and phase velocity c over length L_g. In a practical endfire antenna v would probably equal c at the end of the aperture and be very nearly equal to c over the entire aperture. In this case one would be justified in assuming that both the electric current and

[1] R. S. Elliott, On the Theory of Corrugated Plane Surfaces, *Trans. IRE Antennas Propagation,* **AP-2**(2), April, 1954.

the magnetic current over the aperture travel with phase velocity c. Therefore, let the current sources be

$$K_z = A_1 e^{-jkz'} \qquad 0 \lesssim z' \lesssim L \tag{2-252}$$

and

$$J_z = A_2 e^{-jkz'} \qquad 0 \lesssim z' \lesssim L + L_g \tag{2-253}$$

where A_1 and A_2 are coefficients that take into account the amplitude of J_z with respect to K_z. The two sources are assumed to be in phase at $z' = 0$. The relative far field of K_z is

$$E_1 = A_1 e^{jX_1} \frac{\sin X_1}{X_1} \tag{2-254}$$

where

$$X_1 = \frac{kL}{2} (\cos \theta - 1) \tag{2-255}$$

Similarly the relative far field of J_z is

$$E_2 = A_2 \sin \theta e^{jX_2} \frac{\sin X_2}{X_2} \tag{2-256}$$

where

$$X_2 = \frac{k(L + L_g)}{2} (\cos \theta - 1) \tag{2-257}$$

Field E_1 will have maximum value A_1 at $\theta = 0$. Field E_2 has maximum value

$$E_{2,\text{max}} = A_2 \left| \sin \theta \frac{\sin X_2}{X_2} \right|_{\text{max}} \tag{2-258}$$

which will be a function of θ, L, and L_g. Setting $|E_1|_{\text{max}} = |E_2|_{\text{max}}$ gives

$$A_1 = A_2 \left| \sin \theta \frac{\sin X_2}{X_2} \right|_{\text{max}} \tag{2-259}$$

Thus the total relative far field is

$$E = \left(\left| \sin \theta \frac{\sin X_2}{X_2} \right|_{\text{max}} e^{jX_1} \frac{\sin X_1}{X_1} + \sin \theta e^{jX_2} \frac{\sin X_2}{X_2} \right) A_2 \tag{2-260}$$

A comparison of patterns computed from Eq. (2-260) and measured patterns for a slotted waveguide antenna in a finite ground plane are shown in Fig. 2-25. The actual antenna was known to have aperture fields that were nearly constant in amplitude and with a phase velocity somewhat less than c over most of the aperture.

The principal effect of the finite ground plane on an endfire source is the tilting of the beam away from the ground plane. The beam tilt is a

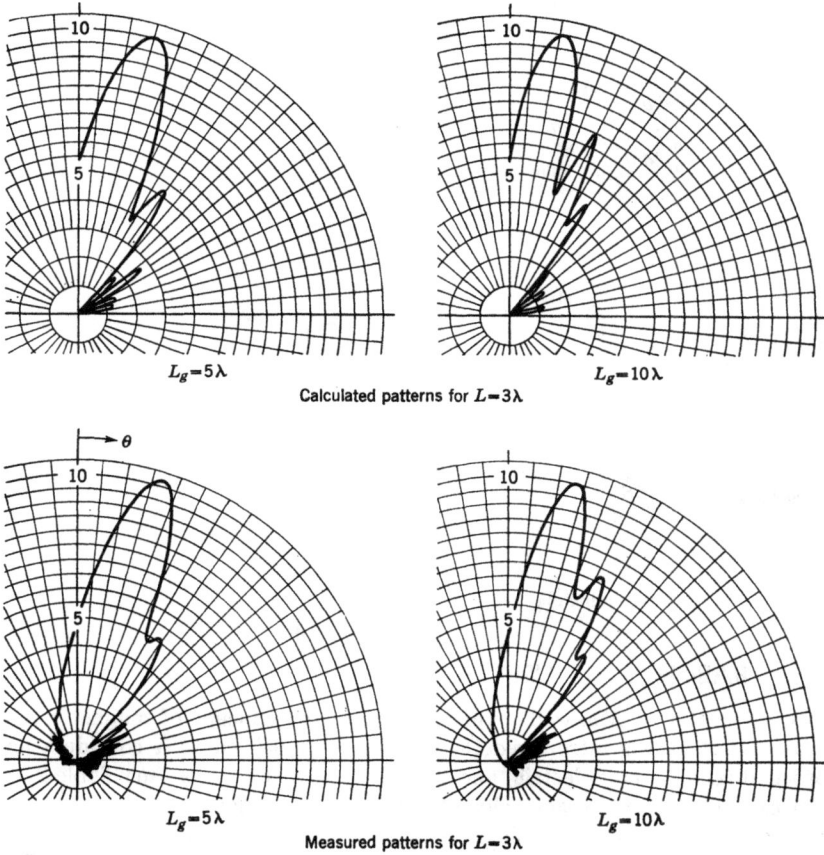

Figure 2-25.   Effect of ground plane length on an endfire antenna.

function of both L and L_g and can be obtained from Eq. (2-260). For small values of beam tilt (say $\theta_t < 20$ deg), Eq. (2-260) gives

$$\theta_t \approx \frac{49}{[(L + L_g)/\lambda]^{\frac{1}{2}}} \tag{2-261}$$

where θ_t is in degrees.

In general the tilt angle depends on the amplitude and phase along the source as well as the lengths L and L_g. The above method can be used to include nonuniform source distributions by the use of the appropriate expressions for E_1 and E_2.

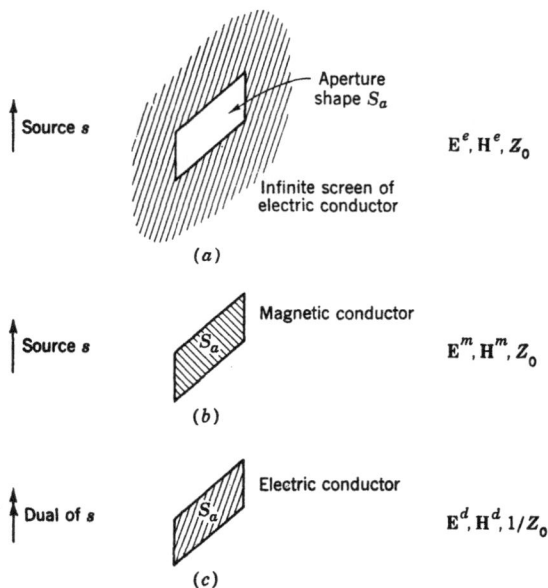

Figure 2-26. Babinet's principle

2-21. BABINET'S PRINCIPLE

The symmetry of Maxwell's equations when the concept of magnetic current and charge is introduced [Eqs. (2-6) to (2-9)] led to a consideration of the dual nature of electric- and magnetic-current sources (see Table 2-1). For example the free-space fields of a line source of magnetic current $\mathbf{M}$ are the same as the free-space fields of an electric-current source $\mathbf{I}$ having the same form as $\mathbf{M}$ when $\mathbf{E}_I$ is replaced by $\mathbf{H}_M$ and $-\mathbf{H}_I$ is replaced by $\mathbf{E}_M$. Physically the line of magnetic current could correspond to a long, narrow slit in a thin conducting plane of infinite extent with an electric field set up across the narrow dimension. The line of electric current would correspond to a thin current-carrying conductor the same width and length as the slit.

The slit and the thin conductor form a dual. They usually are referred to as complementary structures; that is, the two geometries may be fitted together without overlap to form a solid screen. A statement of the duality of electromagnetic fields for complementary screens has been obtained by an extension of Babinet's principle of optics.[1] In optics Babinet's principle states that for the transmission of light through apertures in absorbing screens the superposition of the separate fields on the source-free side of

[1] H. G. Booker, Slot Aerials and Their Relation to Complementary Wire Aerials (Babinet's Principle), *J. Inst. Elec. Engrs.*, **IIIA**:620–626, 1946.

any two complementary absorbing screens will produce the field that would exist without any screen.

Babinet's principle in optics makes no mention of polarization. The more exact statement of the principle by Booker considers polarization and deals with conducting screens, which are usually of more practical interest than absorbing screens. Referring to Fig. 2-26, the source s produces the field $\mathbf{E}^i, \mathbf{H}^i$ with no screen present. In a the source s produces the field $\mathbf{E}^e, \mathbf{H}^e$ to the right of the screen of electric conductor, and in b the source s produces the field $\mathbf{E}^m, \mathbf{H}^m$ to the right of the complementary screen of magnetic conductor. The quantity Z_0 is the characteristic impedance of the medium.

Babinet's principle states that

$$\mathbf{E}^e + \mathbf{E}^m = \mathbf{E}^i$$
$$\mathbf{H}^e + \mathbf{H}^m = \mathbf{H}^i \tag{2-262}$$

A proof of this statement of Babinet's principle is given in Harrington.[1]

If in Fig. 2-26b the original source s is replaced by its dual ($\mathbf{J}$ replaced by $\mathbf{K}$), Z_0 replaced by $1/Z_0$, and the magnetic conductor replaced by electric conductor as shown in c, then $\mathbf{E}^m$ is replaced by $\mathbf{H}^d$ and $\mathbf{H}^m$ by $-\mathbf{E}^d$ (see Table 2-1). Equation 2-262 beomes

$$\mathbf{E}^e + \mathbf{H}^d = \mathbf{E}^i$$
$$\mathbf{H}^e - \mathbf{E}^d = \mathbf{H}^i \tag{2-263}$$

Figure 2-26c represents a more practical complement of Fig. 2-26a.

If the screen and its complement have terminals (see example in Fig. 2-27), it can be shown that[2]

$$Z_1 Z_2 = \frac{Z_0{}^2}{4} \tag{2-264}$$

where Z_1 and Z_2 are the impedances of the two structures.

[1] R. F. Harrington, "Time-Harmonic Electromagnetic Fields," p. 365, McGraw-Hill, New York, 1961.
[2] Booker, *loc. cit.*

Figure 2-27. Impedance relationship between a slot and its complementary dipole.

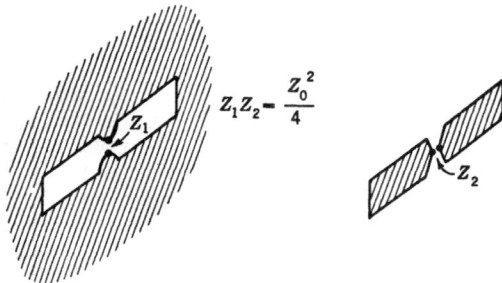

Greatest application of Babinet's principle is found in problems involving scattering and diffraction. However, the impedance relationship, Eq. (2-264), is used frequently in radiation problems. Figure 2-27 illustrates the application of Eq. (2-264). It is necessary to introduce a small gap in the dipole for the impedance Z_2 to have practical significance. Note that, if an antenna is self-complementary, its impedance is constant (i.e., independent of frequency) and equal to $Z_0/2$.

ADDITIONAL REFERENCES

A. Huygens' Principle

1. V. H. Rumsey, Equivalence Theorems Using Normally Directed Dipole and Quadrapole Antennas, URSI Symposium, Toronto, Canada, June, 1959.
2. J. Larmor, *Proc. London Math. Soc., Ser. 2*, **1**:1, January, 1903.
3. A. E. H. Love, *Proc. London Math. Soc., Ser. 2*, **1**:37, January, 1903.
4. H. M. McDonald, *Proc. London Math. Soc.*, **10**:91, 1911; *Phil. Trans. Roy. Soc. London, Ser. A*, 212, 295, 1912.
5. S. A. Schelkunoff, *Bell System Tech. J.*, **15**:92, 1936.
6. J. A. Stratton and L. J. Chu, *Phys. Rev.*, **56**:99, 1939.
7. J. A. Stratton, "Electromagnetic Theory," pp. 460–467, McGraw-Hill, New York, 1941.
8. S. A. Schelkunoff, "Electromagnetic Waves," p. 158, Van Nostrand, Princeton, N.J., 1943.
9. B. B. Baker and E. T. Copson, "The Mathematical Theory of Huygens' Principle," 2d ed., Oxford, Fair Lawn, N.J., 1950.
10. V. H. Rumsey, Huygens' Principle for Electromagnetic Waves, *Ohio State Univ. Antenna Lab. Rept.* 444-15, January, 1954.
11. V. H. Rumsey, Huygens' Principle as an Exact Physical Concept, *Univ. Calif. (Berkeley) Electron. Res. Lab. Rept.*, November, 1954.

B. Fields of Sources

12. S. Silver, "Microwave Antenna Theory and Design," chap. 3, McGraw-Hill, New York, 1949.
13. S. A. Schelkunoff, "Electromagnetic Waves," chap. IX, Van Nostrand, Princeton, N.J., 1943.
14. J. A. Stratton, "Electromagnetic Theory," chap. VIII, McGraw-Hill, New York, 1941.
15. E. C. Jordan, "Electromagnetic Waves and Radiating Systems," chaps. 10 and 15, Prentice-Hall, Englewood Cliffs, N.J., 1950.
16. M. K. Hu, Study of Near Zone Fields of Large Aperture Antennas, *Syracuse Univ. Res. Inst. Rept.* EE 282-574F (1 and 2), April, 1957.
17. R. C. Hansen and L. L. Bailin, Near Field Analysis of Circular Aperture Antennas, *Hughes Aircraft Co. Rept.* AFCRC-TN-59-780, August, 1959.
18. R. F. Goodrich and R. E. Hiatt, On Near Zone Antennas, *Univ. Mich. Dept. Elec. Eng. Radiation Lab. Rept.* 2861-1-F, June, 1959.

19. R. B. Barrar and C. H. Wilcox, On the Fresnel Approximation, *IRE Trans. Antennas Propagation*, **AP-6**:43–48, January, 1958.
20. E. Jacobs, Maximum Power Transfer between Large Aperture Antennas in the Fresnel Region, *Univ. Pennsylvania Inst. Cooperative Res. Rept.* IS-6OUR-6, September, 1960.
21. J. R. Wait and A. M. Conda, Radiation from Flush Mounted Antennas on a Non-ideal Ground Plane, Part 1, *Natl. Bur. Std. Rept.* 5511, Boulder, Colo., August, 1957.
22. L. J. Ricardi, W. C. Danforth, Jr., and M. H. Malone, "Near Field Characteristics of Array Antennas," Lincoln Laboratory, Massachusetts Institute of Technology, October, 1961.
23. D. Dalley, Near-zone Fields of Paraboloid Reflectors, *Rept.* 49 on Office of Naval Research Contract N7onr-29529, Electronics Research Laboratory, University of California, Berkeley, October, 1955.
24. J. Robieux, Near-zone Power Transmission Formulas, *Proc. IRE*, **47**:1161–1163, June, 1959.

C. Stationary Phase

25. S. Silver, "Microwave Antenna Theory and Design," chap. 4, McGraw-Hill, New York, 1949.
26. H. Bremmer, "Terrestrial Radio Waves," p. 20, Elsevier, Amsterdam, 1949.

D. Babinet's Principle

27. E. T. Copson, *Proc. Roy. Soc. London*, **186**:116, 1946.
28. S. Silver, "Microwave Antenna Theory and Design," p. 167, McGraw-Hill, New York, 1949.
29. E. C. Jordan, "Electromagnetic Waves and Radiating Systems," pp. 583–587, Prentice-Hall, Englewood Cliffs, N.J., 1950.
30. J. D. Kraus, "Antennas," pp. 361–364, McGraw-Hill, New York, 1950.
31. V. H. Rumsey, Huygens' Principle for Electromagnetic Waves, *Ohio State Univ. Antenna Lab. Rept.* 444-15, p. 12, January, 1954.
32. R. E. Collin, "Field Theory of Guided Waves," pp. 29–34, McGraw-Hill, New York, 1960.

PROBLEMS

2-1. Verify that ϕ and $\mathbf{A}$ in Eqs. (2-14) and (2-15) are solutions of Eqs. (2-16) and (2-17), respectively.

2-2. Verify Eq. (2-18).

2-3. If only transverse components are considered, show that in the Fresnel and far-field regions,

$$\mathbf{E} = -j\omega\mathbf{A} = -j\omega(\mathbf{\theta}A_\theta + \mathbf{\phi}A_\phi)$$

and

$$\mathbf{H} = \frac{jk}{\mu}(\mathbf{\theta}A_\phi - \mathbf{\phi}A_\theta)$$

2-4. Discuss the significance of the ratio E_θ/H_ϕ in the Fresnel region.

2-5. Show that the usual far-field criterion is obtained for a line source of length L when the maximum phase error contributed by the third term in Eq. (2-78) is $\pi/8$ rad. What is the Fresnel region criterion when the fourth term contributes a maximum error of $\pi/8$ rad? Repeat for a rectangular source of length L and width W. Repeat for a circular source of radius a.

2-6. If the ring radiator in Fig. 2-17 is a slot of width W in an infinite ground plane, show that the far electric field is given by

$$E_\theta(\theta,\phi) = \frac{jke^{-jkr}}{4\pi r} \int_0^{2\pi} \int_{a-W}^{a} [E_y(\rho',\phi') \sin\phi + E_x(\rho',\phi') \cos\phi]\, e^{jk\rho' \sin\theta \cos(\phi-\phi')}$$
$$\times \rho'\, d\rho'\, d\phi'$$

$$E_\phi(\theta,\phi) = \frac{jke^{-jkr}}{4\pi r} \int_0^{2\pi} \int_{a-W}^{a} \cos\theta [E_y(\rho',\phi') \cos\phi - E_x(\rho',\phi') \sin\phi]$$
$$\times \, {}^{jk\rho' \sin\theta \cos(\phi-\phi')} \rho'\, d\rho'\, d\phi'$$

where $E_x(\rho',\phi')$ and $E_y(\rho',\phi')$ are the fields in the slot.

2-7. If W is small enough that radial variation can be neglected in Prob. 2-6, show that the relative far field for a semicircular source extending from $\phi = 0$ to $\phi = \pi$ is given by

$$E_\theta(\theta,\phi) = \int_0^\pi [E_y(\phi') \sin\phi + E_x(\phi') \cos\phi]\, e^{jka \sin\theta \cos(\phi-\phi')}\, d\phi'$$

$$E_\phi(\theta,\phi) = \cos\theta \int_0^\pi [E_y(\phi') \cos\phi - E_x(\phi') \sin\phi]\, e^{jka \sin\theta \cos(\phi-\phi')}\, d\phi'$$

2-8. Find the phase distribution on the source of Prob. 2-7 so that it will produce maximum radiation in the direction $\phi = \pi/2$ and $\theta = \theta_m$. *Hint:* One can picture a planar traveling wave as producing the beam. The phase on a planar source of arbitrary shape can be obtained by sampling the phase of the traveling wave over the region occupied by the source.

2-9. The pattern of a planar source with an arbitrary phase distribution can be found by the methods of this chapter. For the special case where the phase is constant, the beam maximum is broadside to the source. One can concentrate,

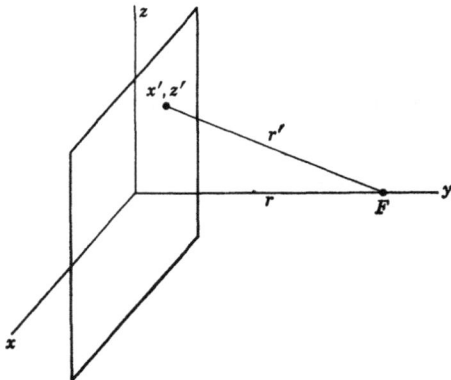

Figure 2-9P.

or focus, the energy from the source to within the Fresnel or near-field region by introducing an appropriate phase taper over the source. This phase taper can be obtained by placing a point source at the desired point of focus and determining the phase distribution it produces over the surface of the source. This is illustrated in Fig. 2-9P, where F is the focal point for on-axis focusing. The relative phase over the source is given by

$$\psi(x',z') = k(r' - r) = kF\left[\left(1 + \frac{x'^2 + z'^2}{F^2}\right)^{\frac{1}{2}} - 1\right]$$

The resulting source distribution for focus at distance F along the y axis is

$$A(x',z') = |A(x',z')|e^{i\psi(x',z')}$$

Show that for the focused condition in the Fresnel region the pattern over the focal plane ($y = F$) and in the vicinity of the y axis is the same as the far field of the unfocused aperture [$\psi(x',z') = 0$].

CHAPTER **3**

SYNTHESIS OF
CONTINUOUS
SOURCE
DISTRIBUTIONS

3-1. INTRODUCTION

An important problem in the theory of radiation from antennas is that of determining a source distribution to produce a given radiation pattern. This is antenna synthesis as contrasted to antenna analysis, which is concerned with finding the fields produced by a given source distribution. Although the general synthesis problem remains unsolved, a substantial body of literature has been written on the subject of antenna synthesis, and a number of methods have been developed for certain restricted classes of patterns and sizes of sources.

A practical antenna may consist of an array of discrete sources, a continuous source distribution, or some combination of the two. Much of the past work on the synthesis problem has been concerned with the determination of an array of discrete sources to produce a given far-field pattern. Good accounts of this work are available in the literature.[1] The interest in aperture antennas during World War II spurred work on the synthesis of continuous sources. In many practical applications a continuous source

---

[1] S. Silver, "Microwave Antenna Theory and Design," chap. 9, McGraw-Hill, New York, 1949; E. C. Jordan, "Electromagnetic Waves and Radiating Systems," chap. 12, Prentice-Hall, Englewood Cliffs, N.J., 1950; J. D. Kraus, "Antennas," chap. 4, McGraw-Hill, New York, 1950.

may be represented by an array of discrete sources having separations of $\lambda/2$ or less (see Chap. 2). Thus synthesis techniques for continuous sources may sometimes be used for discrete sources and vice versa.

It is theoretically possible to obtain an arbitrarily close approximation to any desired pattern from a source of finite size.[1] Experimental observations, however, indicate that there is a practical limit to the gain of a finite source. This important difference between theoretical possibility and experimental actuality has been pointed out in the literature.[2] Any attempt to increase the gain of an antenna beyond a certain value (commonly referred to as supergaining) invariably results in a rapid rate of change of phase and amplitude along the source which produces strong destructive interference in the field of the source. A given field strength can be maintained only by a sharp increase in the magnitude of the source function. It is evident that a radiation pattern having a fixed maximum strength but an increased gain must show a decrease in radiated power. Thus the increase in the source amplitude must arise because of an increase in reactive power rather than radiated power. Large reactive currents result in large ohmic losses, while rapid phase and amplitude variations along the source necessitate precise adjustment of the source function. These are the practical limitations of supergain antennas. The practical synthesis problem is not one of finding a source distribution to produce a specified radiation pattern exactly but rather one of finding a physically realizable source distribution which satisfactorily approximates the specified pattern.

The synthesis problem as considered in this chapter is that of determining a practical source distribution that will give a prescribed far-field pattern. However, this is not the complete synthesis problem. One still has to come up with an antenna geometry that will produce the desired source distribution. This is considered as a problem in antenna design and is discussed in Chap. 4.

3-2. SYNTHESIS OF LINE SOURCES

The relative far-field pattern $E(k\cos\theta)$ of a complex line source $|A(z')|e^{j\psi(z')}$ of length L with element pattern $G(k\cos\theta)$ and lying along the z axis is given by Eq. (2-241), which is rewritten here for convenience as

$$E(k\cos\theta) = G(k\cos\theta)\int_0^L |A(z')|e^{j[\psi(z')+kz'\cos\theta]}\,dz' \qquad (3\text{-}1)$$

[1] C. J. Bouwkamp and N. G. deBruijn, The Problem of Optimum Antenna Current Distributions, *Phillips Res. Rept.*, 1:135–138, January, 1946; P. M. Woodward and J. D. Lawson, The Theoretical Precision with Which an Arbitrary Radiation Pattern May Be Obtained from a Source of Finite Size, *J. Inst. Elec. Engrs.*, Part 3, pp. 363–369, 1948.

[2] Woodward and Lawson, *loc. cit.*; T. T. Taylor, A Discussion of the Maximum Directivity of an Antenna, *Proc. IRE*, 1948, p. 1135; H. J. Riblet, A Note on the Maximum Directivity of an Antenna, *Proc. IRE*, 1958, p. 620; L. J. Chu, Physical Limitations of Omnidirectional Antennas, *J. Appl. Phys.*, 1948, p. 1163.

From this equation one can compute the relative far-field pattern produced by a given line source distribution. Once the source distribution is specified in amplitude and phase, the far-field pattern is determined uniquely in both magnitude and phase. The synthesis problem, that of finding a source function which will produce a given pattern function, is much more difficult, and except for certain special cases an exact solution does not exist. An attempt will be made to present the synthesis methods that have been found to be useful in practice.

A. Fourier Transform Method

Equation (3-1) may be written

$$F(k \cos \theta) = \frac{E(k \cos \theta)}{G(k \cos \theta)} = \int_{-\infty}^{\infty} |A(z')| e^{j[\psi(z') + kz' \cos \theta]} \, dz' \qquad (3\text{-}2)$$

That is, the element pattern $G(k \cos \theta)$ may be absorbed into the field function $F(k \cos \theta)$ where $F(k \cos \theta)$ is simply the far field of a line source in which the elemental radiator is isotropic. The infinite integral will be used, although $A(z')$ will usually be taken as zero outside some finite interval L, where L is the length of the source.

Equation (3-2) will be recognized as one integral of a Fourier transform pair. The other integral of the pair is given by

$$A(z') = \frac{1}{2\pi} \int_{-\infty}^{\infty} F(k \cos \theta) e^{-jz'k \cos \theta} \, d(k \cos \theta)\dagger \qquad (3\text{-}3)$$

Equation (3-3) is the basic equation of antenna synthesis. It permits one to specify a far-field function $F(k \cos \theta)$ and compute the corresponding source function $A(z')$. The requirements on $F(k \cos \theta)$ are that it be at least piecewise continuous and that the integral $\int_{-\infty}^{\infty} |F(k \cos \theta)| d(k \cos \theta)$ exist.

In practice we usually specify only the magnitude of $F(k \cos \theta)$ over a portion of the angular spectrum. Thus we are left with an infinity of possibilities for $A(z')$ depending on the form that we assign to the field function outside the specified interval and the phase that we assign it over the entire interval for which $F(k \cos \theta)$ exists. However, if we impose the condition $A(z') = 0$ outside the finite interval L (the length of the source), then $A(z')$ is unique for a given $F(k \cos \theta)$ if $F(k \cos \theta)$ is known over a continuous but finite range of $k \cos \theta$. If $F(k \cos \theta)$ is zero everywhere outside some finite interval, then $A(z')$ must have an infinite support; i.e., the source must be of infinite extent.

The mathematical problem can be simplified and a satisfactory $A(z')$ can sometimes be obtained by assuming that the phase of $F(k \cos \theta)$ is constant (usually zero). A practical method of selecting the phase of $F(k \cos \theta)$ as well as the amplitude outside a specified interval is described in Sec. 3-2E.

† The factor $1/2\pi$ usually will be disregarded in the discussions that follow, since we are interested primarily in relative distributions.

The evaluation of Eq. (3-3) in some instances can be facilitated by making use of the convolution integral of Fourier transform theory. If the field function $F(k \cos \theta)$ can be expressed as a product

$$F(k \cos \theta) = F_1(k \cos \theta) F_2(k \cos \theta) \tag{3-4}$$

where

$$F(k \cos \theta) = \mathrm{FT}\ A(z')\dagger \tag{3-5}$$
$$F_1(k \cos \theta) = \mathrm{FT}\ A_1(z') \tag{3-6}$$
$$F_2(k \cos \theta) = \mathrm{FT}\ A_2(z') \tag{3-7}$$

then by the convolution theorem,

$$\begin{aligned} A(z') &= \int_{-\infty}^{\infty} A_1(z' - w) A_2(w)\ dw \\ &= \int_{-\infty}^{\infty} A_1(w) A_2(z' - w)\ dw \end{aligned} \tag{3-8}$$

Perhaps the greatest value of the Fourier transform method of antenna synthesis is in recognizing that the source and far-field functions form a transform pair and utilizing the vast amount of information that is available on Fourier transform theory, particularly in circuit theory.

Figure 3-1 lists some of the more common transform pairs. It should be kept in mind that the transformation goes both ways and that each function of the transform pair may represent either the source or the far field. For example a constant source function over a finite interval (1a in Fig. 3-1) gives rise to a $(\sin X)/X$ far field. On the other hand a $(\sin X)/X$ source function (2a in Fig. 3-1) gives rise to a constant far field over a finite range of angles. Shifting the far-field function [for example, the shifted $(\sin X)/X$ in 3b in Fig. 3-1] corresponds to a phase term in the uniform source function which we interpret as a traveling wave. The relationship between the phase velocity of the traveling wave and the angle of maximum radiation was discussed in Chap. 2.

Equation (3-3) enables one to specify a far-field pattern $F(k \cos \theta)$ and compute a source distribution that will produce the specified pattern. However, the source function $A(z')$ as determined by Eq. (3-3) will not, in general, be restricted in length. Utilizing a finite length L of the source function gives an approximate pattern

$$F_a(k \cos \theta) = \int_{-L/2}^{L/2} A(z') e^{jz'k \cos \theta}\ dz' \tag{3-9}$$

which by use of Eq. (3-3) can also be expressed as

$$F_a(k \cos \theta) = \int_{-L/2}^{L/2} \int_{-\infty}^{\infty} F(\xi) e^{jz'(k \cos \theta - \xi)}\ d\xi\ dz' \tag{3-10}$$

where ξ replaces $k \cos \theta$ of Eq. (3-3).

† FT is used as an abbreviation for Fourier transform.

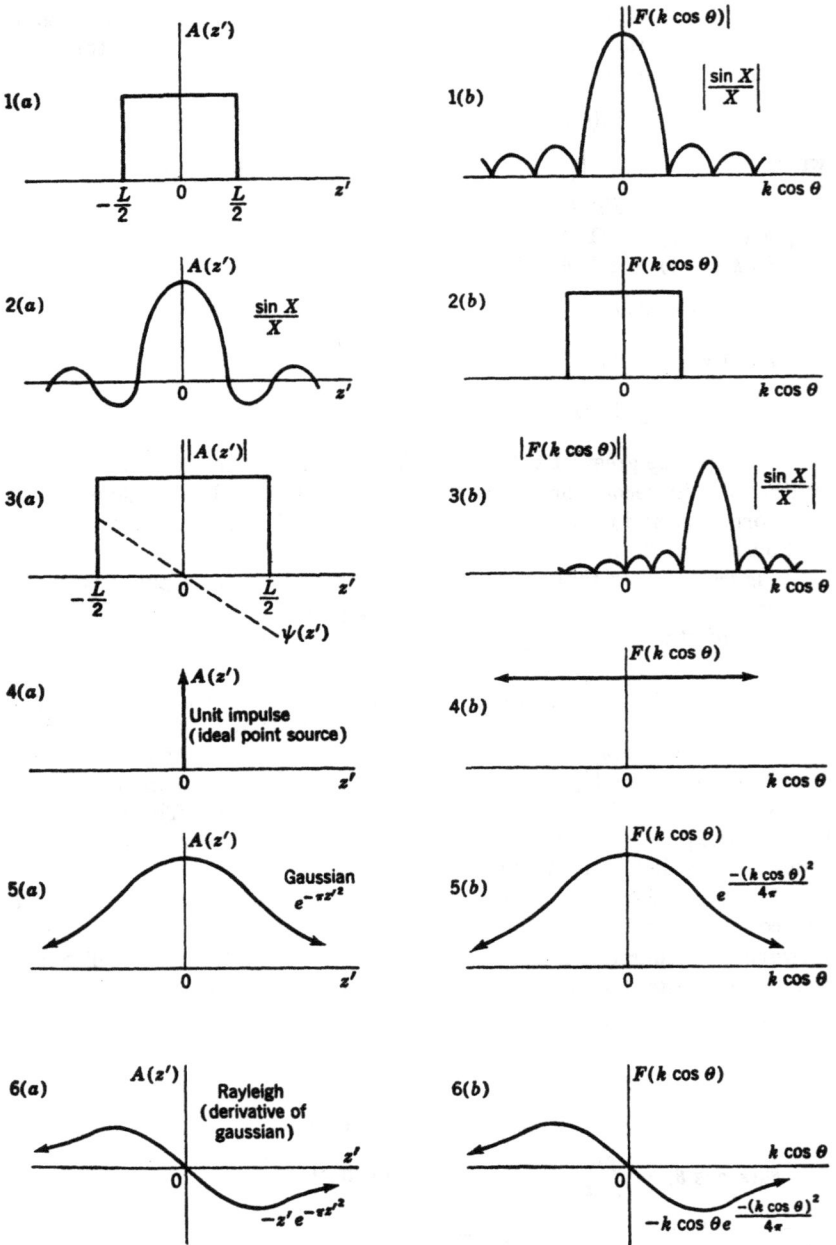

1(a)

1(b) $\left|\dfrac{\sin X}{X}\right|$

2(a) $\dfrac{\sin X}{X}$

2(b)

3(a)

3(b) $\left|\dfrac{\sin X}{X}\right|$

4(a)   Unit impulse (ideal point source)

4(b)

5(a)   Gaussian $e^{-\pi z'^2}$

5(b) $e^{\frac{-(k\cos\theta)^2}{4\pi}}$

6(a)   Rayleigh (derivative of gaussian) $-z'e^{-\pi z'^2}$

6(b) $-k\cos\theta e^{\frac{-(k\cos\theta)^2}{4\pi}}$

Figure 3-1.   Some common Fourier transform pairs.

Interchanging the order of integration in Eq. (3-10) and performing the integration with respect to z' give

$$F_a(k \cos \theta) = L \int_{-\infty}^{\infty} F(\xi) \frac{\sin \dfrac{L(k \cos \theta - \xi)}{2}}{\dfrac{L(k \cos \theta - \xi)}{2}} \, d\xi \tag{3-11}$$

The approximation in Eqs. (3-9) and (3-11) is such that the mean squared error is minimized.[1] The approximation becomes better with increasing source length. Equation (3-11) is known as the Dirichlet formulation of the approximation. It is interpreted as a convolution integral and can be readily evaluated by computer techniques.[2] In some instances, however, it is more convenient to obtain the approximate pattern from Eq. (3-9).

B. The Antenna as a Filter

The Fourier transform techniques provide a means for considering the antenna as a filter. This has been a particularly helpful concept in radio astronomy.[3]

Consider the Fourier transform pair

$$F(\cos \theta) = \frac{1}{2\pi} \int_{-\infty}^{\infty} A(kz') e^{jkz' \cos \theta} \, d(kz') \tag{3-12}$$

$$A(kz') = \int_{-\infty}^{\infty} F(\cos \theta) e^{-jkz' \cos \theta} \, d(\cos \theta) \tag{3-13}$$

In radiometry we are interested in power density. The far-field power density $U(\cos \theta)$ is

$$U(\cos \theta) \sim F(\cos \theta) F^*(\cos \theta) \tag{3-14}$$

or

$$U(\cos \theta) \sim \text{FT } A(kz') \text{ FT } A^*(-kz') \tag{3-15}$$

where the asterisk denotes complex conjugate. The quantity $U(\cos \theta)$ is the power pattern of the antenna. Let the inverse transform of $U(\cos \theta)$ be $P(kz')$. Since $U(\cos \theta)$ is the product of two transformable functions [Eq. (3-15)], the inverse transform gives the convolution integral

$$P(kz') = \int_{-\infty}^{\infty} A^*(w) A(kz' + w) \, dw \tag{3-16}$$

We recognize Eq. (3-16) to be the source autocorrelation function.

[1] J. Ruze, Physical Limitations on Antennas, *MIT Res. Lab. Electron. Tech. Rept.* 248, October, 1952.

[2] F. E. Brammer, A Use of the Analog Computer in Antenna Pattern Studies, *Case Inst. Technol. Sci. Rept.* 6, Cleveland, Ohio, 1959.

[3] J. L. Pawsey and R. N. Bracewell, "Radio Astronomy," International Monographs on Radio, Oxford, Fair Lawn, N.J., 1955.

The power density functions $U(\cos \theta)$ and $P(kz')$ form the Fourier transform pair

$$U(\cos \theta) = \frac{1}{2\pi} \int_{-\infty}^{\infty} P(kz')e^{jkz' \cos \theta} \, d(kz') \tag{3-17}$$

$$P(kz') = \int_{-\infty}^{\infty} U(\cos \theta)e^{-jkz' \cos \theta} \, d(\cos \theta) \tag{3-18}$$

If we let $\cos \theta$ and kz' correspond to t and ω, respectively, then z'/λ corresponds to frequency f. The quantity z'/λ is called *spatial frequency* and is expressed in cycles per radian.

The function $P(kz')$ as obtained from Eq. (3-16) can be thought of as the filter characteristic of the antenna. That is, the antenna filters spatial frequencies just as a network filters conventional frequency.

As an example to illustrate these points consider a uniformly illuminated source of length L as shown in Fig. 3-2a. The far-field pattern is the $(\sin X)/X$ function of Fig. 3-2b. The far-field power pattern is the $(\sin X/X)^2$ function in Fig. 3-2c, and the source autocorrelation function or filter characteristic [from either Eq. (3-16) or (3-18)] is the triangular function in Fig. 3-2d. According to Fig. 3-2d the antenna will not respond to spatial frequencies greater than L/λ.

To summarize, the antenna aperture is viewed as a filter with a certain

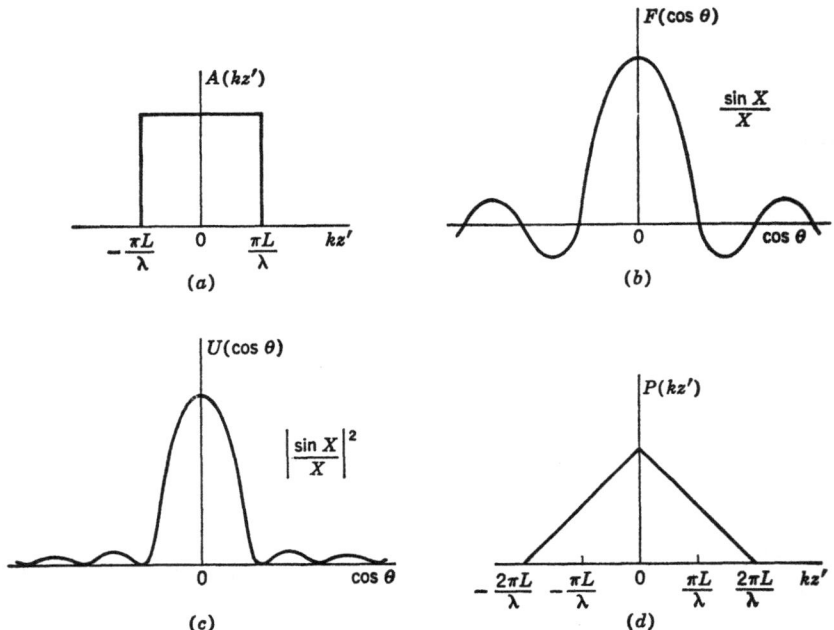

Figure 3-2.   Transform functions for a line source with uniform excitation.

spatial frequency response $P(kz')$ and a certain "waveform response" $U(\cos\theta)$ to a unit impulse input $\delta(\cos\theta)$. The interpretation of $\delta(\cos\theta)$ is that of a point source transmitter at direction θ in the far field of the receiving antenna. For an aperture extending from $-L/2$ to $L/2$ along the z' axis the quantity L/λ is the cutoff frequency of the filter. The antenna will not show any response to spatial frequency components of an incoming signal that are greater than L/λ. To detect fine detail of a far-field distribution of emitters, one may increase the filter bandwidth, i.e., widen the aperture, to cover all significant spatial frequencies in the spectrum of the incoming signal. Alternatively, we can, in principle, measure the finite spatial frequency spectrum detected by our antenna of length L and construct from this the entire spectrum of the far-field distribution of emitters. This is possible in principle because the spatial frequency spectrum will be an analytic function, and if any piece of it is known exactly, it can be mathematically extended to construct the entire function.[1] This is not very feasible in practice, because of the precision with which the output of the antenna would have to be measured and the deteriorating effects of noise.

It should be noted that, when the antenna is considered as a filter, $\cos\theta$ and kz' correspond to t and ω, respectively, whereas in the Fourier transform synthesis method $-k\cos\theta$ and z' correspond to ω and t, respectively. The former analogy provides the concept of spatial frequency. The latter analogy compares phase constant and angular frequency and hence enables us to compare complex frequency and complex propagation constant. This makes network synthesis methods based on the Laplace transform available to the antenna engineer. This is discussed in the next section.

C. Laplace Transform Method

The first significant progress in pattern synthesis for continuous sources occurred when it was recognized that the far-field function and the source function are a Fourier transform pair. However, the source and far-field functions also may be interpreted as a Laplace transform pair. With this interpretation the poles of the pattern function correspond to damped traveling waves at the source just as the poles of the complex frequency response of a network correspond to damped oscillations in the time domain. This approach enables one to apply directly to the antenna problem a considerable amount of network synthesis theory.

The Laplace transform pair from circuit theory is[2]

$$f(t) = \frac{1}{2\pi j}\int_{c-j\infty}^{c+j\infty} g(s)e^{st}\,ds \tag{3-19}$$

$$g(s) = \int_{0}^{\infty} f(t)e^{-st}\,dt \tag{3-20}$$

[1] Y. T. Lo, On the Theoretical Limitation of a Radio Telescope in Determining the Sky Temperature Distribution, *J. Appl. Phys.*, **32**(10): 2052–2054, October, 1961.

[2] See, for example, Gardner and Barnes, "Transients in Linear Systems," chap. 3, Wiley, New York, 1942.

where s is the usual complex frequency $\sigma + j\omega$ (see Fig. 3-3). The quantity $f(t)$ must be single valued almost everywhere in the range $t \gtrless 0$ and must not increase so rapidly as $t \to \infty$ that $e^{-\sigma t}$ will not predominate. The quantity $g(s)$ is defined only in the region of absolute convergence of the integral in Eq. (3-20).

The Laplace transform is a powerful tool in network theory. It may also be used to advantage in antenna theory, in particular pattern synthesis. Let us introduce the complex quantity

$$\gamma = \alpha + j\beta \tag{3-21}$$

where

$$\beta = k \cos \theta \tag{3-22}$$

The quantity γ is the complex propagation constant defined in Chap. 1, and it is analogous to the complex frequency s in network theory. The quantity γ has the dimension of inverse length, whereas s has the dimension of inverse time. The Laplace transform of a function of s results in a function of time, whereas the Laplace transform of a function of γ results in a function of length. A Laplace transform pair involving a field function $F(\gamma)$ and a source function $A(z')$ beginning at the origin may be written as

$$F(\gamma) = \int_0^\infty A(z')e^{\gamma z'}\, dz' \tag{3-23}$$

$$A(z') = \frac{1}{2\pi j} \int_{-c-j\infty}^{-c+j\infty} F(\gamma)e^{-\gamma z'}\, d\gamma \tag{3-24}$$

where $-\gamma$ and z' are analogous to s and t, respectively. Equations (3-23) and (3-24) reduce to the Fourier transform pair (3-2) and (3-3) for $\alpha = 0$.

The profile of $F(\gamma)$ along the $j\beta$ axis in the interval $-k \gtrless \beta \gtrless k$ is the far-field pattern. This is what we see when we plot a measured radia-

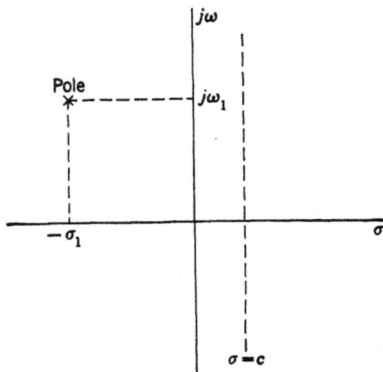

Figure 3-3. Coordinates of the complex s plane.

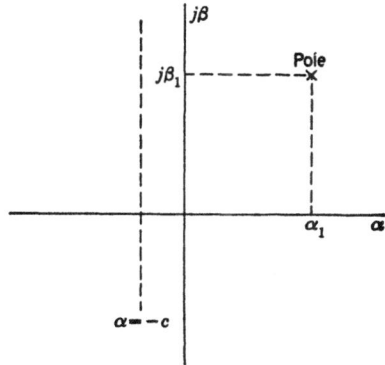

Figure 3-4.   Coordinates of complex γ plane.

tion pattern of an antenna. This interval is called the visible region. The region $|\beta| > k$ is related to the energy stored in the antenna fields. This will be discussed in more detail in Sec. H.

For a line source the synthesis problem is that of finding $A(z')$ when $F(\gamma)$ is specified. The analogous problem in network theory is that of finding the time function corresponding to a given function of complex frequency. Whereas a pole at $s_1 = -\sigma_1 + j\omega_1$ (see Fig. 3-3) corresponds to a damped time function of the form $e^{-\sigma_1 t}e^{j\omega_1 t}$, a pole at $\gamma_1 = \alpha_1 + j\beta_1$ (see Fig. 3-4) corresponds to a traveling wave along the line source of the form $e^{-\alpha_1 z'}e^{-j\beta_1 z'}$. Perhaps the most difficult part of the Laplace transform method of synthesis is determining the poles and zeros of the field function. The details of the method are best illustrated by an example.

An Example—The Butterworth Pattern

As an example of the Laplace transform synthesis procedure, consider the function

$$F(\gamma) = \frac{K}{b_0\gamma^n + b_1\gamma^{n-1} + \cdots + b_n} \tag{3-25}$$

Equation (3-25) with s as the independent variable would represent an n-stage low-pass filter. The magnitude of the antenna pattern is given by

$$|F(\beta)| = \frac{K'}{(\beta^{2n} + A_1\beta^{2n-2} + \cdots + A_n)^{\frac{1}{2}}} \tag{3-26}$$

Setting $d|F(\beta)|/d\beta^2 = 0$ at $\beta = 0$ requires that $A_{n-1} = 0$. Successive differentiation with respect to $\beta^4, \beta^6, \ldots, \beta^{2n-2}$ and requiring the derivatives to be zero at $\beta = 0$ gives

$$|F(\beta)| = \frac{K'}{(\beta^{2n} + A_n)^{\frac{1}{2}}} \tag{3-27}$$

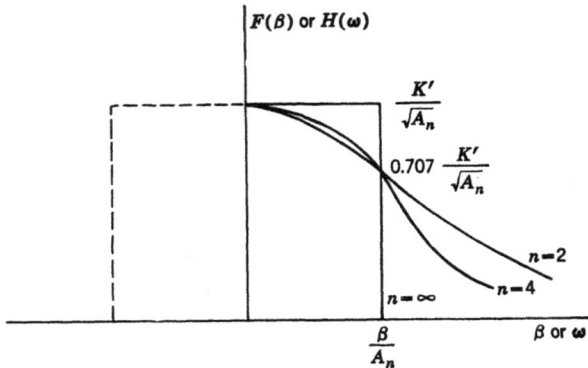

Figure 3-5.   Butterworth response curves.

In filter theory a frequency response

$$H(\omega) = \frac{K'}{(\omega^{2n} + A_n)^{1/2}}$$

(3-28)

is called a maximally flat or Butterworth response.[1] The shape of the spectrum for various values of n is indicated in Fig. 3-5. The larger the value of n, the better approximation Eq. (3-27) or (3-28) becomes to an ideal bandpass filter. As an antenna pattern, the spectrum (pattern) represented by Fig. 3-5 has no sidelobes.

It can be shown that the pole locations for $F(\gamma)$ lie on a circle in the γ plane with center on the $j\beta$ axis at a value $j\beta_m$ that corresponds to the direction of maximum radiation (or frequency of maximum response in the filter case). Furthermore, the half-power points of the response curve are the intersections of the circle and the $j\beta$ axis. The locations of the poles of $F(\gamma)$ are given by the following rules:

1. For odd values of n, a pole is always located on the real axis through the center of the circle. Other poles are displaced from this pole by π/n rad.

2. For even values of n, no poles are located on the real axis through the center of the circle. Poles are located $\pi/2n$ rad from this axis. All poles are displaced π/n rad.

3. In neither case are there poles on the $j\beta$ axis.

Examples of pole locations for $n = 2$ and $n = 3$ are shown in Fig. 3-6.

As a numerical example, consider a broadside Butterworth pattern ($\theta_m = 90$ deg) and a half-power beamwidth of 2 deg. The field function

[1] M. E. Van Valkenburg, "Network Analysis," p. 368, Prentice-Hall, Englewood Cliffs, N.J., 1955.

$F(\gamma)$ is of the form

$$F(\gamma) = \frac{1}{(\gamma - \gamma_1)(\gamma - \gamma_1^*)} \qquad (3\text{-}29)$$

where $\gamma_1 = 0.0124k + j0.0124k$. A partial-fraction expansion of Eq. (3-29) gives

$$F(\gamma) = \frac{K_1}{\gamma - \gamma_1} + \frac{K_1^*}{\gamma - \gamma_1^*} \qquad (3\text{-}30)$$

where

$$K_1 = \frac{1}{\gamma - \gamma_1^*} \Big]_{\gamma = \gamma_1} = \frac{1}{j0.0248k}$$

Therefore

$$F(\gamma) = \frac{1}{j0.0248k} \left(\frac{1}{\gamma - 0.0124k - j0.0124k} \right.$$
$$\left. - \frac{1}{\gamma - 0.0124k + j0.0124k} \right) \qquad (3\text{-}31)$$

The inverse transform gives

$$A(z') = \frac{1}{j0.0248k} \left[e^{(-0.0124 - j0.0124)kz'} - e^{(-0.0124 + j0.0124)kz'} \right] \qquad (3\text{-}32)$$

which except for a constant factor reduces to

$$A(z') = e^{-0.0124kz'} \sin 0.0124kz' \qquad (3\text{-}33)$$

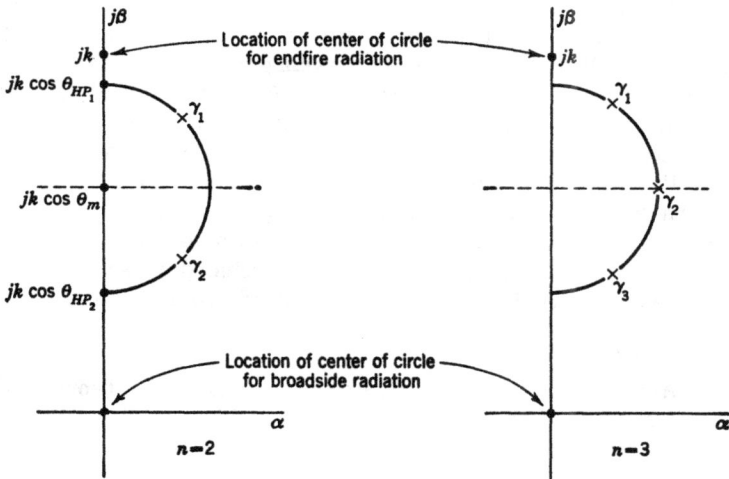

Figure 3-6. Pole locations ($\times$) for Butterworth patterns. θ_m is angle of maximum radiation and ($\theta_{HP_2} - \theta_{HP_1}$) is the half-power beamwidth.

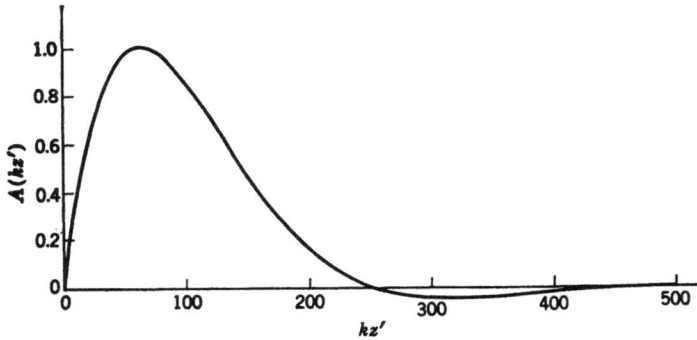

Figure 3-7. Source distribution for an $n = 2$ Butterworth pattern.

A graph of Eq. (3-33) is given in Fig. 3-7. This illustrates the damped sinusoidal amplitude of the source distribution which is analogous to a damped oscillatory time function in a network. If an angle of maximum radiation other than broadside had been chosen, a phase term would be present in Eq. (3-33).

It would require an infinitely long source with the distribution in Eq. (3-33) to give exactly the Butterworth pattern. In practice one may use a finite length of source without seriously distorting the pattern. A nice feature of the Laplace transform method is the relative ease with which the pattern of a source of finite length can be computed. This is due to the partial-fraction expansion of $F(\gamma)$ each term of which under the inverse Laplace transformation results in an exponential function in the form of a traveling wave with constant phase velocity. If the poles of $F(\gamma)$ are in the right half of the γ plane, the traveling waves will be attenuated in the positive z direction. Poles on the $jk \cos \theta$ axis correspond to unattenuated traveling waves. The source equation for the far field of a traveling wave with constant phase velocity can be readily evaluated (see Chap. 2).

Patterns of the source distribution given in Eq. (3-33) or (3-32) are shown in Fig. 3-8 for several source lengths along with the Butterworth pattern ($n = 2$) which would correspond to a source of infinite length. A Butterworth pattern for $n = 4$ is also included to show the significant change in pattern due to the introduction of two more poles.

Chebyshev Pattern

Another pattern shape derivable from network theory is the equal-ripple or Chebyshev response illustrated in Fig. 3-9. For this case the far-field pattern is of the form

$$|F(\beta)| = \frac{1}{[1 + \epsilon T_n{}^2(\beta)]^{1/2}} \tag{3-34}$$

where n is the order of the Chebyshev polynomial T_n and the effect of the constant ϵ is illustrated in Fig. 3-9. The response in Fig. 3-9 is for $n = 4$.

The pole locations for the Chebyshev case are on an ellipse[1] and may be obtained by multiplying the real parts of the poles for the Butterworth case of the same order by tanh a. This results in a projection of the Butterworth poles on a circle of radius cosh a onto an ellipse with semimajor axis of length cosh a and semiminor axis of length sinh a as shown in Fig. 3-10. The center of the ellipse is at $jk \cos \theta_m$, where θ_m is the angle of maximum radiation measured from the axis of the line source.

The Chebyshev pattern shape illustrated in Fig. 3-9 and the corresponding source distribution that would be obtained from the pole configuration in Fig. 3-10 differ entirely from the well-known Dolph-Chebyshev pattern

[1] *Ibid.*, p. 372.

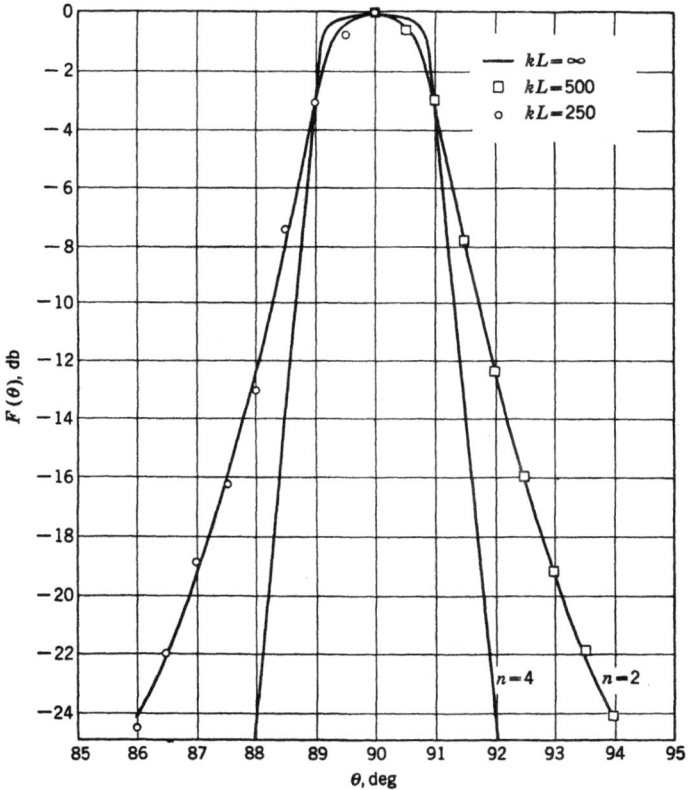

Figure 3-8.   Effect of source length on Butterworth pattern.

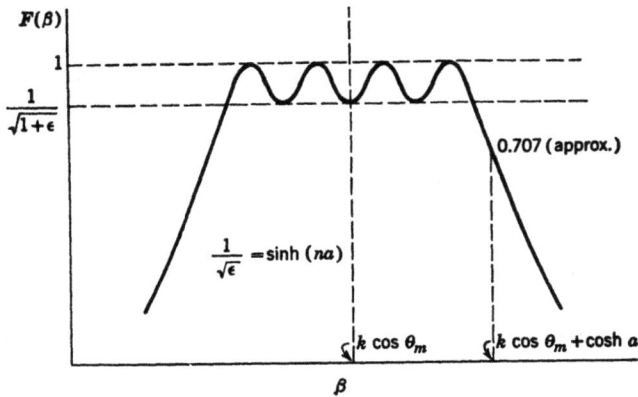

Figure 3-9. Equal-ripple or Chebyshev response curve $(n = 4)$.

and source distribution. In the latter case it is the sidelobe (stop-band) region that has the equal-ripple characteristic of the Chebyshev polynomial (see Prob. 3-5), whereas in the case considered here it is the main-beam (passband) region that has the equal-ripple characteristic.

General Synthesis Procedure

The synthesis of an aperture distribution that will produce a Butterworth pattern shape has been used to illustrate a synthesis procedure based on the Laplace transform.

The first step in the general procedure for a line source is to determine an $F(\gamma)$ that is analytic in the half plane $\alpha < -c$ (see Fig. 3-4) and has the desired $F(j\beta)$ over some specified range of $k \cos \theta$. It is required that the function $F(\gamma)$ be the Laplace transform of a source function $A(z')$. The

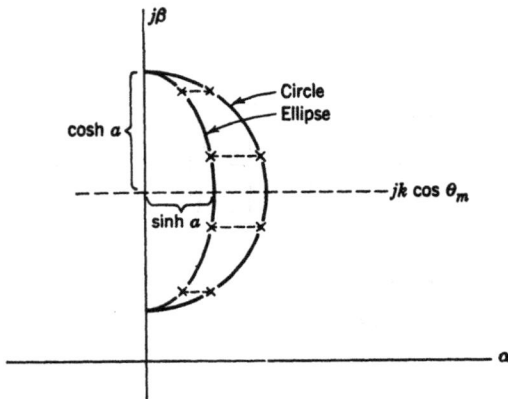

Figure 3-10. Location of poles for Chebyshev case with $n = 4$.

requirements on $A(z')$ are that it be piecewise continuous and that the integral $\int_0^{\infty} |A(z')e^{-cz'}| \, dz'$ exist.

The second step is to determine the poles of $F(\gamma)$. In general it may be necessary to express $F(\gamma)$ as a ratio of polynomials of the form

$$F(\gamma) = \frac{a_0\gamma^m + \cdots + a_m}{b_0\gamma^n + \cdots + b_n} \tag{3-35}$$

and solve for the roots of the polynomial in the denominator. In order for $A(z')$ to be composed only of traveling waves, the order m of the numerator polynomial must be at least one less than the order n of the denominator polynomial.

The third step is to obtain a partial-fraction expansion

$$F(\gamma) = \frac{c_1}{\gamma - \gamma_1} + \frac{c_2}{\gamma - \gamma_2} + \cdots + \frac{c_n}{\gamma - \gamma_n} \tag{3-36}$$

where $\gamma_n = \alpha_n + j\beta_n$. Upon evaluation of the coefficients c_n the source distribution obtained by the inverse Laplace transform is

$$A(z') = c_1 e^{-\gamma_1 z'} + \cdots + c_n e^{-\gamma_n z'} \tag{3-37}$$

A check of how well a finite source approximates the specified pattern is obtained by integrating $A(z')e^{jkz' \cos \theta}$ over some desired length. The approximate pattern for a source of length L is given by Eq. (3-9) with limits of 0 and L.

D. Woodward's Synthesis Method

The far-field pattern of a line source of length L may be expressed as

$$F(k \cos \theta) = \int_{-L/2}^{L/2} A(z')e^{jz'k \cos \theta} \, dz' \tag{3-38}$$

Let $A(z')$ be represented by the following Fourier series:

$$A_N(z') = \sum_{-N}^{N} a_n e^{-j(2n\pi z'/L)} \tag{3-39}$$

The far-field pattern produced by this source function is, from Eq. (3-38),

$$F_N(k \cos \theta) = \sum_{-N}^{N} a_n \int_{-L/2}^{L/2} e^{-jz'[(2n\pi/L) - k \cos \theta]} \, dz' \tag{3-40}$$

Performing the integration in (3-40) we have, except for a constant,

$$F_N(k \cos \theta) = \sum_{-N}^{N} a_n \frac{\sin\left[n\pi - \dfrac{Lk \cos \theta}{2}\right]}{n\pi - \dfrac{Lk \cos \theta}{2}} \tag{3-41}$$

The synthesis problem reduces to a problem of determining the constants a_n in such a way that $F_N(k \cos \theta)$ is a satisfactory approximation to the given pattern $F(k \cos \theta)$ over a specified interval.

Woodward[1] suggested that the coefficients a_n be chosen such that the approximate pattern $F_N(k \cos \theta)$ would fit the given pattern $F(k \cos \theta)$ exactly at the points

$$k \cos \theta = \frac{2n\pi}{L} \qquad n = 0, \pm 1, \pm 2, \ldots, \pm N \qquad (3\text{-}42)$$

where $N = L/\lambda$. The coefficients a_n are determined independently of each other from Eq. (3-41) as

$$a_n = F\left(\frac{2n\pi}{L}\right) \qquad (3\text{-}43)$$

The value of $k \cos \theta$ at which the far field is synthesized can be shifted by changing Eq. (3-43) to

$$a_n = F\left(\frac{2n\pi}{L} \pm k \cos \delta\right) \qquad (3\text{-}43a)$$

where Eq. (3-39) becomes

$$A_N(z') = \sum_{-N}^{N} a_n e^{-j[(2n\pi/L)\pm k \cos \delta]z'} \qquad (3\text{-}39a)$$

The quantity $k \cos \delta$ can be chosen so that the field pattern is fitted exactly at the most desirable points.

The physical meaning of Woodward's method is simple. Each term in the Fourier series representation of $A(z')$ represents a traveling wave of constant amplitude a_n and phase constant $2n\pi/L$ [or $(2n\pi/L) \pm k \cos \delta$]. The far field of each of these waves is of the form $a_n(\sin X_n)/X_n$ and consists of a main lobe, in a direction determined by the phase constant of the wave, together with a number of equally spaced sidelobes. The direction of each main lobe coincides with a null in the pattern of each of the other traveling waves; hence the field in that direction is determined entirely by the magnitude of that lobe.

This is one of the simplest methods to apply in practice. It can be applied as easily when the pattern is given graphically as when given analytically. For large apertures the desired pattern may be fitted exactly at a large number of points. However, for short apertures the number of points is small; thus the accuracy will be correspondingly poorer.

As an example of the Woodward method let us attempt to obtain the pattern indicated by the broken line in Fig. 3-11 from a source 10λ long.

[1] P. M. Woodward, A Method of Calculating the Field Over a Plane Aperture Required to Produce a Given Polar Pattern, *J. Inst. Elec. Engrs.*, Part IIIa, p. 1554, 1946.

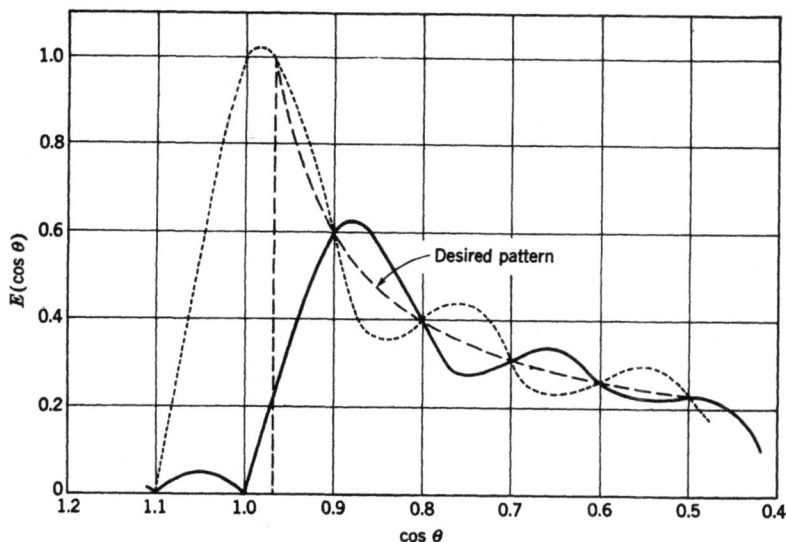

Figure 3-11.   Patterns obtained by Woodward's method compared with the desired pattern.

According to Eq. (3-42) we may fit the desired pattern exactly at the points

$$\cos \theta = \frac{2n\pi}{kL} = \frac{n}{L/\lambda} = \frac{n}{10} \tag{3-44}$$

where $n = 0, \pm 1, \pm 2, \ldots, \pm 10$. Thus the desired pattern in Fig. 3-11 and the approximate pattern will be equal at the values $\cos \theta = 0.5, 0.6, \ldots, 0.9$. In between these values the actual pattern will oscillate about the desired pattern. The actual pattern of the 10λ source is given by

$$E(\cos \theta) = \frac{0.22 \sin (5\pi - 10\pi \cos \theta)}{5\pi - 10\pi \cos \theta} + \frac{0.25 \sin (6\pi - 10\pi \cos \theta)}{6\pi - 10\pi \cos \theta}$$
$$+ \frac{0.32 \sin (7\pi - 10\pi \cos \theta)}{7\pi - 10\pi \cos \theta} + \frac{0.40 \sin (8\pi - 10\pi \cos \theta)}{8\pi - 10\pi \cos \theta}$$
$$+ \frac{0.58 \sin (9\pi - 10\pi \cos \theta)}{9\pi - 10\pi \cos \theta} \tag{3-45}$$

This is shown as the solid curve in Fig. 3-11. In this case the actual pattern was made equal to zero at $\cos \theta = 1$. Some improvement in the pattern in the range $0.97 \gtrless \cos \theta \gtrless 0.5$ can be obtained with values other than zero at $\cos \theta = 1$. However, this may be at the expense of an increase in the field in the region $\cos \theta > 1$ (stored energy region). The dotted curve in Fig. 3-11 shows the effect of making the actual pattern unity at

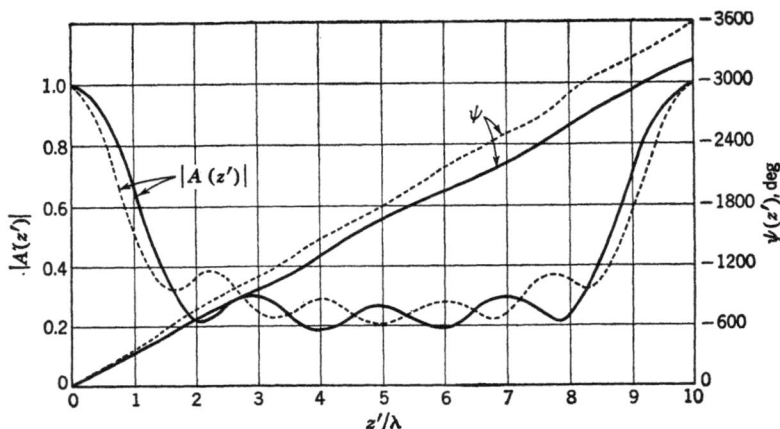

Figure 3-12.   Source distributions for the patterns in Fig. 3-11.

$\cos \theta = 1$. This is accomplished by adding the term $[\sin (10\pi - 10\pi \cos \theta)]/$ $(10\pi - 10\pi \cos \theta)$ to Eq. (3-45). The source distributions for both cases are shown in Fig. 3-12.

In an extension of Woodward's method,[1] specified values can be assigned to the radiation pattern in any finite number of directions by suitably distributing the source function over a finite length. If $E_s(k \cos \theta)$ are N linearly independent patterns obtainable from the given aperture, a composite pattern having specified values in the N directions $k \cos \theta_r$ may be expressed as

$$E(k \cos \theta) = \sum_{s=1}^{N} a_s E_s \qquad (3\text{-}46)$$

where the coefficients a_s are determined from the simultaneous equations

$$E(k \cos \theta_r) = \sum_{s=1}^{N} a_s E_s(k \cos \theta_r) \qquad r = 1, 2, \ldots , N \qquad (3\text{-}47)$$

The component patterns in Eq. (3-46) may be selected arbitrarily provided that they are linearly independent. It is most convenient to select $(a_n \sin X_n)/X_n$ patterns as in Woodward's original method, choose the coefficients as in Eq. (3-43), and in addition introduce other component patterns with main beams in the invisible or stored-energy region ($\cos \theta > 1$). The actual pattern does not fit the specified pattern exactly at any more

[1] Woodward and Lawson, *loc. cit.*; J. L. Yen, On the Synthesis of Line-sources and Infinite Strip-sources, *IRE Trans. Antennas Propagation*, **AP-5**(1):40–46, January, 1957; R. Mittra, On the Synthesis of Strip Sources, *Univ. Illinois Elec. Eng. Res. Lab. Tech. Rept.* 44, December, 1959.

points in this case; however, the overall approximation may be improved in the desired range as indicated by the addition of the component pattern with beam maximum at $\cos \theta = 1$ in the above example of Woodward's method.

In general, the size and extent of the invisible region (which is associated with stored energy) depend on the number of points at which an exact fit is desired as well as the closeness of fit in the intervals between these points. If one keeps the source length fixed and tries to improve the approximation to the desired pattern much beyond that obtained by Woodward's original method, the resultant source distribution will invariably have large variation in amplitude and phase and a large amount of stored energy compared with radiated energy (high Q). More will be said of this in Sec. H on supergain antennas.

E. Optical Synthesis Method

A synthesis method based on geometrical optics is perhaps the most practical method yet devised for a continuous source. The basic idea is to relate the power from an element of the source to that in a wedge-shaped region of space. This implies that the wavelength is either vanishingly small or the source is very large compared with a wavelength. However, the method may be satisfactory for sources as small as $10\,\lambda$. Chu[1] applied the method to curved reflector antennas with very good results for apertures as narrow as $20\,\lambda$. Dunbar[2] and Fry and Goward[3] extended the method to a linear traveling wave source which could be either an actual line source radiator or an equivalent line source representing one dimension of an aperture antenna such as the curved reflector treated by Chu.

By relating the power from an element of source dz' to the power in a wedge-shaped region of space as illustrated in Fig. 3-13, we obtain a one-to-one correspondence between a point on the source and a direction in space. We must be careful, however, not to assign the same direction to nonadjacent elements of the source, since the fields of such elements may add or subtract depending on their relative phases. The optical method gives us no control over the relative phases of nonadjacent elements.

If the far field of the source is independent of ϕ, the power relationship is given by

$$|G(\theta)|^2 |A(z')|^2 \, dz' = a|E(\theta)|^2 \sin \theta \, d\theta \qquad (3\text{-}48)$$

where $|G(\theta)|$ is the far-field pattern of the element dz', $|E(\theta)|^2$ is proportional to the power per unit area in the far field, and a is a constant which may be

[1] L. J. Chu, Microwave Beam Shaping Antennas, *MIT Res. Lab. Electron. Tech. Rept.* 40, June, 1947.

[2] A. S. Dunbar, On the Theory of Antenna Beam Shaping, *J. Appl. Phys.*, 1952, p. 847.

[3] D. W. Fry and F. K. Goward, "Aerials for Centimeter Wavelengths," p. 29, Cambridge, London, 1950.

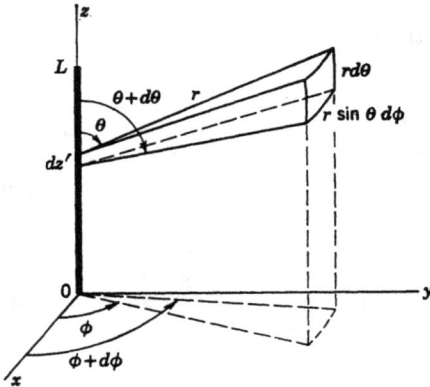

Figure 3-13. Illustrating the optical synthesis method. The power from dz' is restricted to the sector $d\theta$.

adjusted to make Eq. (3-48) dimensionally correct. Let us assume that $G(\theta)$ and $E(\theta)$ have the same dimensions. Then a will have the dimensions of the source amplitude squared. The factor $G(\theta)$ may be absorbed into the field function $F(\theta)$, where

$$F(\theta) = \frac{E(\theta)}{G(\theta)} \tag{3-49}$$

Furthermore, we may use $k \cos \theta$ as the far-field variable rather than θ. Thus Eq. (3-48) can be rewritten as

$$|A(z')|^2 \, dz' = b|F(k \cos \theta)|^2 d(k \cos \theta) \tag{3-50}$$

The power radiated from the beginning of the source to a point z' on the source is given by

$$\int_0^{z'} |A(z')|^2 \, dz' = b \int_{k \cos \theta_1}^{k \cos \theta} |F(k \cos \theta)|^2 d(k \cos \theta) \tag{3-51}$$

where the radiated power is contained in the sector $|\theta - \theta_1|$. The rays representing power from the source in Fig. 3-14 illustrate the effect of interchanging the limits θ_1 and θ_2 in Eq. (3-51). Either arrangement of limits is correct. The choice is usually dictated by the type of antenna structure one plans to use.

The constant b in Eq. (3-51) is determined by requiring all the power radiated from the source to be contained in a sector $|\theta_1 - \theta_2|$. That is,

$$\int_0^L |A(z')|^2 \, dz' = b \int_{k \cos \theta_1}^{k \cos \theta_2} |F(k \cos \theta)|^2 d(k \cos \theta) \tag{3-52}$$

or

$$b = \frac{\displaystyle\int_0^L |A(z')|^2 \, dz'}{\displaystyle\int_{k \cos \theta_1}^{k \cos \theta_2} |F(k \cos \theta)|^2 d(k \cos \theta)} \tag{3-53}$$

Substituting this value of b into Eq. (3-51) gives

$$\frac{\int_0^{z'} |A(z')|^2 \, dz'}{\int_0^L |A(z')|^2 \, dz'} = \frac{\int_{k\cos\theta_1}^{k\cos\theta} |F(k\cos\theta)|^2 d(k\cos\theta)}{\int_{k\cos\theta_1}^{k\cos\theta_2} |F(k\cos\theta)|^2 d(k\cos\theta)} \qquad (3\text{-}54)$$

or

$$\frac{\int_0^{z'} |A(z')|^2 \, dz'}{\int_0^L |A(z')|^2 \, dz'} = \frac{\int_{k\cos\theta_2}^{k\cos\theta} |F(k\cos\theta)|^2 d(k\cos\theta)}{\int_{k\cos\theta_2}^{k\cos\theta_1} |F(k\cos\theta)|^2 d(k\cos\theta)} \qquad (3\text{-}55)$$

depending on which way one chooses the limits. Thus if $|F(k\cos\theta)|$ and $|A(z')|$ are specified, Eq. (3-54) or (3-55) can be solved for $\cos\theta$ as a function of z'. If $|A(z')|^2$ and $|F(k\cos\theta)|^2$ are not readily integrated, one may resort to numerical or graphical techniques. After obtaining $\cos\theta$ as a function of z', it is a simple matter to obtain the phase $\psi(z')$ or the phase velocity $v(z')$ of the source. In applying the optical method the element dz' has been assumed to be large in terms of a wavelength; thus the angle at which the element dz' radiates is given by Eq. (2-249),

$$\cos\theta = \frac{c}{v} \qquad (3\text{-}56)$$

where c is the velocity of light in free space. The phase velocity v is assumed to be constant over dz'. The phase $\psi(z')$ in radians is given by

$$\psi(z') = \psi(0) + k \int_0^{z'} \frac{c}{v}(z') \, dz' \qquad (3\text{-}57)$$

where $k = 2\pi/\lambda$ and $\psi(0)$ is the phase at $z' = 0$, which may be taken as zero.

It should be noted that the optical method is not restricted to the so-called "visible" region of space $(-1 \lesssim \cos\theta \lesssim 1)$. The use of $k\cos\theta$ as the far-field variable permits us to relate a point on the source to a direction in the invisible or stored-energy region of space $(|\cos\theta| > 1)$. This will be illustrated by an example.

Figure 3-14. Illustrating the effect of interchanging limits on Eq. (3-51).

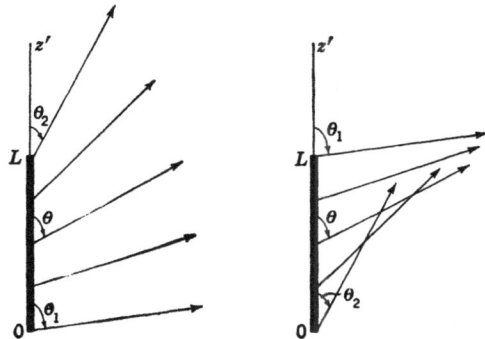

Let us use the optical method to synthesize a source that will produce the csc $\theta \sqrt{\cos \theta}$ far-field pattern shown by the broken curve in Fig. 3-15. The desired pattern extends from $\theta = 15$ deg to $\theta = 60$ deg. These limits in terms of cos θ may be used in Eq. (3-55); however, better results are usually obtained if one overdesigns, so let us use the limits of 5 and 60 deg in Eq. (3-55). Furthermore, let the source have an amplitude distribution $|A(z')| = \sqrt{\cos (\pi z'/2L)}$ and let the source element pattern $G(\theta)$ be isotropic with unit magnitude. The velocity ratio c/v can be computed from Eq. (3-54) and is given as the solid curve in Fig. 3-16. The corresponding far-field pattern for a 10-λ source [computed from Eqs. (3-1) and (3-57)] is given by the solid curve in Fig. 3-15.

The agreement is not very good, but then we have considered the rather extreme case of a source only 10 λ long. Better results would be obtained if we were to use a longer source, of course. However, better results also can be obtained by further overdesign. Using the approximate pattern given by the solid curve in Fig. 3-15 as a guide, let us extend the desired pattern into the invisible region (cos $\theta > 1$) as shown in the same figure. The corresponding velocity ratio c/v can be obtained from Eq. (3-54) by graphical integration and is given by the broken curve in Fig. 3-16. The corresponding pattern is given by the dotted curve in Fig. 3-15 and is substantially better than the solid curve approximation. This procedure

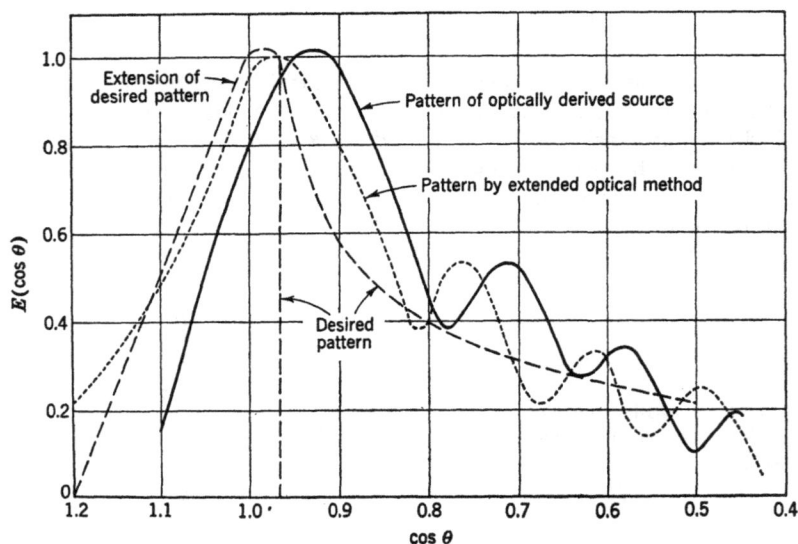

Figure 3-15. Comparison of desired pattern and optically determined patterns for a 10-λ source.

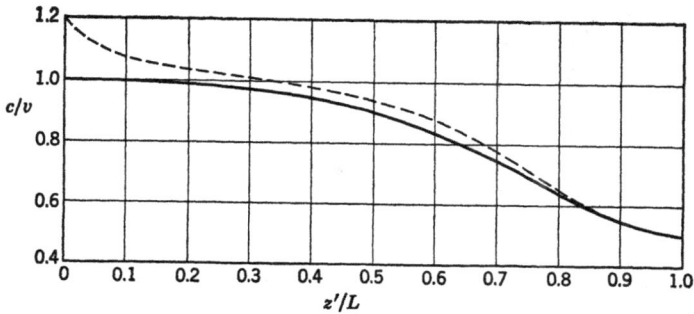

Figure 3-16.   Optically determined c/v necessary to obtain a $\csc^2 \theta \cos \theta$ power pattern from a $\cos (\pi z'/2L)$ power distribution at the source.

may be repeated, and it constitutes an iterative process. More will be said of iterative processes in the next section.

The optical method is quite useful for shaped beam problems and has the advantage over Woodward's method that the source amplitude may be specified as well as the far-field amplitude. For optimum patterns in the sense of minimum sidelobe level for a given half-power beamwidth, best results are obtained by the special distributions discussed in Sec. G.

F. Iterative Methods

It was observed in the last section that the optical method could give improved results if the approximate pattern obtained by specifying a desired pattern in the visible region was used as a guide in specifying a pattern over a somewhat greater range that may extend into the invisible region. This amounts to an iterative process, although there is no guarantee that repeating the process will lead to a pattern that approaches arbitrarily closely to the desired pattern.

An iterative process based on the Fourier transform pair is illustrated in Fig. 3-17. Let $|F(\beta)|$ be the desired pattern shape over the interval $\beta_1 \lessgtr \beta \lessgtr \beta_2$. The source distribution by the Fourier transform method is

$$A(z') = \int_{\beta_1}^{\beta_2} |F(\beta)| e^{-j\beta z'} \, d\beta \tag{3-58}$$

where the phase of $F(\beta)$ has been taken as zero. The source function $A(z')$ will generally have significant value over a greater interval than some desired source length L as illustrated in Fig. 3-18. If $A(z')$ is taken to be zero outside the finite interval L, an approximate pattern $F_a(\beta)$ is obtained as illustrated in Fig. 3-17. The pattern $F_a(\beta)$ corresponds to a source of the desired length L. Let us, therefore, specify a new field function $F'(\beta)$, where

$$
\begin{aligned}
|F'(\beta)| &= |F(\beta)| & \beta_1 \lessgtr \beta \lessgtr \beta_2 \\
F'(\beta) &= F_a(\beta) & \beta_2 < \beta < \beta_1
\end{aligned} \tag{3-59}
$$

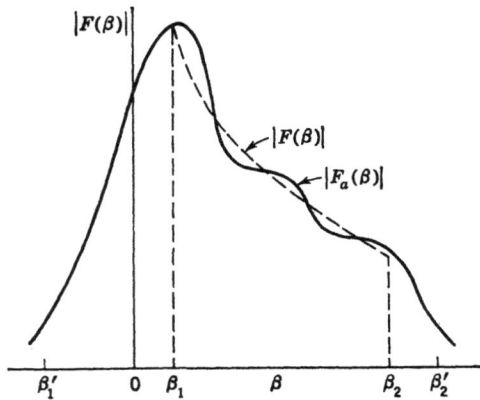

Figure 3-17. Illustrating an iterative process.

The limits β_1 and β_2 may be shifted somewhat, or F_a may be modified slightly to make $F(\beta)$ and $F_a(\beta)$ continuous. The function $F'(\beta)$ may be given the phase of $F_a(\beta)$ over the interval $\beta_1 \lessgtr \beta \lessgtr \beta_2$ or the phase of $F(\beta)$ (which is taken to be zero in this case).

Let us now specify new limits β_1' and β_2' outside which the function $F'(\beta)$ may be neglected. In practice we might use the 20-db level as a criterion. Substituting $F'(\beta)$, β_1', and β_2' for the unprimed quantities in Eq. (3-58) gives

$$A'(z') = \int_{\beta_1'}^{\beta_2'} F'(\beta)e^{-j\beta z'}\,d\beta \qquad (3\text{-}60)$$

If $F'(\beta) = F(\beta)$ over the interval $\beta_1 \lessgtr \beta \lessgtr \beta_2$, then Eq. (3-60) may be written

$$A'(z') = A(z') + \int_{\beta_1'}^{\beta_1} F_a(\beta)e^{-j\beta z'}\,d\beta + \int_{\beta_2}^{\beta_2'} F_a(\beta)e^{-j\beta z'}\,d\beta \qquad (3\text{-}61)$$

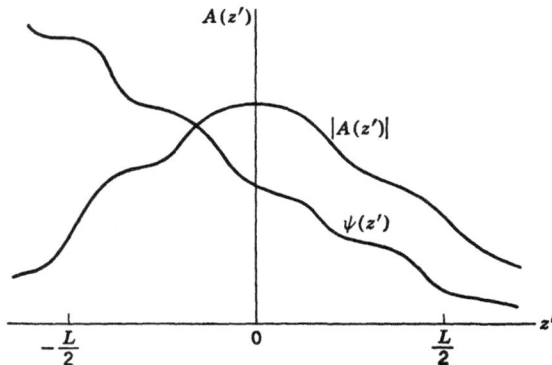

Figure 3-18. Source distribution for the iterative process in Fig. 3-17.

The integrals in Eq. (3-61) show the modification of $A(z')$ due to extending $F(\beta)$ by means of $F_a(\beta)$. This generally results in confining $A'(z')$ more to the interval L than the original function $A(z')$.

The integration

$$F'_a(\beta) = \int_{-L/2}^{L/2} A'(z')e^{j\beta z'}\,dz' \tag{3-62}$$

will show whether or not a satisfactory approximation to $F(\beta)$ in the interval $\beta_1 \lesssim \beta \lesssim \beta_2$ has been obtained by a source of length L. If not, the process may be repeated. Woodward's method can be readily incorporated into this procedure and would replace the integrations in Eqs. (3-60) and (3-62).

An iterative process also may be applied that incorporates the optical method and either the Fourier transform or Woodward's method. The optical method provides a starting point. As described in Sec. E, both source amplitude and far-field amplitude are specified initially. The source phase velocity is computed from Eq. (3-54) or (3-55). From Eq. (3-1) or (3-2) we may compute the amplitude and phase of the far field of the optically derived source distribution. This corresponds to $F_a(\beta)$ in the above discussion on the Fourier transform method, and the same procedure applies. The resulting source distribution can be thought of as an improved distribution in the sense that it will give a pattern more like the desired pattern over its specified range than that of the optically determined source. It will not necessarily be a distribution that is more easily obtained on a practical antenna. In fact it is likely to be less so.

As an example of the latter iterative procedure let us synthesize a $10\text{-}\lambda$ source that will give a $\csc\theta\sqrt{\cos\theta}$ far-field pattern over the range 15 deg $\lesssim \theta \lesssim 60$ deg. The example considered in Sec. E may be used as the first step. The source amplitude is chosen as $\sqrt{\cos(\pi z'/20\lambda)}$ and the far field is taken to be $\csc\theta\sqrt{\cos\theta}$ from 5 deg $\lesssim \theta \lesssim 60$ deg to gain some improvement through overdesign. The source phase velocity as derived by the optical method [Eq. (3-54)] is given by the solid curve in Fig. 3-16.

The second step is to compute the far-field pattern of the optically determined source distribution by means of Eqs. (3-57) and (3-2). The resulting far-field amplitude and phase are given in Fig. 3-19. As we observed in Sec. E, one application of the optical method does not give very satisfactory results in this case.

The next step is to substitute the desired pattern for the approximate pattern over the interval 15 deg $\lesssim \theta \lesssim 60$ deg. Outside this interval one either uses the approximate pattern as it is or uses it as a guide in an extrapolation of the desired pattern. The latter was done in this case in order to keep the half-power beamwidth consistent with the length of the source.

The fourth step is to substitute the far-field phase ϕ and the far-field amplitude given by the broken line in Fig. 3-19 into Eq. (3-3) and compute the corresponding source distribution. Alternatively, one may use Woodward's method taking into account the terms that are contributed by the region $\cos \theta > 1$. The $\times$'s in Fig. 3-19 show the amplitudes and phases that one may use for the coefficients a_n in Eq. (3-42). In this example Eq. (3-3) was used and the resulting source distribution is given in Fig. 3-20. The optically determined source distribution that provided our starting point is also given.

As step five we should compute the far field of the improved source distribution given in Fig. 3-20. This may be done by means of Eq. (3-2). The result is given by the dotted curve in Fig. 3-19. It can be observed that this pattern is significantly better than the optically derived pattern.

It appears that one may continue this process and come arbitrarily close to the specified pattern over some desired range. However, no proof has yet been developed to show that such a process does converge to the desired pattern. The iterative method described here does entail more computation than a direct application of the Fourier transform or Woodward's method. However, in the latter methods one has no control at all over the source distribution whereas when one starts an iterative process as above with the optical method some degree of control is available. Maximum control exists when one performs successive iterations using only the optical method as described in Sec. E. However, convergence to a

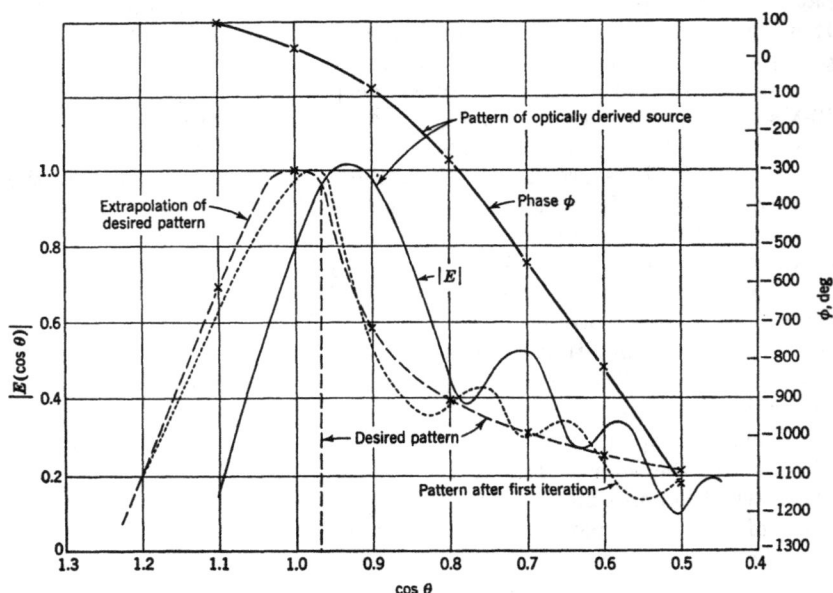

Figure 3-19. Comparison of desired pattern, optically determined pattern, and pattern obtained by iterative process.

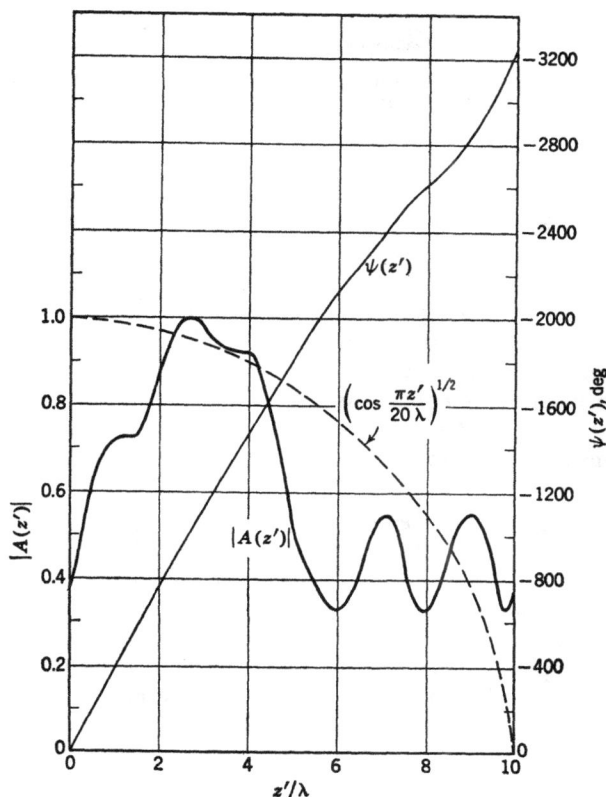

Figure 3-20. The optical source distribution and the distribution after one iteration.

desired pattern appears to be slower and perhaps more questionable in this case than in the case of the combined optical–Fourier transform method.

G. Optimum Design Methods

There are a number of ways in which we can optimize an antenna. When the antenna is considered as part of a system, it can be adjusted to optimize the system in some particular way. For example, in radio astronomy the antenna can be designed to produce an output which maps the region of observation to within some prescribed error (see Sec. 3-2B). The problems that are usually considered, however, involve minimizing beamwidth for a given sidelobe level or maximizing gain. Only the gain and beamwidth problems are considered here.

Minimizing Beamwidth

The synthesis methods discussed in the previous sections are generally satisfactory for finding a source distribution whose pattern will approximate a prescribed pattern of arbitrary shape. However, the problem of optimizing beamwidth, that is, of finding a source to make the beamwidth minimum for a given sidelobe level, is best handled by methods that utilize the constant-ripple characteristic of Chebyshev polynomials.[1] A method by Taylor utilizes an ideal constant sidelobe pattern that is obtained when the number of elements in a Dolph-Chebyshev array is increased indefinitely. One may also specify the far field as a Chebyshev polynomial of order equal to one less than the number of equally spaced elements in an array and determine an approximate source distribution by converting the distribution for the array into a continuous distribution.

The far-field pattern of a Dolph-Chebyshev array of N discrete elements a distance d apart and lying along the z axis is given by

$$E_N = T_{N-1} \left\{ w_0 \cos \left[\frac{kd}{2} \left(\cos \theta - \frac{c}{v} \right) \right] \right\} \qquad (3\text{-}63)$$

where $T_N(w) = \cos (N \cos^{-1} w)$. The beam maximum occurs at $\cos \theta_m = c/v$, and v may be thought of as the phase velocity of a wave that travels along the array and excites each element of the array with the appropriate phase (the phase difference between elements is kdc/v). The parameter w_0 is determined by the sidelobe ratio R (the ratio of main-beam maximum to sidelobe level). For $c/v \lessgtr 1$ we have

$$T_{N-1}(w_0) = R \qquad (3\text{-}64)$$

For $c/v > 1$ we use the relation

$$T_{N-1} \left\{ w_0 \cos \left[\frac{kd}{2} \left(1 - \frac{c}{v} \right) \right] \right\} = R \qquad (3\text{-}65)$$

where the beam maximum in the visible region is at $\theta = 0$. Figure 3-21 shows a Chebyshev polynomial of the seventh degree and the corresponding far-field pattern. In this case the phase delay between elements is given by 180 deg $\cos \theta_m$. If θ_m were chosen to be 90 deg (broadside radiation), the elements would, of course, all be in phase[2] and $\theta = 0$ and $\theta = 180$ deg would both coincide with $w = 0$. The two beams shown in Fig. 3-21 are due to the fact that the three-dimensional beam of a line source is in the shape of a cone. A solid bidirectional beam occurs when $\theta_m = 0$ ($c/v = 1$). That

---

[1] See Kraus, *op. cit.*, p. 105, for a proof that a Dolph-Chebyshev distribution does give minimum beamwidth for a given sidelobe level.

[2] This is the case considered by C. L. Dolph in his paper A Current Distribution for Broadside Arrays Which Optimize the Relationship between Beam Width and Side Lobe Level, *Proc. IRE*, **34**(6):335–348, June, 1946.

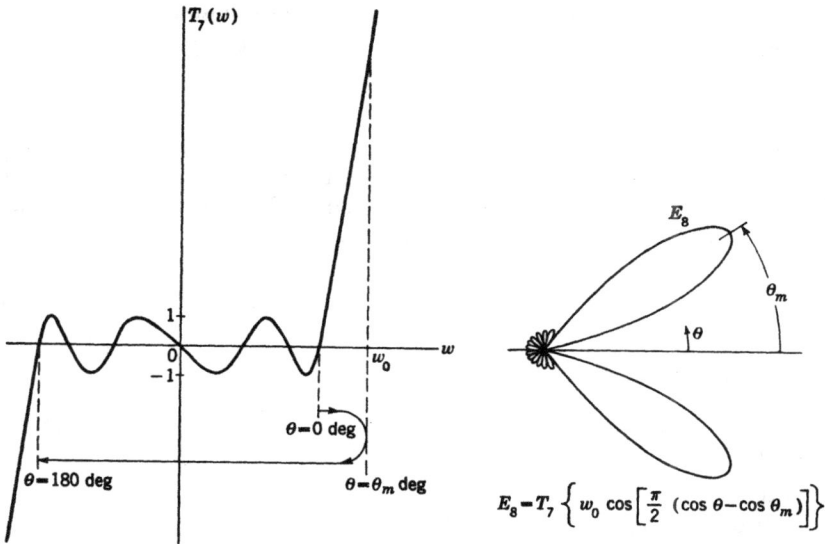

Figure 3-21. Radiation pattern of an eight-element array with $\lambda/2$ spacing between elements. The correspondence between θ and w is indicated.

is, there are beam maxima at $\theta = 0$ deg and $\theta = 180$ deg for $d = \lambda/2$ and $c/v = 1$. A single endfire beam may be obtained by reducing the spacing between elements to something less than $\lambda/2$.

Taylor's Method

It has been shown[1] that, as the number of elements of a Dolph-Chebyshev broadside array of finite length is increased indefinitely ($N \to \infty$), the far-field pattern approaches the function

$$F_0(k \cos \theta, A) = \cos \left[\left(\frac{Lk \cos \theta}{2} \right)^2 - (\pi A)^2 \right]^{1/2} \qquad (3\text{-}66)$$

where L is the length of the array. The quantity A is a real parameter having the property that $\cosh \pi A$ is the ratio of the beam maximum to the sidelobe level (the sidelobe ratio R). The function in Eq. (3-66) is considered to be ideal in the sense that it gives the minimum sidelobe level for a given half-power beamwidth. The sidelobe level is constant and remains so as $|\cos \theta| \to \infty$. On the complex plane ($\gamma = \alpha + jk \cos \theta$) the function in Eq. (3-66) will have zeros on the imaginary axis at points given by

$$(k \cos \theta)_n = \pm \frac{2\pi}{L} [(n - \tfrac{1}{2})^2 + A^2]^{1/2} \qquad n = 1, 2, 3, \ldots, \infty \quad (3\text{-}67)$$

[1] G. J. Van der Maas, A Simplified Calculation for Dolph-Chebyshev Arrays, *J. Appl. Phys.*, **25**(N):121–124, January, 1954.

There is no continuous source distribution that will produce the pattern in Eq. (3-66) because the function does not satisfy the condition that the integral $\int_{-\infty}^{\infty} |F_0(k \cos \theta)| \, d(k \cos \theta)$ exists (see Prob. 3-4). Taylor[1] has shown that the far-field function must decay at least as $\lambda/(L \cos \theta)$ and that the zeros must tend to the integers n instead of $n - \frac{1}{2}$. That is, the function must decay no less rapidly than a $(\sin X)/X$ function. Taylor divides the far-field function into two parts, one an equal sidelobe region and the other a region of decaying sidelobes. In the equal sidelobe region the zeros are given by

$$(k \cos \theta)_n = \pm \frac{2\pi\sigma}{L} [(n - \frac{1}{2})^2 + A^2]^{\frac{1}{2}} \qquad 1 \lesssim n \lesssim \bar{n} \qquad (3\text{-}68)$$

In the tapered sidelobe region

$$(k \cos \theta)_n = \pm \frac{2n\pi}{L} \qquad \text{for } n > \bar{n} \qquad (3\text{-}69)$$

The quantity σ is chosen so that Eqs. (3-68) and (3-69) are identical at an integer n designated by $\bar{n}$. Thus

$$\sigma = \frac{\bar{n}}{[(\bar{n} - \frac{1}{2})^2 + A^2]^{\frac{1}{2}}} \qquad (3\text{-}70)$$

Modifying the ideal function in Eq. (3-66) so that the far-field function has the zeros as given in Eqs. (3-68) and (3-69) results in the function

$$F(k \cos \theta, A, \bar{n}) = C \prod_{n=1}^{\bar{n}-1} \left\{ 1 - \frac{k^2 L^2 \cos^2 \theta}{4\pi^2 \sigma^2 [(n - \frac{1}{2})^2 + A^2]} \right\}$$

$$\times \prod_{n=\bar{n}}^{\infty} \left(1 - \frac{k^2 L^2 \cos^2 \theta}{4\pi^2 n^2} \right) \qquad (3\text{-}71)$$

where the constant C is arbitrary. If C is taken equal to $\cosh \pi A$, then $F_0(0,A) = F(0,A,\bar{n})$ and

$$\lim_{\bar{n} \to \infty} F(k \cos \theta, A, \bar{n}) = F_0(k \cos \theta, A) \qquad (3\text{-}72)$$

Equation (3-71) may be expressed as

$$F(k \cos \theta, A, \bar{n}) = CP(k \cos \theta, A, \bar{n}) Q(k \cos \theta, \bar{n}) \qquad (3\text{-}73)$$

where

$$P = \prod_{n=1}^{\bar{n}-1} \left\{ 1 - \frac{k^2 L^2 \cos^2 \theta}{4\pi^2 \sigma^2 [(n - \frac{1}{2})^2 + A^2]} \right\} \qquad (3\text{-}74)$$

[1] T. T. Taylor, Design of Line Source Antennas for Narrow Beamwidth and Low Side Lobes, *IRE Trans. Antennas Propagation*, **AP-3**(1):16–28, January, 1955.

and

$$Q = \prod_{n=\bar{n}}^{\infty} \left(1 - \frac{k^2 L^2 \cos^2 \theta}{4\pi^2 n^2} \right)$$

$$= \frac{[(\bar{n} - 1)!]^2}{[\bar{n} - 1 + (kL \cos \theta)/2\pi]![\bar{n} - 1 - (kL \cos \theta)/2\pi]!} \tag{3-75}$$

The polynomial P is essentially the pattern function, whereas the function Q introduces a tapering effect which primarily influences the sidelobe region.

The half-power point for the ideal pattern of Eq. (3-66) is found to be

$$(k \cos \theta)_{HP_0} = \frac{2}{L} \left[(\cosh^{-1} R)^2 - \left(\cosh^{-1} \frac{R}{\sqrt{2}} \right)^2 \right]^{\frac{1}{2}} \tag{3-76}$$

The modification of the ideal pattern by Taylor essentially expands the far-field function by the factor σ, thus to a very good approximation the half-power point for the function in Eq. (3-71) is given by

$$(k \cos \theta)_{HP_T} \approx \sigma (k \cos \theta)_{HP_0} \tag{3-77}$$

A graph of the half-power beamwidth as a function of sidelobe level from Eq. (3-76) is given in Fig. 3-22. Recall that θ is measured from the axis of the source (the z axis).

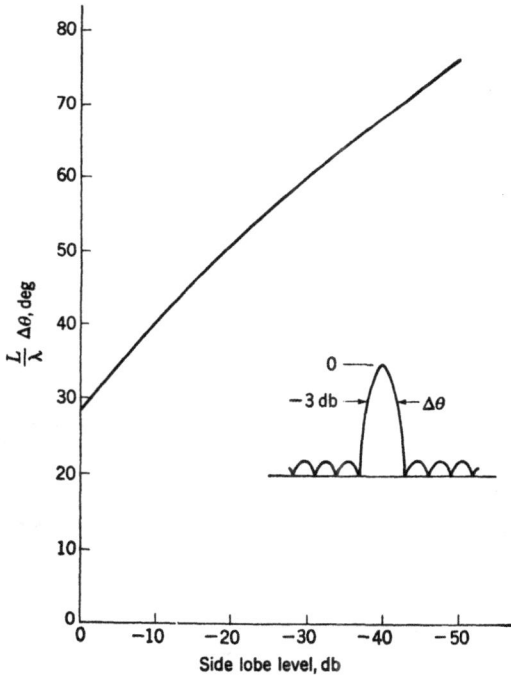

Figure 3-22. Half-power beam-width $\Delta\theta$ versus sidelobe level for broadside Taylor line source of length L.

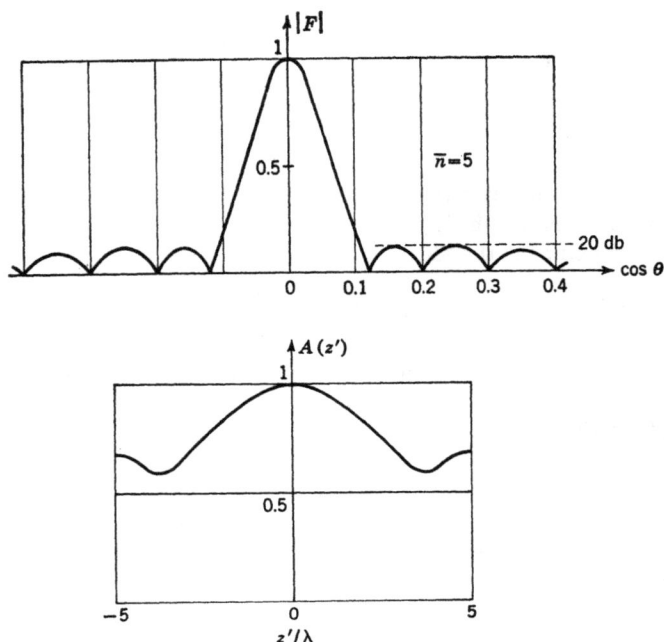

Figure 3-23.   Far field and the corresponding source for an optimum pattern design based on Taylor's method.

In calculating the source function that will produce the field function in Eq. (3-71) it is perhaps most convenient to use Woodward's method (see Sec. D). The source function may be expressed as a sum of uniform traveling waves as follows:

$$A(z') = \sum_m a_m e^{-jm2\pi z'/L} \tag{3-78}$$

The coefficients a_m are given by

$$a_m = F\left(\frac{2m\pi}{L}, A, \bar{n}\right) = P\left(\frac{2m\pi}{L}, A, \bar{n}\right) Q\left(\frac{2m\pi}{L}, \bar{n}\right) \cosh \pi A \tag{3-79}$$

where

$$P\left(\frac{2m\pi}{L}, A, \bar{n}\right) = \prod_{n=1}^{\bar{n}-1} \left\{ 1 - \frac{m^2}{\sigma^2[(n - \frac{1}{2})^2 + A^2]} \right\} \tag{3-80}$$

and

$$Q\left(\frac{2m\pi}{L}, \bar{n}\right) = \frac{[(\bar{n} - 1)!]^2}{(\bar{n} - 1 + m)!(\bar{n} - 1 - m)!} \tag{3-81}$$

Since the far-field function is symmetrical, the source function is given over the interval $-L/2$ to $L/2$ on the z axis as

$$A(z') = F(0,A,\bar{n}) + 2 \sum_{m=1}^{M} F\left(\frac{2m\pi}{L}, A,\bar{n}\right) \cos \frac{2m\pi z'}{L} \qquad (3\text{-}82)$$

An example of Taylor's method is shown in Fig. 3-23.

Taylor's method as outlined above applies to a source with infinite phase velocity, i.e., broadside radiation. The method may be extended to a source with beam maximum at θ_m by substituting $k(\cos\theta - \cos\theta_m)$ for $k \cos\theta$. This amounts to shifting the far-field function [Eq. (3-71)] by $k \cos\theta_m$ units along the $jk \cos\theta$ axis in the complex γ plane.

Dolph-Chebyshev Method Extended to a Continuous Source

Perhaps the simplest way to obtain a line source distribution whose pattern approximates the optimum pattern (minimum beamwidth for a given sidelobe level) is to extend the conventional Dolph-Chebyshev method by converting the source distribution for discrete elements into a continuous distribution.

The far-field pattern of a Dolph-Chebyshev array of N discrete elements spaced a distance d apart and lying along the z axis is given by Eq. (3-63). In general, the far field of a linear array of N elements with excitation coefficients A_l is given by

$$E_{N_e} = 2 \sum_{l=0}^{(N/2)-1} A_l \cos\left[(2l+1)(\cos\theta - \cos\theta_m)\frac{kd}{2} \right] \qquad (3\text{-}83)$$

for N even and

$$E_{N_0} = 2 \sum_{l=0}^{(N-1)/2} A_l \cos\left[lkd(\cos\theta - \cos\theta_m) \right] \qquad (3\text{-}84)$$

for N odd.

The coefficients A_l for the Dolph-Chebyshev array are obtained by equating E_N and T_{N-1} as in Eq. (3-63). Let

$$\psi = kd(\cos\theta - \cos\theta_m) \qquad (3\text{-}85)$$

and relate w and ψ by the equation

$$\cos\frac{\psi}{2} = \frac{w}{w_0} \qquad (3\text{-}86)$$

The parameter w_0 is determined by the sidelobe ratio R as in Eq. (3-64) or (3-65). If E_N in Eq. (3-83) or (3-84) is expanded with w/w_0 substituted for $\cos\psi/2$, we may equate terms of like degree in $E_N(w)$ and $T_{N-1}(w)$ and thus determine the coefficients A_l that will give an optimum far field. More details and proof that the Chebyshev polynomial does give an optimum pattern in the sense of minimum beamwidth for a given sidelobe level are given in Dolph's article[1] and also in Kraus.[2] The literature is liberally sprinkled with methods for computing these excitation coefficients.[3] We shall not discuss these methods here. Our objective is to show that the theory for an array of discrete elements may be used for a continuous source. This is perhaps best accomplished by means of an example.

Let us start with an array of eight elements with $\lambda/2$ spacing between elements. The pattern of such an array is illustrated in Fig. 3-21 and pre-

[1] Dolph, *loc. cit.*
[2] Kraus, *loc. cit.*
[3] D. Barbiere, A Method for Calculating the Current Distribution of Chebyshev Arrays, *Proc. IRE*, **40**:78–82, January, 1952; R. J. Stegen, Excitation Coefficients and Beamwidths of Chebyshev Arrays, *Proc. IRE*, **41**(11):1671–1674, November, 1953.

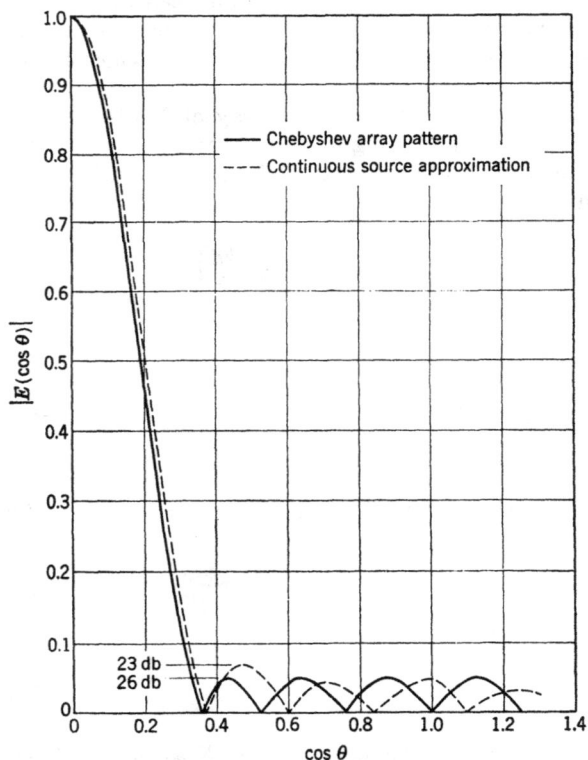

Figure 3-24.   Comparison of patterns for an array and a continuous source.

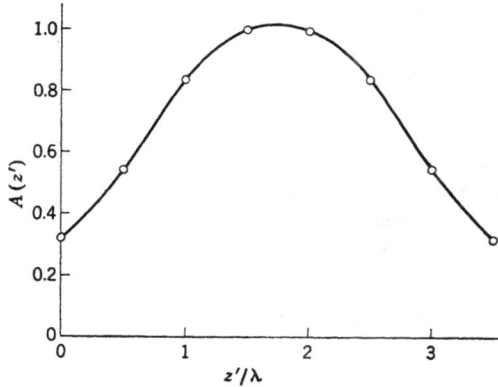

Figure 3-25. Converting Dolph-Chebyshev excitation coefficients into a continuous source.

sented in more detail as the solid curve in Fig. 3-24. The broadside case ($\theta_m = 90$ deg) is considered in this example, but the method applies for any angle of maximum radiation, $0 \lessgtr \theta_m \lessgtr 180$ deg.

The Dolph-Chebyshev excitation coefficients for this array are given by the dots in Fig. 3-25. If we construct a continuous distribution by drawing a smooth curve through the values obtained for the Dolph-Chebyshev coefficients, the result is the curve given in Fig. 3-25.

The far-field pattern of the continuous distribution in Fig. 3-25 was computed by numerical integration and is given by the dashed curve in Fig. 3-24 for comparison with the optimum pattern which is given by the solid line. This and other examples indicate that a good approximation to the Chebyshev pattern can be obtained for a continuous source by simply drawing a smooth curve through the Dolph-Chebyshev excitation coefficients, thus converting an array source into a continuous source.

Optimum Gain

The Dolph-Chebyshev method for an array and Taylor's method for a continuous source constitute optimum methods in the sense that the objective is minimum beamwidth for a given sidelobe level. One can also optimize the gain. This problem has been considered for arrays[1] and also for the continuous source.[2]

For the continuous source it has been proved that any pattern is theoretically realizable, i.e., any shape, any gain, and any sidelobe level (see Sec. 3-1). If certain restrictions are made with regard to the source

[1] A. I. Uzkov, An Approach to the Problem of Optimum Directive Antenna Design, *Compt. Rend.* (Dokl.) *Acad. Sci. URSS*, **L111**(1), 1946; M. Uzsoky and L. Solymár, Theory of Superdirective Linear Arrays, *Acta Phys. Acad. Sci. Hung.*, **6**:185–206, 1956; C. T. Tai, The Optimum Directivity of Uniformly Spaced Broadside Arrays of Dipoles, *IEEE Trans. on Antennas Propagation*, **AP-12** (4): 447–454, July 1964.

[2] L. Solymár, Maximum Gain of a Line Source Antenna if the Distribution Function Is a Finite Fourier Series, *IRE Trans. Antennas Propagation*, **AP-6**:215–219, July, 1958.

distribution, then a distribution that produces optimum gain can be found. For example, Solymár[1] considers the source function to be made up of a finite Fourier series and solves for the coefficients which, for a given super-gain ratio (see next section), will result in the highest directivity.

Following Tai's method for an array, optimizing the gain of a continuous source can be formulated as follows.[2] Assuming no power dissipated in the source (100 per cent efficiency), let a line source of length L on the z axis be composed of a finite sum of waves A_i with coefficients a_i so that the source distribution $A(z)$ is given by

$$A(z) = \sum_{i=1}^{N} a_i A_i(z) \tag{3-87}$$

Let $F_i(k \cos \theta)$ be the pattern of $A_i(z)$. That is,

$$F_i(k \cos \theta) = \int_{-L/2}^{L/2} A_i(z)e^{jkz \cos \theta} \, dz \tag{3-88}$$

Then the pattern resulting from $A(z)$ is

$$F(k \cos \theta) = \int_{-L/2}^{L/2} A(z)e^{jkz \cos \theta} \, dz$$
$$= \sum_{i=1}^{N} a_i F_i(k \cos \theta) \tag{3-89}$$

It will be assumed that a_i and $F_i(k \cos \theta)$ are real and the element dz has an isotropic pattern, but this is no limitation on the method.

The directivity (or gain if the efficiency is 100 per cent) of the line source is given by

$$D = \frac{4\pi \left[\sum_{i=1}^{N} a_i F_i(k \cos \theta_m) \right]^2}{\int_0^{2\pi} \int_0^{\pi} \left[\sum_{i=1}^{N} a_i F_i(k \cos \theta) \right]^2 \sin \theta \, d\theta \, d\phi} \tag{3-90}$$

which can be written as

$$D = \frac{2 \sum_{i=1}^{N} \sum_{j=1}^{N} f_{ij} a_i a_j}{\sum_{i=1}^{N} \sum_{j=1}^{N} g_{ij} a_i a_j} \tag{3-91}$$

[1] *Ibid.*

[2] Actually we shall be optimizing directivity, but the term "optimum gain" is much more commonly used than "optimum directivity." They are the same, of course, if the antenna is 100 per cent efficient.

where

$$f_{ij} = F_i(k \cos \theta_m)F_j(k \cos \theta_m) = f_{ji} \qquad (3\text{-}92)$$

$$g_{ij} = \int_0^\pi F_i(k \cos \theta)F_j(k \cos \theta) \sin \theta \, d\theta = g_{ji} \qquad (3\text{-}93)$$

If we assume that the coefficients a_i are to be adjusted for optimum gain but the functions $A_i(z)$ are fixed, we set

$$\frac{\partial D}{\partial a_n} = 0 \qquad n = 1, 2, \ldots, N \qquad (3\text{-}94)$$

Forming the derivative gives

$$\frac{\partial D}{\partial a_n} = \frac{2 \sum_{i=1}^N \sum_{j=1}^N g_{ij}a_ia_j \left[\sum_{j=1}^N f_{nj}a_j + \sum_{i=1}^N f_{in}a_i \right]}{\left[\sum_{i=1}^N \sum_{j=1}^N g_{ij}a_ia_j \right]^2}$$

$$- \frac{2 \sum_{i=1}^N \sum_{j=1}^N f_{ij}a_ia_j \left[\sum_{j=1}^N g_{nj}a_j + \sum_{i=1}^N g_{in}a_i \right]}{\left[\sum_{i=1}^N \sum_{j=1}^N g_{ij}a_ia_j \right]^2} \qquad (3\text{-}95)$$

Setting the numerator equal to zero and taking advantage of the fact that $f_{ij} = f_{ji}$ and $g_{ij} = g_{ji}$, we have the system of equations

$$\sum_{i=1}^N \sum_{j=1}^N g_{ij}a_ia_j \sum_{m=1}^N f_{nm}a_m - \sum_{i=1}^N \sum_{j=1}^N f_{ij}a_ia_j \sum_{m=1}^N g_{nm}a_m = 0$$

$$n = 1, 2, \ldots, N \qquad (3\text{-}96)$$

If we choose the amplitude of $A_i(z)$ so that $F_i(k \cos \theta_m) = 1$ for all i, then $f_{ij} = 1$ for all i,j. This is no fundamental restriction, and Eq. (3-96) reduces to

$$\sum_{i=1}^N \sum_{j=1}^N g_{ij}a_ia_j \sum_{m=1}^N a_m - \sum_{i=1}^N \sum_{j=1}^N a_ia_j \sum_{m=1}^N g_{nm}a_m = 0 \qquad (3\text{-}97)$$

If $\sum_{m=1}^N a_m$ is divided out, Eq. (3-97) can be expressed as

$$\sum_{m=1}^N g_{nm}a_m = \frac{\sum_{i=1}^N \sum_{j=1}^N g_{ij}a_ia_j}{\sum_{i=1}^N a_i} \qquad n = 1, 2, \ldots, N \qquad (3\text{-}98)$$

Since the quantity on the right side of Eq. (3-98) is independent of n, we may set it equal to a constant K and, using matrix notation, solve the system of equations

$$(g)(a) = (K) \qquad\qquad (3\text{-}99)$$

where

$$(g) = \begin{pmatrix} g_{11} & \cdots & g_{1N} \\ \cdot & & \\ \cdot & & \\ g_{N1} & \cdots & g_{NN} \end{pmatrix} \qquad\qquad (3\text{-}100)$$

$$(a) = \begin{pmatrix} a_1 \\ \cdot \\ \cdot \\ a_N \end{pmatrix} \qquad\qquad (3\text{-}101)$$

$$(K) = \begin{pmatrix} K \\ \cdot \\ \cdot \\ K \end{pmatrix} \qquad\qquad (3\text{-}102)$$

Thus

$$(a) = (g)^{-1}(K)$$
$$\quad = \frac{(b)}{|g|}\,(K) \qquad\qquad (3\text{-}103)$$

where

$$(b) = \begin{pmatrix} b_{11} & \cdots & b_{1N} \\ \cdot & & \\ \cdot & & \\ b_{N1} & \cdots & b_{NN} \end{pmatrix} \qquad\qquad (3\text{-}104)$$

The quantity $|g|$ is the determinant of (g), and (b) is the adjoint of (g) where b_{ij} is the cofactor of g_{ij} (recall that $g_{ij} = g_{ji}$).

Even though K is not actually known, the ratios of the unknown coefficients can be expressed as

$$a_1 : a_2 : \cdots : a_N = \sum_{j=1}^{N} b_{1j} : \sum_{j=1}^{N} b_{2j} : \cdots : \sum_{j=1}^{N} b_{Nj} \qquad (3\text{-}105)$$

This method works only for a finite number of waves $A_i(z)$, and one should choose the functions $A_i(z)$ carefully in order to simplify the computations.

As an example to illustrate the method let us optimize an aperture distribution of the form

$$A(z) = a_1 A_1(z) + a_2 A_2(z) + a_3 A_3(z) \tag{3-106}$$

where

$$A_1(z) = c_1 e^{-j\beta z} \tag{3-107}$$
$$A_2(z) = 1 \tag{3-108}$$
$$A_3(z) = c_3 e^{j\beta z} \tag{3-109}$$

and we utilize only the range $-L/2 \lessgtr z \lessgtr L/2$ in computing $F(k \cos \theta)$ from $A(z)$.

Suppose we try to optimize the gain in the broadside direction ($\theta = 90$ deg). First we compute

$$F_1(k \cos \theta) = c_1 \frac{\sin [(k \cos \theta - \beta)L/2]}{(k \cos \theta - \beta)L/2} \tag{3-110}$$

$$F_2(k \cos \theta) = \frac{\sin [(k \cos \theta)L/2]}{(k \cos \theta)L/2} \tag{3-111}$$

$$F_3(k \cos \theta) = c_3 \frac{\sin [(k \cos \theta + \beta)L/2]}{(k \cos \theta + \beta)L/2} \tag{3-112}$$

We choose c_1 and c_3 so that

$$c_1 \frac{\sin (-\beta L/2)}{-\beta L/2} = 1 \tag{3-113}$$

$$c_3 \frac{\sin (\beta L/2)}{\beta L/2} = 1 \tag{3-114}$$

which makes $f_{ij} = 1$ as mentioned above. Next we compute

$$g_{ij} = \int_0^\pi c_i c_j \frac{\sin [(k \cos \theta + \beta_i)L/2]}{(k \cos \theta + \beta_i)L/2} \frac{\sin [(k \cos \theta + \beta_j)L/2]}{(k \cos \theta + \beta_j)L/2} \sin \theta \, d\theta \tag{3-115}$$

where

$$c_1 = c_3 = \frac{\beta L/2}{\sin (\beta L/2)}$$
$$c_2 = 1$$
$$\beta_1 = -\beta$$
$$\beta_2 = 0 \tag{3-116}$$
$$\beta_3 = \beta$$

If we can determine g_{ij}, we can determine the adjoint of (g). Some elements are, for example,

$$b_{11} = g_{22}g_{33} - g_{32}g_{23}$$
$$b_{12} = -(g_{21}g_{33} - g_{31}g_{23}) \tag{3-117}$$

etc.

Having calculated these numbers, we can determine

$$a_1 : a_2 : a_3 = \sum_{j=1}^{3} b_{1j} : \sum_{j=1}^{3} b_{2j} : \sum_{j=1}^{3} b_{3j} \tag{3-118}$$

The unfinished steps of the example are left as a problem at the end of the chapter.

Hansen-Woodyard Increased Directivity

Perhaps the best known example of optimum gain is the case considered by Hansen and Woodyard.[1] Hansen and Woodyard considered a line source with constant amplitude and phase velocity and adjusted the phase velocity v for optimum directivity.

The relative far-field pattern of a uniform line source of length L and with phase constant β (see Chap. 2) is given by

$$F(k \cos \theta) = \frac{\sin \left([k \cos \theta - \beta) L/2 \right]}{(k \cos \theta - \beta) L/2} \tag{3-119}$$

where θ is measured from the axis of the source (that is, $\theta = 0$ corresponds to the endfire direction).

The directivity for this case is

$$D = \frac{2 \left\{ \dfrac{\sin \left[(k - \beta) L/2 \right]}{(k - \beta) L/2} \right\}^2}{\displaystyle\int_{-1}^{1} \left\{ \dfrac{\sin \left[(k \cos \theta - \beta) L/2 \right]}{(k \cos \theta - \beta) L/2} \right\}^2 d(\cos \theta)} \tag{3-120}$$

Hansen and Woodyard minimized the function

$$\left\{ \frac{k - \beta}{\sin \left[(k - \beta) L/2 \right]} \right\}^2 \int_{-1}^{1} \left\{ \frac{\sin \left[(k \cos \theta - \beta) L/2 \right]}{(k \cos \theta - \beta) L/2} \right\}^2 d(\cos \theta) \tag{3-121}$$

where it was assumed that $L \gg \lambda$. Integration of the quantity (3-121) can be performed in terms of the sine integral giving

$$\frac{2}{kL} \left(\frac{u}{\sin u} \right)^2 \left[\frac{\pi}{2} + \frac{\cos (2u) - 1}{2u} + \text{Si} (2u) \right] \tag{3-122}$$

where

$$u = (k - \beta) \frac{L}{2} \tag{3-123}$$

---

[1] W. W. Hansen and J. R. Woodyard, A New Principle in Antenna Design, *Proc. IRE*, **26**(3):333–345, March, 1938.

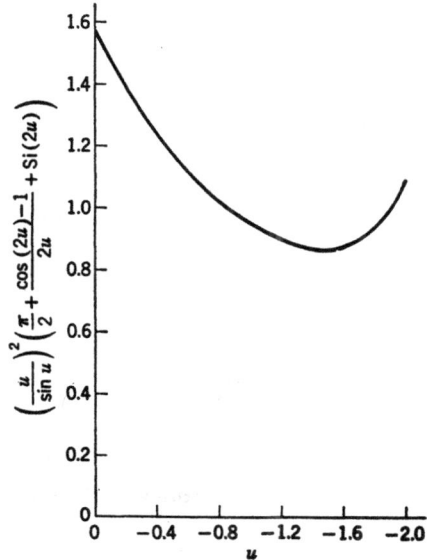

Figure 3-26. Graph of the function (3-122).

The function (3-122) is plotted in Fig. 3-26. The minimum occurs for $u = -1.47$. Thus maximum directivity for a long uniform source results when

$$\beta = k + \frac{2.94}{L} \tag{3-124}$$

or

$$\frac{c}{v} = 1 + \frac{1.47}{\pi L/\lambda} \tag{3-125}$$

This means that the phase delay along the source with increased directivity is 2.94 rad greater than along an ordinary endfire source ($\beta = k$) of the same length. It can be seen from Fig. 3-26 that the region of optimization is quite broad and in practice the commonly used figure is π rather than 2.94. In this case Eq. (3-125) becomes

$$\frac{c}{v} = 1 + \frac{1}{2L/\lambda} \tag{3-126}$$

A plot of relative directivity $\frac{D}{L/\lambda}$ is shown in Fig. 3-27 to illustrate the effects of length and phase velocity on the gain of a line source. Further discussion on the directivity of a line source is given in Sec. 8-3.

H. Supergain Antennas

The supergain antenna is one of the more intriguing problems of antenna theory. Although significant supergaining appears to be impractical at the

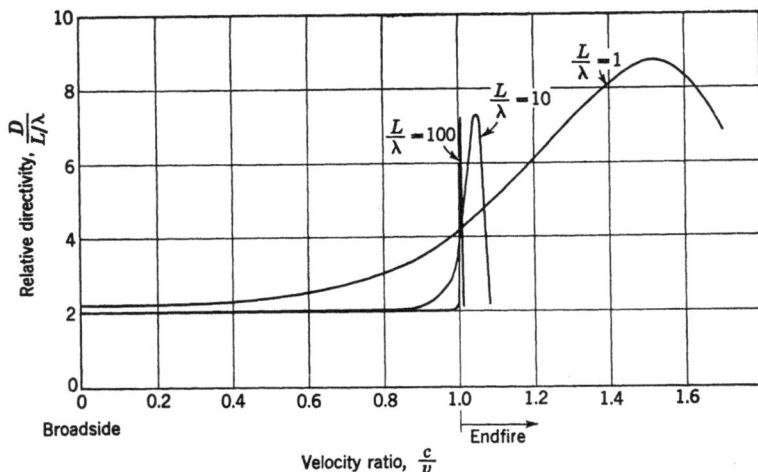

Figure 3-27. Relative directivity as a function of length and phase velocity.

present state of the art, considerable effort has been devoted to its theoretical aspects. A comprehensive study of the subject is beyond the scope of this book. We shall concern ourselves primarily with what we mean by a supergain antenna.

The term supergain can be misleading if antenna efficiency is not taken into account. Because of poor efficiency a practical supergain antenna may have less actual gain than a normal gain antenna of the same size. Superdirectivity is a more exact term; however, the term supergain is in common usage and will be used here. In essence, supergaining is an attempt to increase the gain of an antenna of given size beyond a certain value. This invariably leads to a rapid rate of change of phase and amplitude along the source and an increase in reactive power relative to radiated power (the Q of the antenna increases). Ohmic losses are increased (antenna efficiency drops sharply for conventional conductors), and rapid phase and amplitude variations along the source necessitate precise adjustment of the source function. These are the consequences of supergaining and usually outweigh any advantage of increased resolving power resulting from the increased directivity.

There appears to be no way of getting around the need for precise adjustment of the amplitude and phase of the source. The efficiency problem, however, may be attacked by using superconductive materials wherever relatively large currents are required. Another approach to the efficiency problem involves the use of the negative resistance characteristics of certain solid-state devices integrated into the antenna structure.

Hansen and Woodyard[1] and Schelkunoff[2] were among the first to present evidence that the directivity of an antenna is not completely limited by antenna size. Somewhat later Bouwkamp and de Bruijn[3] proved that theoretically there is no upper limit to the gain of a linear antenna. Riblet[4] showed that a Dolph-Chebyshev source with elements spaced less than $\lambda/2$ could yield as large a directivity as desired. A numerical example of a Dolph-Chebyshev supergain array was carried out in detail by Yaru.[5] He considered a nine-element array $\lambda/4$ in extent which produced a directivity 8.5 times greater than that of a single element. Currents on the order of 14 million amperes were required in the individual elements of the array in order for the array to produce the same maximum radiation intensity as a single element with 19.5 ma. Furthermore, the currents in the individual elements must be adjusted to their correct value to an accuracy of better than 1 part in 10^{11}. Assuming copper half-wave dipoles 1 cm in diameter and a frequency of 10 mc, Yaru found the efficiency of the supergain array to be less than 10^{-14} per cent. This example shows why antenna designers tend to shy away from supergain structures. The efficiency usually decreases at a much greater rate than the rate at which the directivity increases, so that the actual gain of the supergain antenna is well below that of a normal gain antenna.

It was stated earlier that a supergain antenna is a high-Q device; i.e., the ratio of stored to radiated energy is large. The Q of an antenna may be defined as

$$Q = \frac{\text{reactive power in antenna}}{\text{power radiated from antenna}} \tag{3-127}$$

The directivity D of a line source is defined to be the ratio of the power density in the direction of interest (θ_m, ϕ_m) to the average power radiated. This may be expressed mathematically as

$$D = \frac{4\pi P(\theta_m,\phi_m)|F(\cos \theta_m)|^2}{\int_0^{2\pi} \int_0^\pi P(\theta,\phi)|F(\cos \theta)|^2 \sin \theta \, d\theta \, d\phi} \tag{3-128}$$

where $P(\theta,\phi)$ is the power pattern (watts per unit solid angle) of an element of the source. Unless stated otherwise, the directivity is always taken to be for the direction of maximum radiation.

[1] Hansen and Woodyard, *loc. cit.*
[2] S. A. Schelkunoff, A Mathematical Theory of Linear Arrays, *Bell System Tech. J.*, **22**:80–107, 1943.
[3] Bouwkamp and de Bruijn, *loc. cit.*
[4] H. J. Riblet, Discussion on Dolph's Paper, *Proc. IRE*, **35**:489–492, May, 1947.
[5] N. Yaru, A Note on Super-gain Antenna Arrays, *Proc. IRE*, **39**(9):1081–1085, September, 1951.

Chu[1] has considered an antenna under three different conditions of optimization. These are (1) maximum directivity D, (2) minimum Q, and (3) maximum D/Q. It was found that Q increases at a very large rate if one attempts to obtain a directivity greater than $2L/\lambda$, where L is the length of a linear source. Harrington[2] formulated the problem in a manner similar to that used by Chu and found that to avoid supergaining effects the directivity of a source should not exceed $(kR)^2 + 2kR$ where R is the radius of the smallest sphere enclosing the source.

An insight into the nature of supergaining may be obtained by studying the directivity equation (3-128). Usually $P(\theta,\phi)$ is essentially constant over the range of θ where $F(\cos\theta)$ is significant. Thus Eq. (3-128) may be written

$$D = \left[\frac{P(\theta_m,\phi_m)}{\frac{1}{2\pi}\int_0^{2\pi}P(\theta_m,\phi)\,d\phi}\right]\left[\frac{|F(\cos\theta_m)|^2}{\frac{1}{2}\int_0^{\pi}|F(\cos\theta)|^2\sin\theta\,d\theta}\right] \tag{3-129}$$

The second bracketed quantity will be denoted D_i. It is the directivity of the source if it were composed of isotropic elements. Any directivity in the elements of the source contributes to the overall directivity of the source by the factor in the first bracketed quantity. That is, the first bracketed quantity is the directivity of an element of the source. There is no loss in generality of our discussion of supergain if we restrict ourselves to a consideration of the quantity D_i which may be written

$$D_i = \frac{|F(\cos\theta_m)|^2}{\frac{1}{2}\int_{-1}^{1}|F(\cos\theta)|^2 d(\cos\theta)} \tag{3-130}$$

We can bring the length of the source into the directivity equation by changing the variable of integration to $q = (L/\lambda)\cos\theta$ for a source of length L, which gives

$$D_i = \frac{2L}{\lambda}\frac{|F(q)_m|^2}{\int_{-L/\lambda}^{L/\lambda}|F(q)|^2\,dq} \tag{3-131}$$

Taylor[3] has referred to the quantity

$$D_s = \frac{D_i\lambda}{2L} = \frac{|F(q)_m|^2}{\int_{-L/\lambda}^{L/\lambda}|F(q)|^2\,dq} \tag{3-132}$$

as the specific gain. We shall call it specific directivity and refer to it as D_s.

[1] L. J. Chu, Physical Limitations of Omni-directional Antennas, *J. Appl. Phys.*, **19**:1163–1175, December, 1948.

[2] R. F. Harrington, On the Gain and Beamwidth of Directional Antennas, *IRE Trans. Antennas Propagation*, AP-6(3):219–225, July, 1958, and "Time-Harmonic Electromagnetic Fields," pp. 307–311, McGraw-Hill, New York, 1961.

[3] Taylor, *loc. cit.*

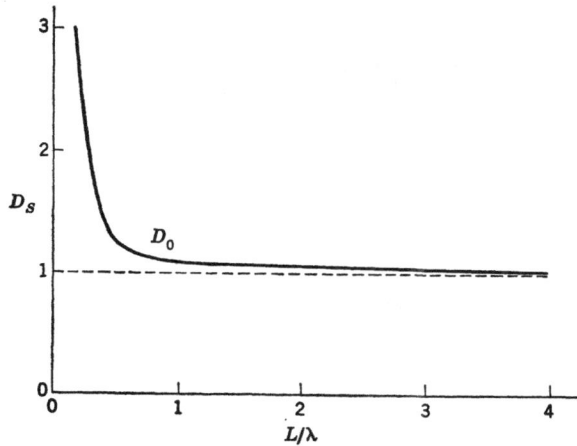

Figure 3-28. Specific directive gain D_0 for a broadside source with uniform amplitude.

The quantity D_s is obviously a function of L/λ. It can be interpreted as directivity per unit length of source. As L/λ increases, D_s decreases quite rapidly for small L/λ and becomes nearly constant as L/λ goes to infinity. This is illustrated in Fig. 3-28 for the special case of a broadside source with constant amplitude, in which case D_s is denoted as D_0. In general the limiting value of D_s as L/λ goes to infinity may be related to the source function by the Parseval formula for Fourier integrals. In our present notation the Fourier transform pair for the far field and the source function is given by

$$F(q) = \int_{-\pi}^{\pi} A(p)e^{jpq}\, dp \tag{3-133}$$

and

$$A(p) = \frac{1}{2\pi} \int_{-\infty}^{\infty} F(q)e^{-jpq}\, dq \tag{3-134}$$

where $p = 2\pi z/L$ for a source of length L centered on the z axis. The specific directivity as L/λ goes to infinity is given by

$$D_{s_\infty} = \frac{|F(q)_m|^2}{\int_{-\infty}^{\infty} |F(q)|^2\, dq} \tag{3-135}$$

By Parseval's formula a function and its Fourier transform are related as

$$2\pi \int_{-\infty}^{\infty} f_1(\tau)f_2^*(\tau)\, d\tau = \int_{-\infty}^{\infty} F_1(\omega)F_2(\omega)\, d\omega \tag{3-136}$$

Applying Eq. (3-136) and the maximum value of $F(q)$ given by

$$|F(q)_m| = \left| \int_{-\pi}^{\pi} A(p)\, dp \right| \tag{3-137}$$

to Eq. (3-135) results in

$$D_{s_\infty} = \frac{\left| \int_{-\pi}^{\pi} A(p)\, dp \right|^2}{2\pi \int_{-\pi}^{\pi} |A(p)|^2\, dp} \tag{3-138}$$

The above discussion is not restricted to a broadside source but applies to a source with phase velocity v given by $c/v = \cos\theta_m$, where θ_m is the angle of maximum radiation. This is illustrated later in Eq. (3-142) for a constant amplitude source.

The quantity D_{s_∞} has a maximum value of unity, and this occurs when $A(p)$ is constant. Thus in this respect a source with a constant amplitude and phase has a specific directivity greater than any other source. However, when L/λ is finite, it is theoretically possible to synthesize sources with arbitrarily high directivity as pointed out earlier.

A ratio of the specific directivity to the limiting specific directivity gives

$$\frac{D_s}{D_{s_\infty}} = \frac{\int_{-\infty}^{\infty} |F(q)|^2\, dq}{\int_{-L/\lambda}^{L/\lambda} |F(q)|^2\, dq} = \text{SGR} \tag{3-139}$$

The ratio D_s/D_{s_∞} is known as the supergain ratio (SGR) and was first defined by Taylor.[1] We shall henceforth use the abbreviation SGR for the ratio given in Eq. (3-139). The SGR may be interpreted as the ratio of the total area under the curve $|F(q)|^2$ to the area under that portion of $|F(q)|^2$ which lies in the visible region ($-1 \lesssim \cos\theta \lesssim 1$). A large SGR will have associated with it large contributions of $F(q)$ in the invisible region. Thus we may encounter supergain not only in highly directive patterns but also when we attempt to fit a "shaped beam" pattern as was evidenced by the examples in the pattern synthesis sections.

A large SGR is also associated with a large ratio of the area under $|A(p)|^2$ to the area under the visible region of $|F(q)|^2$. This may be seen by applying Parseval's equation to the numerator of Eq. (3-139). This gives

$$\text{SGR} = \frac{2\pi \int_{-\pi}^{\pi} |A(p)|^2\, dp}{\int_{-L/\lambda}^{L/\lambda} |F(q)|^2\, dq} \tag{3-140}$$

[1] Taylor, *loc. cit.*, and Taylor, An Antenna Pattern Synthesis Method with Discussion of Energy Storage Considerations, unpublished paper presented at URSI Meeting, Washington, D.C., April, 1951.

Equation (3-140) may be written in terms of the mean square value of $A(p)$ as

$$\text{SGR} = \frac{\dfrac{1}{2\pi}\displaystyle\int_{-\pi}^{\pi}|A(p)|^2\,dp}{\left(\dfrac{1}{2\pi}\right)^2\displaystyle\int_{-L/\lambda}^{L/\lambda}|F(q)|^2\,dq} \tag{3-141}$$

Thus for a given pattern in the visible region of space the mean square value of the source function increases by the same factor as the SGR.

To make the above discussion of supergain more meaningful, let us determine the SGR of a uniformly illuminated traveling wave source. It can be used as a standard against which other sources may be judged. From Eq. (3-139) where

$$F(q) = \frac{\sin \pi(q - q_m)}{\pi(q - q_m)} \tag{3-142}$$

with $q_m = (L/\lambda)\cos \theta_m$, the SGR is given by

$$\text{SGR}_0 = \frac{D_0}{D_{0_\infty}} \tag{3-143}$$

where

$$D_{0_\infty} = 1 \tag{3-144}$$

and

$$D_0 = \frac{1}{\displaystyle\int_{-L/\lambda}^{L/\lambda}\left|\frac{\sin \pi(q - q_m)}{\pi(q - q_m)}\right|^2 dq} \tag{3-145}$$

Equation (3-143) reduces to

$$\text{SGR}_0 = \frac{\pi}{\text{Si}\left[2\pi\left(\dfrac{L}{\lambda} - q_m\right)\right] + \text{Si}\left[2\pi\left(\dfrac{L}{\lambda} + q_m\right)\right] - \dfrac{\sin^2\{\pi[(L/\lambda) - q_m]\}}{\pi[(L/\lambda) - q_m]} - \dfrac{\sin^2\{\pi[(L/\lambda) + q_m]\}}{\pi[(L/\lambda) + q_m]}} \tag{3-146}$$

which reduces to D_0 plotted in Fig. 3-28 for $\cos \theta_m = 0$.

3-3. SYNTHESIS OF TWO-DIMENSIONAL SOURCES

The synthesis methods in the previous section considered only one-dimensional (line) sources. In practice, however, the source may be two-, three-, or even four-dimensional if the source changes with time. Fortunately, most practical antennas can be considered as two-dimensional sources of rectangular or circular shape.

A. Synthesis of Rectangular Sources

Line Source Methods

The synthesis of rectangular sources is generally simplified by assuming that the source distribution is separable. This assumption is quite reasonable, since antennas with rectangular apertures are usually most conveniently fed in this manner. Thus the techniques considered previously for the line source are directly applicable to the rectangular source.

Consider the rectangular source in Fig. 3-29. A typical design problem might call for a beam maximum at $\theta = \theta_m$ and $\phi = 90$ deg with a narrow beam and low sidelobes in the plane containing the x axis and at angle θ_m with respect to the xz plane. The beam in the plane $\phi = 90$ deg may need to be a modified csc θ pattern for a radar application. If the aperture distribution is separable, $A(x',z')$ may be expressed as

$$A(x',z') = A_1(x')A_2(z') \tag{3-147}$$

One could use Taylor's method (Sec. 3-2G) or the Dolph-Chebyshev method (Sec. 3-2G) to synthesize a function $A_1(x')$ for a broadside source of length W to give the required beamwidth and sidelobe level, while Woodward's method (Sec. 3-2D) would be quite appropriate for obtaining $A_2(z')$ of length L to give the shaped beam pattern.

If it is desired that the beam be scanned in a plane parallel to the xz

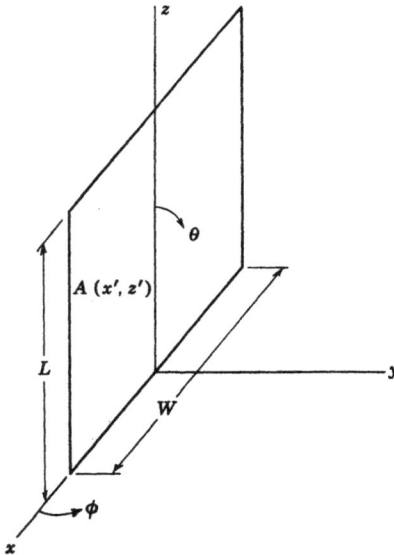

Figure 3-29.   Rectangular source.

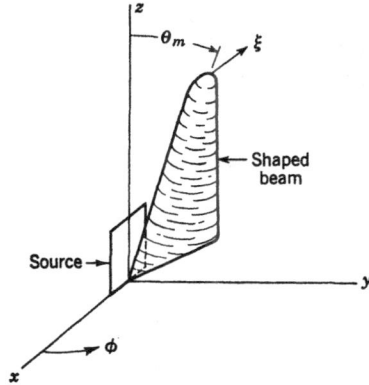

Figure 3-30. Beam scanning.

plane (see Fig. 3-30), this can be accomplished by letting $A_1(x')$ be of the form

$$A_1(x') = |A_1(x')|e^{jkx'(\sin \xi \pm \sin \xi_m)} \tag{3-148}$$

where ξ_m is the angle of maximum radiation measured from the yz plane. Serious distortion of the desired pattern may occur for large ξ_m. A comparable factor may be introduced into $A_2(z')$ to produce movement of the beam in the $\phi = 90$ deg plane; however, beam shape requirements may require the introduction of something other than a simple linear phase variation.

If the desired pattern cannot be separated into two independent patterns (one produced by an x' distribution and one by a z' distribution, for example), the source distribution is not separable and the line source synthesis techniques are not applicable.

Fourier Transform Method

A two-dimensional Fourier transform of the far field can be employed to synthesize a source distribution over a plane, rectangular aperture. This is analogous to the line source case, and except for $1/2\pi$ factors, the pair is given by

$$F(u,v) = \iint\limits_{-\infty}^{\infty} A(x',z')e^{j(ux'+vz')} \, dz' \, dx' \tag{3-149}$$

$$A(x',z') = \iint\limits_{-\infty}^{\infty} F(u,v)e^{-j(ux'+vz')} \, du \, dv \tag{3-150}$$

where $u = k \sin \theta \cos \phi$ and $v = k \cos \theta$.

If the integral $\iint\limits_{-\infty}^{\infty} |F(u,v)|^2 \, du \, dv$ exists, then Eqs. (3-149) and (3-150) constitute a useful Fourier transform pair. In general the integrals may be difficult to evaluate, but for practical distributions they can always be approximated by series where the physical interpretation is that of an array of elements replacing the continuous source (see Sec. 2-18).

As in the line source case, one of the most practical synthesis methods is that of Woodward (see Sec. 3-2D). For the two-dimensional case the field can be expressed as

$$F(u,v) = \sum_n \sum_m a_{mn} \frac{\sin X_m}{X_m} \frac{\sin Z_n}{Z_n} \tag{3-151}$$

where

$$X_m = \left(u - \frac{2\pi m}{W}\right)\frac{W}{2} \tag{3-152}$$

$$Z_n = \left(v - \frac{2\pi n}{L}\right)\frac{L}{2} \tag{3-153}$$

and W and L are the width and length of the source, respectively. The corresponding source distribution [the inverse transform of Eq. (3-151)] is

$$A(x',z') = \sum_n \sum_m a_{mn} e^{-j(2\pi mx'/W)} e^{-j(2\pi nz'/L)} \tag{3-154}$$

If we let $u = 2\pi p/W$ and $v = 2\pi q/L$ where p and q are integers, all the terms of Eq. (3-151) are zero except the one where $p = m$ and $q = n$. That is, it is convenient to select the coefficients a_{mn} such that

$$a_{pq} = F\left(\frac{2\pi p}{W}, \frac{2\pi q}{L}\right) \tag{3-155}$$

This is the procedure that Woodward used for a line source (Sec. 3-2D).

B. Synthesis of Circular Sources

Relatively little work has been done on synthesizing circular sources as compared with rectangular and line sources. Much of the work that has been done was aimed at obtaining a low sidelobe level from a circular source such as a parabolic dish antenna. The far-field patterns of various circular source distributions, such as those given in Sec. 2-7, have been considered in detail. Thus one may "synthesize" a low sidelobe antenna, for example, by choosing a source function that will produce low sidelobes. This, however practical it may be, is not true synthesis. We shall discuss now a synthesis procedure that is based on the Hankel transform pair. This transform pair is analogous to the Fourier transform pair described in Sec. 3-2A.

Hankel Transform Method

Neglecting polarization as in the previous discussions on synthesis, the far-field function $F(k \sin \theta, \phi)$ of a source function $A(\rho', \phi')$ with coordi-

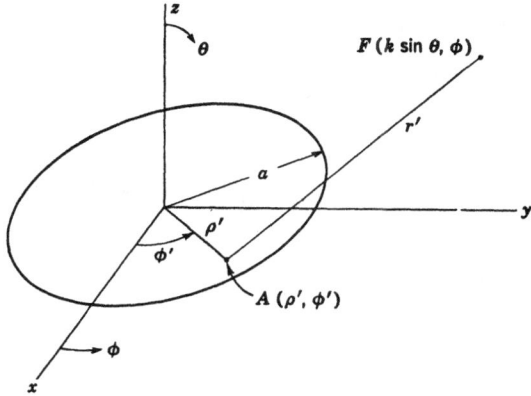

Figure 3-31. Coordinate system for circular source.

nate system as shown in Fig. 3-31 may be obtained from Eq. (2-213) as

$$F(k \sin \theta, \phi) = \int_0^{2\pi} \int_0^a A(\rho', \phi') e^{j\rho'k \sin \theta \cos (\phi-\phi')} \rho' \, d\rho' \, d\phi' \qquad (3\text{-}156)$$

where constant terms are neglected. If the source function is independent of ϕ', the field function becomes

$$F(k \sin \theta) = \int_0^{\infty} A(\rho') \int_0^{2\pi} e^{j\rho'k \sin \theta \cos (\phi-\phi')} \, d\phi' \rho' \, d\rho' \qquad (3\text{-}157)$$

where the infinite limit is used even though the source is of finite extent. Equation (3-157) reduces to

$$F(k \sin \theta) = \int_0^{\infty} \rho' A(\rho') J_0(\rho'k \sin \theta) \, d\rho' \qquad (3\text{-}158)$$

except for a factor of 2π, which will be neglected. In a manner analogous to the Fourier transform case the function $A(\rho')$ is given by the Hankel transform[1] as

$$A(\rho') = \int_0^{\infty} k \sin \theta F(k \sin \theta) J_0(\rho'k \sin \theta) d(k \sin \theta) \qquad (3\text{-}159)$$

As a result of choice of coordinate systems the far-field variable $k \sin \theta$ for the circular source corresponds to the variable $k \cos \theta$ for the rectangular and line sources. The interpretations of visible region $(-1 \lessgtr \sin \theta \lessgtr 1)$ and invisible region $(|\sin \theta| > 1)$ apply as before.

Certain useful transform pairs exist for the circular source. Making use of the results in Sec. 2-16 we obtain the transform pairs given in Fig. 3-32. It should be kept in mind that the transformation goes both ways

[1] I. N. Snedden, "Fourier Transforms," pp. 48–52, McGraw-Hill, New York, 1951.

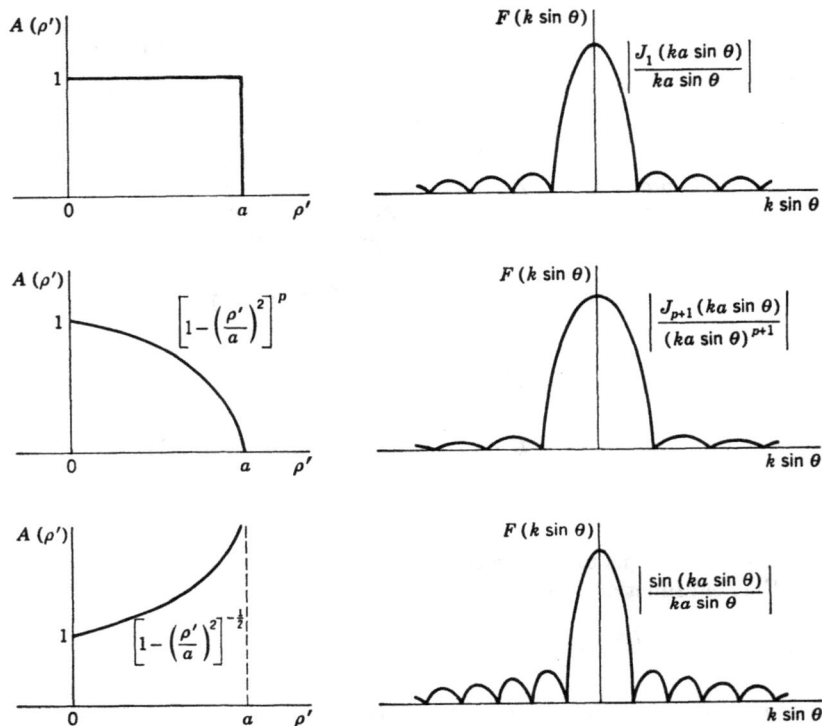

A (ρ')

$\left[1-\left(\frac{\rho'}{a}\right)^2\right]^p$

$F(k\sin\theta)$

$\left|\frac{J_1(ka\sin\theta)}{ka\sin\theta}\right|$

$\left|\frac{J_{p+1}(ka\sin\theta)}{(ka\sin\theta)^{p+1}}\right|$

$\left[1-\left(\frac{\rho'}{a}\right)^2\right]^{-\frac{1}{2}}$

$\left|\frac{\sin(ka\sin\theta)}{ka\sin\theta}\right|$

Figure 3-32.   Some common Hankel transform pairs.

and that each function of the transform pair may represent either the source
or the far field. For example, a step function source distribution gives rise
to a far field of the form $[J_1(ka\sin\theta)]/(ka\sin\theta)$ whereas a source function
of the form $J_1(\rho')/\rho'$ would give rise to a step function in the far field.

Equation (3-159) enables one to specify a far-field pattern $F(k\sin\theta)$
and compute a circularly symmetric source function that will produce the
specified pattern. However, the source function $A(\rho')$ determined by
Eq. (3-159) will not, in general, be restricted to a desired source radius a.
If only the portion of the source out to radius a is to be used, an approximate far-field pattern is obtained as

$$F_a(k\sin\theta) = \int_0^a \rho' A(\rho') J_0(\rho' k\sin\theta)\, d\rho' \tag{3-160}$$

Making use of Eq. (3-159), Eq. (3-160) can be expressed as

$$F_a(k\sin\theta) = \int_0^a \rho'\, d\rho' \int_0^\infty \xi F(\xi) J_0(\rho'\xi) J_0(\rho' k\sin\theta)\, d\xi \tag{3-161}$$

If the inner integral of Eq. (3-161) is uniformly convergent, the order of integration may be interchanged, giving

$$F_a(k \sin \theta) = \int_0^\infty \xi F(\xi) \, d\xi \int_0^a \rho' J_0(\rho'\xi) J_0(\rho'k \sin \theta) \, d\rho' \qquad (3\text{-}162)$$

Performing the integration with respect to ρ' gives

$$\int_0^a \rho' J_0(\rho'\xi) J_0(\rho'k \sin \theta) \, d\rho' = \frac{a}{\xi^2 - (k \sin \theta)^2} [\xi J_1(a\xi) J_0(ak \sin \theta)$$
$$- k \sin \theta J_1(ak \sin \theta) J_0(a\xi)] \qquad (3\text{-}163)$$

Substituting into Eq. (3-162) gives

$$F_a(k \sin \theta) = \int_0^\infty \frac{a\xi F(\xi)}{\xi^2 - (k \sin \theta)^2} [\xi J_1(a\xi) J_0(ak \sin \theta)$$
$$- k \sin \theta J_1(ak \sin \theta) J_0(a\xi)] \, d\xi \qquad (3\text{-}164)$$

This is analogous to the line source case discussed in Sec. 3-2A. The approximate pattern F_a approaches the desired pattern F as a goes to infinity.

Although the Hankel transform method is simple in principle, the integrations involved may be quite formidable, thus limiting the usefulness of the method. It is possible, however, to avoid the direct integration in Eqs. (3-158) and (3-159) by developing an orthonormal set of far-field functions which may be summed to provide an approximation to the desired far-field pattern. Furthermore, a source distribution corresponding to the approximate pattern may be readily obtained by summing the transforms of the far-field functions. The method is somewhat analogous to Woodward's method and appears to have been first used by Johnk.

Johnk's Synthesis Method

Johnk's synthesis method[1] for the circular source is based on a solution of the Hankel transform pair given by Eqs. (3-158) and (3-159); thus the method is restricted to circularly symmetric sources.

Johnk chose the basic transform pair $(1 - r^2)^{m-1}$ and $2^{m-1}(m - 1)! J_m(u)/u^m$, where r is normalized radius ρ'/a and $u = k \sin \theta$. Thus an arbitrary far field may be represented as

$$F(u) = \sum_{m=1}^\infty A_m 2^{m-1}(m - 1)! \frac{J_m(u)}{u^m} \qquad (3\text{-}165)$$

This is analogous to the series of $(\sin X)/X$ terms used by Woodward. The corresponding source distribution for Johnk's method is given by

$$A(r) = \sum_{m=1}^\infty A_m(1 - r^2)^{m-1} \qquad 0 \lesssim r \lesssim 1$$
$$= 0 \qquad\qquad\qquad r > 1 \qquad\qquad (3\text{-}166)$$

[1] C. T. A. Johnk, Synthesis of Aperture Antennas, *Univ. Illinois Eng. Expt. Sta.*, *Antenna Sec.*, *Tech. Rept.* 1, October, 1954.

The problem, now, is one of determining the coefficients A_m of the nonorthogonal series given by Eq. (3-165). By means of an orthogonalization process an orthogonal set may be constructed from a complete nonorthogonal set. The interval of orthogonalization is chosen to be 0 to ∞ in order to make the synthesis method apply to sources of finite extent.

Let the far-field function in Eq. (3-165) be expressed in terms of a nonorthogonal function $h_m(u)$ as

$$F(u) = \sum_m A_m h_m(u) \tag{3-167}$$

where

$$h_m(u) = 2^m m! \frac{J_m(u)}{u^m} \tag{3-168}$$

By means of the Schmidt orthogonalization process[1] an orthonormal set $[G]$ may be constructed with elements given by

$$G_n = \sum_{q=1}^{n} b_{nq} h_q \tag{3-169}$$

Johnk has shown that the set $[G]$ will be incomplete and the sum

$$F_a(u) = \sum_{m=1}^{\infty} A_m G_m \tag{3-170}$$

will oscillate about the desired function $F(u)$. The coefficients A_m in Eq. (3-170) are given by

$$A_m = \int_0^\infty F(u) G_m(u) \, du \tag{3-171}$$

The source distribution given by the Hankel transform as Eq. (3-159) may be written

$$A(r) = \int_0^\infty u F(u) J_0(ur) \, du \tag{3-172}$$

Substituting Eq. (3-170) for $F(u)$ gives

$$A(r) = \int_0^\infty u J_0(ur) \sum_{n=1}^{\infty} A_n G_n \, du \tag{3-173}$$

which may be written

$$A(r) = \sum_{n=1}^{\infty} A_n \int_0^\infty u G_n(u) J_0(ur) \, du \tag{3-174}$$

---

[1] E. T. Whittaker and G. N. Watson, "Modern Analysis," p. 224, Macmillan, New York, 1948.

Substituting G_n from Eq. (3-169) gives

$$A(r) = \sum_{n=1}^{\infty} A_n \sum_{q=1}^{n} b_{nq} \int_0^{\infty} u h_q(u) J_0(ur)\, du \qquad (3\text{-}175)$$

However,

$$\int_0^{\infty} u h_q(u) J_0(ur)\, du = 2q(1 - r^2)^{q-1} \qquad (3\text{-}176)$$

Therefore, Eq. (3-175) becomes

$$A(r) = 2 \sum_{n=1}^{\infty} A_n \sum_{q=1}^{n} q b_{nq}(1 - r^2)^{q-1} \qquad 0 \gtrless r \gtrless 1$$
$$= 0 \qquad\qquad\qquad\qquad\qquad r > 1 \qquad (3\text{-}177)$$

Equation (3-177) may be written

$$A(r) = 2 \sum_{n=1}^{\infty} A_n \mathcal{Q}_n(r) \qquad 0 \gtrless r \gtrless 1$$
$$= 0 \qquad\qquad\qquad r > 1 \qquad (3\text{-}178)$$

where

$$\mathcal{Q}_n(r) = \sum_{q=1}^{n} q b_{nq}(1 - r^2)^{q-1} \qquad (3\text{-}179)$$

The function $\mathcal{Q}_n(r)$ is that source function which produces the orthonormal field function $G_n(u)$. Equation (3-178) gives the required source distribution which will produce the series approximation to $F(u)$ that is given by Eq. (3-170).

A series expression for $G_n(u)$ is developed by Johnk as[1]

$$G_n(u) = N_{nn}^{-\frac{1}{2}} \sum_{q=0}^{\infty} \frac{(-1)^q A_n(q)}{q!} \left(\frac{u}{2}\right)^{2q} \qquad (3\text{-}180)$$

where

$$N_{nn} = R_{nn} \frac{n[(2n-2)!]^2}{(4n-3)!} \qquad (3\text{-}181)$$

$$R_{nn} = \frac{2^{8n-3} n^2 [(n-1)!]^4}{\pi(4n-1)!} \qquad (3\text{-}182)$$

$$A_n(q) = \sum_{m=1}^{n} \frac{m! B_{nm}}{(m+q)!} \qquad (3\text{-}183)$$

$$B_{nm} = \frac{(-1)^{n-m} 2^{4n-4m-2} n^2 m[(n-1)!]^4 (2m)!(2n+2m-2)!}{(2n-1)(4n-3)!(n-m)!(n+m-1)!(m!)^4} \qquad n \gtrless m \quad (3\text{-}184)$$

[1] For full details the reader must refer to Johnk, *loc. cit.*

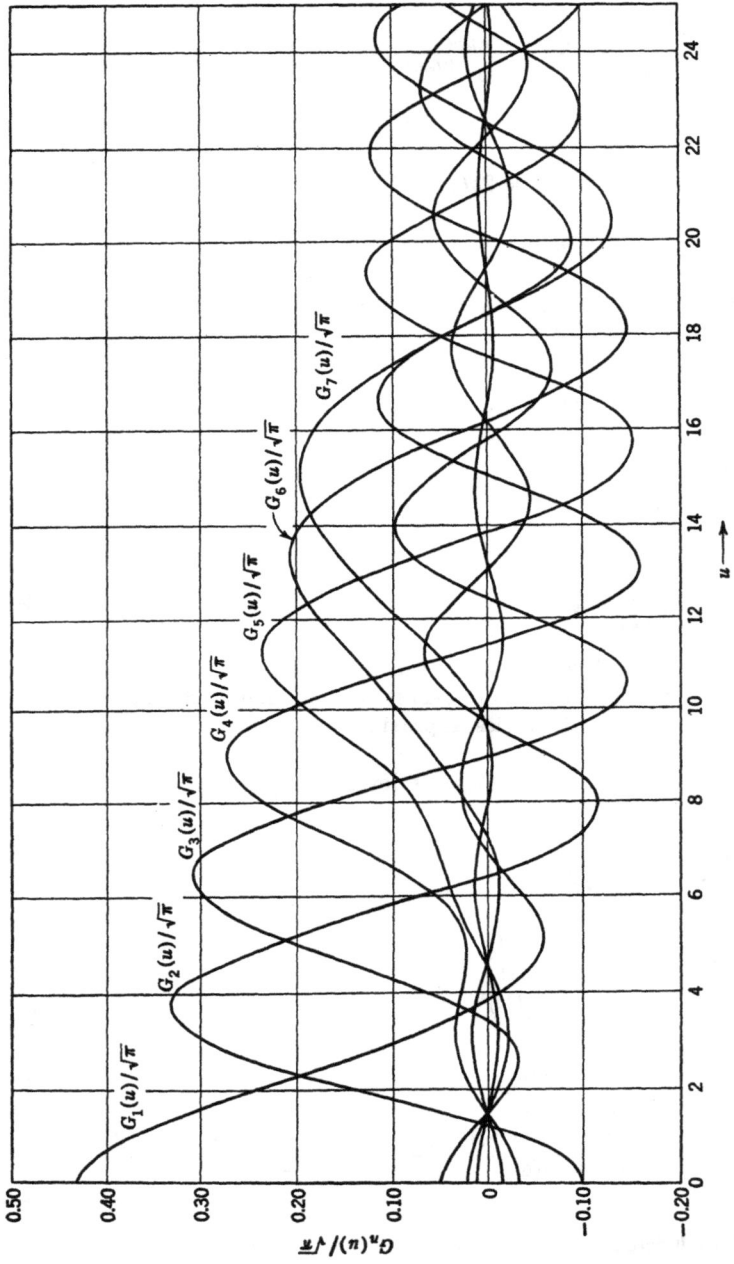

Figure 3-33. Orthonormal functions $G_n(u)$.

136

The coefficients b_{nq} are given by

$$b_{nq} = N_{nn}^{-\frac{1}{2}} B_{nq} \tag{3-185}$$

where q may be substituted for m in Eq. (3-184) to obtain B_{nq}.

The far-field functions $G_n(u)$ for $G_1(u)$ through $G_7(u)$ are plotted in Fig. 3-33, and the source functions $\mathcal{C}_n(r)$ for $\mathcal{C}_1(r)$ through $\mathcal{C}_6(r)$ are plotted in Fig. 3-34.

As an example of Johnk's method consider a gaussian far field of the form

$$F(u) = e^{-0.179u^2} \tag{3-186}$$

for which the 20 per cent amplitude points are at $u = 3$. The coefficients A_m for the series in Eq. (3-170) are perhaps best determined by graphical

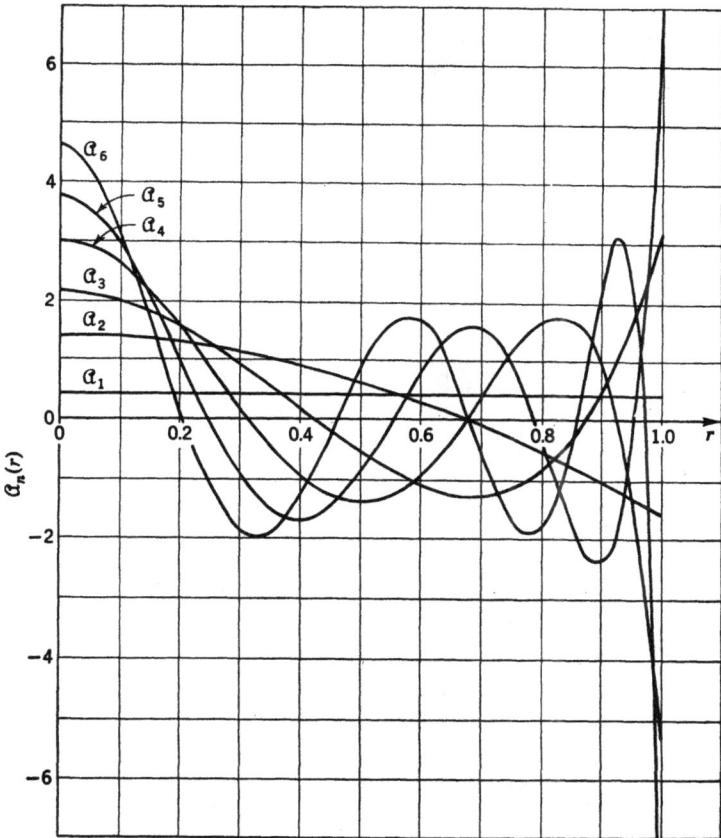

Figure 3-34.   Source functions $\mathcal{C}_n(r)$ corresponding to $G_n(u)$.

integration of Eq. (3-171). Multiplying each $G_n(u)$ in Fig. 3-33 by $F(u)$ in Eq. (3-186) and determining the relative areas under the curves give the coefficients

$$A_1 = 1 \qquad A_5 = 0.0106$$
$$A_2 = 0.116 \qquad A_6 = -0.0141$$
$$A_3 = 0.0423 \qquad A_7 = 0.004$$
$$A_4 = -0.0211$$

Thus the approximation to the gaussian pattern is given as

$$F_a(u) = G_1 + 0.116G_2 + 0.0423G_3 - 0.0211G_4$$
$$+ 0.0106G_5 - 0.0141G_6 + 0.004G_7 \quad (3\text{-}187)$$

A comparison of the curve of Eq. (3-187) with the exact curve given by Eq. (3-186) is shown in Fig. 3-35. The sidelobe level is about 23 db.

The source function $A(r)$ from Eq. (3-178), making use of Fig. 3-34, is given as

$$A(r) = 2\alpha_1 + 0.232\alpha_2 + 0.0846\alpha_3 - 0.0422\alpha_4$$
$$+ 0.0212\alpha_5 - 0.0282\alpha_6 \quad (3\text{-}188)$$

A graph of Eq. (3-188) is shown in Fig. 3-36.

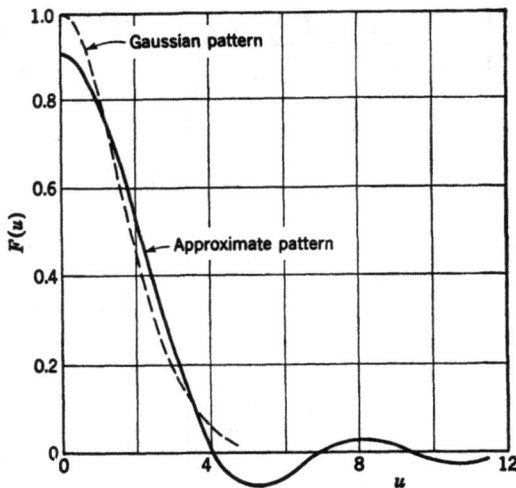

Figure 3-35.   Approximation to a gaussian pattern.

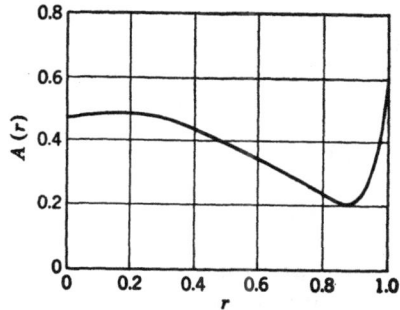

Figure 3-36. Source distribution that gives approximate pattern in Fig. 3-35.

3-4. FRESNEL REGION SYNTHESIS

The previous sections have dealt with the problem of far-field synthesis. In most antenna problems the far field is the region of greatest concern; however, there are some cases where it is desirable or perhaps necessary to find a source that will produce a specified field distribution in the Fresnel region. This section will consider some methods based on Fourier and Fresnel transforms that are quite useful in Fresnel region synthesis.

A. Fresnel Transform Method

A Fresnel transform method as developed by Compton[1] is analogous to the Fourier transform method of far-field synthesis in that the integral equation that gives the Fresnel field of a source may be inverted to give an integral equation for the source distribution.

Consider the electric field of an electric-current line source $I(z')$. An expression that holds in the Fresnel region as well as the far-field region is

$$E_z = \frac{-j\omega\mu \sin^2\theta}{4\pi} \int_{z_1}^{z_2} I(z') \frac{e^{-jkr'}}{r'} \, dz' \tag{3-189}$$

where r' is the distance from the element dz' to the point of observation.

In the far field where $\sin^2\theta \approx 1$, Eq. (3-189) reduces to

$$E_z(k \cos \theta) = \frac{-j\omega\mu e^{-jkr}}{4\pi r} \int_{-\infty}^{\infty} I(z')e^{jkz' \cos \theta} \, dz' \tag{3-190}$$

where r is the distance from the origin to the point of observation and $I(z')$ exists only over a finite interval. The inversion of Eq. (3-190) gives

$$I(z') = \frac{1}{2\pi} \int_{-\infty}^{\infty} \frac{4\pi r e^{jkr} E_z(k \cos \theta)}{-j\omega\mu} e^{-jkz' \cos \theta} d(k \cos \theta) \tag{3-191}$$

[1] R. T. Compton, Jr., Synthesis of Fresnel Zone Antenna Patterns, Master's Thesis, The Ohio State University, 1961.

Except for detail, Eqs. (3-190) and (3-191) are the Fourier transform pair that is used for pattern synthesis in Sec. 3-2A.

Let us examine the meaning of Eqs. (3-190) and (3-191). In differential form (3-190) is written

$$dE_z(k \cos \theta) = \frac{-j\omega\mu}{4\pi} \frac{e^{-jk(r-z' \cos \theta)}}{r} dI(z') \tag{3-192}$$

According to Eq. (3-192), at each point r,θ the field E_z is a sum of contributions from little elements of current $dI(z')$ along the z axis. Each little element $dI(z')$ is multiplied by $e^{-jk(r-z' \cos \theta)}/r$. However, this factor is only an approximation for $e^{-jkr'}/r'$. It would be more accurate to write Eq. (3-192) as

$$dE_z(k \cos \theta) = \frac{-j\omega\mu}{4\pi} \frac{e^{-jkr'}}{r'} dI(z') \tag{3-193}$$

Equation (3-193) will hold in the Fresnel region as well as the far-field region.

Next consider Eq. (3-191). It may be written in differential form as

$$dI(z') = \frac{2re^{jk(r-z' \cos \theta)}}{-j\omega\mu} dE_z(k \cos \theta) \tag{3-194}$$

The interpretation here is that $I(z')$ at a fixed value of z' is obtained by summing contributions $dE_z(k \cos \theta)$ over the range $-\infty < k \cos \theta < \infty$. Each element $dE_z(k \cos \theta)$ is multiplied by $re^{jk(r-z' \cos \theta)}$. It would be more accurate to write Eq. (3-194) as

$$dI(z') = \frac{2r'e^{jkr'}}{-j\omega\mu} dE_z(k \cos \theta) \tag{3-195}$$

Equation (3-195) will hold in the Fresnel region as well as the far-field region.

Equations (3-193) and (3-195) in integral form give

$$E_z(k \cos \theta) = \frac{-j\omega\mu}{4\pi} \int_{z_1}^{z_2} \frac{I(z')e^{-jkr'}}{r'} dz' \tag{3-196}$$

$$I(z') = \frac{2}{-j\omega\mu} \int_{-\infty}^{\infty} r'e^{jkr'} E_z(k \cos \theta) d(k \cos \theta) \tag{3-197}$$

These equations hold in the Fresnel region as well as the far field. In the far field they reduce to Eqs. (3-190) and (3-191). In the Fresnel region they give

$$E_z(k \cos \theta) = \frac{-j\omega\mu e^{-jkr}}{4\pi r} \int_{z_1}^{z_2} I(z')e^{jk\left(z' \cos \theta - \frac{z'^2 \sin^2 \theta}{2r}\right)} dz' \tag{3-198}$$

and

$$I(z') = \frac{2re^{jkr}}{-j\omega\mu} \int_{-\infty}^{\infty} E_z(k \cos \theta)e^{-jk\left(z' \cos \theta - \frac{z'^2 \sin^2 \theta}{2r}\right)} d(k \cos \theta) \tag{3-199}$$

By rearranging terms and completing the square in the exponent, Eq. (3-199) can be written

$$I(z') = \frac{2r\sqrt{\pi k r}\, e^{jkr}}{-j\omega\mu z'} e^{j\frac{\pi}{2}\left(\frac{kr}{\pi} + \frac{kz'^2}{\pi r}\right)} \int_{-\infty}^{\infty} E_z(u) e^{-j\frac{\pi}{2}u^2}\, du \tag{3-200}$$

where

$$u = \frac{z'k\cos\theta}{\sqrt{\pi k r}} + \sqrt{\frac{kr}{\pi}} \tag{3-201}$$

If $E_z(u)$ can be approximated by small segments of constant amplitude, Eq. (3-200) reduces to a sum of Fresnel integrals (see Sec. 2-14).

Equations (3-198) and (3-199) constitute a Fresnel transform pair. In principle one may transform from source to Fresnel pattern or vice versa just as readily as in the far-field case. There is a major difference, however, between the Fresnel transform and the Fourier transform. In the latter the form of the far field does not change with distance; hence a Fourier transform pair is unique. This is not true of the Fresnel transform pair. The Fresnel pattern may change significantly with range; hence the distance between source and pattern must be specified for each Fresnel transform pair.

As an example of Fresnel transform synthesis consider the Fresnel region pattern given in Fig. 3-37. The problem is to find a line source that will produce this pattern 5 λ from the source. From Eq. (3-200)

$$\int_{-\infty}^{\infty} E_z(u) e^{-j(\pi/2)u^2}\, du = \int_{u_1}^{u_2} e^{-j(\pi/2)u^2}\, du \tag{3-202}$$

where

$$u_1 = -0.055z' + 3.16 \tag{3-203}$$
$$u_2 = 0.055z' + 3.16 \tag{3-204}$$

The source $I(z')$ is given by

$$I(z') = K \frac{e^{j\frac{\pi}{5}\left(\frac{z'}{\lambda}\right)^2}}{z'/\lambda} [C(u_2) - C(u_1) - jS(u_2) + jS(u_1)] \tag{3-205}$$

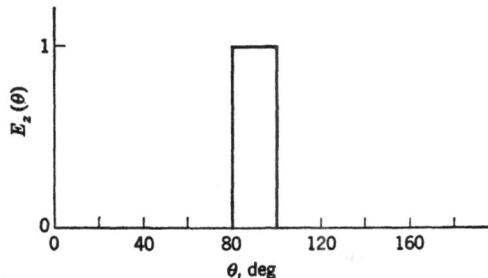

Figure 3-37. Constant-amplitude Fresnel region pattern.

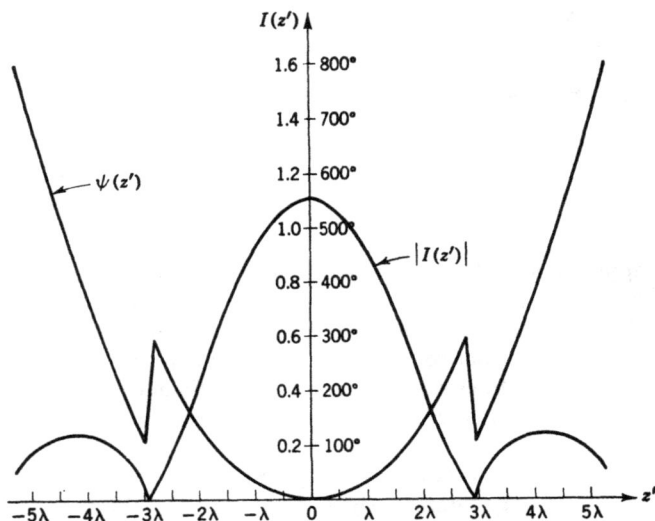

Figure 3-38. Source distribution that gives the pattern in Fig. 3-37 at a range of 5 λ.

where K includes all constants not of interest and C and S are the usual Fresnel integrals (see Sec. 2-14). A graph of Eq. (3-205) is given in Fig. 3-38.

If the rectangular pulse had been specified as a far-field function, the resulting source function would have been of the form $(\sin X)/X$. It is interesting to observe that the Fresnel zone synthesis gives a source amplitude function that is approximately the same as the $(\sin X)/X$ form. In addition, however, there is a quadratic phase term in the Fresnel case that does not arise in the far-field problem.

B. Circular Source Method

If the source distribution is specified on an arc of a circle rather than a straight line, the Fresnel zone field is given by a Fourier integral.[1] In this manner the Fourier transform may be used for Fresnel synthesis.

Consider the geometry in Fig. 3-39. We shall specify $E(z)$ at $y = R_0$ and solve for $I_c(z')$ along the circular arc of radius R_0. $I_c(z')$ is then transformed to $I(z')$ on the z' axis.

The field $E(z)$ at $y = R_0$ is given by

$$E(z) = \frac{-j\omega\mu}{4\pi} \int_{z'} I_c(z') \frac{e^{-jkr'}}{r'} \, dz' \tag{3-206}$$

[1] M. K. Hu, Some New Methods of Analysis and Synthesis of Near Zone Fields, *IRE Intern. Conv. Record*, vol. 8, Part 1, March, 1960.

From the geometry in Fig. 3-39 it can be seen that

$$r'^2 = (z' - z)^2 + (R_0 - a)^2 \tag{3-207}$$

But

$$(R_0 - a)^2 = R_0^2 - z'^2 \tag{3-208}$$

so that

$$r'^2 = z^2 - 2zz' + R_0^2 \tag{3-209}$$

Expanding r' by means of the binomial expansion and using only the first two terms give

$$r' \approx R_0 + \frac{1}{2R_0}(z^2 - 2zz') \tag{3-210}$$

Substituting Eq. (3-210) into Eq. (3-206) gives

$$E(z) \approx \frac{-j\omega\mu}{4\pi R_0} e^{-jkR_0} e^{-jk(z^2/2R_0)} \int_{z'} I_c(z') e^{j(kzz'/R_0)} \, dz' \tag{3-211}$$

Equation (3-211) is a Fourier integral. The inverse equation is

$$I_c(z') = \frac{2R_0 e^{jkR_0}}{-j\omega\mu} \int_{-\infty}^{\infty} E(z) e^{jkz^2/2R_0} e^{-jkzz'/R_0} \, dz \tag{3-212}$$

For simplicity $E(z)$ may be given the phase variation $e^{-jkz^2/2R_0}$ to eliminate the quadratic phase term in Eq. (3-212).

Solution of Eq. (3-212) gives

$$I_c(z') = A(z') e^{j\phi_c(z')} \tag{3-213}$$

Transforming $I_c(z')$ to the z' axis gives

$$I(z') \approx A(z') e^{j[\phi_c(z') + kR_0 - k(R_0^2 - z'^2)^{1/2}]} \tag{3-214}$$

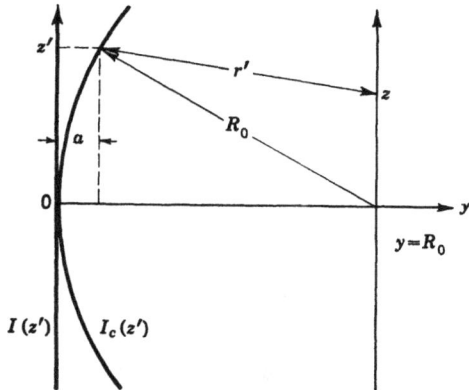

Figure 3-39. Geometry for using Fourier integral in the Fresnel region.

where only phase is considered. Any variation in the amplitude $A(z')$ as a result of the transformation is usually negligible.

The circular source synthesis method is quite useful. It essentially reduces Fresnel synthesis to far-field synthesis for which all the methods in Sec. 3-2 are applicable. Furthermore, it demonstrates that in the plane $y = R_0$ and near the y axis a planar source can produce its far-field pattern in the Fresnel region by means of a quadratic phase distribution over the source. This technique is used in focusing an antenna in its Fresnel zone (see Prob. 2-9).

C. General Fourier Transform Method

The previous section showed a way in which the Fourier transform method of pattern synthesis can be modified to apply in the Fresnel region of an antenna. However, with proper interpretation the Fourier transform method can be exact for any distance from a plane aperture. Consider the Fourier transform pair in Eqs. (3-149) and (3-150). This pair can be rewritten as

$$F(\beta_x,\beta_z) = \iint\limits_{-\infty}^{\infty} A(x',z')e^{j(\beta_x x'+\beta_z z')}\,dz'\,dx' \tag{3-215}$$

$$A(x',z') = \iint\limits_{-\infty}^{\infty} F(\beta_x,\beta_z)e^{-j(\beta_x x'+\beta_z z')}\,d\beta_x\,d\beta_z \tag{3-216}$$

where β_x and β_z have been substituted for u and v, respectively. If $A(x',z')$ represents a field component ($\mathbf{E}$ or $\mathbf{H}$) in a rectangular aperture, then A can be interpreted as a sum of plane waves of the form $F(k)e^{-j\mathbf{k}\cdot\mathbf{r}}$ where

$$\mathbf{k} = \mathbf{x}\beta_x + \mathbf{y}\beta_y + \mathbf{z}\beta_z \tag{3-217}$$

and

$$\mathbf{r} = \mathbf{x}x + \mathbf{y}y + \mathbf{z}z \tag{3-218}$$

With this interpretation a field component at any point x,y,z is

$$A(x,y,z) = \iint\limits_{-\infty}^{\infty} F(\beta_x,\beta_z)e^{-j\beta_y y}e^{-j(\beta_x x+\beta_z z)}\,d\beta_x\,d\beta_z \tag{3-219}$$

where

$$\beta_y^2 = k^2 - \beta_x^2 - \beta_z^2$$

and the transform gives

$$F(\beta_x,\beta_z)e^{-j\beta_y y} = \iint\limits_{-\infty}^{\infty} A(x,y,z)e^{j(\beta_x x+\beta_z z)}\,dx\,dz \tag{3-220}$$

Thus one can specify a field distribution $A_1(x,y,z)$ at $y = y_1$ and find the plane wave spectrum from Eq. (3-220). The inverse transform for $y = 0$

[Eq. (3-216)] gives the source distribution that will produce the specified pattern at $y = y_1$. A practical way of performing the transformations is to utilize what is essentially Woodward's method (see Sec. 3-3A).

ADDITIONAL REFERENCES

1. I. Wolff, Determination of the Radiating System Which Will Produce a Specified Directional Characteristic, *Proc. IRE*, **25**:630, 1937.

2. G. Sinclair and F. V. Cairns, Optimum Patterns for Arrays of Non-isotropic Sources, *Trans. IRE*, **PGAP-1**:50–61, February, 1952.

3. R. H. DuHamel, Optimum Patterns for Endfire Arrays, *Proc. IRE*, **41**(5): 652–659, May, 1953.

4. D. R. Rhodes, The Optimum Linear Array for a Single Main Beam, *Proc. IRE*, **41**(6):793–794, June, 1953.

5. A. Ishimaru and G. Held, Radiation Pattern Synthesis with Sources Located on a Conical Surface, *Univ. Washington Dept. Elec. Eng. Tech. Rept.* 26, April, 1958.

6. A. Ishimaru and G. Held, Optimum Radiation Pattern Synthesis for Circular Aperture, *Univ. Washington Dept. Elec. Eng. Tech. Rept.* 29, October, 1958.

7. A. S. Thomas, Study of Gain Limitations and Pattern Synthesis of Traveling Wave Antennas, *Air Force Cambridge Res. Center Tech. Rept.* AFCRC-TR-58-198, Bedford, Mass., November, 1958.

8. A. Ishimaru and G. Held, Analysis and Synthesis of Radiation Pattern for the Traveling Wave Type Sources on a Circular Aperture, *Univ. Washington Dept. Elec. Eng. Tech. Rept.* 33, March, 1959.

9. E. Spitz, Supergain and Volumetric Antennas, *Air Force Cambridge Res. Center Tech. Rept.* AFCRC-TR-59-194, Bedford, Mass., June, 1959.

10. A. Ksienski, Maximally Flat and Quasi-smooth Sector Beams, *IRE Trans. Antennas Propagation*, **AP-8**(5):476–484, September, 1960.

11. H. E. Shanks, A Geometrical Optics Method of Pattern Synthesis for Linear Arrays, *IRE Trans. Antennas Propagation*, **AP-8**(5):485–490, September, 1960.

12. A. Ksienski, Derivative Control in Shaping Antenna Patterns, *Hughes Aircraft Co. Sci. Rept.* 3508/6, Culver City, Calif., September, 1959.

13. J. H. Harris and H. E. Shanks, A Method for Synthesis of Optimum Directional Patterns from Nonplanar Apertures, *IRE Trans. Antennas Propagation*, **AP-10**(3): 228–236, May, 1962.

14. A. Ksienski, Synthesis of Nonseparable Two-dimensional Patterns by Means of Planar Arrays, *IRE Trans. Antennas Propagation*, **AP-8**(2):224–225, March, 1960.

15. T. T. Taylor, Design of Circular Apertures for Narrow Beamwidth and Low Sidelobes, *IRE Trans. Antennas Propagation*, **AP-8**(1):17–22, January, 1960.

16. R. C. Hansen, Tables of Taylor Distributions for Circular Aperture Antennas, *IRE Trans. Antennas Propagation*, **AP-8**(1):23–26, January, 1960.

17. N. Newman and W. Magnus, Maximal Gains of Antennas, *New York Univ. Inst. Math. Sci. Res. Rept.* EM-142, September, 1959.

18. J. C. Simon, Application of Periodic Functions Approximation to Antenna Pattern Synthesis and Circuit Theory, *IRE Trans. Antennas Propagation*, **AP-4**:429–440, July, 1956.

19. D. R. Rhodes, The Optimum Line Source for the Best Mean-square Approximation to a Given Radiation Pattern, *IRE Trans. Antennas Propagation*, **AP-11**: 440–446, July, 1963.

20. R. Kovács and L. Solymár, Theory of Aperture Aerials Based on the Properties of Entire Functions of the Exponential Type, *Acta Phys. Acad. Sci. Hung.*, **6**: 161–184, 1956.

21. B. M. Minkovich, Symmetrical Diagrams Realized by a Round Radiating Aperture, *Radiotekhnika i Elektronika*, no. 12: 1880–1881 (transl. ed.), December, 1961.

22. B. M. Minkovich, Application of Double Fourier Transform for Design of Plane-Aperture Antennas, *Radiotekhnika i Elektronika*, no. 1: 153–156 (transl. ed.), January, 1962.

23. V. I. Popovkin, Determination of Antenna Length for a Given Radiation Pattern, *Radiotekhnika i Elektronika*, no. 4: 662–666 (transl. ed.), April, 1962.

24. N. G. Ponomarev, Radiation Pattern of Sweeping-Beam Antennas, *Radiotekhnika i Elektronika*, no. 6: 892–904 (transl. ed.), June, 1962.

25. Ye. G. Zelkin, Phase Directivity Pattern and the Problem of Synthesis of an Antenna, *Radiotekhnika i Elektronika*, no. 1: 38–48 (transl. ed.), January, 1963.

26. Ya. N. Fel'd and L. D. Bakhrakh, Present State of Antenna Synthesis Theory, *Radiotekhnika i Elektronika*, no. 2: 163–179 (transl. ed.), February, 1963.

27. V. P. Yakovlev, Synthesis of a Linear Antenna with Current Distribution Represented by a Fourier Series with a Finite Number of Harmonics, *Radiotekhnika i Elektronika*, no. 1: 9–17 (transl. ed.), January, 1964.

28. Ye. G. Zelkin, Synthesis of a Synthesizing Linear Emitter of Arbitrary Form, *Radiotekhnika i Elektronika*, no. 1: 18–25 (transl. ed.), January, 1964.

29. D. R. Rhodes, On a Fundamental Principle in the Theory of Planar Antennas, *Proc. IEEE*, **52**(9):1013–1021, September, 1964.

PROBLEMS

3-1. Show that the approximate pattern function from Eq. (3-9) or (3-11) is an approximation in which the mean squared error is minimized.

3-2. (a) Show that the visible region of an antenna pattern ($|\cos \theta| \lessgtr 1$) corresponds to the power radiated by the source.

(b) Show that the invisible region ($|\cos \theta| > 1$) corresponds to energy stored in the near field of the source.

(c) Find an expression for the Q of an antenna in terms of the far-field function $E(k \cos \theta)$. How is Q related to the supergain ratio SGR?

3-3. Compute the g_{ij} in Eq. (3-115) and determine the coefficients a_1, a_2, and a_3 in Eq. (3-118) to give optimum gain for the source in Eq. (3-106). Determine the gain for $\beta = k = 2\pi/\lambda$ and $L = \lambda/2$.

3-4. Although the far-field function in Eq. (3-66) cannot be produced by a continuous source, show that it has a transform that can be interpreted as a continuous source with a point source at each end. *Hint:* $E_0 = F_1 + F_2$, where

$$F_1 = \cos\left[\left(\frac{Lk \cos \theta}{2}\right)^2 - (\pi A)^2\right]^{1/2} - \cos\frac{kL \cos \theta}{2}$$

$$F_2 = \cos\frac{kL \cos \theta}{2}$$

The transform of F_1 involves a modified Bessel function of order one. The function F_2 is simply the pattern of two point sources separated by distance L.

3-5. Equation (3-34) gives a pattern which has an equal-ripple main beam. Show

that $F(\beta) = \left[\dfrac{\epsilon^2 T_n{}^2(1/\beta)}{1 + \epsilon^2 T_n{}^2(1/\beta)} \right]^{1/2}$ gives a pattern which has equal-ripple sidelobes.

What source distribution will produce this field function?

3-6. What is the SGR for a uniformly illuminated ordinary endfire source of length $10\,\lambda$? Compare this with a uniformly illuminated broadside source of length $10\,\lambda$. Repeat for the Hansen-Woodyard increased directivity case.

3-7. Compare Eq. (3-41) with the Shannon Sampling Theorem.

3-8. Show that for a source of length L on the z axis and a given supergain ratio (SGR) the best mean-square approximation to a far-field function $F(\tfrac{1}{2}kL \cos \theta)$ is given by the source $A(2z/L)$ where

$$F(\tfrac{1}{2}kL \cos \theta) = \sum_{n=0}^{\infty} a_n S_{0n}(kL/2, \cos \theta)$$

$$A(2z/L) = \sum_{n=0}^{\infty} \frac{\pi j^{-n}}{R_{0n}{}^{(1)}(kL/2,\, 1)}\, a_n S_{0n}(kL/2,\, 2z/L)$$

The S_{0n} and $R_{0n}{}^{(1)}$ are angular and radial prolate spheroidal wave functions, respectively. The expansion coefficients are $a_n = b_n/[1 + \mu(\lambda_n{}^{-1} - \text{SGR})]$, where $b_n = \dfrac{1}{k_n} \displaystyle\int_{-1}^{1} F(\tfrac{1}{2}kL \cos \theta) S_{0n}(kL/2, \cos \theta)\, d(\cos \theta),\ \lambda_n = \dfrac{kL}{\pi} [R_{0n}{}^{(1)}(kL/2, 1)]^2$, k_n is a normalizing factor, and μ is a unique positive number satisfying

$$\sum_{n=0}^{\infty} \frac{k_n |b_n|^2/(\lambda_n{}^{-1} - \text{SGR})}{[\mu + 1/(\lambda_n{}^{-1} - \text{SGR})]^2} = 0$$

The pattern function is determined largely by the first $(2L/\lambda) + 1$ terms of its expansion (for details see Additional Reference 19).

CHAPTER **4**

TRAVELING
WAVE
ANTENNA
DESIGN

4-1. INTRODUCTION

By means of the synthesis methods described in Chap. 3 a source distribution may be specified which will give an approximation to the desired pattern. The problem of determining a radiating structure that will give the specified source distribution is defined as antenna design.

An important consideration in the design problem is polarization, which was conveniently neglected in the study of synthesis methods. In practice it is necessary to select a radiating structure which will produce the desired far-field polarization. Usually linear polarization is required, and design data for a number of horizontally and vertically polarized traveling wave structures are given in Chaps. 5, 6, and 8. Circular polarization is also possible and may be obtained from such traveling wave antennas as the helix[1] and the spiral (see Sec. 8-3).

Usually the design data for a traveling wave source are in the form of propagation constant versus source geometry for an antenna with uniform cross section. If the required source distribution has other than a uniform amplitude with linear· phase or an exponential amplitude with linear phase, then the so-called propagation constant γ is no longer constant.

[1] J. D. Kraus, "Antennas," chap. 7, McGraw-Hill, New York, 1950.

148

Both of its components, the attenuation parameter α and the phase parameter β, in general will be functions of position along the source.

If the source is one-dimensional (a line), the distribution is represented as

$$A(z) = |A(z)|e^{j\psi(z)} \tag{4-1}$$

where z is distance along the source. If the source is two-dimensional with separable distribution, the distribution is represented as

$$A(x,z) = |A_1(x)|\,|A_2(z)|e^{j\psi_1(x)}e^{j\psi_2(z)} \tag{4-2}$$

The magnitude and phase of the source function may be related to α and β, respectively, as shown in the following sections, where experimental and analytical methods for determining α and β as functions of source geometry are also described.

4-2. DETERMINING α FROM THE SOURCE AMPLITUDE DISTRIBUTION

From the usual definition of the attenuation parameter α, we have for a line source

$$2\alpha = -\frac{1}{P}\frac{dP}{dz} \tag{4-3}$$

where $P(z)$ is the power flowing along the traveling wave source.

If both ohmic loss and radiation loss are present, the power dissipated per unit length is

$$-\frac{dP(z)}{dz} = p_R + p_L \tag{4-4}$$

where p_R is the power per unit length that is radiated and p_L is the power per unit length that is dissipated in ohmic loss. Equation (4-4) may be written

$$2\alpha = \frac{p_R}{P} + \frac{p_L}{P} = 2\alpha_R + 2\alpha_L \tag{4-5}$$

where

$$2\alpha_R = \frac{p_R}{P} \tag{4-6}$$

and

$$2\alpha_L = \frac{p_L}{P} \tag{4-7}$$

Equations (4-6) and (4-7) define the attenuation parameters due to radiation and ohmic loss, respectively.

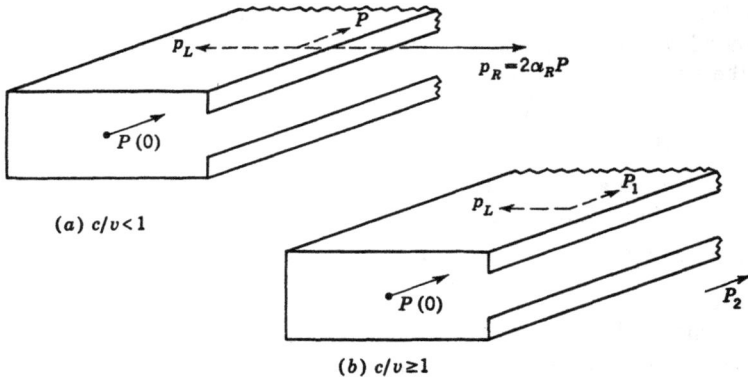

Figure 4-1. Power flow along a traveling wave structure.

For fast wave structures ($c/v < 1$) there is a nonzero component of the Poynting vector that is normal to the structure; that is, there is a flow of power away from the structure as illustrated in Fig. 4-1a. The attenuation parameter α_R is related to the power flowing from the structure by Eq. (4-6). In the case of endfire, or slow wave, structures ($c/v \gtrsim 1$) the attenuation parameter α_R is not so easily interpreted. In this case the power flow both inside and outside the structure is along the structure. There is no component of the Poynting vector normal to the structure (i.e., radiation) except at variations in cross section or discontinuities such as the ends of the structure. Thus it would appear that for endfire structures the attenuation parameter α_R is zero, and in fact, the far-field pattern of a uniform endfire antenna is given fairly accurately by the $(\sin X)/X$ pattern of a constant amplitude source.

It is possible, however, in the endfire case to attach significance to α_R as defined in Eq. (4-6) even though there is apparently no loss along the structure due to radiation. Consider the idealized picture in Fig. 4-1. The slotted waveguide represents a traveling wave antenna. Figure 4-1a illustrates a fast wave structure in which there is a continuous loss of power from the waveguide due to radiation. The attenuation parameter α_R is given by Eq. (4-6). The endfire case is illustrated in Fig. 4-1b. In this case there is no power flow away from the structure in the usual sense, but rather a redistribution of power. That is, if the geometry and/or electrical properties of the structure change with distance along the structure, the distribution of power density over a plane normal to the axis also changes. There is a flow of power out of the guide in the sense that the power density is redistributed over the cross-sectional plane. Thus in Fig. 4-1b if the total input power is $P(0)$, we have

$$P(0) = P_1(z) + P_2(z) + \int_0^z p_L(z)\, dz \qquad (4\text{-}8)$$

where $P_1(z)$ is the total power flowing in the guide and $P_2(z)$ is the total power flowing in the region outside the guide. Both P_1 and P_2 flow parallel to the structure.

Differentiating Eq. (4-8) and dividing by P_1 give

$$\frac{dP_1/dz}{P_1} = \frac{-dP_2/dz}{P_1} - \frac{p_L}{P_1} \qquad (4\text{-}9)$$

Thus the attenuation parameter α_R for the endfire case is given as

$$2\alpha_R = -\frac{dP_1/dz}{P_1} - \frac{p_L}{P_1} \qquad (4\text{-}10)$$

which for $P_1 = P$ is the same expression as for the fast wave case. In the endfire case for a lossless structure, dP_1/dz is produced by a change in cross section and is zero for a uniform structure, whereas in the fast wave case dP/dz is generally nonzero for lossless, uniform structures.

The concept of attenuation parameter for an endfire source must be used with care. The above discussion assumes that a single, nonradiating mode (a surface wave) exists and that the mode is not altered owing to change in cross section. Actually, curvature of the source or a change in cross section will modify the mode and some energy will be converted to radiating modes. If the uniform endfire structure supports a single mode, and if variations in cross section are gradual, the attenuation concept is sometimes useful.

Slow wave sources can be handled by working directly in terms of the amplitude and phase of the appropriate fields or currents along the antenna structure. One can use the natural modes of slow wave structures (see Chaps. 5 and 6) to find these fields. Both amplitude and phase can be controlled by varying geometry (see, for example, the section on modulated slow wave structures in Chap. 8) or by means of coupled line excitation (see Chap. 7). For fast wave sources, however, it is convenient to utilize the attenuation α_R in obtaining a desired amplitude distribution along a source. This is discussed in the following paragraphs.

The radiation pattern of a source is due to an array of infinitesimal elements of the source where the field strength of an individual element is proportional to the current moment of that element. The power radiated from an element is proportional to its current moment squared if the element is radiating by itself. Actually a given element radiates in the presence of the rest of the elements that make up the total source. Consequently, the power radiated from an element is proportional to the current moment squared provided that the variation of source strength with position is gradual and that contributions to the field at the element from neighboring elements are insignificant. These restrictions appear to be approximated in practice even for the case where the phase velocity is equal to or greater than the velocity of light in free space where the latter condition would not be justifiable a priori.

Under conditions where the above provisions are satisfied, it is possible to write for the line source

$$p_R(z) = C|A(z)|^2 \tag{4-11}$$

where $p_R(z)$ is the power radiated per unit length along the traveling wave structure, $|A(z)|$ represents the amplitude distribution of the current moment, and the proportionality constant C may be taken as unity. From Eqs. (4-6) and (4-11) we obtain the relation

$$2\alpha_R(z) = \frac{|A(z)|^2}{P(z)} \tag{4-12}$$

Equation (4-12) determines α_R as a function of z, but it does not give $\alpha_R(z)$ uniquely. If the line source extends from $z = 0$ to $z = L$, the input power $P(0)$ is related to the power $P(L)$ left over at the end of the source by

$$P(0) - P(L) = \int_0^L |A(z)|^2 \, dz + \int_0^L p_L(z) \, dz \tag{4-13}$$

The power $P(z)$ may be expressed as the power yet to be radiated and dissipated from z to L plus the power $P(L)$ remaining at the end of the source; i.e.,

$$P(z) = \int_z^L |A(z)|^2 \, dz + \int_z^L p_L(z) \, dz + P(L) \tag{4-14}$$

Thus Eq. (4-12) may be written

$$2\alpha_R(z) = \frac{|A(z)|^2}{\displaystyle\int_z^L [|A(z)|^2 + p_L(z)] \, dz + \frac{P(L)}{P(0) - P(L)} \times \int_0^L [|A(z)|^2 + p_L(z)] \, dz} \tag{4-15}$$

Equation (4-15) shows that $\alpha_R(z)$ is determined not only by the amplitude distribution $|A(z)|$ and the power dissipated in ohmic loss but also by the fraction of input power which remains at the end of the structure. If ohmic loss in the structure is negligible as is usually the case in practice, Eq. (4-15) reduces to

$$2\alpha(z) = \frac{|A(z)|^2}{\displaystyle\int_z^L |A(z)|^2 \, dz + \frac{P(L)}{P(0) - P(L)} \int_0^L |A(z)|^2 \, dz} \tag{4-16}$$

It can be seen from Eq. (4-16) that, if practically all the power is radiated, α may rise to a very large value near the end of the source, $z = L$. In practice the procedure is to choose some reasonable ratio $P(L)/P(0)$ for a given $|A(z)|$ and then determine $\alpha(z)$ from Eq. (4-16).

If p_R and p_L are known, the efficiency of the source may be determined. The efficiency is defined as

$$\% \text{ efficiency } = \frac{P_R}{P_R + P_L} \times 100 \tag{4-17}$$

where P_R is the total power radiated and P_L is the total power dissipated in ohmic loss. The total power radiated is given by

$$P_R = \int_0^L p_R \, dz \tag{4-18}$$

and the total power dissipated in ohmic loss is

$$P_L = \int_0^L p_L \, dz + P(L) \tag{4-19}$$

where it is assumed that $P(L)$ is absorbed in a load at the end of the source.

For the special case of a uniform structure, α_R and α_L are constant and p_R and p_L are given by

$$p_R = 2\alpha_R P(0)e^{-2(\alpha_R+\alpha_L)z} \tag{4-20}$$

and

$$p_L = 2\alpha_L P(0)e^{-2(\alpha_R+\alpha_L)z} \tag{4-21}$$

Performing the integrations in Eqs. (4-18) and (4-19) gives

$$P_R = \frac{\alpha_R P(0)}{\alpha_R + \alpha_L}[1 - e^{-2L(\alpha_R+\alpha_L)}] \tag{4-22}$$

and

$$P_L = \frac{\alpha_L P(0)}{\alpha_R + \alpha_L}[1 - e^{-2L(\alpha_R+\alpha_L)}] + P(L) \tag{4-23}$$

The efficiency in this case is

$$\% \text{ efficiency } = \frac{100\alpha_R}{\alpha_R + \alpha_L + \dfrac{(\alpha_R + \alpha_L)P(L)}{[1 - e^{-2L(\alpha_R+\alpha_L)}]P(0)}} \tag{4-24}$$

If $P(L)$ may be neglected, Eq. (4-24) reduces to

$$\% \text{ efficiency } = \frac{100\alpha_R}{\alpha_R + \alpha_L} \tag{4-25}$$

As an example of determining α for a given source amplitude, consider a source which has a constant phase velocity with $c/v = 0.939$ (i.e., a beam at 20 deg from the axis of the source). The amplitude is chosen to be

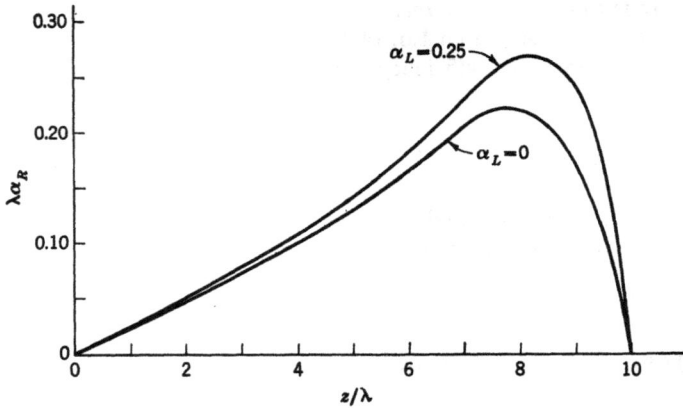

Figure 4-2. Showing the effect of α_L on α_R for a $[\sin(\pi z/10\lambda)]^{\frac{1}{2}}$ amplitude distribution.

$|A(z)|^2 = \sin(\pi z/L)$ and $P(L) = 0.1P(0)$. Neglecting ohmic loss, Eq. (4-16) is used, giving

$$2\alpha(z) = \frac{\pi \sin(\pi z/L)}{L[\cos(\pi z/L) + 1.222]} \tag{4-26}$$

A plot of α from Eq. (4-26) is shown in Fig. 4-2.

If loss in the dielectric is taken into account and the same total loss is maintained, $P(L)$ is modified and becomes

$$P(L) = 0.1P(0) - \int_0^L p_L \, dz \tag{4-27}$$

Assuming a dielectric-filled TM_{11} slotted, square waveguide as illustrated in Fig. 5-10 and a loss tangent of 0.001, the loss attenuation is $\alpha_L \approx 0.252$ neper/λ and Eq. (4-15) gives

$$2\alpha_R(z) \approx$$

$$\frac{\sin(\pi z/L)}{\dfrac{10\lambda}{\pi}\left(\cos\dfrac{\pi z}{L} + 1\right) + 4.78 \times 10^{-8}\dfrac{e^{-2\alpha z - 0.6\alpha}}{\alpha}} \\ + \frac{0.1\alpha - 0.252(1 - e^{-0.6\alpha})}{0.9\alpha + 0.252(1 - e^{-0.6\alpha})}\left(\dfrac{20\lambda}{\pi} + 4.78 \times 10^{-8}\dfrac{1 - e^{-0.6\alpha}}{\alpha}\right) \tag{4-28}$$

where $\alpha = \alpha_L + \alpha_R$. Values of α_R from Eq. (4-28) are also plotted on the graph in Fig. 4-2. The small dielectric loss has caused a significant increase in α_R in order to maintain the same amplitude distribution as in the lossless case.

4-3. DETERMINING β FROM THE SOURCE PHASE DISTRIBUTION

A traveling wave source with constant phase velocity v has a phase constant β given by

$$\beta = \frac{\omega}{v} \tag{4-29}$$

where ω is angular frequency. The phase variation for such a wave traveling along the z coordinate is

$$\psi(z) = -\beta z \tag{4-30}$$

That is, the phase varies linearly with distance. A source distribution of this type is given as

$$A(z) = |A(z)|e^{-j\beta z} \tag{4-31}$$

The minus sign occurs because we have adopted the $e^{j\omega t}$ time convention; thus we associate a phase delay with a wave traveling in the positive direction of the coordinate.

When the phase velocity of a traveling wave is not constant, the phase variation is obtained by summing phase variations of infinitesimal elements of the source. In the limit as the elements shrink to zero, the phase difference between two points z_1 and z_2 (where $z_2 > z_1$) may be written

$$\psi(z) = -\int_{z_1}^{z_2} \beta \, dz \tag{4-32}$$

Thus the phase parameter $\beta(z)$ may be found from the phase $\psi(z)$ by

$$\beta(z) = -\frac{d\psi(z)}{dz} \tag{4-33}$$

In practice it is convenient to work with the velocity ratio c/v where

$$\frac{c}{v} = \frac{\beta}{k} \tag{4-34}$$

Combining Eqs. (4-33) and (4-34) gives

$$\frac{c}{v} = -\frac{1}{k}\frac{d\psi(z)}{dz} \tag{4-35}$$

4-4. RELATING PROPAGATION CONSTANT TO SOURCE CROSS SECTION

The propagation constant γ of a traveling wave source is determined by the geometry and dimensions of the source, the mode of operation, and the electrical characteristics of the materials from which the structure is made. In general any structure that will serve as a transmission line for electromagnetic energy may be modified to operate as a traveling wave antenna. As an ideal transmission line the propagation constant of the structure is imaginary (that is, $\gamma = j\beta$). Modifications that result in a radiating structure are usually such that there is a controlled power loss along the structure due to radiation. This causes the propagation constant to become complex ($\gamma = \alpha + j\beta$). Usually the modification that produces the attenuation quantity also alters the phase constant of the ideal transmission line; however, in some cases α and β may be essentially independent of each other.

In general the desired source distribution does not have a uniform or exponential amplitude and constant phase velocity. When this is the case, α and β are not constant. It will be assumed that all variations in α and β are sufficiently gradual so that the propagation constant γ at a point along a nonuniform traveling wave structure is the same as that for an infinitely long uniform structure of the same cross section as the nonuniform structure at the point that is being considered.

The relationship of γ to the source geometry can be expressed graphically or mathematically and constitutes the design data. The data can be obtained by experimental or analytical methods.

4-5. DETERMINING PROPAGATION CONSTANT BY EXPERIMENT

In many cases the geometry of a traveling wave source is such that an analysis for determining the propagation constant is extremely difficult. In

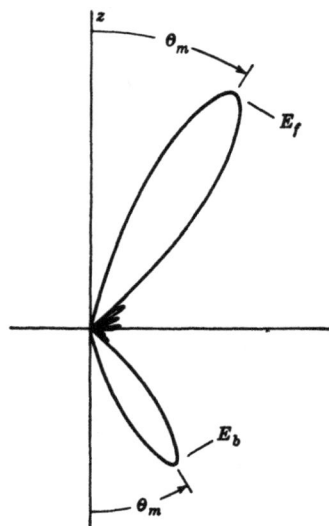

Figure 4-3. Typical pattern for a line source of length L where α is not large.

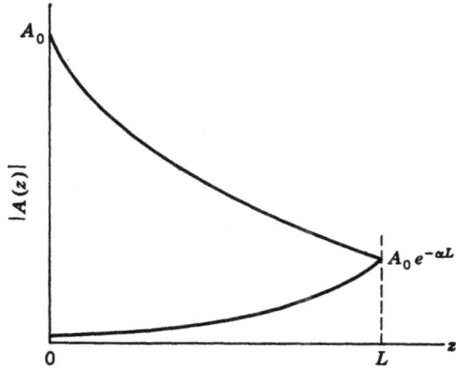

Figure 4-4. Exponential amplitude variation for forward and backward waves along a uniform source with total reflection at $z = L$.

these cases it is most practical to determine α and β by experimental means, and this can be accomplished by either far-field or near-field measurements.

Consider a uniform line source of length L operating in a single mode and oriented along the z axis. If the power $P(L)$ at the end of the source is totally reflected, and if α is small enough that a significant portion of the power is not radiated or dissipated along the source, the far-field pattern in a plane $\phi = \text{const}$ is as shown in Fig. 4-3. The source amplitude for this case would have the form shown in Fig. 4-4. Under the condition of totally reflected power at $z = L$, the magnitude of the forward beam in Fig. 4-3 is

$$E_f = CA_0 \tag{4-36}$$

and that of the back beam is

$$E_b = CA_0 e^{-\alpha L} \tag{4-37}$$

where C is a proportionality constant that is the same for both beams. Combining Eqs. (4-36) and (4-37) gives

$$\frac{E_f}{E_b} = e^{\alpha L} \tag{4-38}$$

The attenuation parameter α may be readily determined from Eq. (4-38). The velocity ratio c/v is given by the relation

$$\frac{c}{v} \approx \cos \theta_m \tag{4-39}$$

where θ_m is the angle of maximum radiation. The phase constant β may be found from Eq. (4-34). Equation (4-39) is a good approximation for $L > 5 \lambda$ and $\theta_m > 20$ deg. For lengths shorter than about 5λ the relatively wide pattern makes it difficult to determine θ_m exactly, and also the pattern of an element of the source begins to have appreciable effect on θ_m. For $\theta_m < 20$ deg the source is approaching endfire operation and θ_m may be greatly influenced by the pattern of an element of the source as well as the extent of a ground plane when one is present. Equation (4-38) is satisfactory for $\alpha < 4$ db/λ.[1] For larger α it may be difficult to measure E_b accurately.

Accurate data for $L < 5\lambda$ and $\alpha > 4$ db/λ can be obtained by near-field measurements. This can be accomplished by direct probing of the fields in the vicinity of the source as illustrated in Fig. 4-5. The diagram shows a reference line for phase measurements. The slotted line probe may be positioned for minimum reading at the meter for each position of the probe at the traveling wave source. The phase between two points along the source is equal to the phase difference between the positions of the slotted line probe that give minimum output at the meter. Distance along the slotted line can be calibrated in degrees or radians. Breaking the circuit at A permits amplitude measurements from which α can be determined.

The phase constant also can be determined from an amplitude measurement. If α is small enough that a well-defined standing wave exists when there is a short at the end of the source, the wavelength along the traveling wave source can be determined from the distance between alternate minima.

The small probe antenna that is used to sample the fields near the source may be a dipole, a loop, or an open-ended waveguide (dielectric loaded to reduce its size).[2]

[1] Although α was defined in Chap. 1 in terms of nepers per unit length, decibels per wavelength (db/λ) is frequently used where attenuation in decibels = 8.686 × attenuation in nepers.

[2] J. H. Richmond and T. E. Tice, Probes for Microwave Near-field Measurements, *IRE Trans. Microwave Theory Tech.*, **MTT-3**(3):32–34, April, 1955.

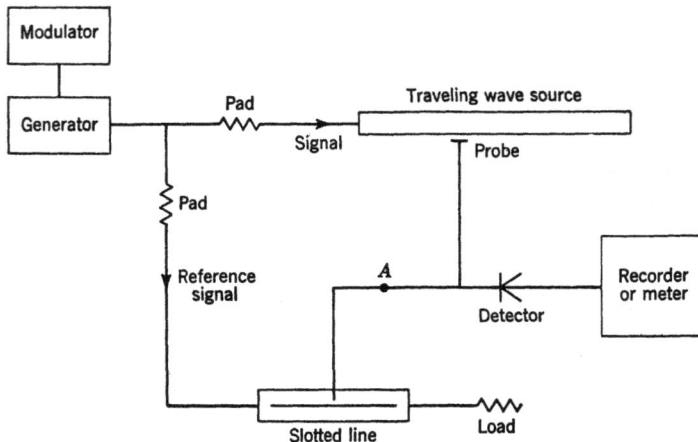

Figure 4-5.   Equipment for direct probing of the near fields of a source.

Nylon string

Dipole scatterer

Traveling wave source

Tuner

Modulator

Generator

Signal E H

Magic T

Detector and recorder

Matched load

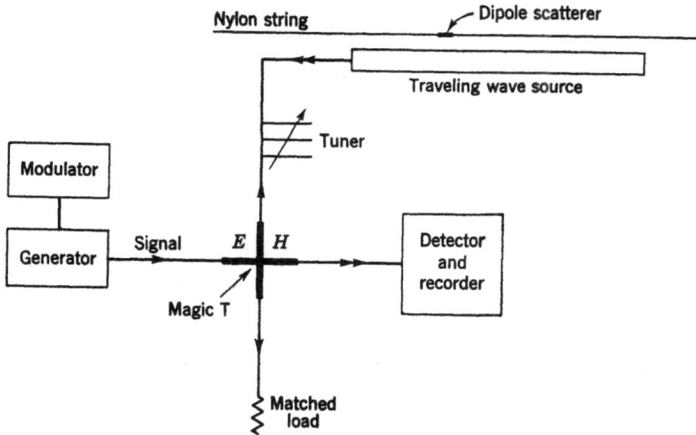

Figure 4-6. Equipment for near-field measurement with scattering technique.

A scattering technique as illustrated in Fig. 4-6 also may be used to measure the amplitude and phase of a source distribution.[1] For amplitude measurements the tuner must be adjusted until isolation between signal source and receiver is about 100 db. Slight detuning provides a reference signal for phase measurements.

Instrumentation for the scattering method is perhaps more difficult than the direct probing method, but it may significantly reduce inaccuracies due to field distortion by the RF cable which is connected to the probe in the direct measuring method.

4-6. DETERMINING PROPAGATION CONSTANT BY ANALYSIS

An analysis of a traveling wave structure to determine the propagation constant as a function of geometry, electrical characteristics, and mode of operation is generally quite difficult. The analysis is simplified considerably if the structure is assumed to be infinitely long with a uniform cross section. The results for a uniform cross section may be applied to a structure with a varying cross section if the variation is sufficiently gradual so that at any cross section the propagation constant is essentially the same as it would be for a uniform structure of that cross section and of infinite length.

The problem of determining the propagation constant for an infinitely long traveling wave structure of uniform cross section may be formulated

[1] J. H. Richmond, A Modulated Scattering Technique for Measurement of Field Distributions, *IRE Trans. Microwave Theory Tech.*, **MTT-3**(4):13–15, July, 1955; R. Justice and V. Rumsey, Measurement of Electric Field Distributions, *IRE Trans. Antennas Propagation*, **AP-3**(4):177–180, October, 1955.

in terms of two general types of solutions to Maxwell's equations. These are the usual TE and TM waves of waveguide theory. These waves are characterized by the fact that each contains an axial (z) component of only one field, electric in the case of TM waves and magnetic in the case of TE waves. The TE and TM waves are sometimes called H waves and E waves, respectively.

Many traveling wave structures can be made to support either TM or TE waves. In some cases, such as the polyrod, both the electric and magnetic fields have axial components and the resultant field configuration may be represented as a sum of TE and TM waves. This is referred to as hybrid mode operation.

Expressions for electric and magnetic fields in terms of vector potential functions were developed in Chap. 2. For convenience the expressions are given again as

$$\mathbf{E} = -\frac{1}{\epsilon} \nabla \times \mathbf{F} - \frac{j}{\omega\mu\epsilon} \nabla \times \nabla \times \mathbf{A} \tag{4-40}$$

and

$$\mathbf{H} = \frac{1}{\mu} \nabla \times \mathbf{A} - \frac{j}{\omega\mu\epsilon} \nabla \times \nabla \times \mathbf{F} \tag{4-41}$$

where $\mathbf{F}$ and $\mathbf{A}$ are the usual electric and magnetic potentials, respectively. For a TM wave propagating in the z direction the axial component of the magnetic field is zero ($H_z = 0$); thus the TM field components can be determined from the magnetic potential

$$\mathbf{A} = \mathbf{z} A_z \tag{4-42}$$

In a similar manner the TE field components can be found from an electric potential function

$$\mathbf{F} = \mathbf{z} F_z \tag{4-43}$$

If the fields of the traveling wave source vary as $e^{j\omega t - \gamma z}$, Eqs. (4-40) and (4-41) give, for rectangular coordinates,

$$
\begin{aligned}
\overset{h}{E}_x &= -\frac{1}{\epsilon}\frac{\partial F_z}{\partial y} \\
\overset{h}{E}_y &= \frac{1}{\epsilon}\frac{\partial F_z}{\partial x} \\
\overset{h}{E}_z &= 0 \\
\overset{h}{H}_x &= -\frac{\gamma}{j\omega\mu\epsilon}\frac{\partial F_z}{\partial x} \\
\overset{h}{H}_y &= -\frac{\gamma}{j\omega\mu\epsilon}\frac{\partial F_z}{\partial y} \\
\overset{h}{H}_z &= -\frac{1}{j\omega\mu\epsilon}\left(\frac{\partial^2 F_z}{\partial x^2} + \frac{\partial^2 F_z}{\partial y^2}\right) = \frac{\beta_x^2 + \beta_y^2}{j\omega\mu\epsilon} F_z
\end{aligned}
\tag{4-44}
$$

for the TE case (H wave) and

$$\overset{e}{E}_x = -\frac{\gamma}{j\omega\mu\epsilon}\frac{\partial A_z}{\partial x}$$

$$\overset{e}{E}_y = -\frac{\gamma}{j\omega\mu\epsilon}\frac{\partial A_z}{\partial y}$$

$$\overset{e}{E}_z = -\frac{1}{j\omega\mu\epsilon}\left(\frac{\partial^2 A_z}{\partial x^2} + \frac{\partial^2 A_z}{\partial y^2}\right) = \frac{(\beta_x^2 + \beta_y^2)A_z}{j\omega\mu\epsilon}$$

$$\overset{e}{H}_x = \frac{1}{\mu}\frac{\partial A_z}{\partial y}$$ (4-45)

$$\overset{e}{H}_y = -\frac{1}{\mu}\frac{\partial A_z}{\partial x}$$

$$\overset{e}{H}_z = 0$$

for the TM case (E wave). The superscripts h and e denote H wave and E wave, respectively. The quantities β_x and β_y may be interpreted as phase constants for waves traveling in the x and y directions, respectively. Solution of the wave equation [see (2-26)] in rectangular coordinates shows that they are related to γ by

$$\beta_x^2 + \beta_y^2 - \gamma^2 = \omega^2\mu\epsilon$$ (4-46)

From Eqs. (4-44) and (4-45) it can be seen that H_z is proportional to F_z and E_z is proportional to A_z. Thus the TE field components can be determined from H_z and the TM field components can be determined from E_z. It would be necessary to find expressions for H_z and E_z that satisfy the two-dimensional wave equation

$$[\nabla_t^2 + (\omega^2\mu\epsilon + \gamma^2)]\begin{pmatrix}E_z\\H_z\end{pmatrix} = 0$$ (4-47)

and the appropriate boundary conditions, where $\nabla_t^2 = (\partial^2/\partial x^2) + (\partial^2/\partial y^2)$ in rectangular coordinates.

In orthogonal cylindrical coordinates (ρ,ϕ,z) Eqs. (4-40) and (4-41) give

$$\overset{h}{E}_\rho = -\frac{1}{\epsilon\rho}\frac{\partial F_z}{\partial\phi}$$

$$\overset{h}{E}_\phi = \frac{1}{\epsilon}\frac{\partial F_z}{\partial\rho}$$

$$\overset{h}{E}_z = 0$$

$$\overset{h}{H}_\rho = -\frac{\gamma}{j\omega\mu\epsilon}\frac{\partial F_z}{\partial\rho}$$ (4-48)

$$\overset{h}{H}_\phi = -\frac{\gamma}{j\omega\mu\epsilon\rho}\frac{\partial F_z}{\partial\phi}$$

$$\overset{h}{H}_z = \frac{\kappa^2}{j\omega\mu\epsilon}F_z$$

for the TE case and

$$\overset{\circ}{E}_\rho = -\frac{\gamma}{j\omega\mu\epsilon}\frac{\partial A_z}{\partial \rho}$$

$$\overset{\circ}{E}_\phi = -\frac{\gamma}{j\omega\mu\epsilon\rho}\frac{\partial A_z}{\partial \phi}$$

$$\overset{\circ}{E}_z = \frac{\kappa^2}{j\omega\mu\epsilon}A_z \tag{4-49}$$

$$\overset{\circ}{H}_\rho = \frac{1}{\mu\rho}\frac{\partial A_z}{\partial \phi}$$

$$\overset{\circ}{H}_\phi = -\frac{1}{\mu}\frac{\partial A_z}{\partial \rho}$$

$$\overset{\circ}{H}_z = 0$$

for the TM case. The quantity κ is interpreted as a transverse phase constant and is related to γ by

$$\kappa^2 - \gamma^2 = \omega^2\mu\epsilon \tag{4-50}$$

Rectangular and cylindrical coordinates have been considered because they are easiest to work with mathematically and they correspond to the cross sections most frequently encountered in practice. Cross sections of two practical traveling wave structures are shown in Fig. 4-7.

The classical method of analysis of a traveling wave structure is to find a field configuration that satisfies Maxwell's equations and the boundary conditions. Exact solutions are readily obtained for structures such as the polyrod and dielectric-coated ground plane. However, for structures such as those in Fig. 4-7 the geometry is such that an exact solution as a

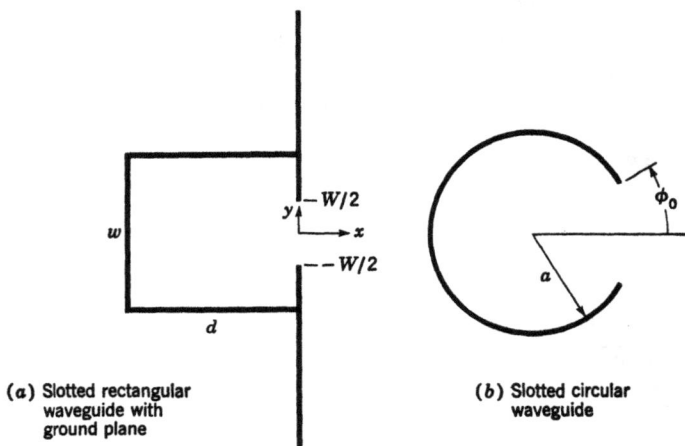

(a) Slotted rectangular waveguide with ground plane

(b) Slotted circular waveguide

Figure 4-7.   Slotted waveguide structures.

boundary value problem is extremely difficult. Therefore, approximate methods such as the variational method and the transverse resonance method are often employed.

4-7. VARIATIONAL METHOD

In some cases where it is not practical to try to solve a field problem by the differential equation approach using Maxwell's equations, an integral equation method called the *variational method* is quite useful.[1] The variational method involves determining a formula that is "stationary" about the correct solution. That is, the formula is relatively insensitive to variations in an assumed field about the correct field.

One of the characteristics of a stationary formulation is that it can be expressed as a ratio of two quantities both of which contain squares of the trial, or assumed, field; thus the amplitude of the trial field will not affect the results.

The usual procedure in establishing a stationary formula is to construct formulas of the proper form and then test them to find one that is stationary. The procedure for testing a variational formulation is illustrated in the following discussion. Harrington[2] gives a general procedure for establishing stationary formulas that is based on the reaction concept[3] (Sec. 2-3).

The variational method has been used by Rumsey[4] and Harrington[5] for a waveguide with a continuous slot parallel to the waveguide axis (z axis). Cross sections of the geometries considered by Rumsey and Harrington are shown in Fig. 4-7. A distribution is assumed for the tangential electric field in the slot, and a stationary expression is derived in which a first-order variation in the assumed field distribution results in a second-order variation in the propagation constant. The air-filled waveguide may be formulated in terms of a single scalar (either a TE or TM mode), whereas the dielectric-filled guide, in general, has to be formulated in terms of two scalars (a hybrid mode).

Rumsey has shown[6] that a stationary expression for the propagation constant γ of the cross section in Fig. 4-7a is

$$\int_{-W/2}^{W/2} [E_y({}^iH_z - {}^eH_z) + E_z({}^iH_y - {}^eH_y)] \, dy = 0 \qquad (4\text{-}51)$$

[1] For a detailed discussion of the variational method as well as numerous applications the reader is referred to R. F. Harrington, "Time-Harmonic Electromagnetic Fields," chap. 7, McGraw-Hill, New York, 1961.
[2] *Ibid.*
[3] V. H. Rumsey, The Reaction Concept in Electromagnetic Theory, *Phys. Rev.*, ser. 2, **94**(6):1483–1491, June 15, 1954.
[4] V. H. Rumsey, Traveling Wave Slot Antennas, *J. Appl. Phys.*, **24**(11), November, 1953.
[5] R. F. Harrington, Propagation along a Slotted Cylinder, *J. Appl. Phys.*, **24**(11), November, 1953.
[6] See footnote 4.

where E_y and E_z are components of an assumed electric field in the slot, ${}^iH$ and ${}^eH$ represent internal and external magnetic fields which fit the assumed E_y and E_z, and the z dependence of the fields is $e^{-\gamma z}$. Equation (4-51) applies to a hybrid mode structure (i.e., one in which both E_z and H_z are present). For a TE wave Eq. (4-51) reduces to

$$\int_{-W/2}^{W/2} E_y({}^iH_z - {}^eH_z)\, dy = 0 \tag{4-52}$$

and for a TM wave,

$$\int_{-W/2}^{W/2} E_z({}^iH_y - {}^eH_y)\, dy = 0 \tag{4-53}$$

For the cylindrical cross section of Fig. 4-7b, the stationary expression for γ is

$$\int_{-\phi_0}^{\phi_0} [E_\phi({}^iH_z - {}^eH_z) + E_z({}^iH_\phi - {}^eH_\phi)]\, d\phi = 0 \tag{4-54}$$

For a TE wave Eq. (4-54) reduces to

$$\int_{-\phi_0}^{\phi_0} E_\phi({}^iH_z - {}^eH_z)\, d\phi = 0 \tag{4-55}$$

and for a TM wave,

$$\int_{-\phi_0}^{\phi_0} E_z({}^iH_\phi - {}^eH_\phi)\, d\phi = 0 \tag{4-56}$$

As in the rectangular case, the components E_ϕ and E_z are assumed and ${}^iH$ and ${}^eH$ are to be determined from the assumed **E** field. In all cases it can be seen that the above expressions are identically zero if the correct electric field is used in determining ${}^eH - {}^iH$.

The stationary properties of Eqs. (4-51) through (4-56) can be readily verified. Consider the TE case for the rectangular guide [Eq. (4-52)]. The magnetic field may be expressed symbolically as

$$ {}^eH_z(x,y,z) = \kappa^2 e^{-\gamma z} \int_{-W/2}^{W/2} E_y(y')\, {}^eh(y,y',\kappa)\, dy' \tag{4-57}$$

and

$$ {}^iH_z(x,y,z) = \kappa^2 e^{-\gamma z} \int_{-W/2}^{W/2} E_y(y')\, {}^ih(y,y',\kappa)\, dy' \tag{4-58}$$

where κ is the transverse (to the z axis) phase constant [see Eq. (4-50)]. We need not concern ourselves with the forms of the functions ${}^eh$ and ${}^ih$. Subtracting Eq. (4-57) from Eq. (4-58) gives

$$ {}^iH_z - {}^eH_z = \kappa^2 e^{-\gamma z} \int_{-W/2}^{W/2} E_y({}^ih - {}^eh)\, dy' \tag{4-59}$$

Thus Eq. (4-52) may be written

$$\int_{-W/2}^{W/2} E_v(y') \, dy' \int_{-W/2}^{W/2} E_v(y) \kappa^2 [{}^ih(y',y,\kappa) - {}^eh(y',y,\kappa)] \, dy = 0 \qquad (4\text{-}60)$$

Now let E_v vary by δE_v from the correct value and denote $\kappa^2({}^ih - {}^eh)$ by Φ. Equation (4-60) becomes

$$\int_{-W/2}^{W/2} [E_v(y') + \delta E_v(y')] \, dy' \int_{-W/2}^{W/2} [E_v(y) + \delta E_v(y)]$$

$$\times \left[\Phi + \delta\kappa \frac{\partial\Phi}{\partial\kappa} \right] dy = 0 \qquad (4\text{-}61)$$

where Φ has been expanded about the true κ by a Taylor expansion and only the first two terms retained. Expanding Eq. (4-61) and retaining only first-order terms give

$$\int_{-W/2}^{W/2} E_v(y') \, dy' \int_{-W/2}^{W/2} E_v(y)\Phi \, dy + \int_{-W/2}^{W/2} E_v(y') \, dy'$$

$$\times \int_{-W/2}^{W/2} E_v(y) \, \delta\kappa \frac{\partial\Phi}{\partial\kappa} \, dy + \int_{-W/2}^{W/2} E_v(y') \, dy' \int_{-W/2}^{W/2} \delta E_v(y)\Phi \, dy$$

$$+ \int_{-W/2}^{W/2} \delta E_v(y') \, dy' \int_{-W/2}^{W/2} E_v(y)\Phi \, dy = 0 \qquad (4\text{-}62)$$

The first term in Eq. (4-62) is obviously zero because it is the same as Eq. (4-52). The fourth term is also zero, since the integral with respect to y contains the correct E_v and κ and hence ${}^iH_z - {}^eH_z$ is zero over the slot. By reciprocity

$$\Phi(y,y',\kappa) = \Phi(y',y,\kappa) \qquad (4\text{-}63)$$

i.e., Φ at y due to a line source at y' and parallel to the z axis is the same as Φ at y' with the line source at y. Thus the third term is equal to the fourth term and is likewise zero. Equation (4-62) reduces to

$$\delta\kappa \int_{-W/2}^{W/2} E_v(y') \, dy' \int_{-W/2}^{W/2} E_v(y) \frac{\partial\Phi}{\partial\kappa} \, dy = 0 \qquad (4\text{-}64)$$

from which we conclude that $\delta\kappa$ is zero. Thus we see that Eq. (4-52) is a stationary expression for κ. A first-order variation in the electric field of the slot produces a second-order variation in κ and hence in the propagation constant γ. The same arguments apply to Eqs. (4-51) and (4-53) to (4-56).

Although the stationary expressions [Eqs. (4-51) to (4-56)] are simple in form, the solution for κ (or γ) is generally quite involved because of the rather complicated equations that arise in determining ${}^eH$ and ${}^iH$ from the assumed electric field in the slot.

To illustrate the variational method we shall determine γ for the slotted circular waveguide in Fig. 4-7b. Let the waveguide be excited in the TE_{11} mode so that the electric field is as shown in Fig. 4-8 and Eq. (4-55)

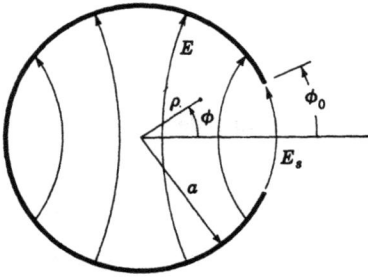

Figure 4-8. Field configuration in a TE_{11} slotted circular waveguide.

applies. For a TE mode the field components can be determined from F_s (or H_z) by Eq. (4-48). In the internal region ($\rho < a$), H_z must satisfy the two-dimensional wave equation [see Eqs. (4-47) and (4-50)]

$$\nabla_t^2 H_z + \kappa^2 H_z = 0 \tag{4-65}$$

and be finite at the origin. Therefore, we may express ^iH_z in terms of Bessel functions as

$$^iH_z = \sum_{n=0}^{\infty} a_n J_n(\kappa\rho) \cos n\phi \tag{4-66}$$

Only cosine terms are used for the ϕ variation because the field configuration is assumed to be symmetrical about $\phi = 0$. In the external region ($\rho > a$) the field is assumed to be an outward traveling wave; thus the Hankel function of the second kind is used and

$$^\bullet H_z = \sum_{n=0}^{\infty} b_n H_n^{(2)}(\kappa\rho) \cos n\phi \tag{4-67}$$

By Eq. (4-48) the electric field E_ϕ is given as

$$E_\phi = \frac{j\omega\mu}{\kappa} \frac{\partial H_z}{\partial(\kappa\rho)} \tag{4-68}$$

thus

$$^iE_\phi = \frac{j\omega\mu}{\kappa} \sum_{n=0}^{\infty} a_n J_n'(\kappa\rho) \cos n\phi \tag{4-69}$$

and

$$^\bullet E_\phi = \frac{j\omega\mu}{\kappa} \sum_{n=0}^{\infty} b_n H_n^{(2)\prime}(\kappa\rho) \cos n\phi \tag{4-70}$$

where the prime on Bessel and Hankel functions denotes the derivative with respect to $\kappa\rho$.

The coefficients a_n and b_n can be determined from the boundary condition at $\rho = a$, which is

$$E_\phi = E_s \tag{4-71}$$

where E_s is the field that is assumed to exist in the slot. Applying the boundary condition to Eqs. (4-69) and (4-70) gives

$$E_s = \frac{j\omega\mu}{\kappa} \sum_{n=0}^{\infty} a_n J_n'(\kappa a) \cos n\phi \tag{4-72}$$

and

$$E_s = \frac{j\omega\mu}{\kappa} \sum_{n=0}^{\infty} b_n H_n^{(2)'}(\kappa a) \cos n\phi \tag{4-73}$$

Utilizing the orthogonality of the cosine function the coefficients a_n and b_n are found to be

$$\begin{aligned} a_n &= \frac{\kappa}{j\omega\mu\pi J_n'(\kappa a)} \int_{-\phi_0}^{\phi_0} E_s \cos n\phi'\, d\phi' \qquad n \neq 0 \\ a_0 &= \frac{\kappa}{2j\omega\mu\pi J_0'(\kappa a)} \int_{-\phi_0}^{\phi_0} E_s\, d\phi' \end{aligned} \tag{4-74}$$

and

$$\begin{aligned} b_n &= \frac{\kappa}{j\omega\mu\pi H_n^{(2)'}(\kappa a)} \int_{-\phi_0}^{\phi_0} E_s \cos n\phi'\, d\phi' \qquad n \neq 0 \\ b_0 &= \frac{\kappa}{2j\omega\mu\pi H_0^{(2)'}(\kappa a)} \int_{-\phi_0}^{\phi_0} E_s\, d\phi' \end{aligned} \tag{4-75}$$

The field H_z at the slot may now be expressed as

$${}^{i}H_z = \frac{\kappa J_0(\kappa a)}{2j\omega\mu\pi J_0'(\kappa a)} \int_{-\phi_0}^{\phi_0} E_s\, d\phi' + \sum_{n=1}^{\infty} \frac{\kappa J_n(\kappa a) \cos n\phi}{j\omega\mu\pi J_n'(\kappa a)} \\ \times \int_{-\phi_0}^{\phi_0} E_s \cos n\phi'\, d\phi' \tag{4-76}$$

and

$${}^{e}H_z = \frac{\kappa H_0^{(2)}(\kappa a)}{2j\omega\mu\pi H_0^{(2)'}(\kappa a)} \int_{-\phi_0}^{\phi_0} E_s\, d\phi' + \sum_{n=1}^{\infty} \frac{\kappa H_n^{(2)}(\kappa a) \cos n\phi}{j\omega\mu\pi H_n^{(2)'}(\kappa a)} \\ \times \int_{-\phi_0}^{\phi_0} E_s \cos n\phi'\, d\phi' \tag{4-77}$$

Subtracting (4-77) from (4-76) gives

$${}^{i}H_z - {}^{e}H_z = \frac{\kappa}{2j\omega\mu\pi} \left(\frac{J_0(\kappa a)}{J_0'(\kappa a)} - \frac{H_0^{(2)}(\kappa a)}{H_0^{(2)'}(\kappa a)} \right) \int_{-\phi_0}^{\phi_0} E_s\, d\phi' \\ + \sum_{n=1}^{\infty} \frac{\kappa \cos n\phi}{j\omega\mu\pi} \left(\frac{J_n(\kappa a)}{J_n'(\kappa a)} - \frac{H_n^{(2)}(\kappa a)}{H_n^{(2)'}(\kappa a)} \right) \int_{-\phi_0}^{\phi_0} E_s \cos n\phi'\, d\phi' \tag{4-78}$$

Substituting (4-78) into (4-55) gives

$$\int_{-\phi_0}^{\phi_0} E_s \left[\frac{1}{2}\left(\frac{J_0(\kappa a)}{J_0'(\kappa a)} - \frac{H_0^{(2)}(\kappa a)}{H_0^{(2)'}(\kappa a)} \right) \int_{-\phi_0}^{\phi_0} E_s \, d\phi' \right.$$
$$+ \sum_{n=1}^{\infty} \cos n\phi \left(\frac{J_n(\kappa a)}{J_n'(\kappa a)} - \frac{H_n^{(2)}(\kappa a)}{H_n^{(2)'}(\kappa a)} \right)$$
$$\left. \times \int_{-\phi_0}^{\phi_0} E_s \cos n\phi' \, d\phi' \right] d\phi = 0 \quad (4\text{-}79)$$

For the TE case we shall assume that E_s is constant with ϕ and take it to be unity. Thus the integrations in Eq. (4-79) can be readily performed, giving

$$2\phi_0^2 \left(\frac{J_0(\kappa a)}{J_0'(\kappa a)} - \frac{H_0^{(2)}(\kappa a)}{H_0^{(2)'}(\kappa a)} \right)$$
$$+ \sum_{n=1}^{\infty} \left(\frac{2 \sin n\phi_0}{n} \right)^2 \left(\frac{J_n(\kappa a)}{J_n'(\kappa a)} - \frac{H_n^{(2)}(\kappa a)}{H_n^{(2)'}(\kappa a)} \right) = 0 \quad (4\text{-}80)$$

If we introduce the Neumann number ϵ_n where

$$\epsilon_n = 1 \qquad n = 0$$

and $\qquad\qquad\qquad\qquad\qquad\qquad\qquad\qquad\qquad\qquad\qquad (4\text{-}81)$

$$\epsilon_n = 2 \qquad n \neq 0$$

Eq. (4-80) may be written

$$\sum_{n=0}^{\infty} \epsilon_n \left[\frac{J_n(\kappa a)}{J_n'(\kappa a)} - \frac{H_n^{(2)}(\kappa a)}{H_n^{(2)'}(\kappa a)} \right] \left(\frac{2 \sin n\phi_0}{n} \right)^2 = 0 \quad (4\text{-}82)$$

Equation (4-82) gives κ as a function of a and ϕ_0 for a TE-excited slotted circular waveguide. The same procedure may be followed for the TM case, giving

$$\sum_{n=0}^{\infty} \epsilon_n \left[\frac{J_n'(\kappa a)}{J_n(\kappa a)} - \frac{H_n^{(2)'}(\kappa a)}{H_n^{(2)}(\kappa a)} \right] \left[\frac{4\pi\phi_0 \cos n\phi_0}{\pi^2 - (2n\phi_0)^2} \right]^2 = 0 \quad (4\text{-}83)$$

where E_z in the slot is assumed to be

$$E_s = \cos \frac{\pi\phi}{2\phi_0} \quad (4\text{-}84)$$

Equations (4-82) and (4-83) were first obtained by Harrington.[1] They give κ as a function of a and ϕ_0 and the type of excitation (TE or TM). Actually, Eqs. (4-82) and (4-83) have an infinite number of solutions, each

[1] Harrington, *op. cit.*, 1953.

corresponding to a slotted waveguide mode similar to waveguide modes in a closed circular waveguide. The slotted cylinder modes are classified according to the mode of the unslotted cylinder. The dominant mode solution of Eq. (4-82) is thus referred to as the TE_{11} slotted waveguide mode.

In the above analysis we are not restricted to one slot. Multiple slots may be taken into account by introducing the appropriate intervals of integration in Eq. (4-79). If the slot or slots are small enough that κ for the slotted waveguide mode is in the vicinity of κ_0 for the closed waveguide, κa may be written

$$\kappa a = \kappa_0 a + \delta \tag{4-85}$$

where δ is a small complex number. The solution of Eqs. (4-82) and (4-83) can be simplified by expanding these equations about $\kappa_0 a$ and solving for δ.

Consider $J_n(\kappa a)$ in Eq. (4-82). It may be expanded into a Taylor series as

$$J_n(\kappa a) = J_n(\kappa_0 a) + \delta J_n'(\kappa_0 a) + \frac{\delta^2}{2} J_n''(\kappa_0 a) + \cdots \tag{4-86}$$

The remaining Bessel and Hankel functions may be expanded in a similar manner. Keeping only terms in δ through the second power, Eq. (4-82) or (4-83) reduces to

$$a_2 \delta^2 + a_1 \delta + a_0 = 0 \tag{4-87}$$

For the TE_{11} case

$$a_2 = \left(\frac{H_1''}{H_1'} + \frac{J_1'''}{2J_1''}\right) b_0 - b_1 \tag{4-88}$$

$$a_1 = b_0 \tag{4-89}$$

$$a_0 = (2 \sin \phi_0)^2 \tag{4-90}$$

and

$$b_1 = J_1'' H_1' \sum_{\substack{n=0 \\ n \neq 1}}^{\infty} \epsilon_n \left[\frac{J_n'' H_n' + J_n' H_n''}{(J_n' H_n')^2}\right] \left(\frac{2 \sin n\phi_0}{n}\right)^2 \tag{4-91}$$

$$b_0 = J_1'' H_1' \sum_{\substack{n=0 \\ n \neq 1}}^{\infty} \frac{\epsilon_n}{J_n' H_n'} \left(\frac{2 \sin n\phi_0}{n}\right)^2 \tag{4-92}$$

with

$$\kappa_0 a = 1.841 \tag{4-93}$$

For simplicity the superscript 2 has been deleted from the Hankel functions and the argument $\kappa_0 a$ has been deleted from both Hankel and Bessel functions.

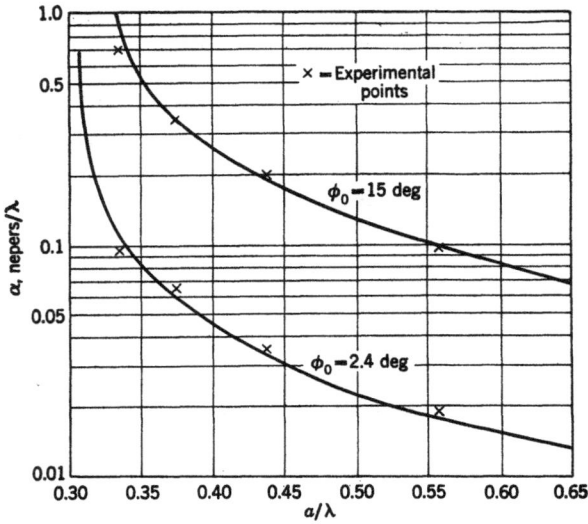

(a) α as a function of a/λ and ϕ_0

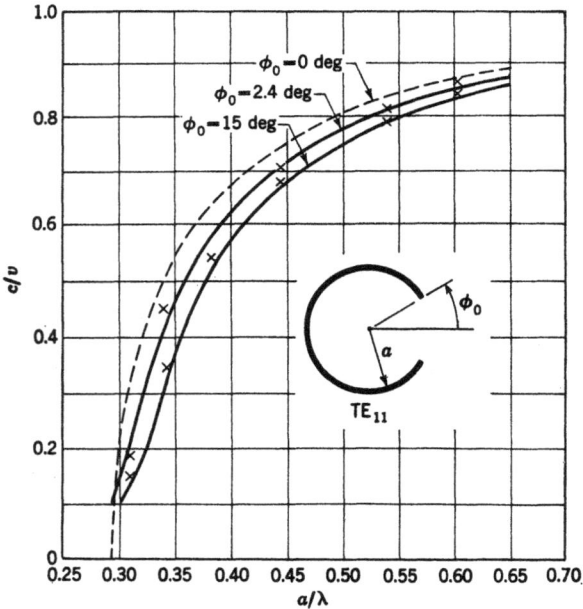

(b) c/v as a function of a/λ and ϕ_0

Figure 4-9. Theoretical and experimental values of α and c/v for the TE_{11} slotted cylinder.

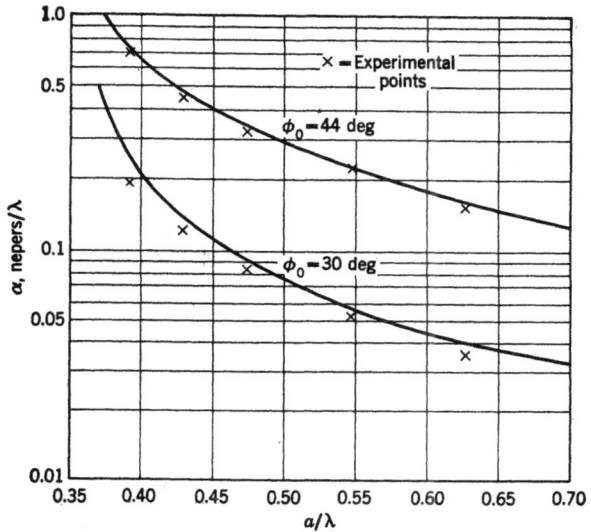

(a) α as a function of a/λ and ϕ_0

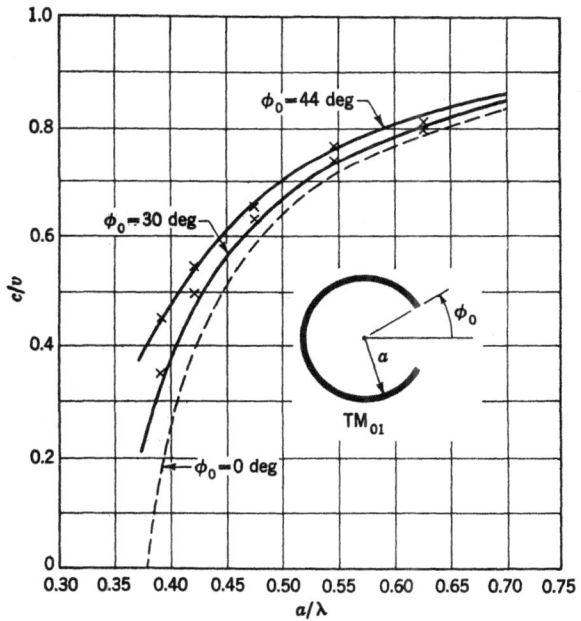

(b) c/v as a function of a/λ and ϕ_0

Figure 4-10. Theoretical and experimental values of α and c/v for the TM_{01} slotted cylinder.

For the TM_{01} case

$$a_2 = \left(\frac{H_0'}{H_0} + \frac{J_0''}{2J_0'}\right) c_0 - c_1 \qquad (4\text{-}94)$$

$$a_1 = c_0 \qquad (4\text{-}95)$$

$$a_0 = \left(\frac{4\phi_0}{\pi}\right)^2 \qquad (4\text{-}96)$$

and

$$c_1 = 2J_0'H_0 \sum_{n=1}^{\infty} \frac{J_n'H_n + J_nH_n'}{(J_nH_n)^2} \left[\frac{4\pi\phi_0 \cos n\phi_0}{\pi^2 - (2n\phi_0)^2}\right]^2 \qquad (4\text{-}97)$$

$$c_0 = 2J_0'H_0 \sum_{n=1}^{\infty} \frac{1}{J_nH_n} \left[\frac{4\pi\phi_0 \cos n\phi_0}{\pi^2 - (2n\phi_0)^2}\right]^2 \qquad (4\text{-}98)$$

with

$$\kappa_0 a = 2.405 \qquad (4\text{-}99)$$

Determination of γ from κ is straightforward. Substituting Eq. (4-85) into (4-50) gives

$$\gamma^2 = (\kappa_0 a + \delta)^2 - k^2 \qquad (4\text{-}100)$$

where $\gamma = \alpha + j\beta$ and $k = 2\pi/\lambda$. Theoretical results from Eqs. (4-82) and (4-83) are given in Figs. 4-9 and 4-10 in terms of α and c/v, where γ is expressed as

$$\gamma = \alpha + jk\frac{c}{v} \qquad (4\text{-}101)$$

with α in nepers per wavelength. Experimental verification of the theory is illustrated by the crosses, which represent measured values of α and c/v obtained by the methods outlined in Sec. 4-5.

4-8. TRANSVERSE RESONANCE METHOD

The propagation constant of many practical traveling wave structures may be found by a method of transverse resonance such as used by Marcuvitz[1] for the analysis of composite waveguide structures. Transverse resonance was perhaps first applied to a traveling wave antenna by Rotman.[2] The method also has been used quite successfully by Honey,[3] Zucker,[4] and

[1] N. Marcuvitz (ed.), "Waveguide Handbook," chap. 8, McGraw-Hill, New York, 1951.

[2] W. Rotman, The Channel Guide Antenna, *Air Force Cambridge Res. Center, Rept.* E5054, Cambridge, Mass., January, 1950.

[3] R. C. Honey, Horizontally Polarized Long Slot Array, *Stanford Res. Inst. Tech. Rept.* 47, Menlo Park, Calif., August, 1954.

[4] F. J. Zucker, The Guiding and Radiation of Surface Waves, *Proc. Symp. Mod. Advan. Microwave Tech.*, Polytechnic Institute of Brooklyn, November, 1954.

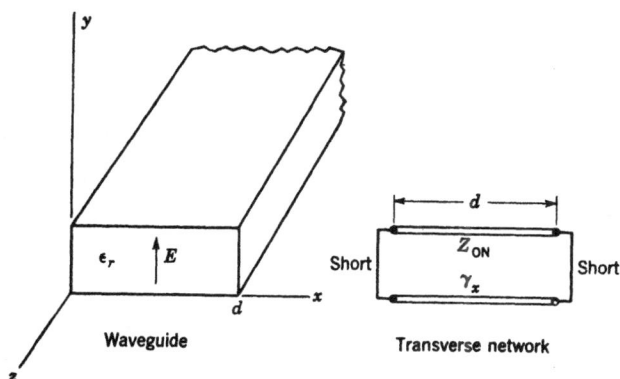

Figure 4-11. Transverse equivalent network of TE_{0N} waveguide.

Goldstone and Oliner.[1] Goldstone and Oliner have applied the method to the largest number of traveling wave antenna structures and appear to be the first to have applied the method to structures having circular cross sections.

We observed in the last section that a uniform traveling wave structure with fields having z variation of the form $e^{-\gamma z}$ could be considered as a two-dimensional problem in the plane transverse to the direction of propagation (xy plane). The fields of such a structure must satisfy the transverse wave equation [see Eq. (4-47)]; hence there is wave propagation in the transverse direction with phase constant κ, where $\kappa^2 = \omega^2 \mu \epsilon + \gamma^2$.

In the transverse resonance method the cross section of a traveling wave structure is represented as a transmission line network. The resonances of this transverse network yield values of κ and hence γ for the traveling wave structure. As a simple example to illustrate the method consider the conventional TE_{0N} waveguide in Fig. 4-11. For this structure a TE wave travels in the x direction with propagation constant γ_x. The characteristic impedance of the transverse line is expressed as Z_{0N}. It is a function of the mode of operation of the guide. At any point along the transverse network the impedance of the line looking in the positive x direction is equal and opposite to that looking in the negative x direction (the same applies to admittance). This follows from the continuity of the components of **E** and **H** that are tangential to a plane orthogonal to the transverse transmission line ($x = $ const plane in Fig. 4-11). Another way of stating the impedance relationship is that the sum of the two impedances that are observed by looking in opposite directions from a point on the line must add

[1] L. O. Goldstone and A. A. Oliner, Leaky Wave Antennas I: Rectangular Waveguides, *IRE Trans. Antennas Propagation*, **AP-7**:307–319, October, 1959; and Leaky Wave Antennas II: Circular Waveguides, *IRE Trans. Antennas Propagation*, **AP-9**: 280–290, May, 1961.

up to zero. At $x = d$ in Fig. 4-11 the impedance looking in the positive x direction is zero owing to the short circuit of the waveguide wall. Looking in the negative x direction at $x = d$ the impedance is

$$Z = Z_{0N} \tanh \gamma_x d \qquad (4\text{-}102)$$

where γ_x is the propagation constant for the TE wave in the x direction and Z_{0N} is the characteristic impedance in the x direction. The two impedances add up to zero, giving

$$Z_{0N} \tanh \gamma_x d = 0 \qquad (4\text{-}103)$$

Solution of Eq. (4-103) gives permissible values of γ_x for TE_{0N} modes. These solutions for the lossless case are

$$\gamma_x = j\beta_x = j\frac{N\pi}{d} \qquad N = 1, 2, 3, \ldots \qquad (4\text{-}104)$$

Starting with the wave equation in rectangular coordinates it can be shown that

$$\gamma_x{}^2 + \gamma_y{}^2 + \gamma^2 + \omega^2\mu\epsilon = 0 \qquad (4\text{-}105)$$

For this case ($\gamma_y = 0$ and no losses) Eq. (4-105) can be written

$$\beta_x{}^2 + \beta^2 = \epsilon_r k^2 \qquad (4\text{-}106)$$

where k is the free-space phase constant and ϵ_r is the relative permittivity (dielectric constant) of the material in the waveguide. Thus we arrive at the usual expression for the phase constant of a TE_{0N} waveguide:

$$\beta = \left[\epsilon_r k^2 - \left(\frac{N\pi}{d}\right)^2 \right]^{\frac{1}{2}} \qquad (4\text{-}107)$$

Equation (4-107) can be rewritten as

$$\beta = k \left[\epsilon_r - \left(\frac{N\lambda}{2d}\right)^2 \right]^{\frac{1}{2}} \qquad (4\text{-}108)$$

Application of the transverse resonance method to a structure such as Fig. 4-7a is somewhat more involved. In this case one needs to know the impedance (or admittance) of the slot as well as the impedance of the line. If the impedance of a slot of width W in an infinite ground plane for the TE_{01} mode is Z_s, then the transverse resonance equation for TE_{01} excitation ($\mathbf{E} = \mathbf{y}E_y$) of the slotted guide may be written

$$Z_{01} \tanh \gamma_x d = -Z_s{}' = -R_s - jX_s \qquad (4\text{-}109)$$

This is a complex transcendental equation and an exact solution cannot readily be obtained. Solution of Eq. (4-109) is facilitated by a perturbation method which will be described in the next example (slotted cylindrical guide).

From the above discussion it is clear that one needs to know both the impedance of the equivalent line and the aperture impedance in order to apply transverse resonance. Both of these impedances can be determined by a modal analysis of the type used by Marcuvitz.[1] A brief account of modal analysis applied to radial transmission lines is given in the following paragraphs to illustrate the method. The results are used in the analysis of the TE$_{11}$ slotted cylinder that was treated in the last section by the variational method.

Radial Transmission-line Modes[2]

In general the fields of a waveguide mode can be expressed as a product of two functions. One factor, called the modal amplitude, is a function only of the direction along which one desires to use a transmission line representation. The remaining portion of the field expression is called the mode function.

In Marcuvitz' treatment[3] the mode functions are independent of the modal amplitude coordinate, hence are functions only of the coordinates transverse to the modal amplitude, or transmission line, coordinate. This is suitable for slotted rectangular guides.[4]

For a radial transmission-line representation in a circular guide, Goldstone and Oliner[5] have modified the mode function concept to permit the mode functions to be a function of the radial transmission-line coordinate ρ as well as the transverse coordinate ϕ where z variation of the form $e^{-\gamma z}$ is assumed. This permits a vector modal representation of the fields in a slotted circular guide (Fig. 4-7b) with orthogonal properties over cylindrical surfaces $\rho = $ const.

From Maxwell's equations (source-free case) the transverse (to ρ) components E_z, E_ϕ, H_z, H_ϕ can be written for cylindrical coordinates as

$$(k^2 + \gamma^2)H_\phi = -j\omega\epsilon \frac{\partial E_z}{\partial \rho} - \frac{\gamma}{\rho}\frac{\partial H_z}{\partial \phi} \qquad (4\text{-}110)$$

$$(k^2 + \gamma^2)E_\phi = -j\omega\mu \frac{\partial H_z}{\partial \rho} - \frac{\gamma}{\rho}\frac{\partial E_z}{\partial \phi} \qquad (4\text{-}111)$$

where z dependence is $e^{-\gamma z}$. The components E_z and H_z also must satisfy the wave equation

$$\left[\frac{1}{\rho}\frac{\partial}{\partial \rho}\left(\rho\frac{\partial}{\partial \rho}\right) + \frac{\partial^2}{\rho^2 \partial \phi^2} + (k^2 + \gamma^2)\right]\begin{Bmatrix}E_z\\H_z\end{Bmatrix} = 0 \qquad (4\text{-}112)$$

---

[1] Marcuvitz, *op. cit.*, chap. 1.
[2] This treatment follows that of Goldstone and Oliner, *op. cit.*, 1961.
[3] Marcuvitz, *loc. cit.*
[4] Goldstone and Oliner, *op. cit.*, 1959.
[5] Goldstone and Oliner, *op. cit.*, 1961.

The axial field E_z is associated with TM fields ($H_z = 0$), and H_z is associated with TE fields ($E_z = 0$). The axial field components can be represented as

$$E_z(\rho,\phi) = \sum_n \overset{e}{V}_n(\rho)\overset{e}{e}_{zn}(\phi) \qquad \text{TM fields} \tag{4-113}$$

$$H_z(\rho,\phi) = \sum_n \overset{h}{I}_n(\rho)\overset{h}{h}_{zn}(\phi) \qquad \text{TE fields} \tag{4-114}$$

From the wave equation (4-112) the mode functions $\overset{e}{e}_{zn}$ and $\overset{h}{h}_{zn}$ must satisfy the equation

$$\left(\frac{d^2}{d\phi^2} + n^2\right) \left\{\begin{matrix}\overset{e}{e}_{zn}\\[4pt]\overset{h}{h}_{zn}\end{matrix}\right\} = 0 \tag{4-115}$$

The boundary conditions on the mode functions require them to be periodic in ϕ; hence n will be an integer.

Also from the wave equation (4-112) the modal amplitudes I_n and V_n satisfy Bessel's equation

$$\left[\frac{1}{\rho}\frac{d}{d\rho}\left(\rho\frac{d}{d\rho}\right) + \kappa_n^2\right]\left\{\begin{matrix}\overset{h}{I}_n\\[4pt]\overset{e}{V}_n\end{matrix}\right\} = 0 \tag{4-116}$$

where

$$\kappa_n^2 = \kappa^2 - \frac{n^2}{\rho^2} \tag{4-117}$$

with

$$\kappa^2 = k^2 + \gamma^2 \tag{4-118}$$

Furthermore the modal amplitudes satisfy the transmission-line equations

$$\frac{dV_n}{d\rho} = -j\kappa_n Z_{0n} I_n \tag{4-119}$$

$$\frac{dI_n}{d\rho} = -j\kappa_n Y_{0n} V_n \tag{4-120}$$

where the characteristic impedance and admittance are given by

$$\overset{e}{Y}_{0n} = \frac{1}{\overset{e}{Z}_{0n}} = \frac{\epsilon\omega\rho\kappa_n}{\kappa^2} \tag{4-121}$$

$$\overset{h}{Y}_{0n} = \frac{1}{\overset{h}{Z}_{0n}} = \frac{\kappa^2}{\omega\mu\rho\kappa_n} \tag{4-122}$$

From Eqs. (4-110), (4-111), (4-113), and (4-114) a general modal representation of the transverse (to ρ) fields may be written as

$$\mathbf{E}_t(\rho,\phi) = \sum_n [\overset{e}{V}_n(\rho)\overset{e}{\mathbf{e}}_n(\rho,\phi) + \overset{h}{V}_n(\rho)\overset{h}{\mathbf{e}}_n(\rho,\phi)] \tag{4-123}$$

$$\mathbf{H}_t(\rho,\phi) = \sum_n [\overset{e}{I}_n(\rho)\overset{e}{\mathbf{h}}_n(\rho,\phi) + \overset{h}{I}_n(\rho)\overset{h}{\mathbf{h}}_n(\rho,\phi)] \tag{4-124}$$

In terms of e_{zn} and h_{zn} the remaining mode functions are found from Eqs. (4-110) and (4-111) to be

$$\overset{e}{h}_{\phi n} = -\frac{\overset{e}{e}_{zn}}{\rho} \tag{4-125}$$

$$\overset{e}{e}_{\phi n} = \frac{\gamma}{\kappa^2}\frac{1}{\rho}\frac{d\overset{e}{e}_{zn}}{d\phi} \tag{4-126}$$

$$\overset{h}{h}_{\phi n} = \frac{\gamma}{\kappa^2}\frac{1}{\rho}\frac{d\overset{h}{h}_{zn}}{d\phi} \tag{4-127}$$

$$\overset{h}{e}_{\phi n} = \frac{\overset{h}{h}_{zn}}{\rho} \tag{4-128}$$

The mode functions defined above possess orthogonality properties over cylindrical surfaces transverse to ρ and may be normalized such that

$$\int_0^{2\pi} \overset{e}{\mathbf{h}}_n^* \times \boldsymbol{\varrho} \cdot \overset{e}{\mathbf{e}}_m \rho\, d\phi = \int_0^{2\pi} \overset{h}{\mathbf{h}}_n^* \times \boldsymbol{\varrho} \cdot \overset{h}{\mathbf{e}}_m \rho\, d\phi = \delta_{mn} \tag{4-129}$$

$$\int_0^{2\pi} \overset{e}{\mathbf{h}}_n^* \times \boldsymbol{\varrho} \cdot \overset{h}{\mathbf{e}}_m \rho\, d\phi = \int_0^{2\pi} \overset{h}{\mathbf{h}}_n^* \times \boldsymbol{\varrho} \cdot \overset{e}{\mathbf{e}}_m \rho\, d\phi = 0 \tag{4-130}$$

where $\boldsymbol{\varrho}$ is a unit vector in the radial direction. The orthogonality properties are useful in the determination of aperture admittance in Sec. 4-9.

For the slotted cylinders considered here the mode functions are even functions with respect to ϕ; hence the mode functions become for TM modes

$$\overset{e}{e}_{zn} = \sqrt{\frac{\epsilon_n}{2\pi}}\cos n\phi \tag{4-131}$$

$$\overset{e}{h}_{zn} = 0 \tag{4-132}$$

$$\overset{e}{e}_{\phi n} = \frac{-\gamma}{\kappa^2}\frac{n}{\rho}\frac{\sin n\phi}{\sqrt{\pi}} \tag{4-133}$$

$$\overset{e}{h}_{\phi n} = -\sqrt{\frac{\epsilon_n}{2\pi}}\frac{\cos n\phi}{\rho} \tag{4-134}$$

and for TE modes

$$\overset{h}{h}_{zn} = \sqrt{\frac{\epsilon_n}{2\pi}} \cos n\phi \tag{4-135}$$

$$\overset{h}{e}_{zn} = 0 \tag{4-136}$$

$$\overset{h}{h}_{\phi n} = \frac{\gamma}{\kappa^2} \frac{n}{\rho} \frac{\sin n\phi}{\sqrt{\pi}} \tag{4-137}$$

$$\overset{h}{e}_{\phi n} = \sqrt{\frac{\epsilon_n}{2\pi}} \frac{\cos n\phi}{\rho} \tag{4-138}$$

where $n = 0, 1, 2, \ldots$, and

$$\epsilon_n = 1 \qquad n = 0$$
$$\epsilon_n = 2 \qquad n > 0$$

Transverse Resonance Applied to Slotted Cylindrical Guide

As another example to illustrate the transverse resonance method let us consider the TE slotted cylindrical waveguide that was analyzed by the variational method in the last section. The network representation of the cross section in Fig. 4-8 is shown in Fig. 4-12. A radial transmission line of length a is terminated in the admittance $\overset{h}{Y}_s$ which is due to the slot. The effect of the slot may be presumed to be capacitive for the TE case because of electric field across the slot.

If $\overset{h}{Y}_1(\kappa\rho)$ is the admittance of the radial transmission line in Fig. 4-12, the transverse resonance equation may be written

$$\overset{h}{\overset{\leftarrow}{Y}}_1(\kappa a) + \overset{h}{\vec{Y}}_s = 0 \tag{4-139}$$

where an arrow to the left means that the admittance (or impedance) is viewed in the direction of decreasing ρ and an arrow to the right means in the direction of increasing ρ.

The admittance $\overset{h}{\overset{\leftarrow}{Y}}_1$ is

$$\overset{h}{\overset{\leftarrow}{Y}}_1 = -\frac{\overset{h}{I}_1}{\overset{h}{V}_1} \tag{4-140}$$

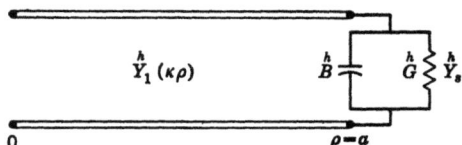

Figure 4-12. Transverse equivalent network of TE₁₁ slotted cylinder.

where $\overset{h}{I}_1$ and $\overset{h}{V}_1$ are the current and voltage, respectively, along the radial transmission line that represents the cross section of the slotted cylinder with TE_{11} excitation. The superscript h denotes an H wave (TE mode). A similar expression may be written for TM excitation (E wave).

For the present example $\overset{h}{I}_1$ is found from Eq. (4-116) to be

$$\overset{h}{I}_1 = A_1 J_1(\kappa\rho) \qquad \rho \lesssim a \tag{4-141}$$

From Eqs. (4-120) and (4-122)

$$\overset{h}{V}_1 = \frac{j\omega\mu\rho}{\kappa} A_1 J_1'(\kappa\rho) \qquad \rho \lesssim a \tag{4-142}$$

From Eqs. (4-141) and (4-142) the admittance $\overset{h}{\overset{\leftarrow}{Y}}_1(\kappa a)$ is found to be

$$\overset{h}{\overset{\leftarrow}{Y}}_1(\kappa a) = -\frac{\kappa}{j\omega\mu a} \frac{J_1(\kappa a)}{J_1'(\kappa a)} \tag{4-143}$$

Equation (4-143) may be substituted into Eq. (4-139), and if $\overset{h}{Y}_s$ is known, one may solve for κ and hence γ. Solution of the complex transcendental equation (4-139) is usually extremely tedious, involving a trial-and-error procedure. If the frequency of operation of the traveling wave structure is well above cutoff, explicit expressions for α and β may be obtained from Eq. (4-139) by the following perturbation method.

As in Sec. 4-7 let κa be expressed as

$$\kappa a = \kappa_0 a + \delta \tag{4-144}$$

where κ_0 is the transverse phase constant of the unperturbed structure and δ is a small complex number. The propagation constant γ from Eq. (4-100) is given approximately as

$$\gamma \approx \left(\kappa_0{}^2 - k^2 + 2\frac{\kappa_0\delta}{a}\right)^{\frac{1}{2}} \tag{4-145}$$

where $(\kappa_0{}^2 - k^2)^{\frac{1}{2}}$ may be identified as the propagation constant γ_0 of the unperturbed waveguide. Expanding Eq. (4-145) gives

$$\gamma \approx \gamma_0\left(1 + \frac{\kappa_0\delta}{\gamma_0{}^2 a} + \cdots\right) \tag{4-146}$$

where only the first two terms are retained. For a lossless guide,

$$\gamma_0 = j\beta_0 = \frac{j2\pi}{\lambda_{g0}} \tag{4-147}$$

hence Eq. (4-146) becomes

$$\alpha + j\beta \approx j\frac{2\pi}{\lambda_{g0}}\left(1 - \frac{\kappa_0\lambda_{g0}^2\delta}{4\pi^2 a}\right) \tag{4-148}$$

Equating real and imaginary parts of Eq. (4-148) gives

$$\alpha \approx \frac{\kappa_0\lambda_{g0}\delta_i}{2\pi a} \tag{4-149}$$

and

$$\beta \approx \frac{2\pi}{\lambda_{g0}}\left(1 - \frac{\kappa_0\lambda_{g0}^2\delta_r}{4\pi^2 a}\right) \tag{4-150}$$

where

$$\delta = \delta_r + j\delta_i \tag{4-151}$$

Equations (4-149) and (4-150) are approximate expressions for α and β in terms of δ. Greater accuracy could be obtained by retaining δ^2 in Eq. (4-145), but the resulting expressions for α and β become much more complicated. Equations (4-149) and (4-150) are found to give good results for c/v greater than approximately 0.3.

The complex number δ may be determined by expanding the transverse resonance equation about κ_0. Because of difficulties due to $\overset{h}{Y}_1(\kappa_0 a)$ being infinite, it is more convenient to expand the transverse resonance equation when it is expressed as an impedance of the form

$$\overset{h}{\overleftrightarrow{Z}}(\kappa a) = 0 \tag{4-152}$$

where

$$\overset{h}{\overleftrightarrow{Z}}(\kappa a) = \overset{h}{\overleftarrow{Z}}_1(\kappa a) + \overset{h}{R}_s - j\overset{h}{X}_s \tag{4-153}$$

and

$$\overset{h}{\overleftarrow{Z}}_1(\kappa a) = \frac{\omega\mu a}{j\kappa}\frac{J_1'(\kappa a)}{J_1(\kappa a)} \tag{4-154}$$

$$\overset{h}{R}_s = \frac{\overset{h}{G}}{\overset{h}{G^2} + \overset{h}{B^2}} \tag{4-155}$$

$$\overset{h}{X}_s = \frac{\overset{h}{B}}{\overset{h}{G^2} + \overset{h}{B^2}} \tag{4-156}$$

In normalized form Eq. (4-153) may be expressed as

$$\overset{h}{\overleftrightarrow{Z}}(\kappa a) = \overset{h}{\overleftarrow{Z}}(\kappa a) + \overset{h}{\mathfrak{R}_s} - j\overset{h}{\mathfrak{X}_s} = 0 \tag{4-157}$$

where each term has been divided by $\omega\mu a/\kappa$.

Expanding Eq. (4-157) about κ_0 gives

$$\overset{h}{\overleftrightarrow{Z}}(\kappa a) \approx \overset{h}{\overleftrightarrow{Z}}(\kappa_0 a) + \delta \overset{h}{\overleftrightarrow{Z}}'(\kappa_0 a) = 0 \tag{4-158}$$

where the prime denotes differentiation with respect to κa. Solving for δ gives

$$\delta = - \frac{\overset{h}{\overleftrightarrow{Z}}(\kappa_0 a)}{\overset{h}{\overleftrightarrow{Z}}'(\kappa_0 a)} \tag{4-159}$$

Evaluation of $\overset{h}{\overleftrightarrow{Z}}(\kappa_0 a)$ and $\overset{h}{\overleftrightarrow{Z}}'(\kappa_0 a)$ gives

$$\overset{h}{\overleftrightarrow{Z}}(\kappa_0 a) = \overset{h}{\mathfrak{R}_s}(\kappa_0 a) - j\overset{h}{\mathfrak{X}_s}(\kappa_0 a) \tag{4-160}$$

and

$$\overset{h}{\overleftrightarrow{Z}}'(\kappa_0 a) \approx \frac{1 - (\kappa_0 a)^2}{j(\kappa_0 a)^2} \tag{4-161}$$

where it is recalled that $J_1'(\kappa_0 a)$ is zero and the derivatives of $\mathfrak{R}$ and $\mathfrak{X}$ are neglected. Thus δ for the TE_{11} mode reduces to

$$\delta = \frac{(\kappa_0 a)^2 [\overset{h}{\mathfrak{X}_s}(\kappa_0 a) + j\overset{h}{\mathfrak{R}_s}(\kappa_0 a)]}{(\kappa_0 a)^2 - 1} \tag{4-162}$$

If the above perturbation method is applied to a TM_{01} slotted cylinder, α and β are given by Eqs. (4-149) and (4-150) with

$$\delta = j\overset{e}{\mathfrak{R}_s} - \overset{e}{\mathfrak{X}_s} \tag{4-163}$$

The transverse equivalent network for this case is shown in Fig. 4-13.

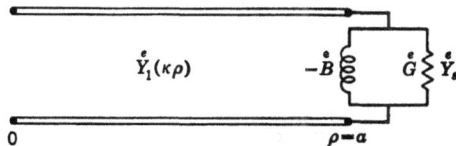

Figure 4.13. Transverse network for slotted cylinder with TM_{01} excitation.

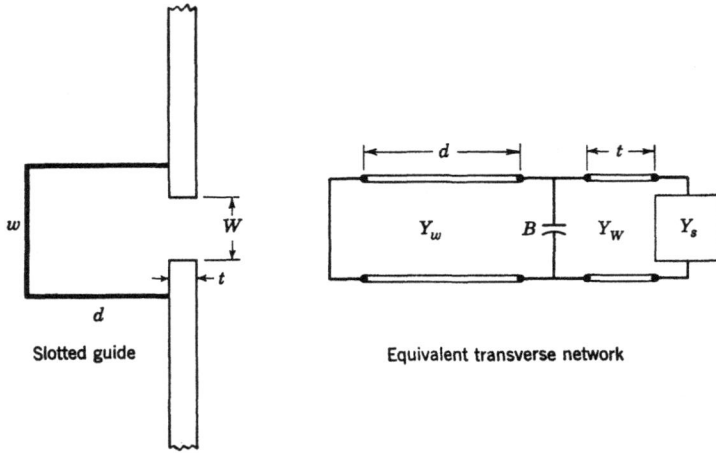

Figure 4-14. Slot in a thick wall.

The above methods have been applied to rectangular structures,[1] and the results are given in the next chapter.

The transverse resonance method is quite practical for waveguides with radiating apertures because it gives good approximations for α and β in explicit and relatively simple form in terms of the aperture admittance (or impedance). Aperture admittances for many structures are available in the literature, and a number of them are given in the next chapter. If it is necessary to solve for the aperture admittance, however, the computations are likely to be as involved as (and perhaps similar to) the computations for γ in the variational method. An example will be given in the next section.

Before leaving the transverse resonance method it should be noted that the method may be applied to structures with significant wall thicknesses as in Fig. 4-14. The wall thickness t is treated as a transmission line of length t and characteristic admittance Y_W which is determined by the mode of operation and the dimension W. The admittance $Y(d + t)$ is obtained by transforming the admittance at d,

$$Y(d) = jB - jY_W \cot \kappa d \tag{4-164}$$

through the length of line t. The susceptance B is associated with energy stored in higher-order modes at the discontinuity created by the junction of the two sections of transmission line. The susceptance B may be evaluated by the methods of Marcuvitz.[2]

An accurate determination of B_s in this case will be complicated by

[1] Goldstone and Oliner, *op. cit.*, 1959.
[2] Marcuvitz, *op. cit.*, chap. 5, sec. 26.

the interior geometry. If $W \ll 2\pi/\kappa$ and t is on the order of W or larger, it may be permissible to neglect the contribution to B_s from the interior region. However, B should be included in this case. As $t \to 0$, B is replaced by the part of B_s that is due to energy stored in the interior region. A method for determining B_s is discussed in the next section.

4-9. DETERMINING APERTURE ADMITTANCE

In general the determination of aperture admittance (or impedance) is quite involved. The admittance depends on the mode of excitation as well as the geometry of the structure, and because of the difficulties encountered in attempting an exact solution, approximate methods (such as the variational method) are usually employed.

Let us determine the admittance $\overset{h}{Y}_s$ that terminates the radial transmission line for the TE_{11} slotted cylinder that was considered in the previous section. Assuming that all losses are due to radiation, the conductance of the slot may be determined from the radial real power per unit length and the mode voltage $\overset{h}{V}_1$ as

$$\overset{h}{G}_s = \frac{\text{Re} \int_{\text{slot}} \mathbf{E} \times \mathbf{H}^* \cdot \varrho a \, d\phi}{\left| \overset{h}{V}_1 \right|^2} \tag{4-165}$$

For this case the radial real power is given by

$$\text{Re} \int_{-\phi_0}^{\phi_0} \mathbf{E} \times \mathbf{H}^* \cdot \varrho a \, d\phi = \text{Re} \int_{-\phi_0}^{\phi_0} E_s H_z^* a \, d\phi \tag{4-166}$$

where E_s is the electric field in the slot. The magnetic field H_z may be represented in terms of the mode functions introduced in the preceding section [see Eq. (4-114)] as

$$H_z = \sum_{n=0}^{\infty} \overset{h}{I}_n \overset{h}{h}_{zn} \tag{4-167}$$

where the modal current I_n is related to modal voltage by

$$\overset{h}{I}_n = \overset{h}{Y}_n \overset{h}{V}_n \tag{4-168}$$

With the use of the orthogonality relations in Eqs. (4-129) and (4-130) the voltage $\overset{h}{V}_n$ is obtained from Eq. (4-113) as

$$\overset{h}{V}_n = \int_{-\phi_0}^{\phi_0} \overset{h}{h}_{zn} E_s a \, d\phi \tag{4-169}$$

Using Eqs. (4-167) to (4-169), Eq. (4-165) can be written

$$\overset{h}{G_s} = \frac{\mathrm{Re} \sum_{n=0}^{\infty} \overset{h}{\overrightarrow{Y}_n}(a) \int_{-\phi_0}^{\phi_0} E_s(\phi)\overset{h}{h_{zn}}(\phi)\, d\phi \int_{-\phi_0}^{\phi_0} E_s(\phi')\overset{h}{h_{zn}}(\phi')\, d\phi'}{\left| \int_{-\phi_0}^{\phi_0} \overset{h}{h_{z1}}E_s(\phi)\, d\phi \right|^2} \tag{4-170}$$

It can be shown that $\overset{h}{G_s}$ in Eq. (4-170) is stationary with respect to a variation of E_s about its correct value.

The modal admittance for the external region is found in the same
manner as for the internal region [see Eq. (4-143)]. It is found to be

$$\overset{h}{\overrightarrow{Y}_n}(a) = \frac{\kappa}{j\omega\mu a} \frac{H_n^{(2)}(\kappa a)}{H_n^{(2)\prime}(\kappa a)} \tag{4-171}$$

The real part of Eq. (4-171) is

$$\mathrm{Re}\,\overset{h}{\overrightarrow{Y}_n}(a) = \frac{1}{2}\left[\overset{h}{\overrightarrow{Y}_n}(a) + \overset{h}{\overrightarrow{Y}_n^*}(a) \right] \tag{4-172}$$

Making use of the Wronskian of solutions of Bessel's equation,

$$J_n(\kappa a)H_n^{(2)\prime}(\kappa a) - J_n'(\kappa a)H_n^{(2)}(\kappa a) = \frac{-j2}{\pi \kappa a} \tag{4-173}$$

Eq. (4-172) reduces to

$$\mathrm{Re}\,\overset{h}{\overrightarrow{Y}_n}(a) = \frac{2}{\pi\omega\mu a^2} \frac{1}{|H_n^{(2)\prime}(\kappa a)|^2} \tag{4-174}$$

The quantity $\overset{h}{h_{zn}}$ in Eq. (4-170) is found from Eq. (4-115) to be

$$\overset{h}{h_{zn}} = a_n \cos n\phi \tag{4-175}$$

where only $\cos n\phi$ terms are used because of symmetry.

Substituting Eqs. (4-175) and (4-128) into (4-129) gives

$$a_n = \sqrt{\frac{\epsilon_n}{2\pi}} \tag{4-176}$$

where $\epsilon_n = 2$ for $n > 0$ and $\epsilon_n = 1$ for $n = 0$.

Equation (4-170) can now be solved for $\overset{h}{G_s}$ if E_s is specified. For $E_s = 1$,
Eq. (4-170) gives

$$\overset{h}{G_s} = \frac{1}{\pi\omega\mu a^2} \sum_{n=0}^{\infty} \frac{\epsilon_n}{|H_n^{(2)\prime}(\kappa a)|^2}\left(\frac{\sin n\,\phi_0}{n \sin \phi_0} \right)^2 \tag{4-177}$$

For

$$E_s = \frac{1}{[1 - (\phi/\phi_0)^2]^{1/2}} \tag{4-178}[1]$$

Eq. (4-170) gives

$$\overset{h}{G_s} = \frac{1}{\pi\omega\mu a^2} \sum_{n=0}^{\infty} \frac{\epsilon_n}{|H_n^{(2)'}(\kappa a)|^2} \left[\frac{J_0(n\phi_0)}{J_0(\phi_0)}\right]^2 \tag{4-179}$$

The susceptance $\overset{h}{B_s}$ due to the slot is associated with stored energy in both the internal and external regions of the slotted cylinder. The susceptance due to stored energy in the internal region may be expressed in variational form as

$$j\overset{ih}{B_s} = \sum_{n=2}^{\infty} \overset{h}{Y_n} \frac{|V_n|^2}{|V_1|^2} \tag{4-180}[2]$$

where only nonpropagating modes are considered (the waveguide is assumed to be small enough to support only the TE_{11} mode). The quantities $\overset{h}{Y_n}$ and V_n are given by

$$\overset{h}{Y_n} = \frac{\kappa}{j\omega\mu a} \frac{J_n(\kappa a)}{J_n'(\kappa a)} \tag{4-181}$$

and

$$V_n = \int_{-\phi_0}^{\phi_0} E_s \overset{h}{h_{zn}} a \, d\phi \tag{4-182}$$

Equation (4-180) thus becomes

$$j\overset{ih}{B_s} = \sum_{n=2}^{\infty} \frac{\kappa}{j\omega\mu a} \frac{J_n(\kappa a)}{J_n'(\kappa a)} \frac{\left(\int_{-\phi_0}^{\phi_0} E_s \cos n\phi \, d\phi\right)^2}{\left(\int_{-\phi_0}^{\phi_0} E_s \cos \phi \, d\phi\right)^2} \tag{4-183}$$

For $E_s = 1$, Eq. (4-183) reduces to

$$j\overset{ih}{B_s} = \sum_{n=2}^{\infty} \frac{\kappa}{j\omega\mu a} \frac{J_n(\kappa a)}{J_n'(\kappa a)} \left(\frac{\sin n\phi_0}{n \sin \phi_0}\right)^2 \tag{4-184}$$

For $E_s = \dfrac{1}{[1 - (\phi/\phi_0)^2]^{1/2}}$, Eq. (4-183) reduces to

$$j\overset{ih}{B_s} = \sum_{n=2}^{\infty} \frac{\kappa}{j\omega\mu a} \frac{J_n(\kappa a)}{J_n'(\kappa a)} \left(\frac{J_0(n\phi_0)}{J_0(\phi_0)}\right)^2 \tag{4-185}$$

---

[1] This is the form chosen by Goldstone and Oliner, *op. cit.*, 1961.
[2] This is the form used by Marcuvitz, *op. cit.*, chap. 3, in treating windows in waveguides.

The susceptance associated with energy stored in the external region may be found from

$$\overset{eh}{B_s} = \text{Im} \sum_{n=0}^{\infty} \overset{h}{Y}_n \frac{|V_n|^2}{|V_1|^2} \tag{4-186}$$

where

$$\overset{h}{Y}_n = \frac{\kappa}{j\omega\mu a} \frac{H_n^{(2)}(\kappa a)}{H_n^{(2)'}(\kappa a)} \tag{4-187}$$

Alternatively, $\overset{eh}{B_s}$ may be written

$$\overset{eh}{B_s} = \frac{-\text{Im} \int_{-\phi_0}^{\phi_0} E_s H_z^* a \, d\phi}{|V_1|^2} \tag{4-188}$$

Equations (4-186) and (4-188) reduce to

$$\overset{eh}{B_s} = \frac{\text{Im} \sum_{n=0}^{\infty} \epsilon_n \overset{h}{Y}_n \left(\int_{-\phi_0}^{\phi_0} E_s \cos n\phi \, d\phi \right)^2}{\left(\int_{-\phi_0}^{\phi_0} E_s \cos \phi \, d\phi \right)^2} \tag{4-189}$$

The imaginary part of $\overset{h}{Y}_n$ may be found from

$$\text{Im} \overset{h}{Y}_n = \frac{1}{2j} (\overset{h}{Y}_n - \overset{h}{Y}_n^*) \tag{4-190}$$

and is given by

$$\text{Im} \overset{h}{Y}_n = \frac{-\kappa}{\omega\mu a} \frac{J_n(\kappa a)J_n'(\kappa a) + N_n(\kappa a)N_n'(\kappa a)}{|H_n^{(2)'}(\kappa a)|^2} \tag{4-191}$$

For $E_s = 1$, Eq. (4-189) reduces to

$$\overset{eh}{B_s} = \frac{-\kappa}{\omega\mu a} \sum_{n=0}^{\infty} \frac{\epsilon_n [J_n(\kappa a)J_n'(\kappa a) + N_n(\kappa a)N_n'(\kappa a)]}{|H_n^{(2)'}(\kappa a)|^2} \left(\frac{\sin n\phi_0}{n \sin \phi_0} \right)^2 \tag{4-192}$$

and for $E_s = \dfrac{1}{[1 - (\phi/\phi_0)^2]^{1/2}}$, Eq. (4-189) reduces to

$$\overset{eh}{B_s} = \frac{-\kappa}{\omega\mu a} \sum_{n=0}^{\infty} \frac{\epsilon_n [J_n(\kappa a)J_n'(\kappa a) + N_n(\kappa a)N_n'(\kappa a)]}{|H_n^{(2)'}(\kappa a)|^2} \left[\frac{J_0(n\phi_0)}{J_0(\phi_0)} \right]^2 \tag{4-193}$$

The determination of slot admittance for the slotted cylinder with TE$_{11}$ excitation illustrates a general method. The same procedure may be followed for slotted cylinders with TM excitation as well as for slotted rectangular waveguides with either TE or TM excitation.

In addition to the variational method one may employ such methods as the integral equation method, the equivalent static method, and the transform method as used in the determination of the lumped circuit parameters of window and obstacle discontinuities in waveguide structures.[1] Fortunately, the aperture admittances for many waveguides with radiating apertures have already been determined, and a number of these will be given in the next chapter.

ADDITIONAL REFERENCES

1. R. C. Honey, A Flush-mounted Leaky-wave Antenna with Predictable Patterns, *IRE Trans. Antennas Propagation*, **AP-7**:320–329, October, 1959.
2. H. Jasik (ed.), "Antenna Engineering Handbook," chap. 16, McGraw-Hill, New York, 1961.
3. *Proc. Symp. Mod. Advan. Microwave Tech.*, pp. 403–435, Polytechnic Institute of Brooklyn, 1955.

PROBLEMS

4-1. Derive Eq. (4-51).
4-2. Derive Eqs. (4-119) and (4-120).
4-3. Go through the transmission-line development starting on page 175 for the case of a rectangular waveguide.
4-4. Show that Eq. (4-170) is stationary with respect to a variation of E_s about its correct value.
4-5. Show that Eqs. (4-180) and (4-186) are stationary with respect to a variation of E_s about its correct value.

[1] Marcuvitz, *op. cit.*, sec. 3.5.

5-1. INTRODUCTION

It was pointed out in Chap. 4 that design data for traveling wave radiators can be conveniently presented in terms of attenuation constant α and phase constant β (or α and the velocity ratio c/v). In this chapter design data for many practical line sources are summarized. The data are presented in the form of α and β (or α and c/v) versus geometry and mode of excitation. Equations are given whenever possible. In some cases only experimental data are available, and these are presented in graphical form. In many instances data are presented graphically to show agreement between theory and experiment. For waveguide structures with radiating apertures the graphical display of the data shows the amount of perturbation due to the addition of the aperture.

The data are classified according to mode of excitation. This gives three categories: TE structures, TM structures, and hybrid (TE + TM) structures. Experimental evidence shows that mode purity is preserved except for some hybrid structures in which TE or TM excitation may give hybrid mode operation. In this case the above classification is somewhat ambiguous. Attention is called to such situations when they arise.

5-2. TE LINE SOURCE RADIATORS

TE structures lying along the z axis are characterized by $E_z = 0$. Only rectangular and circular cross sections having dominant TE mode excitation are considered. Other geometries are possible as well as higher-order mode operation, but these are not needed in most practical applications.

188

A. TE_{01} Slotted Rectangular Guide

The TE_{01} slotted rectangular guide shown in Fig. 5-1 has been studied extensively. It appears to have been first conceived by Hansen[1] in 1939 and has been studied in detail by many investigators,[2] most of whom have considered the special case where the slot is the full width of the guide $(W = w)$.

Transverse resonance (see Sec. 4-8) gives

$$j \cot \kappa_1 d - \overset{h}{\mathcal{G}}(\kappa_1) - j\overset{h}{\mathcal{B}}(\kappa_1) = 0 \tag{5-1}$$

where the aperture admittance has been normalized with respect to $\kappa_1/\omega\mu$. For $W \ll w$, an approximate solution of Eq. (5-1) based on a perturbation about the closed guide $(W = 0)$ gives

$$\alpha \approx \frac{\lambda_{g0}}{2d^2} \frac{\overset{h}{\mathcal{G}}}{\overset{h}{\mathcal{G}^2} + \overset{h}{\mathcal{B}^2}} \tag{5-2}$$

$$\frac{c}{v} \approx \frac{\lambda}{\lambda_{g0}}\left(1 - \frac{\lambda_{g0}^2}{4\pi d^2} \frac{\overset{h}{\mathcal{B}}}{\overset{h}{\mathcal{G}^2} + \overset{h}{\mathcal{B}^2}}\right) \tag{5-3}$$

[1] W. W. Hansen, Radiating Electromagnetic Waveguide, U.S. Patent 2402622 (1940).

[2] H. G. Booker, Girders and Trenches as End-fire Aerials, *Telecommunications Res. Estab. Rept.* 30, Swanage, England, 1941; A. L. Cullen, On the Channel Section Waveguide Radiator, *Phil. Mag.*, **40**:417–428, April, 1949; W. Rotman, The Channel Guide Antenna, *Air Force Cambridge Res. Center Rept.* E5054, Cambridge, Mass., January, 1950; V. H. Rumsey, Traveling Wave Slot Antennas, *J. Appl. Phys.*, **24**:1358–1365, November, 1953; J. N. Hines, V. H. Rumsey, and C. H. Walter, Traveling Wave Slot Antennas, *Proc. IRE*, **41**:1624–1631, November, 1953; L. O. Goldstone and A. A. Oliner, Leaky Wave Antennas I: Rectangular Waveguides, *IRE Trans. Antennas Propagation*, **AP-7**:307–319, October, 1959.

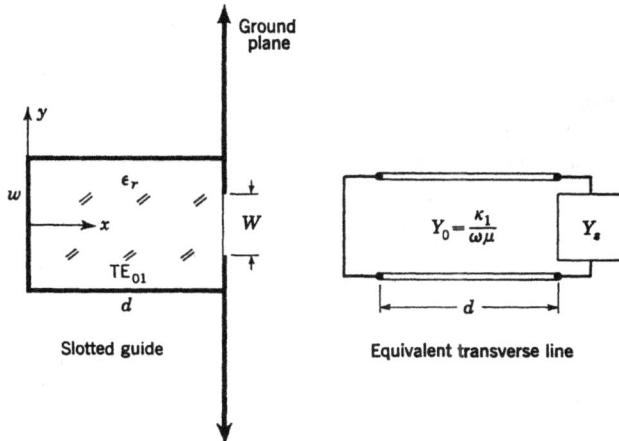

Figure 5-1. Slotted rectangular guide.

where[1]

$$\lambda_{g0} = \frac{\lambda}{[\epsilon_r - (\lambda/2d)^2]^{1/2}} \tag{5-4}$$

$$\overset{h}{\mathcal{G}} \approx \frac{w[\pi^2 - k^2 d^2(\epsilon_r - 1)]}{2\pi d} \tag{5-5}$$

$$\overset{h}{\mathcal{B}} \approx \frac{w}{d} \ln\left(\csc\frac{\pi W}{2w}\right) + \frac{w[\pi^2 - k^2 d^2(\epsilon_r - 1)]}{\pi^2 d}$$
$$\times \ln\frac{2.718\pi d}{1.781 W[\pi^2 - k^2 d^2(\epsilon_r - 1)]^{1/2}} \tag{5-6}$$

For $W = w$, transverse resonance with a solution based on a perturbation about a closed guide of width $2d$ gives (see Prob. 5-1)

$$\alpha \approx \frac{\lambda_{g0}\kappa_0 \overset{h}{\mathcal{G}}(\kappa_0)}{2\pi d \, \csc^2(\kappa_0 d)} \tag{5-7}$$

$$\frac{c}{v} \approx \left(\epsilon_r - \frac{\lambda^2 \kappa_0{}^2}{4\pi^2}\right)^{1/2} \tag{5-8}$$

where

$$\kappa_0 \approx \frac{\pi}{2d} - \frac{1}{d}\overset{h}{\mathcal{B}}\left(\frac{\pi}{2d}\right) \tag{5-9}$$

$$\overset{h}{\mathcal{G}}(\kappa_0) \approx \frac{w[\kappa_0{}^2 - k(\epsilon_r - 1)]}{2\kappa_0} \tag{5-10}$$

$$\overset{h}{\mathcal{B}}\left(\frac{\pi}{2d}\right) \approx \frac{w[\pi^2 - 4d^2 k^2(\epsilon_r - 1)]}{2\pi^2 d} \ln\frac{5.436\pi d}{1.781 W[\pi^2 - 4d^2 k^2(\epsilon_r - 1)]^{1/2}} \tag{5-11}$$

Graphs of α and c/v from Eqs. (5-7) and (5-8) are shown in Figs. 5-2 and 5-3 along with experimental data. Because of the approximations in the perturbation solution of the transverse resonance equation (see Sec. 4-8), Eqs. (5-2), (5-3), (5-7), and (5-8) give satisfactory results only when the frequency is such that the slotted waveguide is well above cutoff.

For operation at frequencies near cutoff one should determine an exact solution for the transverse resonance equation. Alternatively, one may utilize the variational solution of Rumsey.[2]

The data in Figs. 5-2 and 5-3 show that the propagation constant for the TE$_{01}$ slotted rectangular guide can be varied over quite a wide range. The quantities α and c/v are not independent of each other, but within limits one can adjust the geometry to hold one quantity essentially constant while varying the other.

To get small values of α for certain applications such as extremely long antennas, one is forced to endfire operation or to a very small W with

---

[1] $\mathcal{G}$ and $\mathcal{B}$ are different from the quantities used by Goldstone and Oliner, *op. cit.*, to the extent that dielectric loading is taken into account here.

[2] Rumsey, *op. cit.*

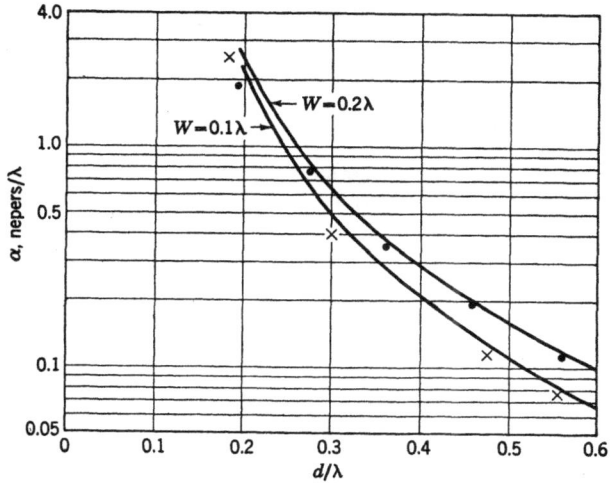

Figure 5-2.   Attenuation constant α for TE_{01} slotted rectangular guide $(W = w)$. (Points are measured data taken from Hines, Rumsey, and Walter, Traveling Wave Slot Antennas, *Proc. IRE*, **41**: 1624–1631, 1953, for $\epsilon_r = 1$.)

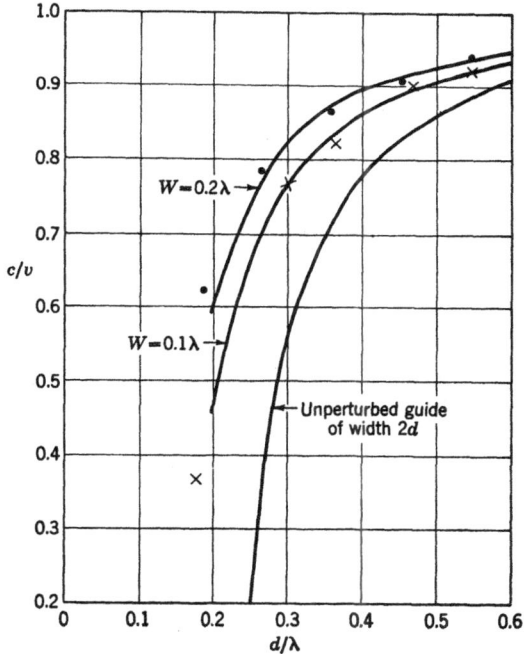

Figure 5-3.   Velocity ratio c/v for TE_{01} slotted rectangular guide $(W = w)$. (Points are measured data taken from Hines, Rumsey, and Walter, Traveling Wave Slot Antennas, *Proc. IRE*, **41**:1624–1631, November, 1953, for $\epsilon_r = 1$.)

perhaps a thick wall. Small W, however, will limit power-handling capability. When very small values of α are required, it is better to consider the following holey waveguide structure.

B. TE₀₁ Holey Rectangular Guide

A rectangular guide periodically loaded with holes as shown in Fig. 5-4 has been considered by several investigators[1] Transverse resonance gives

$$j \cot \kappa_1 d - \overset{h}{\mathcal{G}}(\kappa_1) + j\overset{h}{\mathcal{B}}(\kappa_1) = 0 \tag{5-12}$$

An approximate solution based on a perturbation about the closed guide results in

$$\alpha \approx \frac{\lambda_{g0}}{2d^2} \frac{\overset{h}{\mathcal{G}}}{\overset{h}{\mathcal{G}}{}^2 + \overset{h}{\mathcal{B}}{}^2} \tag{5-13}$$

$$\frac{c}{v} \approx \frac{\lambda}{\lambda_{g0}} \left(1 + \frac{\lambda_{g0}^2}{4\pi d^2} \frac{\overset{h}{\mathcal{B}}}{\overset{h}{\mathcal{G}}{}^2 + \overset{h}{\mathcal{B}}{}^2} \right) \tag{5-14}$$

[1] Goldstone and Oliner, *op. cit.*; J. N. Hines and J. R. Upson, A Wide Aperture Tapered-depth Scanning Antenna, *Ohio State Univ. Res. Found. Rept.* 667-7, Columbus, December, 1957.

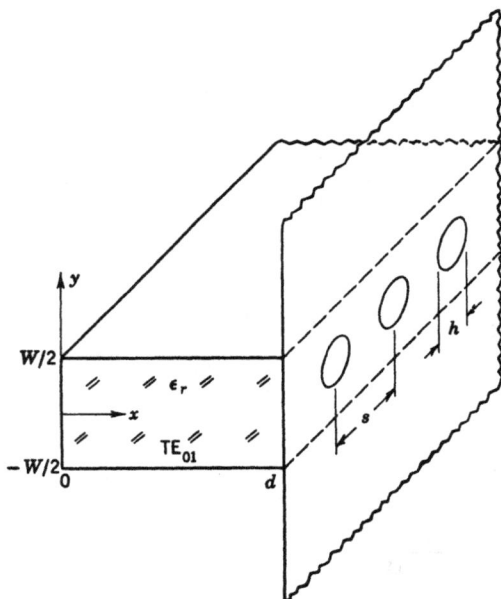

Figure 5-4. Holey rectangular guide.

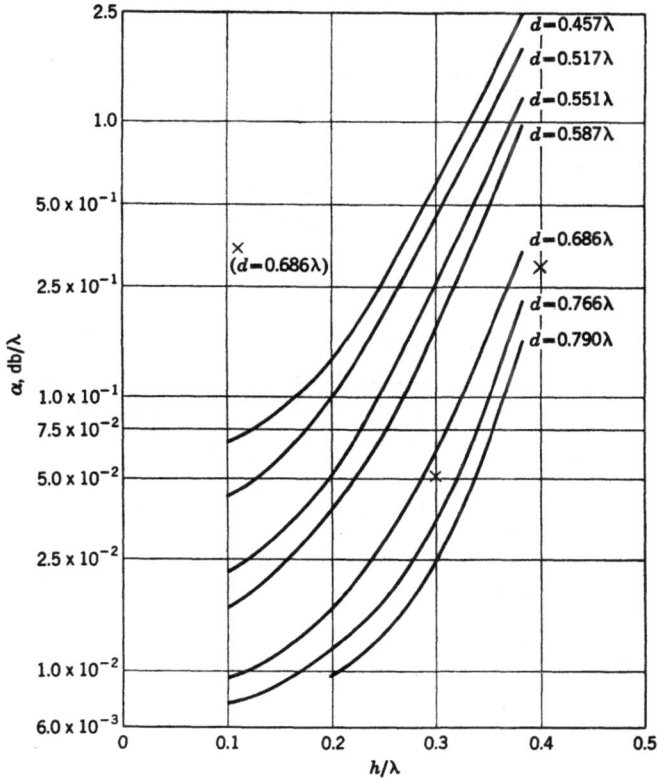

Figure 5-5. Measured α for holey RG 91/U waveguide. Points are obtained from Eq. (5-13). Spacing $s = 0.437\lambda$. (After Hines and Upson, A Wide Aperture Tapered-Depth Scanning Antenna, *Ohio State Univ. Res. Found. Rept.*, 667-7, Columbus, December, 1957.) Note that α, db/λ = 8.686 α, nepers/λ.

where[1]

$$\lambda_{g0} = \frac{\lambda}{[\epsilon_r - (\lambda/2d)^2]^{1/2}} \tag{5-15}$$

$$\frac{h}{G} \approx \frac{w[\pi^2 - k^2d^2(\epsilon_r - 1)]}{2\pi d} \tag{5-16}$$

$$\frac{h}{\mathcal{B}} \approx \frac{6dws}{\pi h^3} \tag{5-17}$$

Equations (5-13) and (5-14) become very poor approximations for frequencies near cutoff. For operation near cutoff the transverse resonance

[1] The functions $\mathcal{G}$ and $\mathcal{B}$ are taken from Goldstone and Oliner, *op. cit.*, and modified to include $\epsilon_r > 1$.

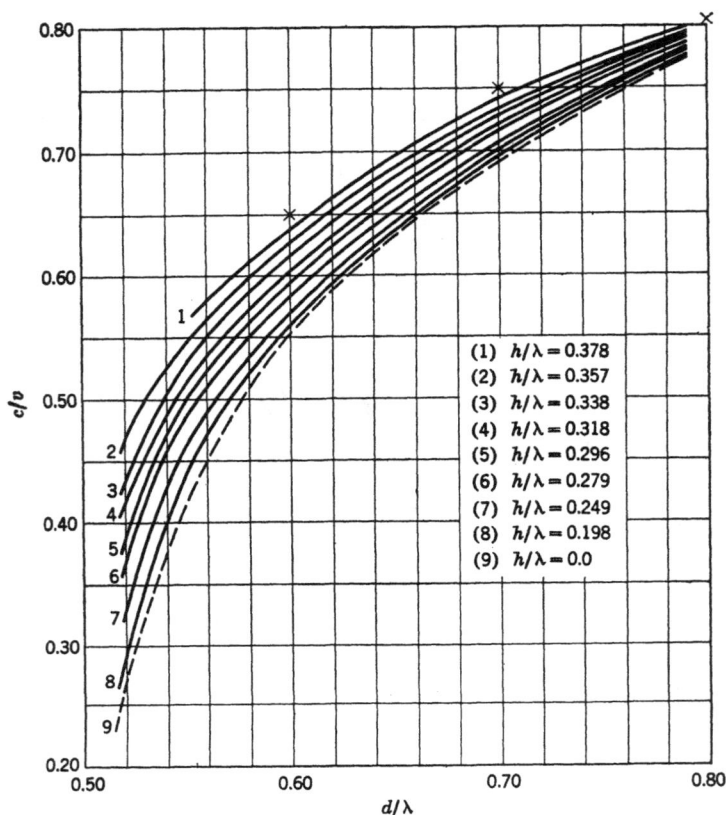

Figure 5-6.   Measured c/v for holey RG 91/U waveguide. Points are obtained from Eq. (5-14) for $h = 0.378\lambda$. Spacing $s = 0.437\lambda$. (After Hines and Upson, A Wide Aperature Tapered-Depth Scanning Antenna, *Ohio State Univ. Res. Found. Rept.*, 667–7, Columbus, December, 1957.)

equation (5-12) should be solved exactly. In all cases the use of the above equations is restricted to $s < \lambda/2$ and $h < w$. The accuracy of Eqs. (5-13) and (5-14) is indicated by the points in Figs. 5-5 and 5-6 where experimental data of Hines and Upson[1] are presented.

The data in Fig. 5-5 show that extremely low rates of attenuation can be obtained for the holey waveguide. The data in Fig. 5-5 are presented in decibels per wavelength. The conversion to nepers per wavelength may be made by noting that decibels per wavelength = 8.686 × nepers per wavelength.

[1] Hines and Upson, *op. cit.*

C. TE_{11} Slotted Cylinder

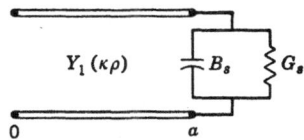

The TE_{11} slotted cylindrical guide shown in Fig. 5-7 was considered in Sec. 4-8 to illustrate a transverse resonance method of determining propagation constant, and it was also considered in Sec. 4-7 to illustrate a variational method. A comparison of some measured data and theoretical results by the variational method has been presented in Fig. 4-8. In this section we shall present approximate equations obtained by a perturbation solution of the transverse resonance equation.

Transverse resonance gives (see Sec. 4-8)

$$j \frac{J_1(\kappa a)}{J_1'(\kappa a)} + \mathcal{G} + j\mathcal{B} = 0 \tag{5-18}$$

where for convenience the superscript h has been dropped and $\mathcal{G}$ and $\mathcal{B}$ are G_s and B_s, respectively, normalized with respect to $\kappa/\omega\mu a$. Solution by a perturbation about the closed guide yields

$$\alpha \approx \frac{(\kappa_0 a)^3 \lambda_{g0}}{2\pi a^2[(\kappa_0 a)^2 - 1]} \frac{\mathcal{G}}{\mathcal{G}^2 + \mathcal{B}^2} \tag{5-19}$$

$$\frac{c}{v} \approx \frac{\lambda}{\lambda_{g0}} \left\{ 1 - \frac{(\kappa_0 a)^3 \lambda_{g0}^2}{4\pi^2 a^2[(\kappa_0 a)^2 - 1]} \frac{\mathcal{B}}{\mathcal{G}^2 + \mathcal{B}^2} \right\} \tag{5-20}$$

where

$$\lambda_{g0} = \frac{2\pi a}{[\epsilon_r(ka)^2 - (\kappa_0 a)^2]^{1/2}} \tag{5-21}$$

$$\mathcal{G} \approx \frac{1}{\pi\kappa_0 a} \sum_n \frac{\epsilon_n}{|H_n^{(2)'}\{[(\kappa_0 a)^2 - (ka)^2(\epsilon_r - 1)]^{1/2}\}|^2} \left[\frac{J_0(n\phi_0)}{J_0(\phi_0)} \right]^2 \tag{5-22}$$

$$\mathcal{B} \approx \frac{2[(\kappa_0 a)^2 - (ka)^2(\epsilon_r - 1)]}{\kappa_0 a} \ln \frac{2}{\phi_0} \tag{5-23}$$

and $\kappa_0 a = 1.841$. The quantity ϵ_n is the Neumann nomber.

The above equations are found to give results in close agreement with the results of the variational method shown in Fig. 4-8.

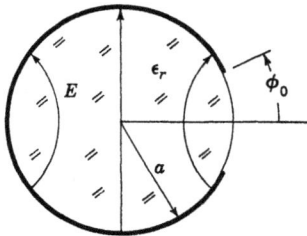

TE$_{11}$ slotted cylinder

Equivalent transverse network

Figure 5-7.   TE$_{11}$ slotted cylinder.

D. Yagi-Uda Array

An array of short, parallel rods or pins may be excited at one end of the structure to form an endfire traveling wave antenna. Such a structure is commonly referred to as a Yagi array, although Yagi-Uda array is more appropriate. This type of array was originally described by Professor S. Uda[1] in Japanese. The first English version was that of H. Yagi.[2] Hence more publicity has been given the latter. The phase velocity of the traveling wave is a function of the geometry of the parallel rods, and it can be determined either experimentally or analytically. The polarization of the radiation pattern is that of an electric dipole transverse to the axis of the antenna.

Consider the array of perfectly conducting rods in Fig. 5-8. We may think of each rod as a short-circuited, center-fed dipole. The elements of the array are parasitically excited, and the terminal voltage of each shorted dipole element will be zero. A superposition of the voltage produced by the feed plus the voltages induced by the currents in all the elements gives

$$V_0 = V_0(F) + \sum_n I_n Z_{0n} = 0 \qquad (5\text{-}24)$$

where V_0 is the terminal voltage at the zeroth element, $V_0(F)$ is the direct contribution of the feed, I_n is the terminal current of the nth element, and Z_{0n} is the mutual impedance between elements 0 and n.

Let us assume that the array is very long, that the zeroth element is near the center of the array, and that the geometry is such that a slow wave $(c/v > 1)$ can be supported by the structure. At a sufficiently large distance from the source the guided-wave field will be much larger than

---

[1] S. Uda, Wireless Beam of Short Electric Waves, *J. Inst. Elec. Engrs.* (*Japan*), No. 452, pp. 273–282, March, 1926; No. 472, pp. 1209–1219, November, 1927 (in Japanese).

[2] H. Yagi, Beam Transmission of Ultra Short Waves, *Proc. IRE*, **16**:715–741, June, 1928.

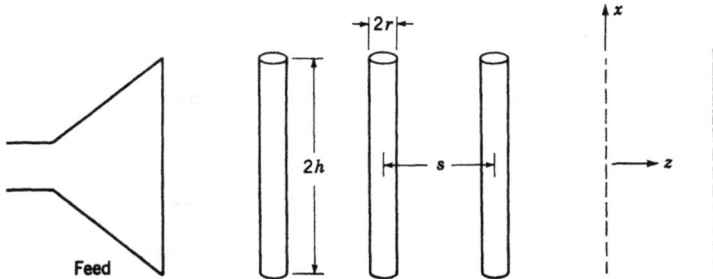

Figure 5-8.   Uniform array of parasitic dipoles.

the source field; hence the latter can be neglected in Eq. (5-24), giving

$$V_0 = \sum_n Z_{0n}I_n = 0 \tag{5-25}$$

For a uniform structure the guided wave will be of the form

$$I = I_0 e^{-\gamma z} \tag{5-26}$$

but $z = ns$; therefore

$$I_n = I_0 e^{-n\gamma s} \tag{5-27}$$

where $\gamma = \alpha + j\beta$ is the usual complex propagation constant. Substituting Eq. (5-27) into (5-25) and assuming a lossless structure of infinite extent give

$$\sum_{-\infty}^{\infty} Z_{0n}e^{-jn\beta s} = 0 \tag{5-28}$$

Equation (5-28) is the fundamental equation from which β can be determined, but computations are formidable. Equation (5-28) has been solved by numerical methods by Serracchioli and Levis[1] using formulas for self- and mutual impedances by Brown and King.[2] Calculated curves are given in Fig. 5-9 and some measured results are shown for comparison.[3]

In addition to the TE structures described above, dielectric rods and dielectric-coated metal rods also may be used as TE radiators. Expressions for the propagation constants in these two cases follow easily from the analysis of TM modes; hence the results for TE dielectric rods are given with the results for TM structures in the next section.

5-3. TM LINE SOURCE RADIATORS

The TE structures described in the previous section are characterized by a zero longitudinal electric field ($E_z = 0$), and the transverse electric field in the aperture is equivalent to a longitudinal (z-directed) magnetic current. In contrast, a TM line source is characterized by zero longitudinal magnetic field ($H_z = 0$) and is equivalent to a longitudinal electric current. TM excitation, however, does not always ensure the radiation characteristics

[1] F. Serracchioli and C. Levis, The Calculated Phase Velocity of Long End-fire Uniform Dipole Arrays, *IRE Trans. Antennas Propagation*, **AP-7**:S424–S434, December, 1959 (special supplement).

[2] G. Brown and R. King, High Frequency Models in Antenna Investigations, *Proc. IRE*, **22**:957–980, April, 1934.

[3] H. W. Ehrenspeck and H. Poehler, A New Method for Obtaining Maximum Gain from Yagi Antennas, *IRE Trans. Antennas Propagation*, **AP-7**(4):379–386, October, 1959.

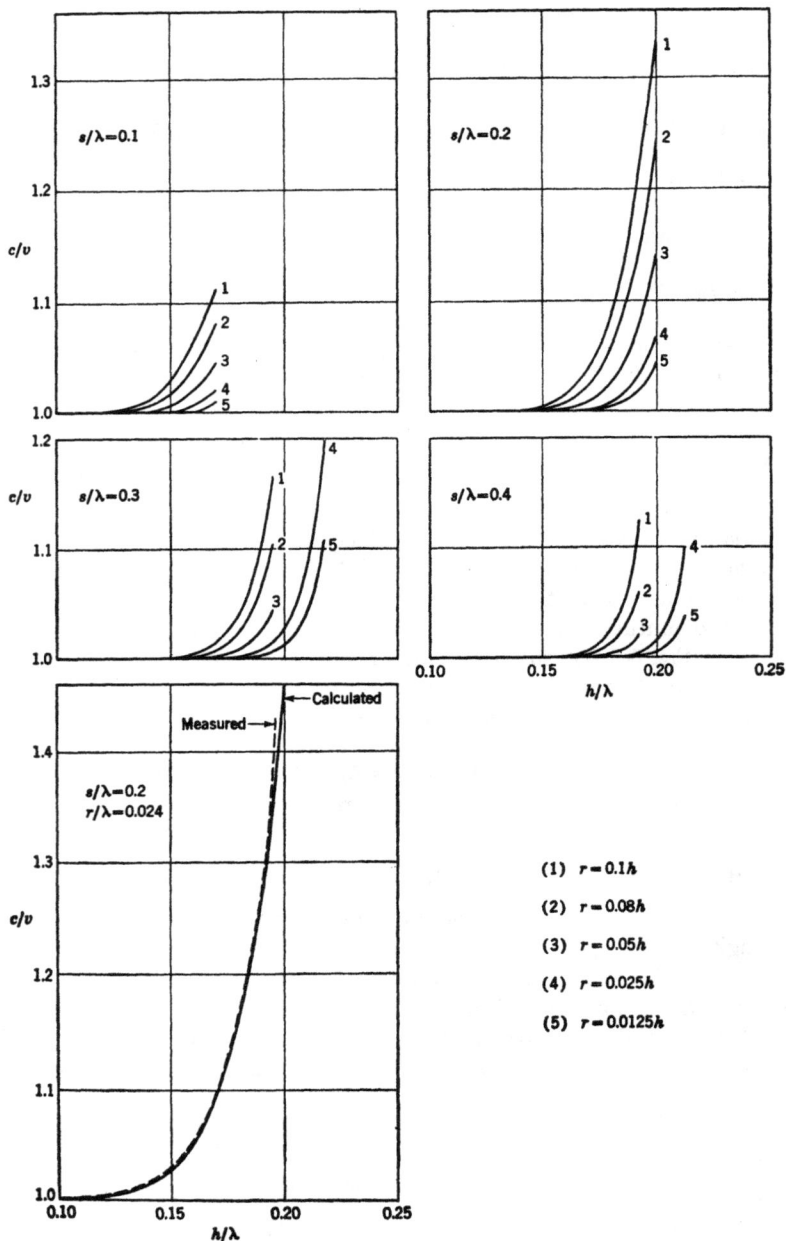

Figure. 5-9. Velocity ratio c/v for long Yagi-Uda array. (After Serracchioli and Levis, The Calculated Phase Velocity of Long End-fire Uniform Dipole Arrays, *IRE Trans. Antennas Propagation*, **AP-7**: S424–S434, December, 1959, Special Supplement.)

of a TM source. The pattern characteristics of a longitudinal electric-current element are exhibited by such structures as TM air-filled slotted guides, dielectric cylinders, and dielectric-coated metal cylinders. However, TM-excited dielectric-filled slotted guides for $\epsilon_r > 2$ are found to have an aperture distribution equivalent to transverse magnetic-current elements[1] (i.e., the electric field is essentially longitudinal). This is interpreted as hybrid mode operation,[1] but such structures will be included in this section and analyzed as TM dielectric-filled slotted guides.

A. TM Slotted Rectangular Guide

Slotted waveguide radiators with TM excitation have not been studied as extensively as has the case of TE excitation; however, significant work has been done on the cross section shown in Fig. 5-10.[1,2]

Useful results may be obtained from the transverse resonance method. Application of transverse resonance for the TM_{11} mode gives

$$j \cot \kappa d - \mathcal{G}(\kappa) + j\mathcal{B}(\kappa) = 0 \tag{5-29}$$

where the aperture admittance has been normalized with respect to the characteristic admittance of the line. For $W \ll w$ an approximate solution of Eq. (5-29) based on a perturbation about $\kappa_c = \pi/d$ for the closed guide gives

$$\alpha \approx \frac{\lambda_{gt}}{2d^2} \frac{\mathcal{G}}{\mathcal{G}^2 + \mathcal{B}^2} \tag{5-30}$$

$$\frac{c}{v} \approx \frac{\lambda}{\lambda_{g0}} \left[1 - \frac{\lambda_{g0}^2}{4\pi d^2} \frac{\mathcal{B}}{\mathcal{G}^2 + \mathcal{B}^2} \right] \tag{5-31}$$

[1] Hines, Rumsey, and Walter, *op. cit.*
[2] Goldstone and Oliner, *op. cit.*

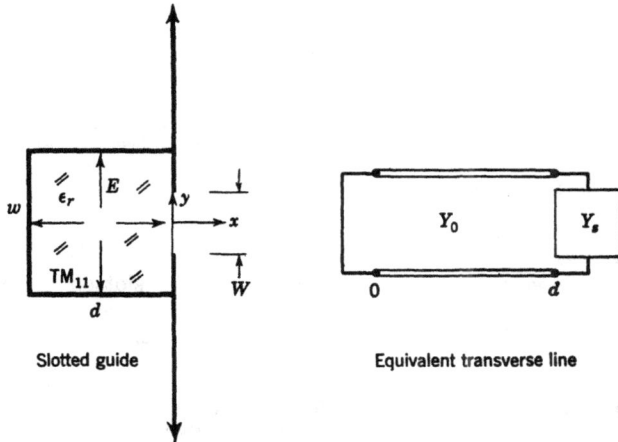

Figure 5-10.   Slotted rectangular guide with TM_{11} excitation.

where[1]

$$\lambda_{g0} = \frac{\lambda}{[\epsilon_r - (\lambda/2d)^2 - (\lambda/2w)^2]^{\frac{1}{2}}} \tag{5-32}$$

$$G \approx \frac{0.570d}{W} \left\{ \frac{\pi^2 w[1 - (W/w)^2]^2}{16W \cos^2 (\pi W/2w)} \right\} \tag{5-33}$$

$$\mathcal{B} \approx \frac{0.312d}{W} \left\{ \frac{\pi^2 w[1 - (W/w)^2]^2}{16W \cos^2 (\pi W/2w)} \right\} + \frac{d}{w} \cot^2 \frac{\pi W}{2w} \tag{5-34}$$

As W approaches w, the effect of inductive loading by the slot is overcome by an increase in the effective size of the guide. In this case good results may be obtained by first formulating a resonance problem involving only the susceptance to obtain an improvement in κ_0. Thus

$$\cot \kappa_0 d = -\mathcal{B}(\kappa_0) \tag{5-35}$$

A solution of Eq. (5-35) for a perturbation about the closed guide gives

$$\kappa_0 \approx \frac{\pi}{d} - \frac{1}{d} \frac{1}{\mathcal{B}(\pi/d)} \tag{5-36}$$

Following the method in Sec. 4-8 a perturbation solution of Eq. (5-29) about κ_0 in Eq. (5-36) gives

$$\alpha \approx \frac{\lambda_{g0}\kappa_0}{2\pi d \sec^2 \kappa_0 d} \frac{G}{G^2 + \mathcal{B}^2} \tag{5-37}$$

$$\frac{c}{v} \approx \frac{\lambda}{\lambda_{g0}} \left(1 + \frac{\kappa_0 \lambda_{g0}^2}{4\pi^2 d} \frac{\mathcal{X} + \tan \kappa_0 d}{\sec^2 \kappa_0 d} \right) \tag{5-38}$$

where

$$\lambda_{g0} = \frac{\lambda}{[\epsilon_r - (\lambda/2w)^2 - (\lambda\kappa_0/2\pi)^2]^{\frac{1}{2}}} \tag{5-39}$$

$$G \approx \frac{0.570\pi}{\kappa_0 W} \left\{ \frac{\pi^2 w[1 - (W/w)^2]^2}{16W \cos^2 (\pi W/2w)} \right\} \tag{5-40}$$

$$\mathcal{B} \approx \frac{0.312\pi}{\kappa_0 W} \left\{ \frac{\pi^2 w[1 - (W/w)^2]^2}{16W \cos^2 (\pi W/2w)} \right\} + \frac{\pi}{\kappa_0 w} \cot^2 \frac{\pi W}{2w} \tag{5-41}$$

$$\mathcal{X} = \frac{\mathcal{B}}{G^2 + \mathcal{B}^2} \tag{5-42}$$

and κ_0 is given by Eq. (5-36) using $\mathcal{B}$ from Eq. (5-34).

Measured data for an air-filled TM_{11} slotted square guide are shown in Figs. 5-11 and 5-12. Theoretical results from Eqs. (5-37) and (5-38) for $W/w = 0.56$ are given by the broken curves.

[1] The quantities G and $\mathcal{B}$ are the same as those of Goldstone and Oliner, *op. cit.* They were obtained by modifying the results of N. Marcuvitz, "Waveguide Handbook," pp. 184 and 191, McGraw-Hill, New York, 1951. Marcuvitz gives the impedance of an aperture where $W = w$. The modification by Goldstone and Oliner takes into account $W < w$.

Figure 5-11. Attenuation constant α for an air-filled, slotted square guide with TM_{11} excitation. Solid lines represent measured data. Broken lines are theoretical.

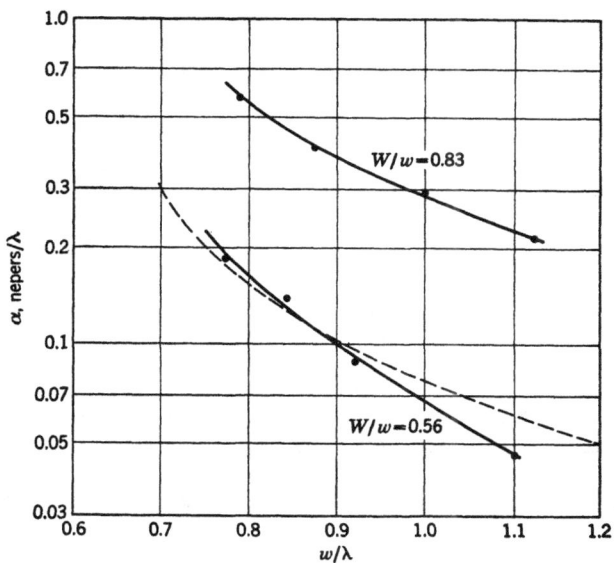

Figure 5-12. Velocity ratio c/v for an air-filled, slotted square guide with TM_{11} excitation. Solid lines represent measured data. Broken lines are theoretical.

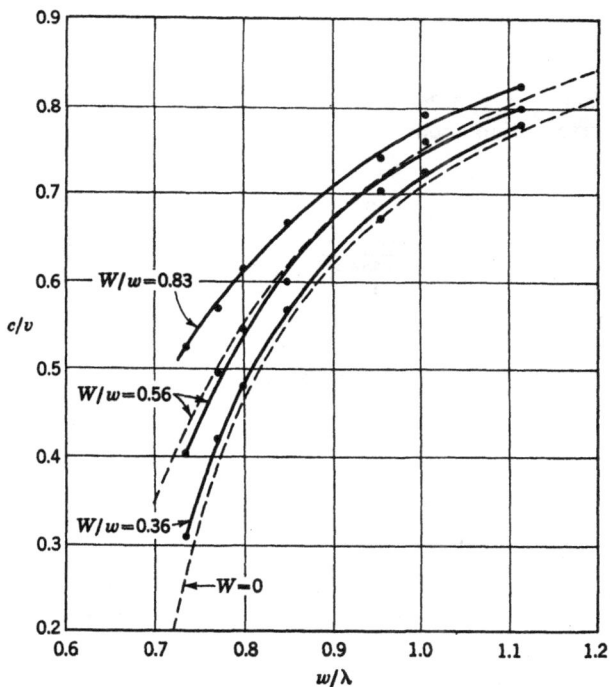

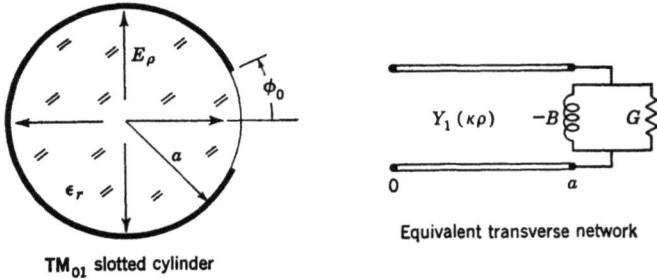

TM$_{01}$ slotted cylinder

Equivalent transverse network

Figure 5-13. TM$_{01}$ slotted cylinder.

B. TM Slotted Cylinder

The cross section of a TM$_{01}$ slotted cylinder is shown in Fig. 5-13. This case has been studied rather extensively[1] and may be analyzed by the variational method in Sec. 4-7 or the transverse resonance method in Sec. 4-8. The transverse resonance method will be used here since it leads to relatively simple equations which give good results.

Transverse resonance gives for this case

$$\frac{jJ_0'(\kappa a)}{J_0(\kappa a)} + \mathcal{G} - j\mathcal{B} = 0 \qquad (5\text{-}43)$$

where $\mathcal{G}$ and $\mathcal{B}$ have been normalized with respect to $\omega\epsilon a/\kappa$. Solution by a perturbation about the closed guide yields

$$\alpha \approx \frac{\kappa_0 a \lambda_{g0}}{2\pi a^2}\frac{\mathcal{G}}{\mathcal{G}^2 + \mathcal{B}^2} \qquad (5\text{-}44)$$

$$\frac{c}{v} \approx \frac{\lambda}{\lambda_{g0}}\left(1 + \frac{\kappa_0 a \lambda_{g0}^2}{4\pi^2 a^2}\frac{\mathcal{B}}{\mathcal{G}^2 + \mathcal{B}^2}\right) \qquad (5\text{-}45)$$

where[2]

$$\lambda_{g0} = \frac{2\pi a}{[\epsilon_r(ka)^2 - (\kappa_0 a)^2]^{1/2}} \qquad (5\text{-}46)$$

$$\mathcal{G} \approx \frac{8\kappa_0 a}{\epsilon_r \pi[(\kappa_0 a)^2 - (ka)^2(\epsilon_r - 1)]}$$

$$\times \sum_{n=0}^{\infty}\frac{\epsilon_n}{|H_n^{(2)}\{[(\kappa_0 a)^2 - (ka)^2(\epsilon_r - 1)]^{1/2}\}|^2}\left[\frac{J_1(n\phi_0)}{n\phi_0}\right]^2 \qquad (5\text{-}47)$$

---

[1] Hines, Rumsey, and Walter, *op. cit.*; R. F. Harrington, Propagation along a Slotted Cylinder, *J. Appl. Phys.*, **24**(11), November, 1953; L. O. Goldstone and A. A. Oliner, Leaky Wave Antennas II: Circular Waveguides, *IRE Trans. Antennas Propagation*, **AP-9**:280–290, May, 1961.

[2] The functions $\mathcal{G}$ and $\mathcal{B}$ are taken from Goldstone and Oliner, *op. cit.*, 1961, and modified to take into account dielectric loading.

with

$$\epsilon_n = 1 \quad n = 0$$
$$\epsilon_n = 2 \quad n > 0$$
$$\kappa_0 a = 2.405$$
$$k = \frac{2\pi}{\lambda}$$

and

$$\mathfrak{G} \approx \frac{8(ka)(\kappa_0 a)}{\epsilon_r \phi_0^2 [(\kappa_0 a)^2 - (ka)^2 (\epsilon_r - 1)]} \tag{5-48}$$

Equations (5-44) and (5-45) give results in close agreement with the results of the variational method shown in Fig. 4-9 except in the region near cutoff, where Eq. (5-43) should be solved exactly.

C. TM Dielectric Cylinder

Dielectric cylinders are well known as transmission lines, and the earliest work on this subject appears to be that of Hondros and Debye[1]. Not much interest was shown in dielectric lines until just before and during World War II when Bell Laboratories did considerable work on dielectric rods as transmission lines[2] and as "polyrod" antennas.[3]

In general a dielectric rod (solid or hollow) may support a TM, TE, or a hybrid (TM + TE) mode. In all cases a finite length of dielectric rod transmission line may be used as an antenna (usually endfire) wherein the radiation characteristics can be controlled to a certain extent by varying the cross section and/or dielectric constant along the structure. The usual polyrod antenna utilizes a hybrid mode which will be discussed in the next section. The simplest dielectric rod modes from the standpoint of analysis are the circularly symmetric TM and TE modes for a solid circular rod.

[1] Hondros and Debye, Electromagnetische Wellen an dielektrischen Drahten, *Ann. Physik*, **32**:465–476, 1910.

[2] J. R. Carson, S. P. Mead, and S. A. Schelkunoff, Hyper-Frequency Wave Guides—Mathematical Theory, *Bell System Tech. J.*, **15**:310–333, 1936.

[3] G. E. Mueller and W. A. Tyrrell, Polyrod Antennas, *Bell System Tech. J.*, **26**:837–851, 1947.

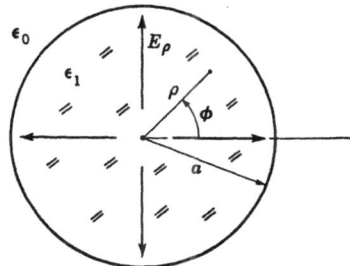

Figure 5-14. Circular dielectric rod with radially symmetric TM mode.

The TM mode will be discussed here, and results will be given for the corresponding TE case.

The TM mode is characterized by zero longitudinal magnetic field. For the cross section shown in Fig. 5-14 the fields can be determined from the electric field E_z [see Eq. (4-49)] as

$$E_\rho = -\frac{\gamma}{\kappa^2}\frac{\partial E_z}{\partial \rho} \tag{5-49}$$

$$E_\phi = \frac{-\gamma}{\kappa^2 \rho}\frac{\partial E_z}{\partial \phi} \tag{5-50}$$

$$H_\rho = \frac{j\omega\epsilon}{\kappa^2 \rho}\frac{\partial E_z}{\partial \phi} \tag{5-51}$$

$$H_\phi = \frac{-j\omega\epsilon}{\kappa^2}\frac{\partial E_z}{\partial \rho} \tag{5-52}$$

where E_z is a solution of the two-dimensional wave equation (4-47) and is given by

$$E_z = A_n e^{jn\phi} J_n(\kappa_1\rho) \qquad \rho \lessgtr a \tag{5-53}$$
$$E_z = B_n e^{jn\phi} H_n{}^{(2)}(\kappa_0\rho) \qquad \rho \gtrless a \tag{5-54}$$

The quantities A_n and B_n are unknown constants, and E_z is understood to vary as $e^{j\omega t - \gamma z}$.

Applying the boundary conditions to tangential **E** and **H** at $\rho = a$ gives three equations in the two unknowns A_n and B_n except when $n = 0$. For $n = 0$, E_ϕ is zero and the remaining two equations from E_z and H_ϕ serve to eliminate A_0 and B_0. Thus the only TM modes that can exist on the dielectric cylinder are those with radial symmetry.

Matching E_z at $\rho = a$ for this case gives

$$A_0 J_0(\kappa_1 a) = B_0 \epsilon_0 H_0{}^{(2)}(\kappa_0 a) \tag{5-55}$$

and matching H_ϕ at $\rho = a$ gives

$$\frac{A_0 \epsilon_1}{\kappa_1} J_1(\kappa_1 a) = \frac{B_0 \epsilon_0}{\kappa_0} H_1{}^{(2)}(\kappa_0 a) \tag{5-56}$$

Combining Eqs. (5-55) and (5-56) gives

$$\frac{\kappa_1 J_0(\kappa_1 a)}{\epsilon_1 J_1(\kappa_1 a)} = \frac{\kappa_0 H_0{}^{(2)}(\kappa_0 a)}{\epsilon_0 H_1{}^{(2)}(\kappa_0 a)} \tag{5-57}$$

Alternatively one could use transverse resonance, which gives

$$\overset{\leftarrow}{Z}_1(\kappa_1 a) + \vec{Z}_0(\kappa_0 a) = 0 \tag{5-58}$$

where

$$\overleftarrow{Z}_1(\kappa_1 a) = \frac{\kappa_1}{j\omega\epsilon_1} \frac{J_0(\kappa_1 a)}{J_1(\kappa_1 a)} \tag{5-59}$$

and

$$\overrightarrow{Z}_0(\kappa_0 a) = \frac{-\kappa_0}{j\omega\epsilon_0} \frac{H_0^{(2)}(\kappa_0 a)}{H_1^{(2)}(\kappa_0 a)} \tag{5-60}$$

Equation (5-58) gives the same result as (5-57) as it should, since matching tangential E and H at the boundary is the same as matching impedance.

The radially symmetric TE case may be solved as above, or it may be noted that it is the dual of the radially symmetric TM case where the dual of Eq. (5-57) is given by

$$\frac{\kappa_1 J_0(\kappa_1 a)}{\mu_1 J_1(\kappa_1 a)} = \frac{\kappa_0 H_0^{(2)}(\kappa_0 a)}{\mu_0 H_1^{(2)}(\kappa_0 a)} \tag{5-61}$$

The quantities κ_0 and κ_1 in Eqs. (5-57) and (5-61) are interrelated by

$$\kappa_1^2 - \gamma^2 = k^2\epsilon_r \tag{5-62}$$

and

$$\kappa_0^2 - \gamma^2 = k^2 \tag{5-63}$$

where it is assumed that $\mu_1 = \mu_0$. Combining Eqs. (5-62) and (5-63) gives

$$\kappa_1^2 - \kappa_0^2 = k^2(\epsilon_r - 1) \tag{5-64}$$

Equations (5-57) and (5-64) for the TM case [or Eqs. (5-61) and (5-64) for the TE case] can be solved for either κ_1 or κ_0. The propagation constant is then determined from Eq. (5-62) or (5-63).

Solving for κ_1 (or κ_0) is complicated by the transcendental equation [Eq. (5-57) or (5-61)]. The solution is simplified when $\epsilon_1 > \epsilon_0$ and there are no dielectric losses. In this case κ_1 is real and κ_0 is imaginary. The TM or TE wave will be guided by the rod with the fields decreasing exponentially in the radial direction but having no attenuation in the axial direction. Cutoff occurs when $\kappa_0 = 0$. This corresponds to $c/v = 1$.

Lossy Dielectric Cylinder

If the dielectric cylinder in Fig. 5-14 has a conductivity σ_1, E_z for radially symmetric TM modes is again given by Eqs. (5-53) and (5-54) with $n = 0$. Equation (5-52) may be generalized to give

$$H_\phi = - \frac{(\sigma + j\omega\epsilon)}{\kappa} \frac{\partial E_z}{\partial(\kappa\rho)} \tag{5-65}$$

Matching E_z at $\rho = a$ gives Eq. (5-55). Matching H_ϕ at $\rho = a$ gives

$$\frac{A_0(\sigma_1 + j\omega\epsilon_1)}{j\omega\kappa_1} J_1(\kappa_1 a) = \frac{B_0\epsilon_0}{\kappa_0} H_1^{(2)}(\kappa_0 a) \tag{5-66}$$

Combining Eqs. (5-55) and (5-66) gives

$$\frac{\kappa_0 H_0^{(2)}(\kappa_0 a)}{\epsilon_0 H_1^{(2)}(\kappa_0 a)} = \frac{j\omega\kappa_1 J_0(\kappa_1 a)}{(\sigma_1 + j\omega\epsilon_1)J_1(\kappa_1 a)} \tag{5-67}$$

Conducting Cylinder (Sommerfeld Wave)

For $\omega\epsilon_1 \ll \sigma_1$, Eq. (5-67) reduces to

$$\frac{\kappa_0 H_0^{(2)}(\kappa_0 a)}{\epsilon_0 H_1^{(2)}(\kappa_0 a)} = \frac{j\omega\kappa_1 J_0(\kappa_1 a)}{\sigma_1 J_1(\kappa_1 a)} \tag{5-68}$$

Equation (5-68) gives the propagation constant for a radially symmetric TM wave along an imperfect conductor. This case was first considered by Sommerfeld[1] and later by Goubau.[2]

Coaxial Dielectric Cylinder

The coaxial dielectric cylinder in Fig. 5-15, although more complicated than the solid dielectric cylinder, also can be analyzed by applying Maxwell's equations and the appropriate boundary conditions.[3] For radially symmetric TM modes Eqs. (5-49) to (5-52) apply with

$$E_{z1} = \frac{Av^2 J_0(v\rho)e^{-j\beta z}}{j\omega\epsilon_0\epsilon_r} \qquad \text{region 1} \tag{5-69}$$

$$E_{z2} = \frac{-[B_1 I_0(u\rho) + B_2 K_0(u\rho)]u^2 e^{-j\beta z}}{j\omega\epsilon_0} \qquad \text{region 2} \tag{5-70}$$

$$E_{z3} = \frac{[C_1 J_0(v\rho) + C_2 N_0(v\rho)]v^2 e^{-j\beta z}}{j\omega\epsilon_0\epsilon_r} \qquad \text{region 3} \tag{5-71}$$

$$E_{z4} = \frac{-DK_0(u\rho)u^2 e^{-j\beta z}}{j\omega\epsilon_0} \qquad \text{region 4} \tag{5-72}$$

where J_0 and N_0 are zero-order Bessel functions of the first and second kinds, respectively, and I_0 and K_0 are zero-order modified Bessel functions.[4] The modified Bessel functions are used to represent evanescent waves. In terms of the conventional Bessel and Hankel functions the modified

[1] A. Sommerfeld, Fortpflanzung Electrodynamischer Wellen an einem zylindrischen Leiter, *Ann. Physik Chem.*, **67**:233, 1899.
[2] G. Goubau, Surface Waves and Their Application to Transmission Lines, *J. Appl. Phys.*, **21**:1119, 1950.
[3] The analysis is given in complete detail by R. Barnett, "An All Dielectric Coaxial Waveguide and Antenna," Ph.D. Dissertation, The Ohio State University, Columbus, 1963.
[4] For a discussion of Bessel functions see, for example, R. F. Harrington, "Time-Harmonic Electromagnetic Fields," chap. 5, McGraw-Hill, New York, 1961.

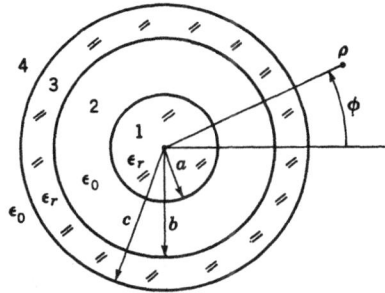

Figure 5-15. Cross section of dielectric coaxial waveguide.

Bessel functions are defined by

$$I_n(\kappa\rho) = j^n J_n(-j\kappa\rho) \tag{5-73}$$

$$K_n(\kappa\rho) = \frac{\pi}{2}(-j)^{n+1}H_n^{(2)}(-j\kappa\rho) \tag{5-74}$$

where n is the order of the function and $H_n^{(2)}$ is the Hankel function of the second kind.

The magnetic field for the radially symmetric TM modes is given by

$$H_{\phi 1} = -AJ_0'(v\rho)ve^{-j\beta z} \tag{5-75}$$
$$H_{\phi 2} = -[B_1 I_0'(u\rho) + B_2 K_0'(u\rho)]ue^{-j\beta z} \tag{5-76}$$
$$H_{\phi 3} = -[C_1 J_0'(v\rho) + C_2 N_0'(v\rho)]ve^{-j\beta z} \tag{5-77}$$
$$H_{\phi 4} = -DK_0'(u\rho)ue^{-j\beta z} \tag{5-78}$$

where the prime denotes differentiation with respect to the argument.

The quantities u and v in the above equations are related to the usual phase constants by

$$u^2 = -\kappa_2^2 = -\kappa_4^2 = \beta^2 - k^2 \tag{5-79}$$
$$v^2 = \kappa_1^2 = \kappa_3^2 = \epsilon_r k^2 - \beta^2 \tag{5-80}$$

Requiring E_z and H_ϕ to be continuous at the boundaries $\rho = a$, $\rho = b$, and $\rho = c$ in Fig. 5-15 results in six equations with the six unknown coefficients A, B_1, B_2, C_1, C_2, and D. In order for a solution to exist, the determinant of these coefficients must be zero. The resulting determinant is

$$\begin{vmatrix} \dfrac{v^2}{\epsilon_r}J_0(va) & u^2 I_0(ua) & u^2 K_0(ua) & 0 & 0 & 0 \\ vJ_0'(va) & -uI_0'(ua) & -uK_0'(ua) & 0 & 0 & 0 \\ 0 & u^2 I_0(ub) & u^2 K_0(ub) & \dfrac{v^2}{\epsilon_r}J_0(vb) & \dfrac{v^2}{\epsilon_r}N_0(vb) & 0 \\ 0 & uI_0'(ub) & uK_0'(ub) & -vJ_0'(vb) & -vN_0'(vb) & 0 \\ 0 & 0 & 0 & \dfrac{v^2}{\epsilon_r}J_0(vc) & \dfrac{v^2}{\epsilon_r}N_0(vc) & u^2 K_0(uc) \\ 0 & 0 & 0 & vJ_0'(vc) & vN_0'(vc) & -uK_0'(uc) \end{vmatrix} = 0 \tag{5-81}$$

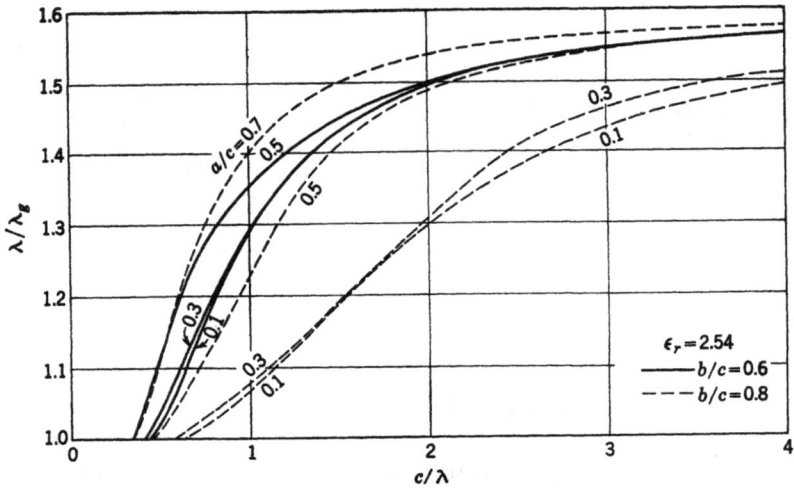

Figure 5-16. Wavelength ratio λ/λ_g for the TM$_{01}$ mode on the dielectric. The wavelength ratio λ/λ_g is equal to the velocity ratio c/v.

For radially symmetric TE modes the determinant is

$$\begin{vmatrix} v^2 J_0(va) & u^2 I_0(ua) & u^2 K_0(ua) & 0 & 0 & 0 \\ v J'_0(va) & -u I'_0(ua) & -u K'_0(ua) & 0 & 0 & 0 \\ 0 & u^2 I_0(ub) & u^2 K_0(ub) & v^2 J_0(vb) & v^2 N_0(vb) & 0 \\ 0 & u I'_0(ub) & u K'_0(ub) & -v J'_0(vb) & -v N'_0(vb) & 0 \\ 0 & 0 & 0 & v^2 J_0(vc) & v^2 N_0(vc) & u^2 K_0(uc) \\ 0 & 0 & 0 & v J'_0(vc) & v N'_0(vc) & -u K'_0(uc) \end{vmatrix} = 0$$

$$(5\text{-}82)$$

The oscillatory nature of the Bessel function means that higher-order radially symmetric modes can exist. Some results for the dominant TM mode (designated TM$_{01}$) are given in Fig. 5-16 and for the dominant TE mode in Fig. 5-17. The quantity λ_g is the wavelength of the waveguide mode.

D. TM Dielectric-coated Metal Rod

A TM dielectric-coated perfectly conducting cylinder (see Fig. 5-18) was first considered by Harms.[1] Later, Goubau[2] studied the TM coated cylinder and showed it to be quite practical as a single-conductor transmission line.

For use as a low-loss transmission line the dielectric constant of the coating material must be greater than that of free space and the TM wave must be efficiently launched at the transmitting end and efficiently

[1] F. Harms, Electromagnetische Wellen an einem Draht mit Isolierender Zylindischer Hülle, *Ann. Physik*, **23**:44, 1907.

[2] G. Goubau, *op. cit.*, and Single Conductor Surface-wave Transmission Lines, *Proc. IRE*, **39**:619, June, 1951.

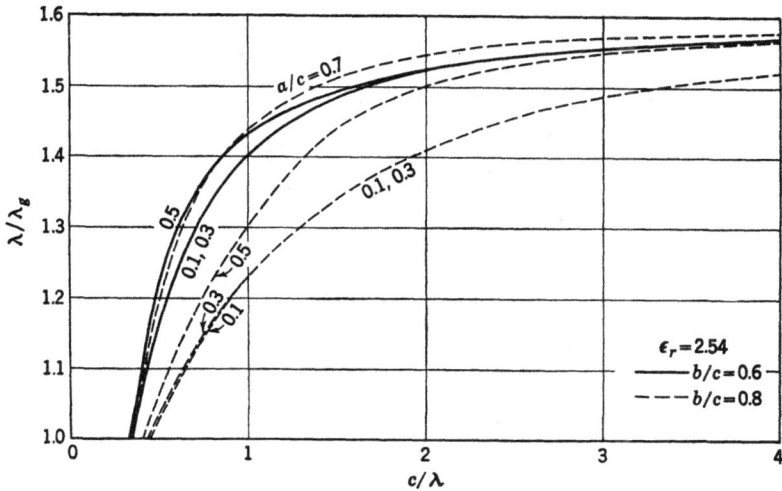

Figure 5-17.   Wavelength ratio λ/λ_g for the TE_{01} mode on the dielectric coaxial waveguide in Fig. 5-15. The wavelength ratio λ/λ_g is equal to the velocity ratio c/v.

recovered at the receiving end. When a section of the line is used as an antenna, the receiving end is usually abruptly terminated. If end effects are neglected, the phase and amplitude along the structure when used as an antenna can be controlled by the cross-sectional dimensions and the dielectric material. The radiation characteristics of the TM coated cylinder are similar to those of a conventional wire antenna; however, the dielectric coating permits greater pattern control.

For TM modes, $H_z = 0$ and the fields can be determined from the z component of electric field by means of Eqs. (5-49) to (5-52). For radially symmetric modes

$$E_z = C_1 J_0(\kappa_1\rho) + C_2 N_0(\kappa_1\rho) \qquad a \lesssim \rho \lesssim b \tag{5-83}$$

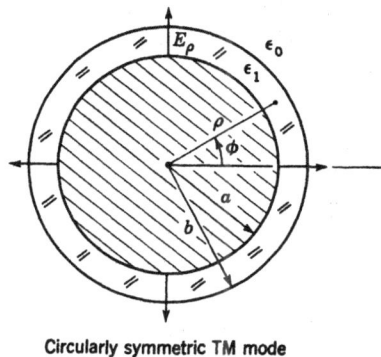

Figure 5-18.   Dielectric-coated metal cylinder.

Circularly symmetric TM mode

and

$$E_z = C_0 H_0^{(2)}(\kappa_0 \rho) \qquad \rho \gtrless b \tag{5-84}$$

where it is understood that all fields vary as $e^{j\omega t - \gamma z}$.

The propagation constant γ may be obtained by matching tangential **E** and **H** (E_z and H_ϕ) at the boundary $\rho = b$ and requiring E_z to be zero at $\rho = a$. The latter condition eliminates one of the unknowns in Eq. (5-83), giving

$$E_z = \frac{C_1}{N_0(\kappa_1 a)} [N_0(\kappa_1 a) J_0(\kappa_1 \rho) - J_0(\kappa_1 a) N_0(\kappa_1 \rho)] \tag{5-85}$$

From Eq. (5-52)

$$H_\phi = \frac{j\omega \epsilon_1 C_1}{\kappa_1 N_0(\kappa_1 a)} [N_0(\kappa_1 a) J_1(\kappa_1 \rho) - J_0(\kappa_1 a) N_1(\kappa_1 \rho)] \qquad a \lessgtr \rho \lessgtr b \tag{5-86}$$

$$H_\phi = \frac{j\omega \epsilon_0 C_0}{\kappa_0} H_1^{(2)}(\kappa_0 \rho) \qquad \rho \gtrless b \tag{5-87}$$

Matching E_z at $\rho = b$ gives

$$\frac{C_1}{N_0(\kappa_1 a)} [N_0(\kappa_1 a) J_0(\kappa_1 b) - J_0(\kappa_1 a) N_0(\kappa_1 b)] = C_0 H_0^{(2)}(\kappa_0 b) \tag{5-88}$$

and matching H_ϕ at $\rho = b$ gives

$$\frac{\epsilon_1 C_1}{\kappa_1 N_0(\kappa_1 a)} [N_0(\kappa_1 a) J_1(\kappa_1 b) - J_0(\kappa_1 a) N_1(\kappa_1 b)] = \frac{\epsilon_0 C_0}{\kappa_0} H_1^{(2)}(\kappa_0 b) \tag{5-89}$$

Combining Eqs. (5-88) and (5-89) gives

$$\frac{\kappa_1}{\epsilon_1} \frac{N_0(\kappa_1 a) J_0(\kappa_1 b) - J_0(\kappa_1 a) N_0(\kappa_1 b)}{N_0(\kappa_1 a) J_1(\kappa_1 b) - J_0(\kappa_1 a) N_1(\kappa_1 b)} = \frac{\kappa_0}{\epsilon_0} \frac{H_0^{(2)}(\kappa_0 b)}{H_1^{(2)}(\kappa_0 b)} \tag{5-90}$$

The propagation constant γ may be found from Eqs. (5-62), (5-63), and (5-90).

Equation (5-90) also may be obtained by matching radial impedance at $\rho = b$ (transverse resonance). In this case

$$\overleftarrow{Z}_b = -\overrightarrow{Z}_b \tag{5-91}$$

where

$$\overleftarrow{Z}_b = \frac{\kappa_1}{j\omega \epsilon_1} \frac{N_0(\kappa_1 a) J_0(\kappa_1 b) - J_0(\kappa_1 a) N_0(\kappa_1 b)}{N_0(\kappa_1 a) J_1(\kappa_1 b) - J_0(\kappa_1 a) N_1(\kappa_1 b)} \tag{5-92}$$

and

$$\overrightarrow{Z}_b = \frac{-\kappa_0 H_0^{(2)}(\kappa_0 b)}{j\omega \epsilon_0 H_1^{(2)}(\kappa_0 b)} \tag{5-93}$$

Solution of Eq. (5-90) is generally quite difficult, especially if ϵ_1 is less than unity or is complex. In some instances a perturbation solution may be useful. Such a solution may be obtained by viewing the dielectric-coated conducting rod as a perturbation of a solid dielectric cylinder.

Let

$$\overleftrightarrow{Z} = \frac{\kappa_0}{\epsilon_0}\frac{H_0^{(2)}(\kappa_0 b)}{H_1^{(2)}(\kappa_0 b)} - \frac{\kappa_1}{\epsilon_1}\frac{N_0(\kappa_1 a)J_0(\kappa_1 b) - J_0(\kappa_1 a)N_0(\kappa_1 b)}{N_0(\kappa_1 a)J_1(\kappa_1 b) - J_0(\kappa_1 a)N_1(\kappa_1 b)} = 0 \qquad (5\text{-}94)$$

where

$$\kappa_1 = \kappa_{01} + \delta \qquad (5\text{-}95)$$

and κ_{01} is a solution to Eq. (5-94) for $a = 0$. Expanding $\overleftrightarrow{Z}$ about κ_{01} and keeping only the first two terms give

$$\overleftrightarrow{Z}(\kappa_1,\kappa_0) = \overleftrightarrow{Z}(\kappa_{01},\kappa_{00}) + \delta\frac{\partial\overleftrightarrow{Z}}{\partial\kappa_1}\bigg]_{\kappa_{01},\kappa_{00}} = 0 \qquad (5\text{-}96)$$

Solving for δ gives

$$\delta = \frac{-\overleftrightarrow{Z}(\kappa_{01},\kappa_{00})}{\partial\overleftrightarrow{Z}/\partial\kappa_1\Big]_{\kappa_{01},\kappa_{00}}} \qquad (5\text{-}97)$$

where κ_{01} and κ_{00} are solutions of

$$\frac{\kappa_{00}H_0^{(2)}(\kappa_{00}b)}{\epsilon_0 H_1^{(2)}(\kappa_{00}b)} - \frac{\kappa_{01}J_0(\kappa_{01}b)}{\epsilon_1 J_1(\kappa_{01}b)} = 0 \qquad (5\text{-}98)$$

If one attempts to solve Eq. (5-92) exactly it is convenient to rewrite the expression as

$$\kappa_1 b\left[\frac{N_0(\eta\kappa_1 b)J_0(\kappa_1 b) - J_0(\eta\kappa_1 b)N_0(\kappa_1 b)}{N_0(\eta\kappa_1 b)J_1(\kappa_1 b) - J_0(\eta\kappa_1 b)N_1(\kappa_1 b)}\right] = \epsilon_r\kappa_0 b\frac{H_0^{(2)}(\kappa_0 b)}{H_1^{(2)}(\kappa_0 b)} \qquad (5\text{-}99)$$

where $\eta = a/b$ and $\epsilon_r = \epsilon_1/\epsilon_0$. Equation (5-99) is of the form

$$f_1(\eta,\kappa_1 b) = f_2(\epsilon_r,\kappa_0 b) \qquad (5\text{-}100)$$

For ϵ_r real and greater than unity, κ_1 will be real and κ_0 imaginary and a solution may be obtained for a given η and ϵ_r by plotting Eqs. (5-100) and (5-64) on the $\kappa_1 b,\kappa_0 b$ plane. If Eq. (5-64) is rewritten as

$$(\kappa_1 b)^2 - (\kappa_0 b)^2 = (kb)^2(\epsilon_r - 1) \qquad (5\text{-}101)$$

it gives a circle of radius $kb\sqrt{\epsilon_r - 1}$ on the $\kappa_1 b,\kappa_0 b$ plane. Intersection of this circle with a plot of Eq. (5-100) gives either $\kappa_1 b$ or $\kappa_0 b$ from which γb can be determined by means of Eq. (5-62) or (5-63). If ϵ_r is less than unity or is complex, κ_1 and κ_0 generally will be complex. Thus it is necessary to solve Eq. (5-99) with complex arguments.

TE Modes

Radially symmetric TE modes also may exist on the coated cylinder in Fig. 5-18. The propagation constant may be found in the same manner as in the TM case. The equation for the TE case, corresponding to Eq. (5-90) for the TM case, is

$$\frac{\kappa_1}{\mu_1} \frac{N_1(\kappa_1 a)J_0(\kappa_1 b) - J_1(\kappa_1 a)N_0(\kappa_1 b)}{N_1(\kappa_1 a)J_1(\kappa_1 b) - J_1(\kappa_1 a)N_1(\kappa_1 b)} = \frac{\kappa_0}{\mu_0} \frac{H_0^{(2)}(\kappa_0 b)}{H_1^{(2)}(\kappa_0 b)} \tag{5-102}$$

E. TM Corrugated Cylinder

Earliest work on the propagation of an electromagnetic wave along a corrugated surface appears to be that of Goldstein[1] and Cutler.[2] Applications of the transmission and filter properties of corrugated structures have been made to traveling wave tubes,[3] linear accelerators,[4] and microwave transmission lines and antennas.[5]

A corrugated cylinder having the longitudinal section shown in Fig. 5-19 has a propagating wave that is a radially symmetric TM mode. This mode has the usual TM components E_z, E_ρ, and H_ϕ, and an approximate solution for the propagation constant can be obtained by assuming that only the propagating TM mode exists and matching radial impedance at the boundary $\rho = b$ (transverse resonance). This procedure usually gives satisfactory results if there are five or more corrugations per wavelength.

[1] H. Goldstein, "Theoretical Analysis of Corrugated Surfaces," Ph.D. Thesis, Massachusetts Institute of Technology, 1943, and The Theory of Corrugated Transmission Lines and Waveguides, *MIT Radiation Lab. Rept.* 494, April, 1944.

[2] C. C. Cutler, "Electromagnetic Waves Guided by Corrugated Conducting Surfaces," Bell Telephone Laboratories, Inc., Whippany, N.J., October, 1944.

[3] J. R. Pierce, Traveling Wave Tubes, *Bell System Tech. J.*, **29**:189, chaps. IV–VI, April, 1950.

[4] R. B. R. Shersby-Harvie, Traveling Wave Linear Accelerators, *Proc. Phys. Soc.*, **61**(3):255, September, 1948; L. Brillouin, Waveguides for Slow Waves, *J. Appl. Phys.*, **19**:1023, November, 1948; E. L. Chu and W. W. Hansen, The Theory of Disk Loaded Waveguides, *J. Appl. Phys.*, **18**:996, 1947.

[5] Goldstein, *op. cit.*, 1943 and 1944; Cutler, *op. cit.*; W. Rotman, A Study of Single-surface Corrugated Guides, *Proc. IRE*, **39**:952, August, 1951.

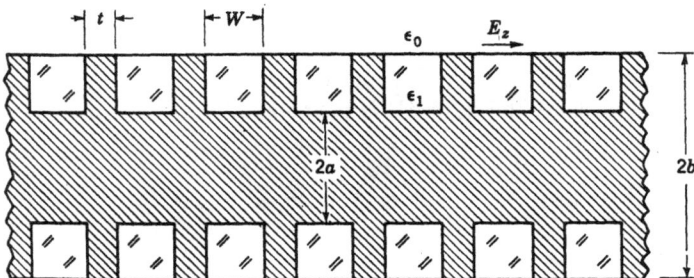

Figure. 5-19.   Longitudinal section of corrugated cylinder.

Matching impedance at $\rho = b$ requires that

$$\overleftarrow{Z}_b + \overrightarrow{Z}_b = 0 \tag{5-103}$$

The impedance $\overrightarrow{Z}_b$ is the same as for the dielectric-coated cylinder and is given by Eq. (5-93). The impedance $\overleftarrow{Z}_b$, however, is taken to be zero over the interval t, and over the interval W, $\overleftarrow{Z}_b$ is taken to be the same as for the dielectric-coated cylinder (Eq. 5-92). The average impedance over the interval $W + t$ is

$$\overleftarrow{Z}_b = \frac{W}{W + t} \frac{\kappa_1}{j\omega\epsilon_1} \frac{N_0(\kappa_1 a)J_0(\kappa_1 b) - J_0(\kappa_1 a)N_0(\kappa_1 b)}{N_0(\kappa_1 a)J_1(\kappa_1 b) - J_0(\kappa_1 a)N_1(\kappa_1 b)} \tag{5-104}$$

Substituting Eqs. (5-93) and (5-104) into (5-103) gives

$$\frac{W\kappa_1}{(W + t)\epsilon_1} \frac{N_0(\kappa_1 a)J_0(\kappa_1 b) - J_0(\kappa_1 a)N_0(\kappa_1 b)}{N_0(\kappa_1 a)J_1(\kappa_1 b) - J_0(\kappa_1 a)N_1(\kappa_1 b)} = \frac{\kappa_0}{\epsilon_0} \frac{H_0^{(2)}(\kappa_0 b)}{H_1^{(2)}(\kappa_0 b)} \tag{5-105}$$

Solution of Eq. (5-105) is simplified by observing that $\gamma = 0$ for $\rho < b$; thus $\kappa_1 = k\sqrt{\epsilon_r}$. Because of the corrugations only transverse propagation exists for $\rho < b$. Equation (5-105) may be written

$$\frac{Wkb}{(W + t)\sqrt{\epsilon_r}} \frac{N_0(\eta kb \sqrt{\epsilon_r})J_0(kb \sqrt{\epsilon_r}) - J_0(\eta kb \sqrt{\epsilon_r})N_0(kb \sqrt{\epsilon_r})}{N_0(\eta kb \sqrt{\epsilon_r})J_1(kb \sqrt{\epsilon_r}) - J_0(\eta kb \sqrt{\epsilon_r})N_1(kb \sqrt{\epsilon_r})}$$
$$= \frac{\kappa_0 b H_0^{(2)}(\kappa_0 b)}{H_1^{(2)}(\kappa_0 b)} \tag{5-106}$$

where $\eta = a/b$ and $\epsilon_r = \epsilon_1/\epsilon_0$. Equation (5-106) can be solved for κ_0, which may then be used to determine γ by means of Eq. (5-63). Cutoff occurs for $\kappa_0 = 0$.

5-4. HYBRID LINE SOURCES

It has been pointed out that a TE source is equivalent to a longitudinal magnetic current and a TM source is equivalent to a longitudinal electric current. It is also possible to excite certain structures with a field configuration corresponding to transverse rather than longitudinal currents. In this case longitudinal field components are present and the wave is no longer TE or TM. A wave of this type can be resolved into a sum of TE and TM waves and is called a hybrid mode. Hybrid mode structures are useful as low loss transmission lines and as endfire antennas (see Chap. 8).

A. Dielectric Cylinder

Perhaps the best known hybrid mode structure is the dielectric cylinder operating in what is commonly called the "dipole" mode. A sketch of the

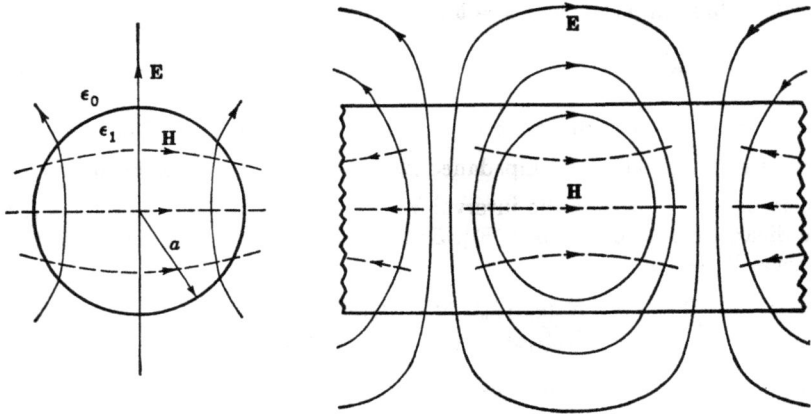

Figure 5-20.   Dielectric rod with dipole mode.

field configuration is shown in Fig. 5-20. Although the dielectric rod was studied as a transmission line as early as 1910,[1] it was not until the war years (World War II) and immediately afterward that the dielectric rod became important as a transmission line[2] and as an antenna.[3]

The analysis to determine the field configurations and propagation constants for hybrid modes on the circular dielectric rod follows that of the TM dielectric cylinder (see Sec. 5-3C). The nth order hybrid mode is denoted as HE_{nm} and is characterized by the longitudinal fields

$$E_z = A_n J_n(\kappa_1 \rho) \cos n\phi \qquad \rho \lessgtr a \qquad (5\text{-}107)$$
$$E_z = B_n H_n^{(2)}(\kappa_0 \rho) \cos n\phi \qquad \rho \gtrless a \qquad (5\text{-}108)$$

and

$$H_z = C_n J_n(\kappa_1 \rho) \sin n\phi \qquad \rho \lessgtr a \qquad (5\text{-}109)$$
$$H_z = D_n H_n^{(2)}(\kappa_0 \rho) \sin n\phi \qquad \rho \gtrless a \qquad (5\text{-}110)$$

where variation of the form $e^{j\omega t - \gamma z}$ is assumed. The subscript n is the number of cycles in the ϕ variation, and m is the number of zeros included in $J_n(\kappa_1 \rho)$ for $0 \lessgtr \rho \lessgtr a$.

Equations (5-107) and (5-108) correspond to the TM portion of the hybrid mode, and Eqs. (5-109) and (5-110) correspond to the TE portion. The remaining components may be determined by means of Eqs. (4-48) and (4-49), respectively, and an expression for the propagation constant

[1] Hondros and Debye, *op. cit.*

[2] R. M. Whitmer, Fields in Nonmetallic Waveguides, *Proc. IRE*, **36**:1105, 1948; C. H. Chandler, An Investigation of Dielectric Rod as Wave Guide, *J. Appl. Phys.*, **20**:1188, 1949; W. M. Elsasser, Attenuation in a Dielectric Circular Rod, *J. Appl. Phys.*, **20**:1193, 1949.

[3] Mueller and Tyrrell, *op. cit.*; R. B. Watson and C. W. Horton, The Radiation Patterns of Dielectric Rods—Experiment and Theory, *J. Appl. Phys.*, **19**:661, 1948.

may be obtained by matching tangential components at $\rho = a$. Therefore, we need determine only E_ϕ and H_ϕ, where

$$E_\phi = \frac{j\omega\mu}{\kappa}\frac{\partial H_z}{\partial(\kappa\rho)} - \frac{\gamma}{\kappa^2\rho}\frac{\partial E_z}{\partial\phi} \tag{5-111}$$

and

$$H_\phi = \frac{-\gamma}{\kappa^2\rho}\frac{\partial H_z}{\partial\phi} - \frac{j\omega\epsilon}{\kappa}\frac{\partial E_z}{\partial(\kappa\rho)} \tag{5-112}$$

Equations (5-111) and (5-112) give

$$E_\phi = \left[\frac{j\omega\mu_1 C_n J_n'(\kappa_1\rho)}{\kappa_1} + \frac{n\gamma A_n J_n(\kappa_1\rho)}{\kappa_1^2\rho}\right]\sin n\phi \qquad \rho \lessgtr a \tag{5-113}$$

$$E_\phi = \left[\frac{j\omega\mu_0 D_n H_n^{(2)\prime}(\kappa_0\rho)}{\kappa_0} + \frac{\gamma n B_n H_n^{(2)}(\kappa_0\rho)}{\kappa_0^2\rho}\right]\sin n\phi \qquad \rho \gtrless a \tag{5-114}$$

and

$$H_\phi = \left[\frac{-\gamma n C_n J_n(\kappa_1\rho)}{\kappa_1^2\rho} - \frac{j\omega\epsilon_1 A_n J_n'(\kappa_1\rho)}{\kappa_1}\right]\cos n\phi \qquad \rho \lessgtr a \tag{5-115}$$

$$H_\phi = \left[\frac{-\gamma n D_n H_n^{(2)}(\kappa_0\rho)}{\kappa_0^2\rho} - \frac{j\omega\epsilon_0 B_n H_n^{(2)\prime}(\kappa_0\rho)}{\kappa_0}\right]\cos n\phi \qquad \rho \gtrless a \tag{5-116}$$

Matching the tangential components at the interface $\rho = a$ gives the four equations

$$J_n(\kappa_1 a)A_n - H_n^{(2)}(\kappa_0 a)B_n = 0 \tag{5-117}$$

$$J_n(\kappa_1 a)C_n - H_n^{(2)}(\kappa_0 a)D_n = 0 \tag{5-118}$$

$$\frac{\gamma n J_n(\kappa_1 a)}{\kappa_1^2 a}A_n - \frac{\gamma n H_n^{(2)}(\kappa_0 a)}{\kappa_0^2 a}B_n$$
$$+ \frac{j\omega\mu_1 J_n'(\kappa_1 a)}{\kappa_1}C_n - \frac{j\omega\mu_0 H_n^{(2)\prime}(\kappa_0 a)}{\kappa_0}D_n = 0 \tag{5-119}$$

$$\frac{j\omega\epsilon_1 J_n'(\kappa_1 a)}{\kappa_1}A_n - \frac{j\omega\epsilon_0 H_n^{(2)\prime}(\kappa_0 a)}{\kappa_0}B_n$$
$$+ \frac{\gamma n J_n(\kappa_1 a)}{\kappa_1^2 a}C_n - \frac{\gamma n H_n^{(2)}(\kappa_0 a)}{\kappa_0^2 a}D_n = 0 \tag{5-120}$$

The primes denote the derivative with respect to $\kappa\rho$.

The determinant of these four equations must vanish. Setting the determinant equal to zero gives the characteristic equation for the HE_{nm} mode as

$$\frac{\omega^2\mu_0\epsilon_0}{\kappa_0^2}\left[\frac{H_n^{(2)\prime}(\kappa_0 a)}{H_n^{(2)}(\kappa_0 a)}\right]^2 + \frac{\omega^2\mu_1\epsilon_1}{\kappa_1^2}\left[\frac{J_n'(\kappa_1 a)}{J_n(\kappa_1 a)}\right]^2$$
$$- \left(\frac{\omega^2\mu_1\epsilon_0}{\kappa_0\kappa_1} + \frac{\omega^2\mu_0\epsilon_1}{\kappa_0\kappa_1}\right)\frac{J_n'(\kappa_1 a)H_n^{(2)\prime}(\kappa_0 a)}{J_n(\kappa_1 a)H_n^{(2)}(\kappa_0 a)}$$
$$= -\gamma^2 n^2\left(\frac{1}{\kappa_1^2 a} - \frac{1}{\kappa_0^2 a}\right)^2 \tag{5-121}$$

Eliminating ω^2 and γ^2 by means of the equations

$$\omega^2 = \frac{\kappa_1{}^2 - \kappa_0{}^2}{\mu_1 \epsilon_1 - \mu_0 \epsilon_0} \tag{5-122}$$

and

$$\gamma^2 = \frac{\kappa_0{}^2 \mu_1 \epsilon_1 - \kappa_1{}^2 \mu_0 \epsilon_0}{\mu_1 \epsilon_1 - \mu_0 \epsilon_0} \tag{5-123}$$

gives

$$\frac{1}{\kappa_0{}^2 a^2}\left[\frac{H_n{}^{(2)\prime}(\kappa_0 a)}{H_n{}^{(2)}(\kappa_0 a)}\right]^2 + \frac{\epsilon_r}{\kappa_1{}^2 a^2}\left[\frac{J_n'(\kappa_1 a)}{J_n(\kappa_1 a)}\right]^2 - \left(\frac{\epsilon_r+1}{\kappa_0 \kappa_1 a^2}\right)\frac{J_n'(\kappa_1 a)H_n{}^{(2)\prime}(\kappa_0 a)}{J_n(\kappa_1 a)H_n{}^{(2)}(\kappa_0 a)}$$
$$= n^2\left(\frac{\epsilon_r}{\kappa_1{}^2 a^2} - \frac{1}{\kappa_0{}^2 a^2}\right)\left(\frac{1}{\kappa_1{}^2 a^2} - \frac{1}{\kappa_0{}^2 a^2}\right) \tag{5-124}$$

where $\mu_0 = \mu_1$ and $\epsilon_r = \epsilon_1/\epsilon_0$. The mode usually employed is the principal, or dipole, mode (HE$_{11}$) which is sketched in Fig. 5-20.

It is interesting to note that for $n = 0$, Eq. (5-124) reduces to

$$\left[\frac{1}{\kappa_0 a}\frac{H_0{}^{(2)\prime}(\kappa_0 a)}{H_0{}^{(2)}(\kappa_0 a)} - \frac{\epsilon_r}{\kappa_1 a}\frac{J_0'(\kappa_1 a)}{J_0(\kappa_1 a)}\right]\left[\frac{1}{\kappa_0 a}\frac{H_0{}^{(2)\prime}(\kappa_0 a)}{H_0{}^{(2)}(\kappa_0 a)}\right.$$
$$\left. - \frac{1}{\kappa_1 a}\frac{J_0'(\kappa_1 a)}{J_0(\kappa_1 a)}\right] = 0 \tag{5-125}$$

from which

$$\frac{1}{\kappa_0 a}\frac{H_0{}^{(2)\prime}(\kappa_0 a)}{H_0{}^{(2)}(\kappa_0 a)} - \frac{\epsilon_r}{\kappa_1 a}\frac{J_0'(\kappa_1 a)}{J_0(\kappa_1 a)} = 0 \tag{5-126}$$

is the characteristic equation for the symmetrical TM mode [see Eq. (5-57)] and

$$\frac{1}{\kappa_0 a}\frac{H_0{}^{(2)\prime}(\kappa_0 a)}{H_0{}^{(2)}(\kappa_0 a)} - \frac{1}{\kappa_1 a}\frac{J_0'(\kappa_1 a)}{J_0(\kappa_1 a)} = 0 \tag{5-127}$$

is the characteristic equation for the symmetrical TE mode [see Eq. (5-61) for $\mu_0 = \mu_1$].

Equation (5-124) can be solved for specific values of n, ϵ_r, and $\kappa_1 a$. The corresponding value of $\kappa_0 a$ is found by successive approximation. For ϵ_r real and greater than unity a graph of $\kappa_1 a$ versus $\kappa_0 a$ may be plotted on the $\kappa_1 a, \kappa_0 a$ plane along with the circle given by Eq. (5-101) (with a substituted for b). The intersection of the two curves gives the correct $\kappa_1 a$ (or $\kappa_0 a$) which may be used to determine γa by means of Eq. (5-62) or (5-63). If ϵ_r is less than unity or is complex, κ_1 and κ_0 generally will be complex and it will be necessary to solve Eqs. (5-124) and (5-101) with complex arguments. Cutoff occurs for $\kappa_0 = 0$.

An interesting modification of the HE$_{11}$ dielectric rod is the dielectric image line[1] in which the bottom half of the rod is replaced by a ground

[1] D. D. King, Dielectric Image Line, *J. Appl. Phys.*, **23**:699, 1952.

Figure 5-21.   Image line.

ϵ_0

ϵ_1

Semicircular dielectric rod

Ground plane

plane. This is illustrated in Fig. 5-21. In this form support and shielding problems are greatly simplified.

Some of the advantages of the HE_{11} mode are:

1. It has no cutoff for $\epsilon_r > 1$. This is also true for the dominant TM mode on the dielectric-coated rod.

2. It has low loss over a wider range of frequencies and diameters than any other dielectric rod mode.

3. It can be launched directly from a TE_{01} waveguide, which is of considerable practical importance.

Attenuation along a dielectric rod due to losses in the dielectric may be determined experimentally or analytically. Experimentally one may probe the fields along the rod (see Sec. 4-5) to determine the rate of attenuation, but for low rates of attenuation it has been found to be more accurate to use a resonator method.[1] In this method a section of dielectric rod is terminated at each end by a large metal plate. A small loop may be used to couple into the resonator at each plate or the dielectric rod may pass through a hole in each plate to provide coupling. A response curve of the resonator may be plotted as frequency is varied through resonance and Q may be found from the usual definition

$$Q = \frac{f}{\Delta f} \tag{5-128}$$

where f is the resonant frequency and Δf is the bandwidth between the 3-db points of the response curve. The attenuation constant α of the rod may be found from the relation[1]

$$Q \approx \frac{\beta}{2[\alpha + (1 - R)/L]} \tag{5-129}$$

where L is the distance between resonator plates, R is the reflection coefficient at either end of the resonator (ends assumed to be identical), and β is the phase constant of the rod which may be found by the methods of Sec. 4-5. The coefficient R may be eliminated by taking measurements for two different values of L.

Analytically, the attenuation along a dielectric rod may be obtained by solving the characteristic equation (5-124) for ϵ_r complex. For low loss,

[1] Chandler, *op. cit.*, E. H. Schiebe, B. G. King, and D. L. Van Zeeland, Loss Measurements of Surface Wave Transmission Lines, *J. Appl. Phys.*, **25**:790, 1954.

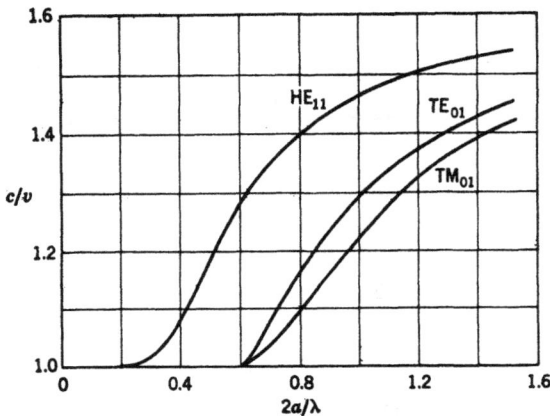

Fig. 5-22. Velocity ratio c/v for polystyrene rod ($\epsilon_r =$ 2.56). (From data obtained from Elsasser, Attenuation in a Dielectric Circular Rod, *J. Appl. Phys.*, **20**:1193, 1949.)

however, the usual procedure is to assume that the fields are of the same form as in the lossless case and find the attenuation constant from the power dissipated per unit length along the rod.

The loss is associated with the conduction-current density

$$\mathbf{J} = \sigma \mathbf{E} \tag{5-130}$$

where σ is the conductivity of the dielectric. The conductivity is related to loss tangent, $\tan \delta$, by

$$\sigma = \omega \epsilon_1 \tan \delta \tag{5-131}$$

If power lost per unit length along a dielectric rod is small compared with the power transmitted along the rod, the Poynting vector $(\mathbf{E} \times \mathbf{H})$ may be taken as parallel to the axis of the rod. Let $P(z)$ be the power along the rod where

$$P(z) = \tfrac{1}{2}\text{Re} \int_{\Sigma_1 + \Sigma_0} [\mathbf{E} \times \mathbf{H}^*]_z \rho \, d\rho \, d\phi \tag{5-132}$$

The cross section Σ_1 is that of the rod, and Σ_0 that of the region outside the rod. The field components in Eq. (5-132) are taken to be those for the rod with no loss.

The attenuation constant due to loss [see Eq. (4-3)] is

$$\alpha = -\frac{1}{2} \frac{1}{P(z)} \frac{dP(z)}{dz} \tag{5-133}$$

where $P(z)$ is given by Eq. (5-132) and $[dP(z)]/dz$ may be written in terms of σ as

$$\frac{dP(z)}{dz} = \frac{1}{2} \sigma \int_{\Sigma_1} |\mathbf{E}|^2 \rho \, d\rho \, d\phi \tag{5-134}$$

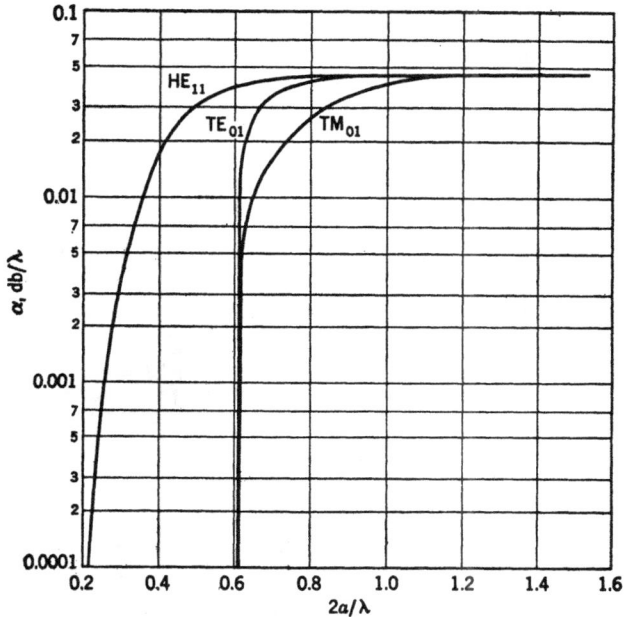

Figure 5-23. Attenuation in decibels per wavelength for polystyrene rod ($\epsilon_r = 2.56$ and tan $\delta = 0.001$). (From data obtained from Elsasser, Attenuation in a Dielectric Circular Rod, *J. Appl. Phys.*, **20**:1193, 1949.) Note that α, db/λ = 8.686 α, nepers/λ.

Disregarding sign, Eq. (5-133) may be written

$$2\alpha = \frac{\sigma \int_{\Sigma_1} |\mathbf{E}|^2 \rho \, d\rho \, d\phi}{\text{Re} \int_{\Sigma_1 + \Sigma_0} [\mathbf{E} \times \mathbf{H}^*]_z \rho \, d\rho \, d\phi} \qquad (5\text{-}135)$$

Graphs of c/v and attenuation for a polystyrene rod ($\epsilon_r = 2.56$ and tan $\delta = 0.001$) are shown in Figs. 5-22 and 5-23 for the HE_{11} mode and the lowest-order radially symmetric TM and TE modes (TM_{01} and TE_{01}).

Dielectric Coaxial Waveguide

Hybrid HE_{nm} modes also can exist on the dielectric coaxial waveguide of Fig. 5-15.[1] Assuming sinusoidal variation in ϕ as in the case of the solid dielectric rod and requiring continuity of E_ϕ, H_ϕ, E_z, and H_z at the three boundaries result in 12 linear homogeneous equations. A solution exists if the determinant of the coefficients is zero. The resulting twelfth-order determinant is

[1] Barnett, *op. cit.*

$$
\begin{vmatrix}
QJ_n(va) & PJ_n(va) & SI_n(ua) & SK_n(ua) & 0 & 0 & 0 & 0 & 0 & 0 & 0 & 0 \\[4pt]
0 & RJ_n(va) & 0 & 0 & SI_n(ua) & SK_n(ua) & PJ_n(ub) & QN_n(ub) & PJ_n(ub) & PN_n(ub) & 0 & 0 \\[4pt]
\dfrac{nYI_n(va)}{c\,a/c} & \dfrac{-nXJ_n(va)}{a/c} & \dfrac{-nYI_n(ua)}{a/c} & \dfrac{-nYK_n(ua)}{a/c} & -WI_n'(ua) & -WK_n'(ua) & 0 & \dfrac{-nYN_n(ub)}{c\,b/c} & -RJ_n'(ub) & -RN_n'(ub) & 0 & 0 \\[8pt]
-2\pi RJ_n'(va) & 0 & 2\pi WI_n'(ua) & 2\pi WK_n'(ua) & \dfrac{nXI_n(ua)}{a/c} & \dfrac{nXK_n(ua)}{a/c} & QJ_n(ub) & -2\pi RN_n'(ub) & \dfrac{-nXJ_n(ub)}{b/c} & \dfrac{-nXN_n(ub)}{b/c} & 0 & 0 \\[8pt]
0 & 0 & SI_n(ub) & SK_n(ub) & SI_n(ub) & SK_n(ub) & \dfrac{-nYJ_n(ub)}{c\,b/c} & QN_n(uc) & 0 & 0 & SK_n(uc) & SK_n(uc) \\[8pt]
0 & 0 & \dfrac{nYI_n(ub)}{b/c} & \dfrac{nYK_n(ub)}{b/c} & WI_n'(ub) & WK_n'(ub) & -2\pi RJ_n'(ub) & \dfrac{nYN_n(uc)}{c_r} & PJ_n(uc) & PN_n(uc) & 0 & -WK_n'(uc) \\[8pt]
0 & 0 & 2\pi WI_n'(ub) & 2\pi WK_n'(ub) & \dfrac{nXI_n(ub)}{b/c} & \dfrac{nXK_n(ub)}{b/c} & QJ_n(uc) & 2\pi RN_n'(uc) & RJ_n'(uc) & RN_n'(uc) & -nYK_n(uc) & -nXK_n(uc) \\[8pt]
0 & 0 & 0 & 0 & 0 & 0 & \dfrac{nYJ_n(uc)}{c_r} & & nXJ_n(uc) & nXN_n(uc) & -2\pi WK_n'(uc) &
\end{vmatrix}
= 0
$$

$$(5\text{-}136)$$

where

$$P = \epsilon_r - \left(\frac{\lambda}{\lambda_g}\right)^2$$

$$R = \sqrt{P}$$

$$Q = \frac{P}{\epsilon_r}$$

$$S = \left(\frac{\lambda}{\lambda_g}\right)^2 - 1 \qquad (5\text{-}137)$$

$$W = \sqrt{S}$$

$$X = \frac{\lambda/\lambda_g}{120\pi(c/\lambda)}$$

$$Y = \frac{60(\lambda/\lambda_g)}{c/\lambda}$$

Barnett[1] has programmed Eqs. (5-81), (5-82), and (5-136) for the IBM 7090. Some of the results for the TM_{01} and TE_{01} modes are given in Figs. 5-16 and 5-17, respectively. Some results for the HE_{11} mode from Eq. (5-136) are given in Fig. 5-24.[2] The ratio of free space to guide wavelength λ/λ_g is presented as a function of geometry. Note that the ratio λ/λ_g is the same as the velocity ratio c/v used in previous traveling wave structures.

[1] Barnett, *op. cit.*
[2] For extensive curves of λ_g/λ for TM_{01}, TE_{01}, HE_{11}, HE_{12}, and HE_{21} modes as well as a description of the computer program the reader is referred to *ibid.*

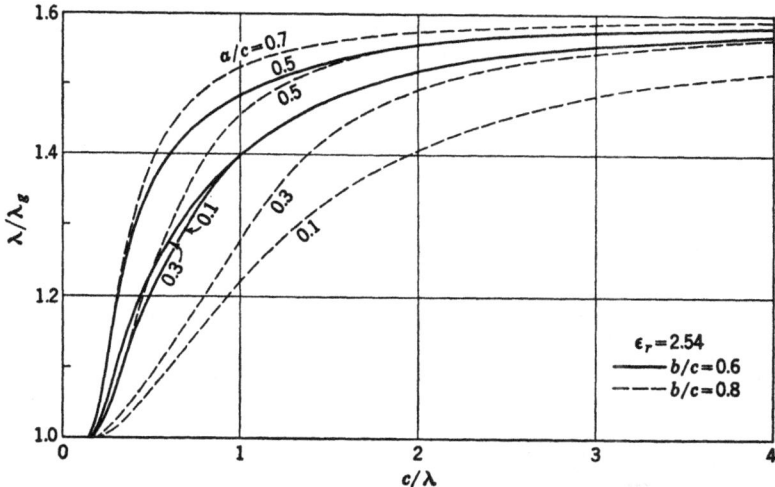

Figure 5-24. Wavelength ratio λ/λ_g for HE_{11} mode on the coaxial dielectric guide of Fig. 5-15. The wavelength ratio λ/λ_g is equal to the velocity ratio c/v.

For the special case where $a = 0$, the coaxial line reduces to a hollow dielectric tube.[1]

Cutoff for the coaxial dielectric waveguides occurs when $\lambda_g = \lambda$. That is, the TE, TM, or HE mode ceases to exist when the phase velocity is equal to the velocity of light in free space. In the above notation

$$u = k\left[\left(\frac{\lambda}{\lambda_g}\right)^2 - 1\right]^{\frac{1}{2}} \to 0$$

$$v = k\left[\epsilon_r - \left(\frac{\lambda}{\lambda_g}\right)^2\right]^{\frac{1}{2}} \to k(\epsilon_r - 1)^{\frac{1}{2}}$$

(5-138)

as $\lambda_g \to \lambda$. The value of λ (from $k = 2\pi/\lambda$) in v of Eqs. (5-138) is denoted λ_c for the cutoff wavelength. It can be found from curves such as Figs. 5-16, 5-17, and 5-24 at the point where $\lambda_g = \lambda$, or λ_c can be found by solving Eqs. (5-81), (5-82), and (5-136) with u and v as given in Eqs. (5-138). In general the dominant (TE$_{01}$, TM$_{01}$, and HE$_{11}$) modes do not have a cutoff as long as any dielectric is present. Cutoff for these modes occurs when the dielectric thickness goes to zero and the modes degenerate into a plane wave in free space.

B. Dielectric-coated Metal Rod

A "dipole" type of hybrid mode also may propagate along a dielectric-coated metal rod. The field configuration is indicated in Fig. 5-25. A structure of this type may be used as a reinforced polyrod antenna, and it may be fed by a coaxial line propagating the lowest-order TE wave having circumferential variation.

[1] D. G. Kiely, "Dielectric Aerials," Methuen, London, 1953; R. E. Beam, Wave Propagation in Dielectric Tubes, *Final Rept.*, Contract DA 36-039-sc-5397, Northwestern University, Evanston, Ill., 1952.

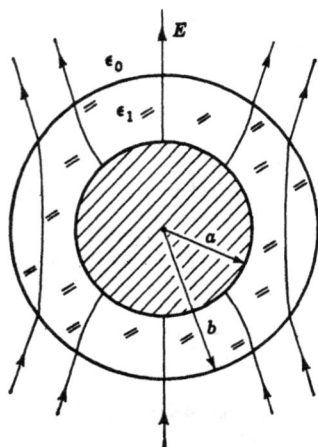

Figure 5-25.   Dipole mode on coated metal cylinder.

The characteristic equation for TE waves in a coaxial line of inner radius a and outer radius b is[1]

$$\frac{N_n'(\kappa_1 a)}{J_n'(\kappa_1 a)} = \frac{N_n'(\kappa_1 b)}{J_n'(\kappa_1 b)} \tag{5-139}$$

In general for the nth-order TE wave with circumferential variation, a good approximation to κ_1 for b/a up to 5 is

$$\kappa_1 \approx \frac{2n}{a+b} \qquad n = 1, 2, 3, \ldots \tag{5-140}$$

The characteristic equation for the dielectric-coated rod in the dipole mode is considerably more complicated than Eq. (5-139) for the TE coaxial line modes. The analysis, however, is straightforward and follows that of the solid dielectric rod in the previous section.

The nth-order hybrid mode is again denoted as HE_{nm} and is characterized by the longitudinal fields

$$E_z = [A_n J_n(\kappa_1 \rho) + E_n N_n(\kappa_1 \rho)] \cos n\phi \qquad a \lessgtr \rho \lessgtr b \tag{5-141}$$
$$E_z = B_n H_n^{(2)}(\kappa_0 \rho) \cos n\phi \qquad \rho \gtrless b \tag{5-142}$$

and

$$H_z = [C_n J_n(\kappa_1 \rho) + F_n N_n(\kappa_1 \rho)] \sin n\phi \qquad a \lessgtr \rho \lessgtr b \tag{5-143}$$
$$H_z = D_n H_n^{(2)}(\kappa_0 \rho) \sin n\phi \qquad \rho \gtrless b \tag{5-144}$$

The transverse fields E_ϕ and H_ϕ may be obtained from Eqs. (5-111) and (5-112). The radial fields E_ρ and H_ρ are not needed in determining the characteristic equation.

Setting E_z equal to zero at the conducting surface ($\rho = a$) eliminates one of the coefficients in Eq. (5-141), giving

$$E_z = \frac{A_n}{N_n(\kappa_1 a)} [N_n(\kappa_1 a) J_n(\kappa_1 \rho) - J_n(\kappa_1 a) N_n(\kappa_1 \rho)] \cos n\phi \tag{5-145}$$

Setting E_ϕ from Eq. (5-111) equal to zero at $\rho = a$ gives

$$F_n = -\frac{C_n J_n'(\kappa_1 a)}{N_n'(\kappa_1 a)} \tag{5-146}$$

therefore Eq. (5-143) becomes

$$H_z = \frac{C_n}{N_n'(\kappa_1 a)} [N_n'(\kappa_1 a) J_n(\kappa_1 \rho) - J_n'(\kappa_1 a) N_n(\kappa_1 \rho)] \sin n\phi \tag{5-147}$$

From Eqs. (5-111), (5-112), (5-142), (5-144), (5-145), and (5-147) the

[1] See, for example, S. Ramo and J. R. Whinnery, "Fields and Waves in Modern Radio," p. 333, Wiley, New York, 1944.

transverse components E_ϕ and H_ϕ are

$$E_\phi = \frac{j\omega\mu_1}{\kappa_1} \frac{C_n}{N'_n(\kappa_1 a)} [N'_n(\kappa_1 a)J'_n(\kappa_1\rho) - J'_n(\kappa_1 a)N'_n(\kappa_1\rho)] \sin n\phi$$

$$+ \frac{\gamma n}{\kappa_1^2\rho} \frac{A_n}{N_n(\kappa_1 a)} [N_n(\kappa_1 a)J_n(\kappa_1\rho) - J_n(\kappa_1 a)N_n(\kappa_1\rho)] \sin n\phi$$

$$a \lessgtr \rho \lessgtr b \quad (5\text{-}148)$$

$$E_\phi = \frac{j\omega\mu_0}{\kappa_0} D_n H_n^{(2)}(\kappa_0\rho) \sin n\phi + \frac{\gamma n}{\kappa_0^2\rho} B_n H_n^{(2)}(\kappa_0\rho) \sin n\phi \quad \rho \gtrless b$$

$$(5\text{-}149)$$

and

$$H_\phi = \frac{-j\omega\epsilon_1}{\kappa_1} \frac{A_n}{N_n(\kappa_1 a)} [N_n(\kappa_1 a)J'_n(\kappa_1\rho) - J_n(\kappa_1 a)N'_n(\kappa_1\rho)] \cos n\phi$$

$$- \frac{\gamma n}{\kappa_1^2\rho} \frac{C_n}{N'_n(\kappa_1 a)} [N'_n(\kappa_1 a)J_n(\kappa_1\rho) - J'_n(\kappa_1 a)N_n(\kappa_1\rho)] \cos n\phi$$

$$a \lessgtr \rho \lessgtr b \quad (5\text{-}150)$$

$$H_\phi = \frac{-j\omega\epsilon_0}{\kappa_0} B_n H_n^{(2)\prime}(\kappa_0\rho) \cos n\phi - \frac{\gamma n}{\kappa_0^2\rho} D_n H_n^{(2)}(\kappa_0\rho) \cos n\phi \quad \rho \gtrless b$$

$$(5\text{-}151)$$

Matching the tangential fields at $\rho = b$ gives the four equations

$$[N_n(\kappa_1 a)J_n(\kappa_1 b) - J_n(\kappa_1 a)N_n(\kappa_1 b)] \frac{A_n}{N_n(\kappa_1 a)} - H_n^{(2)}(\kappa_0 b)B_n = 0 \quad (5\text{-}152)$$

$$[N'_n(\kappa_1 a)J_n(\kappa_1 b) - J'_n(\kappa_1 a)N_n(\kappa_1 b)] \frac{C_n}{N'_n(\kappa_1 a)} - H_n^{(2)}(\kappa_0 b)D_n = 0 \quad (5\text{-}153)$$

$$\frac{\gamma n}{\kappa_1^2 b} [N_n(\kappa_1 a)J_n(\kappa_1 b) - J_n(\kappa_1 a)N_n(\kappa_1 b)] \frac{A_n}{N_n(\kappa_1 a)}$$

$$- \frac{\gamma n}{\kappa_0^2 b} H_n^{(2)}(\kappa_0 b)B_n + \frac{j\omega\mu_1}{\kappa_1} [N'_n(\kappa_1 a)J'_n(\kappa_1 b) - J'_n(\kappa_1 a)N'_n(\kappa_1 b)]$$

$$\times \frac{C_n}{N'_n(\kappa_1 a)} - \frac{j\omega\mu_0}{\kappa_0} H_n^{(2)\prime}(\kappa_0 b)D_n = 0 \quad (5\text{-}154)$$

and

$$\frac{j\omega\epsilon_1}{\kappa_1} [N_n(\kappa_1 a)J'_n(\kappa_1 b) - J_n(\kappa_1 a)N'_n(\kappa_1 b)] \frac{A_n}{N_n(\kappa_1 a)} - \frac{j\omega\epsilon_0}{\kappa_0} H_n^{(2)\prime}(\kappa_0 b)B_n$$

$$+ \frac{\gamma n}{\kappa_1^2 b} [N'_n(\kappa_1 a)J_n(\kappa_1 b) - J'_n(\kappa_1 a)N_n(\kappa_1 b)] \frac{C_n}{N'_n(\kappa_1 a)}$$

$$- \frac{\gamma n}{\kappa_0^2 b} H_n^{(2)}(\kappa_0 b)D_n = 0 \quad (5\text{-}155)$$

Setting the determinant of the above equations equal to zero as in the previous section gives the characteristic equation for the HE_{nm} mode as

$$\frac{1}{\kappa_0^2 b^2} \left[\frac{H_n^{(2)\prime}(\kappa_0 b)}{H_n^{(2)}(\kappa_0 b)}\right]^2 - \frac{1}{\kappa_1\kappa_0 b^2} \frac{g'H_n^{(2)\prime}(\kappa_0 b)}{hH_n^{(2)}(\kappa_0 b)} - \frac{\epsilon_r}{\kappa_0\kappa_1 b^2} \frac{iH_n^{(2)\prime}(\kappa_0 b)}{gH_n^{(2)}(\kappa_0 b)}$$

$$+ \frac{\epsilon_r}{\kappa_1^2 b^2} \frac{g'i}{gh} = n^2 \left(\frac{\epsilon_r}{\kappa_1^2 b^2} - \frac{1}{\kappa_0^2 b^2}\right)\left(\frac{1}{\kappa_1^2 b^2} - \frac{1}{\kappa_0^2 b^2}\right) \quad (5\text{-}156)$$

where

$$g = N_n(\kappa_1 a) J_n(\kappa_1 b) - J_n(\kappa_1 a) N_n(\kappa_1 b) \qquad (5\text{-}157)$$
$$g' = N'_n(\kappa_1 a) J'_n(\kappa_1 b) - J'_n(\kappa_1 a) N'_n(\kappa_1 b) \qquad (5\text{-}158)$$
$$h = N'_n(\kappa_1 a) J_n(\kappa_1 b) - J'_n(\kappa_1 a) N_n(\kappa_1 b) \qquad (5\text{-}159)$$
$$i = N_n(\kappa_1 a) J'_n(\kappa_1 b) - J_n(\kappa_1 a) N'_n(\kappa_1 b) \qquad (5\text{-}160)$$

For $a = 0$, Eq. (5-156) reduces to the HE_{nm} mode for the dielectric rod. For $n = 0$, Eq. (5-156) gives the characteristic equation for symmetrical TM and TE modes [Eqs. (5-90) and (5-102) with $\mu_0 = \mu_1$] in Sec. 5-3D.

C. Rectangular Dielectric Slab between Parallel Plates

A rectangular dielectric slab between parallel metal plates is illustrated in Fig. 5-26. A structure of this type is sometimes called an H guide.[1] An exact analysis of this structure is straightforward when the plates are of infinite extent, and the analysis for infinite plates is a good approximation for finite plates.

As we shall see in the analysis that follows, certain modes have E_y and E_z components that are zero at $x = 0$. In this case a perfectly conducting plate may be inserted at $x = 0$, forming the structure in Fig. 5-27a. As before, the analysis for infinite plates is a good approximation for finite plates. The structures in Figs. 5-26 and 5-27a are useful as low loss transmission lines. With the side plates folded to form a ground plane as in Fig. 5-27b the structure is useful as a flush-mounted transmission line or antenna. The analysis for the infinite plates also serves as a good approximation for the structure in Fig. 5-27b.

Three types of modes can be supported by the structure in Fig. 5-26. These are conventional TE modes ($\mathbf{E} = \mathbf{y}E_y$), hybrid modes characterized

[1] F. J. Tischer, The H-Guide, a Waveguide for Microwaves, *IRE Natl. Conv. Record*, part 5, p. 44, 1956.

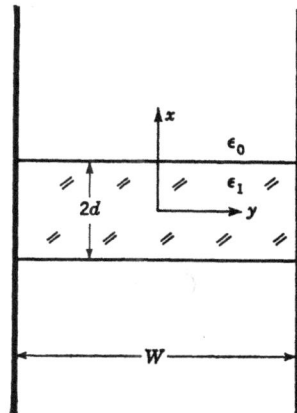

Figure 5-26. Rectangular dielectric slab between parallel metal plates.

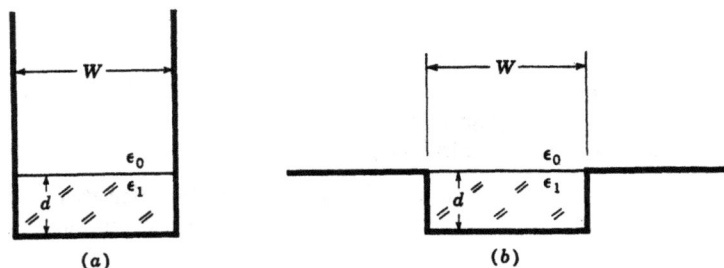

Figure 5-27.   Bisected parallel plate structure.

by $H_x = 0$, and hybrid modes characterized by $E_x = 0$. In the hybrid modes longitudinal components of both **E** and **H** are present; however, $H_x = 0$ modes may be considered as TM modes with respect to the x direction. The $E_x = 0$ modes on the other hand may be considered as TE modes with respect to the x direction.

The dominant mode for the structure is TE in the usual sense, in that it has no longitudinal component of electric field.

$H_x = 0$ Modes

These are perhaps the most practical modes for the structures in Figs. 5-26 and 5-27. The lowest-order mode may be excited by TE_{01} waveguide, and the polarization is such that a flush-mounted antenna with the cross section of Fig. 5-27b gives good endfire radiation.[1]

The fields for the $H_x = 0$ modes may be found from a vector potential $\mathbf{x}A_x$ by means of Eqs. (4-40) and (4-41). The components are found to be

$$E_x = \frac{-1}{j\omega\mu\epsilon}\left(\frac{\partial^2 A_x}{\partial y^2} + \frac{\partial^2 A_x}{\partial z^2}\right) \tag{5-161}$$

$$E_y = \frac{1}{j\omega\mu\epsilon}\frac{\partial^2 A_x}{\partial x\,\partial y} \tag{5-162}$$

$$E_z = \frac{1}{j\omega\mu\epsilon}\frac{\partial^2 A_x}{\partial x\,\partial z} \tag{5-163}$$

$$H_y = \frac{1}{\mu}\frac{\partial A_x}{\partial z} \tag{5-164}$$

$$H_z = -\frac{1}{\mu}\frac{\partial A_x}{\partial y} \tag{5-165}$$

The component E_x (and also A_x) must satisfy the wave equation

$$[\nabla^2 + (\omega^2\mu\epsilon - \kappa^2)]E_x = 0 \tag{5-166}$$

[1] B. T. Stephenson and C. H. Walter, Endfire Slot Antennas, *IRE Trans. Antennas Propagation*, **AP-3**(2), April, 1955.

A solution is

$$E_z = A_1 \cos\left(\kappa_1 x - \frac{n\pi}{2}\right)\cos\frac{m\pi y}{W} \qquad -d \lessgtr x \lessgtr d \qquad (5\text{-}167)$$

$$E_z = (-1)^n A_0 e^{j\kappa_0 x}\cos\frac{m\pi y}{W} \qquad x < -d \qquad (5\text{-}168)$$

$$E_z = A_0 e^{-j\kappa_0 x}\cos\frac{m\pi y}{W} \qquad x > d \qquad (5\text{-}169)$$

where

$$0 \lessgtr \kappa_1 x - \frac{n\pi}{2} \lessgtr \frac{\pi}{2} \qquad n = 0, 1, 2, \ldots$$

and κ is the phase constant (or wave number) for the x direction. The subscripts 1 and 0 on κ denote the dielectric and air regions, respectively, and variation of the form $e^{j\omega t - \gamma z}$ is assumed. Note that for $m = 0$ the field E_z is zero because of the boundary conditions at the conducting surfaces. Therefore permissible values of m are $m = 1, 3, 5, \ldots$

Equating tangential components at either $x = d$ or $x = -d$ for the above equations gives the characteristic equation

$$\epsilon_0 \kappa_1 \tan\left(\kappa_1 d - \frac{n\pi}{2}\right) = j\kappa_0 \epsilon_1 \qquad n = 0, 1, 2, \ldots \qquad (5\text{-}170)$$

Equation (5-170) in conjunction with

$$\kappa_1{}^2 + \left(\frac{m\pi}{W}\right)^2 - \gamma^2 = \frac{\epsilon_1}{\epsilon_0}k^2 \qquad (5\text{-}171)$$

$$\kappa_0{}^2 + \left(\frac{m\pi}{W}\right)^2 - \gamma^2 = k^2 \qquad (5\text{-}172)$$

may be solved for γ according to the methods of the previous sections. Solutions for n even $(0, 2, 4, \ldots)$ apply to the structures in Fig. 5-27 where a metal plate exists at $x = 0$.

The mode that is generally used in practice has $n = 0$ and $m = 1$. It corresponds somewhat to the TE_{01} mode of rectangular guide and may be efficiently launched from a TE_{01} guide. Because it is a hybrid mode, we shall denote it HE_{01}. A graph of c/v versus d/λ for several values of W is given in Fig. 5-28 for polystyrene dielectric $(\epsilon_r = 2.56)$. The ratio c/v approaches $\sqrt{\epsilon_r}$ for large d/λ and W/λ.

The attenuation along the structure in Fig. 5-26 can be computed from Eq. (5-133). Calculations are greatly simplified if the structure is designed to carry most of the power in the air region. If the dielectric has low loss in this case, the attenuation may be attributed to conductor losses only. Under these conditions an H-guide transmission line may have significantly less attenuation than a conventional rectangular waveguide of the same type of metal.

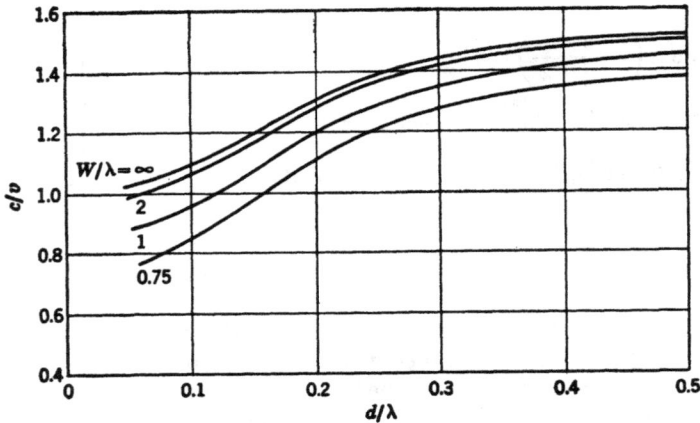

Figure 5-28. Velocity ratio c/v for HE_{01} mode in dielectric-loaded parallel plate guide. Dielectric is polystyrene ($\epsilon_r = 2.56$).

$E_x = 0$ Modes

The hybrid modes associated with $E_x = 0$ may be denoted EH_{nm} to distinguish them from the $H_x = 0$ modes. The fields may be found from a vector potential xF_x by means of Eqs. (4-40) and (4-41) as before. The components are found to be

$$H_x = \frac{-1}{j\omega\mu\epsilon}\left(\frac{\partial^2 F_x}{\partial y^2} + \frac{\partial^2 F_x}{\partial z^2}\right) \tag{5-173}$$

$$H_y = \frac{1}{j\omega\mu\epsilon}\frac{\partial^2 F_x}{\partial x \partial y} \tag{5-174}$$

$$H_z = \frac{1}{j\omega\mu\epsilon}\frac{\partial^2 F_x}{\partial x \partial z} \tag{5-175}$$

$$E_y = -\frac{1}{\epsilon}\frac{\partial F_x}{\partial z} \tag{5-176}$$

$$E_z = \frac{1}{\epsilon}\frac{\partial F_x}{\partial y} \tag{5-177}$$

The component H_x (and also F_x) satisfies the wave equation

$$[\nabla^2 + (\omega^2\mu\epsilon - \kappa^2)]H_x = 0 \tag{5-178}$$

A solution to Eq. (5-178) is

$$H_x = B_1 \cos\left(\kappa_1 x - \frac{n\pi}{2}\right)\sin\frac{m\pi y}{W} \qquad -d \lesseqgtr x \lesseqgtr d \tag{5-179}$$

$$H_x = (-1)^n B_0 e^{j\kappa_0 x}\sin\frac{m\pi y}{W} \qquad x < -d \tag{5-180}$$

$$H_x = B_0 e^{-j\kappa_0 x}\sin\frac{m\pi y}{W} \qquad x > d \tag{5-181}$$

where

$$0 \lesssim \kappa x - \frac{n\pi}{2} \lesssim \frac{\pi}{2} \qquad\qquad n = 0, 1, 2, \ldots$$

Equating tangential components at either $x = d$ or $x = -d$ gives the characteristic equation

$$\kappa_1 \mu_0 \tan\left(\kappa_1 d - \frac{n\pi}{2}\right) = j\kappa_0 \mu_1 \qquad\qquad (5\text{-}182)$$

Equation (5-182) in conjunction with Eqs. (5-171) and (5-172) may be solved for γ. Solutions for n odd apply to the structures in Fig. 5-27 where a metal plate exists at $x = 0$. Here, again, $m = 0$ is a trivial case.

In general for a given structure the EH_{nm} mode will have a smaller wavelength than the HE_{nm} mode for the same frequency of operation.

TE Modes

Transverse electric modes for the structure in Fig. 5-26 may be found in the usual manner from an electric vector potential function zF_z (or the magnetic field component H_z which is proportional to F_z). If only a y component of E is assumed to exist for TE modes, then the remaining fields are most readily found from E_y.

From Maxwell's equations we have

$$H_x = \frac{1}{j\omega\mu} \frac{\partial E_y}{\partial z} \qquad\qquad (5\text{-}183)$$

$$H_z = \frac{-1}{j\omega\mu} \frac{\partial E_y}{\partial x} \qquad\qquad (5\text{-}184)$$

and

$$H_y = 0 \qquad\qquad (5\text{-}185)$$

A solution for E_y that satisfies Maxwell's equation and the boundary conditions is

$$E_y = A_1 \cos\left(\kappa_1 x - \frac{n\pi}{2}\right) \qquad -d \lesssim x \lesssim d \quad n = 0, 1, 2, \ldots \quad (5\text{-}186)$$

$$E_y = A_0 e^{-j\kappa_0 x} \qquad\qquad x > d \qquad\qquad (5\text{-}187)$$

$$E_y = (-1)^n A_0 e^{j\kappa_0 x} \qquad\qquad x < -d \qquad\qquad (5\text{-}188)$$

where variation of the form $e^{j\omega t - \gamma z}$ is assumed. Note that there can be no variation in the y direction for TE operation.

Equating tangential components at $x = d$ or $x = -d$ gives the characteristic equation

$$\kappa_1 \mu_0 \tan\left(\kappa_1 d - \frac{n\pi}{2}\right) = j\kappa_0 \mu_1 \qquad\qquad (5\text{-}189)$$

which is the same as for the EH_{nm} modes in the previous section. Equation (5-189) in conjunction with

$$\kappa_1{}^2 - \gamma^2 = k^2\epsilon_r \tag{5-190}$$

and

$$\kappa_0{}^2 - \gamma^2 = k^2 \tag{5-191}$$

may be solved for γ. Solutions for n odd apply to the structure in Fig. 5-27 where a conducting plate exists at $x = 0$.

Cutoff Conditions

The TE_n modes with components given by Eqs. (5-183) to (5-188) are the dominant modes for the structure in Fig. 5-26. None of the components depends on y; thus W may be made sufficiently small so that all hybrid modes are suppressed and only the TE_n modes may propagate.

A cutoff criterion for hybrid modes may be derived from Eqs. (5-182), (5-171), and (5-172). The critical distance W_c which suppresses all hybrid modes (that is, $\gamma = 0$) may be expressed as

$$\frac{W_c}{\lambda} = \frac{1}{2}\left(\frac{1 + \tan^2 \kappa_1 d}{1 + \epsilon_r \tan^2 \kappa_1 d}\right)^{1/2} \tag{5-192}$$

For $W < W_c$ only TE_n modes may propagate. The dominant TE mode (TE_0) has no cutoff at all (i.e., it can exist down to zero frequency). Higher-order TE modes ($n = 1, 2, \ldots$) have a cutoff condition in the sense that κ_0 may be made zero; hence the wave degenerates into a TEM wave with the velocity of light in the medium μ_0, ϵ_0 (taken to be free space). This is also true for the hybrid modes when $W > W_c$. The condition for cutoff in this sense ($\kappa_0 = 0$) for both TE and hybrid modes is

$$\tan\left(\kappa_1 d - \frac{n\pi}{2}\right) = 0 \tag{5-193}$$

from which

$$\kappa_1 d = \frac{n\pi}{2} \tag{5-194}$$

For the TE modes Eq. (5-194) along with Eqs. (5-190) and (5-191) (for $\kappa_0 = 0$) give

$$2d/\lambda = \frac{n}{2\sqrt{\epsilon_r - 1}} \tag{5-195}$$

as the critical thickness for cutoff for the nth mode.

For the hybrid modes HE_{nm} and EH_{nm}, Eq. (5-194) and Eqs. (5-171) and (5-172) (for $\kappa_0 = 0$) give the same result as Eq. (5-195).

Transmission Losses

In practice the conductor portion of a traveling wave structure will have large but finite conductivity and the dielectric portion will have small but finite conductivity. If the losses are small, the method of Sec. 5-4A involving Eq. (5-133) may be used and wall losses and dielectric losses may be considered separately.

The wall loss per unit length along the structure of Fig. 5-26 is

$$\frac{dP}{dz} = R_s \int_{-\infty}^{\infty} |H|^2_{W/2} \, dx \qquad (5\text{-}196)$$

where

$$R_s = \sqrt{\frac{\omega\mu_0}{2\sigma_W}} = \text{surface resistance} \qquad (5\text{-}197)$$

$$|H|^2_{W/2} = (|H_x|^2 + |H_z|^2)_{y=W/2} \qquad (5\text{-}198)$$

and σ_W is the conductivity of the metal walls. The total power $P(z)$ transmitted is given by Eq. (5-131). Thus the attenuation in nepers per unit length due to wall losses is

$$\alpha_W = \frac{R_s \int_{-\infty}^{\infty} |H|^2_{W/2} \, dx}{\text{Re} \int_{-\infty}^{\infty} \int_{-W/2}^{W/2} (\mathbf{E} \times \mathbf{H}^*)_z \, dy \, dx} \qquad (5\text{-}199)$$

where the field components are those of the particular mode of interest as determined for the lossless case. For the TE modes, for example, the components would be E_y, H_x, and H_z as given by Eqs. (5-183) to (5-188). If one considers the structure in Fig. 5-27a, then the integration on x is from zero to infinity.

The dielectric loss will be associated with region 1 in Fig. 5-26 if we let region 0 be free space. In this case the power lost per unit length is

$$\frac{dP}{dz} = \frac{\sigma_e}{2} \int_{-d}^{d} \int_{-W/2}^{W/2} |E|^2 \, dy \, dx \qquad (5\text{-}200)$$

where

$$|E|^2 = |E_x|^2 + |E_y|^2 + |E_z|^2 \qquad (5\text{-}201)$$

$$\sigma_e = \omega\epsilon_1 \tan \delta \qquad (5\text{-}202)$$

and $\tan \delta$ is the loss tangent of the dielectric. The attenuation in nepers per unit length due to dielectric loss is, therefore,

$$\alpha_e = \frac{\sigma_e \int_{-d}^{d} \int_{-W/2}^{W/2} |E|^2 \, dy \, dx}{2 \, \text{Re} \int_{-\infty}^{\infty} \int_{-W/2}^{W/2} (\mathbf{E} \times \mathbf{H}^*)_z \, dy \, dx} \qquad (5\text{-}203)$$

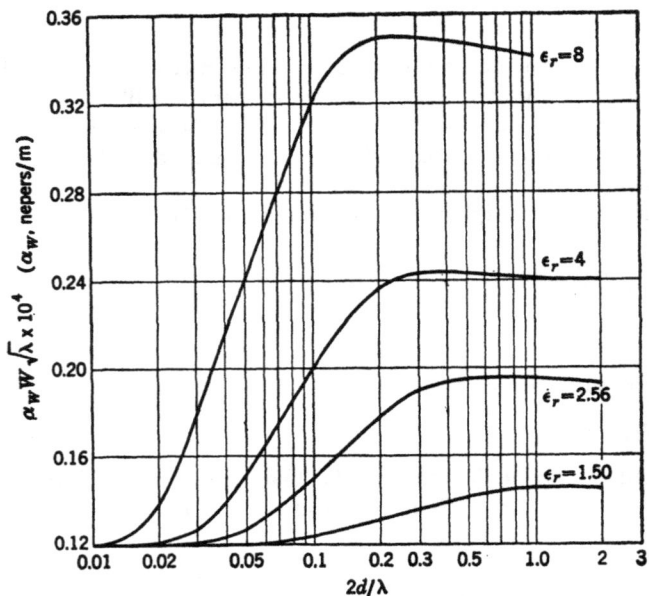

Figure 5-29.   Attenuation due to wall loss for TE_0 mode. Wall material is copper ($\sigma_w = 5.80 \times 10^7$ mhos/m). (From Cohn, Propagation in a Dielectric Loaded Parallel Plane Waveguide, *IRE Trans. Microwave Theory Tech.*, April, 1959, pp. 202–208.)

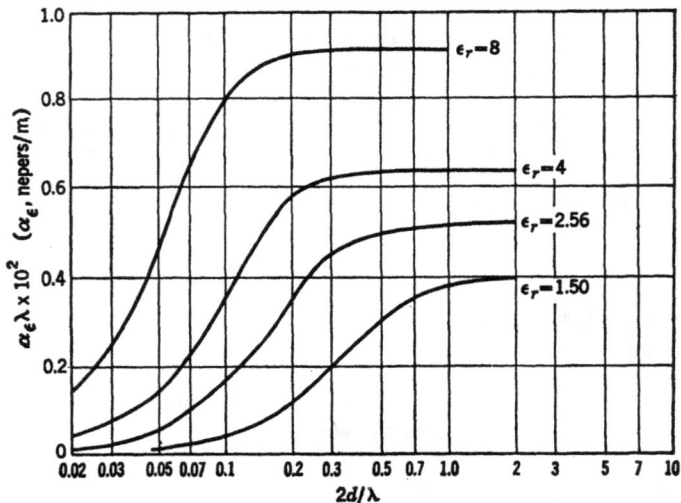

Figure 5-30.   Attenuation due to dielectric loss for TE_0 mode (tan $\delta = 0.001$). (From Cohn, Propagation in a Dielectric Loaded Parallel Plane Waveguide, *IRE Trans. Microwave Theory Tech.*, April, 1959, pp. 202–208.)

Again the field components are those of the particular mode of interest as determined for the lossless case.

Some graphs of wall loss and dielectric loss for the TE_0 mode[1] are shown in Figs. 5-29 and 5-30 for a structure with copper walls and with dielectric material with tan $\delta = 0.001$.

ADDITIONAL REFERENCES

Slotted and Holey Waveguides

1. H. Jasik (ed.), "Antenna Engineering Handbook," chap. 16, McGraw-Hill, New York, 1961.
2. S. Nishida, Theory of Thin Dielectric Cover on Slotted Rectangular Waveguide Antenna, *Res. Rept.* R-754-59 PIB-682, Polytechnic Institute of Brooklyn, July, 1959.

Yagi-Uda Arrays

3. F. Serracchioli, The Bandwidth of Uniform Arrays of Parasitic Dipoles, *Ohio State Univ. Res. Found. Antenna Lab. Rept.* 662-33, June, 1959.
4. D. K. Reynolds, Broadband Traveling Wave Antenna, *IRE Natl. Conv. Record,* Part I, pp. 99–107, 1957.
5. D. L. Sengupta, On Uniform and Linearly Tapered Long Yagi Antennas, *IRE Trans. Antennas Propagation,* **AP-8**(1):11–17, January, 1960.
6. H. Jasik (ed.), "Antenna Engineering Handbook," chap. 5, McGraw-Hill, New York, 1961.

Dielectric Rod Antenna

7. S. A. Schelkunoff, "Electromagnetic Waves," pp. 425–428, Van Nostrand, Princeton, N.J., 1943.
8. G. C. Southworth, "Principles and Applications of Waveguide Transmission," Van Nostrand, Princeton, N.J., 1950.
9. H. Jasik (ed.), "Antenna Engineering Handbook," chap. 16, McGraw-Hill, New York, 1961.

PROBLEMS

5-1. Verify Eqs. (5-7) and (5-8). *Hint:* For a wide slot the field in the slotted waveguide of Fig. 5-1 approximates that of a closed guide of width $2d$. Good results are obtained in this case by first formulating a transverse resonance problem for a short-circuited line of length d and terminated in susceptance B (given as $\mathcal{B}$ when normalized to Y_0 of the line). A perturbation solution of the problem about the transverse phase constant $\pi/2d$ of the closed guide gives an improved

---

[1] These data were taken from M. Cohn, Propagation in a Dielectric Loaded Parallel Plane Waveguide, *IRE Trans. Microwave Theory Tech.*, April, 1959, pp. 202–208.

234   Traveling Wave Antennas

phase constant [Eq. (5-9)] that can be used in the transverse resonance problem of the complete structure which includes normalized conductance $\mathcal{G}$ as well as susceptance $\mathcal{B}$.

5-2. Verify Eqs. (5-59) and (5-60).

5-3. Verify Eqs. (5-128) and (5-129).

5-4. Derive Eq. (5-192).

5-5. Verify Eq. (5-196).

5-6. A relative dielectric constant ϵ_r less than unity can be achieved in a plasma. Discuss the TM circular dielectric rod [(Eqs. (5-57), (5-62), (5-63), and (5-64)] for $0 \lesssim \epsilon_r < 1$ and compare it with the slotted TM_{01} cylinder in Fig. 5-6P. The

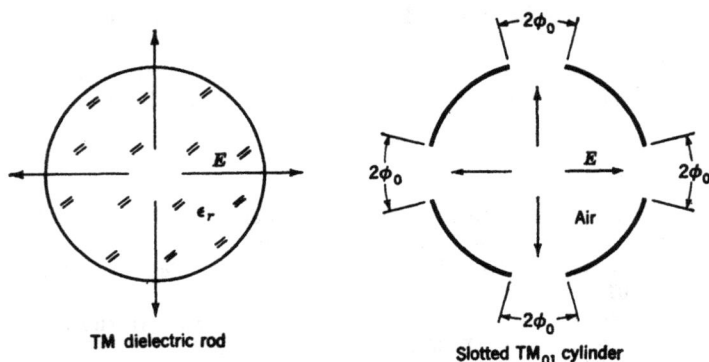

TM dielectric rod

Slotted TM_{01} cylinder

Figure 5-6P.

slotted cylinder is a fast wave structure and acts like a dielectric medium with index between 0 and 1. It can be analyzed by the transverse resonance method or the variational method of Chap. 4. Find γ, κ_1, and κ_0 when $\epsilon_r = 0$. What kind of waves can exist on the dielectric rod for $\epsilon_r < 0$?

5-7. Determine the elements of the determinant in Eq. (5-81).

5-8. Determine the elements of the determinant in Eq. (5-82).

5-9. Determine the elements of the determinant in Eq. (5-136).

CHAPTER **6**

DESIGN DATA

—PLANAR SOURCES

6-1. INTRODUCTION

The structures in Chap. 5 are usually less than a wavelength in extent in cross section; hence they have been classified as line sources. They are useful as transmission lines, line radiators, and elements in antenna arrays.

In this chapter we consider propagation along planar structures many wavelengths wide, wide enough, in fact, to permit analysis on the basis of a structure of infinite extent. Structures capable of both fast and slow wave propagation are considered. As in the previous chapter the classification is with respect to mode of operation, that is, TM, TE, or hybrid (TM + TE).

The planar structures described in the following pages are well suited for use as large, two-dimensional, flush-mounted antennas. The structures having isotropy in the plane of the structure may also be used as surface wave lenses (see Chap. 8). When properly excited and terminated, the slow wave planar structures also may be used as transmission lines.

As in the previous chapter on line sources, the analysis is based on structures of infinite length. For a given mode of operation a practical structure of finite length may be assumed to have the same propagation constant as the infinitely long structure with the same cross section.

Methods of exciting desired modes of operation for planar structures are mentioned briefly. A more detailed discussion of traveling wave excitation is given in the next chapter.

6-2. TE PLANAR SOURCES

The planar structures will be taken to lie in the yz plane with propagation in the z direction. Thus for TE operation the field E_z will be zero. Some

235

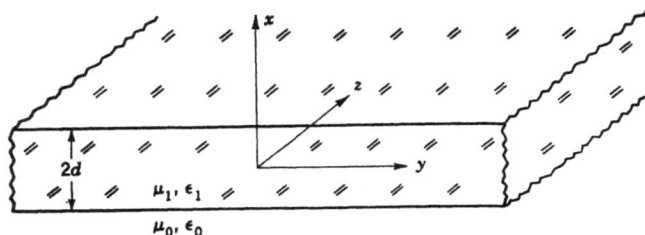

Figure 6-1.   Infinite sheet of homogeneous, isotropic dielectric material.

practical structures which will support TE waves are dielectric sheet, dielectric sheet on a ground plane, and various parallel plate structures with slots or holes in one plate to form a radiating slow or fast wave structure.

A. TE Dielectric Sheet

A dielectric sheet structure is illustrated in Fig. 6-1. The natural modes of operation of such a structure include both TE and TM modes. This is also true for the case where a conducting sheet is placed at $x = 0$. TE and TM surface wave propagation along a dielectric sheet on a ground plane has been the subject of a good many investigations. Perhaps the earliest was that of Attwood.[1] Later, Zucker[2] gave an excellent summary of surface wave propagation on corrugated and dielectric-coated conductors and discussed the use of transverse resonance for determining the propagation constants for TE and TM waves. Transverse resonance is extremely useful for determining the propagation constant of a planar structure if the surface impedance is known; however, it does not give the field configurations. To find the fields for a particular mode it is necessary to apply Maxwell's equations and the appropriate boundary conditions.

The structure in Fig. 6-1 may be analyzed by applying the results for the dielectric slab between parallel plates from Sec. 5-4C. For $W = \infty$, the dielectric slab between parallel plates is simply the infinite sheet in Fig. 6-1.

With propagation in the z direction and no variation in the y direction the results from Sec. 5-4C for the TE case are directly applicable and are repeated here for convenience. The field components are

$$E_y = A_1 \cos\left(\kappa_1 x - \frac{n\pi}{2}\right) \qquad -d \lesseqgtr x \lesseqgtr d \qquad (6\text{-}1)$$

$$E_y = A_0 e^{-j\kappa_0 x} \qquad\qquad x > d \qquad\qquad (6\text{-}2)$$

$$E_y = (-1)^n A_0 e^{j\kappa_0 x} \qquad x < -d \qquad\qquad (6\text{-}3)$$

[1] S. S. Attwood, Surface Wave Propagation over a Coated Plane Conductor, *J. Appl. Phys.*, **22**:504, 1951.

[2] F. J. Zucker, The Guiding and Radiation of Surface Waves, *Proc. Symp. Mod. Advan. Microwave Tech.*, pp. 403–436, Polytechnic Institute of Brooklyn, New York, November, 1954.

$$H_x = \frac{-\gamma}{j\omega\mu} E_y \tag{6-4}$$

$$H_z = \frac{-1}{j\omega\mu} \frac{\partial E_y}{\partial x} \tag{6-5}$$

where

$$0 \lesssim \kappa_1 x - \frac{n\pi}{2} \lesssim \frac{\pi}{2} \qquad n = 0, 1, 2, \ldots$$

and variation of the form $e^{j\omega t - \gamma z}$ is assumed.

Equating tangential components at $x = d$ or $x = -d$ gives the characteristic equation for TE_n modes as

$$\kappa_1 \mu_0 \tan\left(\kappa_1 d - \frac{n\pi}{2}\right) = j\kappa_0 \mu_1 \tag{6-6}$$

Equation (6-6) in conjunction with

$$\kappa_1{}^2 - \gamma^2 = k^2 \epsilon_r \tag{6-7}$$

and

$$\kappa_0{}^2 - \gamma^2 = k^2 \tag{6-8}$$

may be solved for γ. If the material is lossless, γ will be imaginary ($\gamma = j\beta$). Solutions for n odd apply to the structure in Fig. 6-2 where a conducting sheet exists at $x = 0$. For $n = 1$, for example, Eq. (6-6) gives

$$-\mu_0 \kappa_1 \cot \kappa_1 d = j\kappa_0 \mu_1 \tag{6-9}$$

This is the lowest-order TE mode (TE_1) for a dielectric sheet on a ground plane.

Equation (6-9) may be obtained by simply matching impedance at $x = d$. That is,

$$\overrightarrow{Z}\Big]_{x=d} = -\overleftarrow{Z}\Big]_{x=d} \tag{6-10}$$

where the arrow to the right denotes that the impedance is that seen in the positive x direction and the arrow to the left denotes the impedance

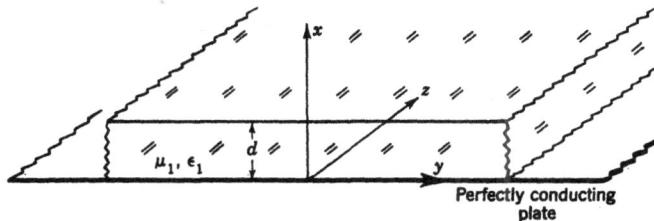

Figure 6-2.   Dielectric sheet on ground plane.

seen in the negative x direction. This was referred to as transverse resonance in Chap. 4. It corresponds in circuitry to a determination of the free oscillations of a resonant circuit.

For the TE case the impedance $\overrightarrow{Z}\Big]_d$ is given by

$$\overrightarrow{Z}\Big]_d = \frac{E_y}{H_z} = \frac{\omega\mu_0}{\kappa_0} \tag{6-11}$$

and the impedance $\overleftarrow{Z}\Big]_d$ for a perfectly conducting sheet at $x = 0$ is

$$\overleftarrow{Z}\Big]_d = \frac{j\omega\mu_1}{\kappa_1} \tan \kappa_1 d \tag{6-12}$$

Substituting Eqs. (6-11) and (6-12) into (6-10) gives Eq. (6-9). For a perfect dielectric μ_1 is equal to μ_0, the permeability of free space.

Equations (6-7) and (6-8) may be combined and used with the transcendental equation (6-6) in a graphical solution for γ. Equations (6-7) and (6-8) give the equation of a circle in the $\kappa_1 d$, $\kappa_0 d$ plane. For a lossless dielectric with $\epsilon_r > 1$, κ_1 is real and κ_0 is imaginary and the equation for the circle is

$$(\kappa_1 d)^2 + |\kappa_0|^2 d^2 = (kd)^2(\epsilon_r - 1) \tag{6-13}$$

Intersection of the curve from Eq. (6-6) (with both sides multiplied by d) and the circle of Eq. (6-13) gives κ_0 (or κ_1) from which γ may be obtained by Eq. (6-8) [or (6-7)]. For the lossless case, $\gamma = j\beta$ and the velocity ratio c/v is given in terms of κ_0 by

$$\frac{c}{v} = \frac{\beta}{k} = \left[1 - \left(\frac{\kappa_0 d}{kd}\right)^2\right]^{1/2} \tag{6-14}$$

Some curves of c/v versus d/λ are given in Fig. 6-3 for the TE_0 and TE_1 modes. The TE_0 mode is the dominant mode for the plane dielectric sheet of thickness $2d$, and the TE_1 mode is the dominant mode for a sheet of thickness d on a ground plane.

The TE_0 mode in the dielectric sheet has no cutoff, as shown in Fig. 6-3. Higher-order modes (TE_1, etc.), however, have cutoff frequencies. This occurs at $c/v = 1$ (that is, $\kappa_0 = 0$). For d/λ smaller than the value at which $c/v = 1$, the mode will not propagate. The cutoff condition for TE_n modes was derived in Sec. 5-4C. The critical thickness for cutoff for the nth mode is

$$\frac{2d}{\lambda} = \frac{n}{2\sqrt{\epsilon_r - 1}} \qquad \epsilon_r > 1 \tag{6-15}$$

For lossy dielectrics one can apply the above analysis for a complex permittivity

$$\epsilon_1 = \epsilon_1' - j\epsilon_1'' \tag{6-16}$$

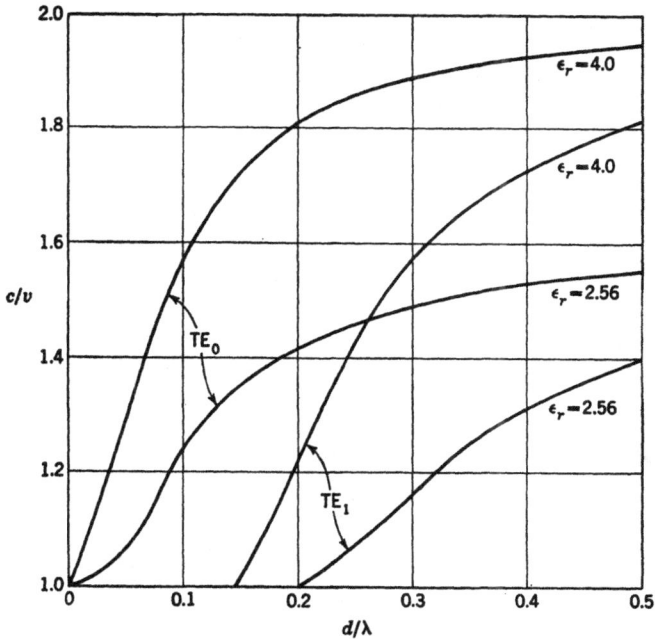

Figure 6-3. Velocity ratio c/v for an infinite sheet of perfect dielectric of thickness d on a lossless ground plane (TE$_1$ case) or thickness $2d$ with no ground plane (TE$_0$ case and TE$_1$ case).

However, it is simpler for low loss dielectric to use the method of Sec. 5-4C that results in the expression for attenuation constant given by Eq. (5-203), which is directly applicable here.

The analysis for TE modes in a dielectric slab that has been presented is based on a homogeneous, isotropic dielectric slab with relative dielectric constant ϵ_r and $\mu_1 = \mu_0$. The analysis is also applicable to an isotropic, homogeneous sheet with relative permeability μ_r.

Although the exact analysis applies to a sheet of infinite width or a sheet of finite width between conducting plates of infinite extent, it is a good approximation for sheets more than several wavelengths wide without side plates.

Excitation of TE waves on sheet structures such as shown in Figs. 6-1 and 6-2 but of finite width is easily accomplished in practice. For propagation in the z direction all that is needed is a source with principal component E_y and, for best results, extending the width of the sheet. For example, a sectoral horn with polarization parallel to the sheet makes an excellent feed. For a sheet on a ground plane it is also possible to excite a TE wave through slots in the conducting sheet.

Layered Dielectric

The analysis of traveling wave propagation over layers of dielectric sheets is complicated by the additional boundary conditions. For TE propagation in the three-layer structure of Fig. 6-4 the field E_y is given by

$$E_y = A_0 e^{-\gamma_0 x} e^{-\gamma z} \qquad \text{in region 0} \qquad (6\text{-}17)$$
$$E_y = (A_1 e^{-\gamma_1 x} + B_1 e^{\gamma_1 x}) e^{-\gamma z} \qquad \text{in region 1} \qquad (6\text{-}18)$$

$$E_y = A_2 \cosh\left(\gamma_2 x - j\frac{n\pi}{2}\right) e^{-\gamma z} \qquad \text{in region 2} \qquad (6\text{-}19)$$

where propagation constant $\gamma_m = \alpha_m + j\beta_m$ (for $m = 0, 1, 2$) is used for the x direction. If region 0 is taken to be free space and regions 1 and 2 are lossless and isotropic, then from the relations

$$\gamma_0^2 - \beta^2 = -\omega^2 \mu_0 \epsilon_0 = -k^2 \qquad (6\text{-}20)$$
$$\gamma_1^2 - \beta^2 = -\omega^2 \mu_1 \epsilon_1 = -k^2 \mu_{1r} \epsilon_{1r} \qquad (6\text{-}21)$$
$$\gamma_2^2 - \beta^2 = -\omega^2 \mu_2 \epsilon_2 = -k^2 \mu_{2r} \epsilon_{2r} \qquad (6\text{-}22)$$

we find that $\beta_0 = 0$ and γ_1 and γ_2 can assume only real or imaginary values and γ_1 and γ_2 cannot both be real [see Eqs. (6-24) to (6-26)].

Using Eq. (6-5) for H_z and requiring E_y and H_z to be continuous at the boundaries gives a system of four equations. Setting the determinant equal to zero gives the transcendental equation

$$\tanh\left(\gamma_2 d_2 - j\frac{n\pi}{2}\right) = \frac{-\mu_2\gamma_1}{\mu_1\gamma_2} \frac{\mu_0\gamma_1 \tanh \gamma_1 d_1 + \mu_1\alpha_0}{\mu_1\alpha_0 \tanh \gamma_1 d_1 + \mu_0\gamma_1}$$
$$n = 0, 1, 2, \ldots \qquad (6\text{-}23)$$

Note that for $d_1 = 0$, $\alpha_0 = j\kappa_0$, and $\gamma_2 = j\kappa_1$, Eq. (6-23) reduces to Eq. (6-6).

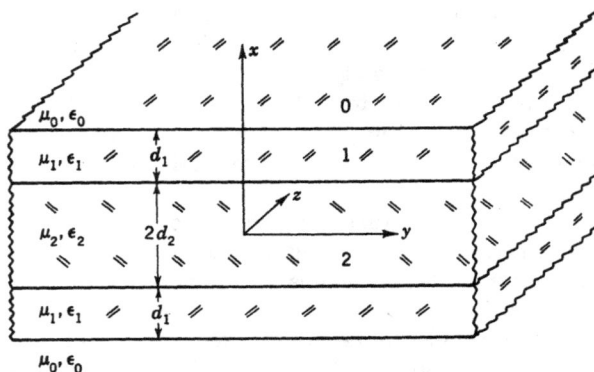

Figure 6-4.   Three-layer dielectric sheet structure.

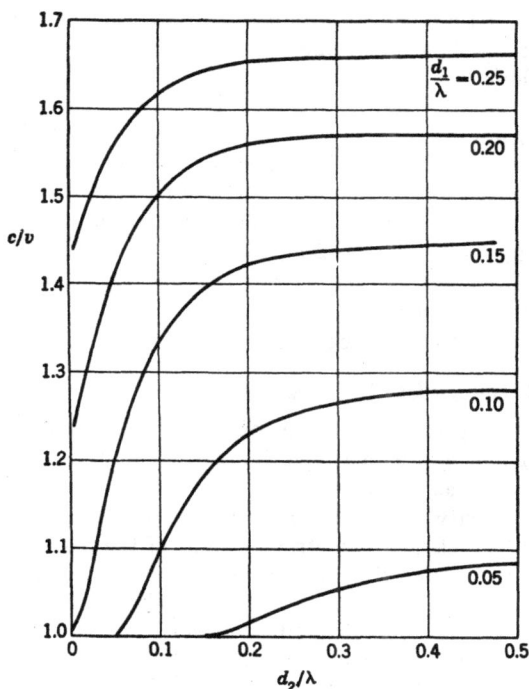

Figure 6-5. Velocity ratio c/v versus air-layer thickness for dielectric layer ($\epsilon_1/\epsilon_0 = 4$) above a ground plane and with TE_1 surface wave propagation.

For the three-layer structure in Fig. 6-4, Richmond[1] has shown that three cases cover the possible forms of γ_1 and γ_2 for lossless, isotropic regions and where $\gamma_0 = \alpha_0$. These are

$$\left.\begin{array}{l} \gamma_0 = \alpha_0 \\ \gamma_1 = j\beta_1 \\ \gamma_2 = \alpha_2 \end{array}\right\} \quad \text{case 1} \qquad (6\text{-}24)$$

$$\left.\begin{array}{l} \gamma_0 = \alpha_0 \\ \gamma_1 = \alpha_1 \\ \gamma_2 = j\beta_2 \end{array}\right\} \quad \text{case 2} \qquad (6\text{-}25)$$

$$\left.\begin{array}{l} \gamma_0 = \alpha_0 \\ \gamma_1 = j\beta_1 \\ \gamma_2 = j\beta_2 \end{array}\right\} \quad \text{case 3} \qquad (6\text{-}26)$$

The fields are evanescent in the x direction in region 0, and only slow wave propagation ($c/v > 1$) exists in the z direction.

Equation (6-23) along with Eqs. (6-20) to (6-22) determine the phase constant β for TE modes on the structure in Fig. 6-4. For n odd ($n = 1, 3$,

[1] J. H. Richmond, Reciprocity Theorems and Plane Surface Waves, *Ohio State Univ. Eng. Expt. Sta. Bull.* 176, Columbus, July, 1959.

. . .) the field E_y is zero at $x = 0$; hence a ground plane may be inserted at $x = 0$, and the above results apply to a two-layer structure on a ground plane.

Cutoff for the above modes (which may be designated TE_n modes) occurs when $\alpha_0 = 0$. It can be seen from Eq. (6-23) that there is no cutoff for the dominant TE mode ($n = 0$). As d_1 and d_2 go to zero, the TE_0 mode degenerates into a plane wave in free space.

A case of considerable practical importance (see Sec. 8-3) occurs in Fig. 6-4 for region 2 equal to free space (that is, $\mu_2 = \mu_0$ and $\epsilon_2 = \epsilon_0$) and TE_1 operation with a ground plane at $x = 0$. Velocity ratio c/v as a function of layer thickness for this case is given in Fig. 6-5 for a relative dielectric constant of 4 in region 1.

B. Holey Plate Structure

A holey plate structure consisting of a parallel plate waveguide with holes in one of the plates is illustrated in Fig. 6-6. When such a structure is excited as a TE_{0n} waveguide, the plate spacing may be adjusted to provide a velocity ratio c/v from nearly zero to a value approaching the square root of the relative dielectric constant of the material between the plates. As an antenna this structure gives horizontally polarized radiation (electric field parallel to the surface). If both sheets are holey, bidirectional radiation ($c/v < 1$) or endfire radiation ($c/v \gtrless 1$) may be obtained.

An expression for the propagation constant may be obtained by assuming that the holey plate structure is composed of stacked TE_{0n} wave-guides with holes in a narrow wall of each guide. Since the electric field would be everywhere normal to the common walls ($\mathbf{E} = \mathbf{y}E_y$ in this case), the walls may be removed to form the structure in Fig. 6-6. Thus the analysis of the structure in Fig. 6-6 reduces to that of a holey waveguide, which has been treated in Sec. 5-2B. There is a difference, however, in that the

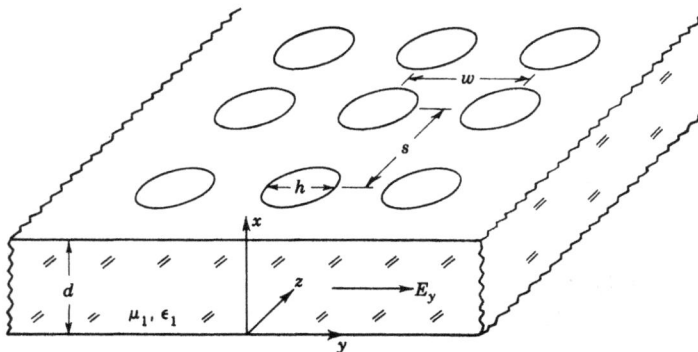

Figure 6-6.   TE holey plate structure.

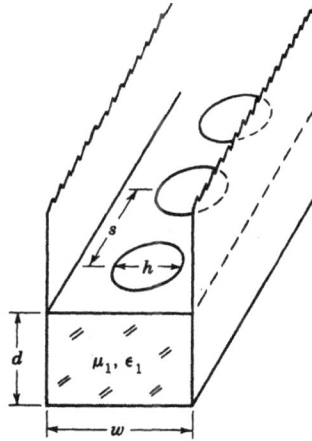

Figure 6-7.   An element of the holey plate structure.

holey waveguide that makes up the parallel plate structure radiates into a parallel plate region rather than half space as in Sec. 5-2B. The structure to be considered here is shown in Fig. 6-7. The characteristic equation for the structure in Fig. 6-7 may be obtained by applying the transverse resonance procedure as in Sec. 5-2B. The equivalent transverse circuit for Fig. 6-7 is illustrated in Fig. 6-8 in which a transmission line with propagation constant κ_1 and length d is shorted at one end and terminated in admittance $G - jB$ at the other end. Transverse resonance requires that

$$\overrightarrow{Y}\bigg]_d = -\overleftarrow{Y}\bigg]_d \qquad (6\text{-}27)$$

which gives

$$-j\frac{\kappa_1}{\omega\mu_1}\cot\kappa_1 d = -G + jB \qquad (6\text{-}28)$$

where

$$G = \frac{\kappa_0}{\omega\mu_0} \qquad (6\text{-}29)$$

and

$$B \approx \frac{6sw}{\omega\mu h^3} \qquad (6\text{-}30)$$

Figure 6-8.   Equivalent transverse network for holey plate structure.

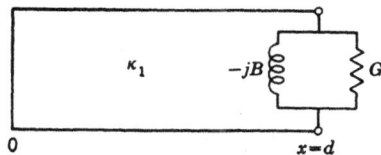

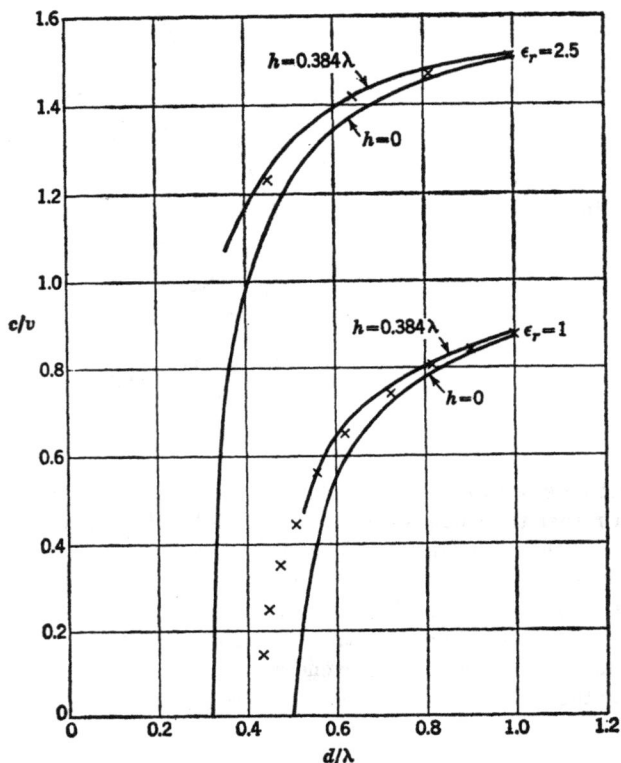

Figure 6-9.   Velocity ratio c/v for the holey plate structure in Fig. 6-6 with $w = s = 0.423\lambda$. The $\times$'s are measured points at a wavelength of 3.1 cm for holes in a $\frac{1}{16}$-in. plate. Solid lines are computed for zero plate thickness.

The conductance G is simply the admittance of a TE_{01} wave propagating between infinite parallel plates. The susceptance B is that of a circular aperture in an iris of zero thickness in a TE_{01} parallel plate waveguide.[1]

An approximate solution of Eq. (6-28) may be obtained by the perturbation method described in Sec. 4-8. Such a solution is generally adequate if the traveling wave structure is not operated near cutoff.

Solving for the propagation constant $\gamma = \alpha + j\beta$ yields

$$\alpha \approx \frac{\lambda_{g0}}{2d^2} \frac{G}{G^2 + B^2} \qquad (6\text{-}31)$$

$$\beta \approx \frac{2\pi}{\lambda_{g0}} \left(1 + \frac{\lambda_{g0}^2}{4\pi d^2} \cdot \frac{B}{G^2 + B^2}\right) \qquad (6\text{-}32)$$

[1] N. Marcuvitz, Waveguide Circuit Theory: Coupling of Waveguides by Small Apertures, *Polytech. Inst. Brooklyn Rept.* R-157-47, PIB-106, 1947.

where

$$\lambda_{g0} = \frac{\lambda}{[\epsilon_r - (\lambda/2d)^2]^{\frac{1}{2}}} \tag{6-33}$$

$$\mathcal{G} \approx \frac{[\pi^2 - (kd)^2(\epsilon_r - 1)]^{\frac{1}{2}}}{\pi} \tag{6-34}$$

$$\mathcal{B} \approx \frac{6wsd}{\pi h^3} \tag{6-35}$$

The quantities $\mathcal{G}$ and $\mathcal{B}$ are G and B, respectively, normalized with respect to $\kappa_1/\omega\mu_1$. In the perturbation solution the transverse propagation constant for the unperturbed guide ($\kappa_1 = \pi/d$) is used and μ_1 is taken to be equal to μ_0.

Graphs of α and c/v for the dominant TE mode in a holey plate structure are shown in Figs. 6-9 and 6-10. Computed curves were obtained from Eqs. (6-31) and (6-32). Some measured data are included for comparison to indicate the accuracy of the approximate analysis. The results indicate that the attenuation constant α is significantly influenced by the thickness of the holey plate. The effect of the plate thickness can be taken into account in the transverse resonance method (see Sec. 4-8).

It may be noted that the holey plate effectively forms an inductive sheet. An inductive grid structure made up of thin conducting strips is described in the next section.

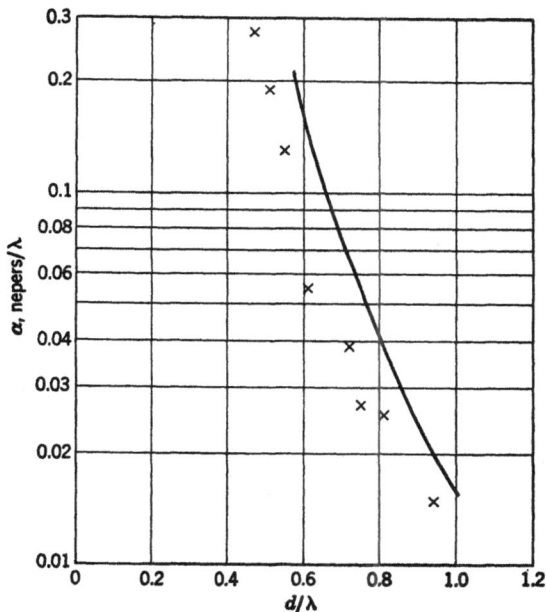

Figure 6-10. Attenuation constant α for the $\epsilon_r = 1$ case in Fig. 6-9.

C. TE Inductive Grid Structure

A parallel plate structure consisting of a perfectly conducting sheet and an inductive grid as shown in Fig. 6-11 is useful as a traveling wave antenna. The inductive grid can be formed by round conducting wires or flat conducting strips. The polarization is parallel to the inductive grid ($\mathbf{E} = \mathbf{y}E_y$) as in the holey plate case. The inductive grid as shown in Fig. 6-11 cannot be made isotropic in the plane of the structure, whereas the holey plate structure can when $s = w$ (see Fig. 6-6). However, the inductive grid structure is simple and provides a wide range of propagation constants. Virtually independent control of the amplitude and the phase of the illumination over an inductive grid aperture may be obtained by varying strip sizes and spacings.

An inductive grid structure consisting of thin flat strips has been studied by Honey.[1] The characteristic equation for such a structure may be obtained by the transverse resonance procedure. Applying transverse resonance one obtains

$$\frac{j\kappa_1}{\omega\mu_1} \cot \kappa_1 d = \left. \vec{Y} \right]_d \tag{6-36}$$

where $\left. \vec{Y} \right]_d$ is the admittance of the inductive grid in free space for a TE wave at normal incidence. The equivalent circuit for $\left. \vec{Y} \right]_d$ is shown in Fig. 6-12.

Equation (6-36) may be written in normalized form as

$$j \cot \kappa_1 d = \frac{\kappa_0 \mu_1}{\kappa_1 \mu_0} - j \frac{B\omega\mu_1}{\kappa_1}$$
$$= \mathcal{G} - j\mathcal{B} \tag{6-37}$$

[1] R. C. Honey, A Flush Mounted Leaky Wave Antenna with Predictable Patterns, *IRE Trans. Antennas Propagation*, **AP-7**:320–329, October, 1959.

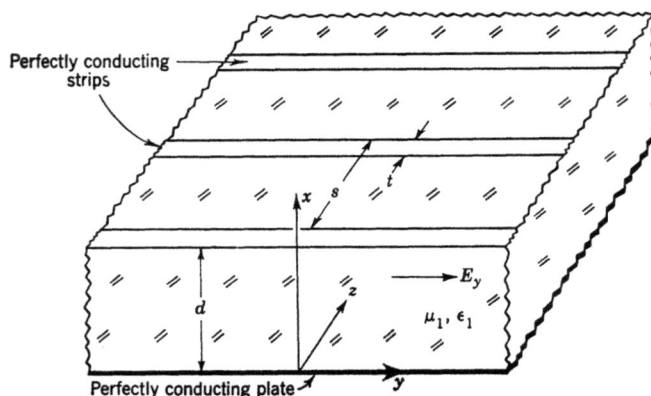

Figure 6-11. Inductive grid traveling wave structure.

Figure 6-12. Equivalent circuit for inductive grid of thin conducting strips.

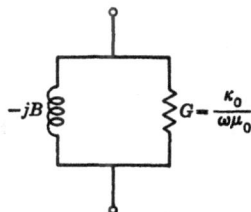

The usual perturbation solution gives

$$\alpha \approx \frac{\lambda_{g0}}{2d^2}\left(\frac{\mathcal{G}}{\mathcal{G}^2 + \mathcal{B}^2}\right) \tag{6-38}$$

$$\frac{c}{v} \approx \frac{\lambda}{\lambda_{g0}}\left(1 + \frac{\lambda_{g0}^2}{4\pi d^2}\frac{\mathcal{B}}{\mathcal{G}^2 + \mathcal{B}^2}\right) \tag{6-39}$$

where

$$\lambda_{g0} = \frac{\lambda}{[\epsilon_r - (\lambda/2d)^2]^{1/2}} \tag{6-40}$$

$$\mathcal{G} \approx \frac{\mu_1[\pi^2 - (kd)^2(\epsilon_r - 1)]^{1/2}}{\pi\mu_0} \tag{6-41}$$

For a grid of thin, perfectly conducting strips of width t and center-to-center separation s the normalized susceptance $\mathcal{B}$ is given by[1]

$$\mathcal{B} \approx \frac{2\,d\mu_1}{\mu_0 s \ln\left[\csc\left(\pi t/2s\right)\right]} \qquad s \ll \lambda \tag{6-42}$$

where κ_1 has been taken to be that of the closed guide, that is, $\kappa_1 = \pi/d$.

For a grid of perfectly conducting wires of radius r and center-to-center separation s the equivalent circuit[2] is shown in Fig. 6-13. A perturbation solution gives

$$\alpha \approx \frac{\lambda_{g0}}{2d^2}\frac{Z_0 X_a^2}{Z_0^2 + (X_a - X_b)^2} \tag{6-43}$$

$$\frac{c}{v} \approx \frac{\lambda}{\lambda_{g0}}\left[1 + \frac{\lambda_{g0}^2}{4\pi d^2}\frac{(X_a - X_b)(Z_0^2 - 2X_a X_b - X_b^2)}{Z_0^2 + (X_a - X_b)^2}\right] \tag{6-44}$$

[1] N. Marcuvitz (ed.), "Waveguide Handbook," p. 284, McGraw-Hill, New York, 1951.
[2] Ibid., p. 286.

Figure 6-13. Equivalent circuit for inductive grid of wires.

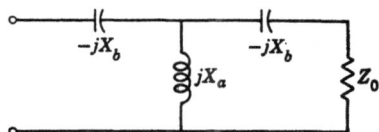

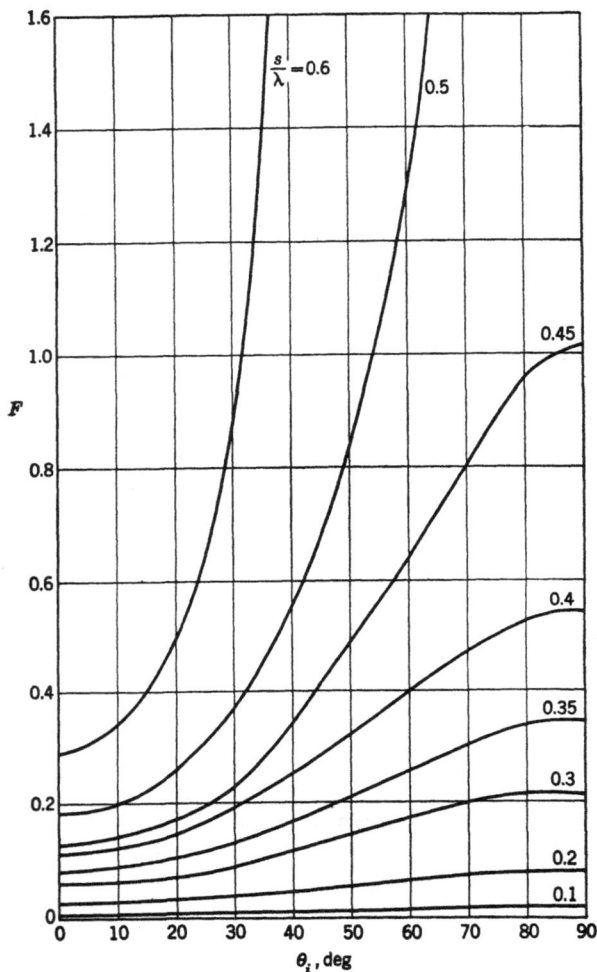

Figure 6-14. Correction term F for a wire grid as a function of spacing and angle of incidence.

where λ_{g0} is given by Eq. (6-40) and

$$Z_0 = \frac{1}{\mathcal{G}} \approx \frac{\pi\mu_0}{\mu_1[\pi^2 - (kd)^2(\epsilon_r - 1)]^{1/2}} \tag{6-45}$$

$$X_a \approx \frac{\mu_0 s \ln{(s/2\pi r)}}{2\,d\mu_1} \qquad s \ll \lambda^{\cdot} \tag{6-46}$$

$$X_b \approx \frac{2\mu_0\pi^2 r^2}{d\mu_1 s} \tag{6-47}$$

and κ_1 has been taken to be that for the closed guide, that is, π/d.

For $t \ll s$, the normalized shunt susceptance for the conducting strips [Eq. (6-42)] becomes

$$\mathfrak{B} \approx \frac{2\,d\mu_1}{\mu_0 s\,\ln\,(2s/\pi t)} \tag{6-48}$$

whereas for the conducting wire case [Eq. (6-47)] X_b may be neglected and from Eq. (6-46)

$$\mathfrak{B} = \frac{1}{X_a} \approx \frac{2\,d\mu_1}{\mu_0 s\,\ln\,(2s/4\pi r)} \tag{6-49}$$

Comparing Eqs. (6-48) and (6-49) we see that for $t \ll s$ and $s \ll \lambda$ the results for the conducting strip case apply to the conducting wire case if we let t represent twice the wire diameter instead of the strip width.

If the spacing s is not sufficiently small in terms of the wavelength, a correction term F must be added to the susceptance $\mathfrak{B}$ in Eq. (6-42). The correction term F from MacFarlane[1] is shown in Fig. 6-14 as a function of

[1] G. G. MacFarlane, Surface Impedance of an Infinite Parallel Wire Grid at Oblique Angles of Incidence, *J. Inst. Elec. Engrs.* (*London*), **93**(III-A, 10): 1523–1527, 1946.

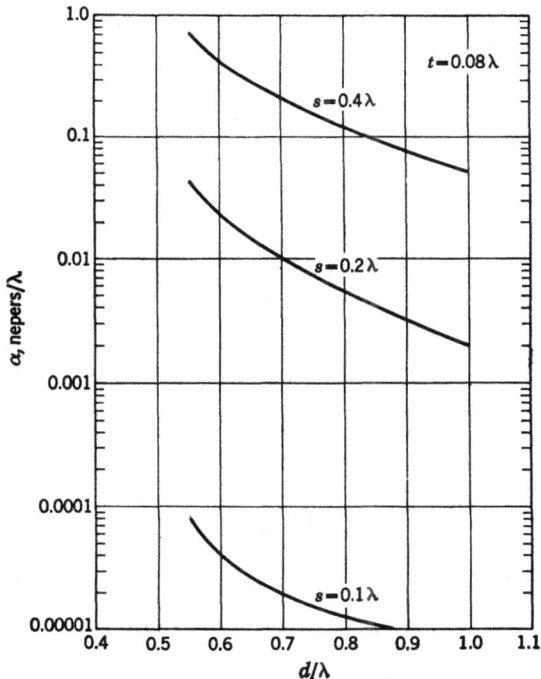

Figure 6-15. Attenuation constant α for grid of conducting strips calculated by Eq. (6-38) for $\epsilon_r = \mu_r = 1$. MacFarlane's correction term has been included.

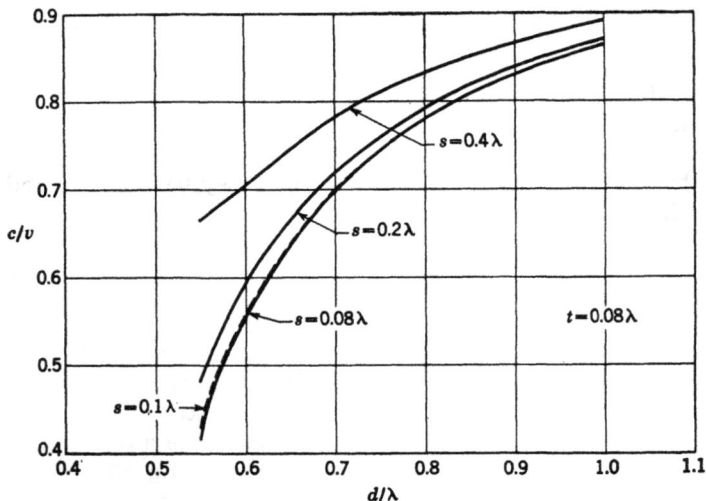

Figure 6-16. Velocity ratio c/v for grid of conducting strips calculated by Eq. (6-39) for $\epsilon_r = \mu_r = 1$. MacFarlane's correction term has been included.

incidence angle θ_i and spacing s for a wire grid.[1] The angle θ_i may be found approximately from

$$\theta_i \approx \sin^{-1}\frac{\lambda}{\lambda_{g0}} \tag{6-50}$$

where θ_i is measured from the normal to the grid. Including the correction term the normalized susceptance of the grid of conducting strips becomes

$$\mathcal{B} \approx \frac{2\pi\mu_1}{\kappa_1\mu_0 s\{\ln\,[\csc\,(\pi t/2s)] + F\}} \tag{6-51}$$

Some calculated curves of α and c/v as functions of grid spacing s and plate separation d are given in Figs. 6-15 and 6-16.

D. TE Capacitive Grid Structure

A parallel plate structure consisting of a perfectly conducting sheet and a capacitive grid as shown in Fig. 6-17 is also useful as a traveling wave antenna.[2] This is essentially an array of the TE slotted guides considered

---

[1] A more exact correction term that takes into account the presence of the conducting back wall has been derived by J. R. Wait in his paper Reflection from a Wire Grid Parallel to a Conducting Plane, *Can. J. Phys.*, **32**:571–579, September, 1954.

[2] R. C. Honey, Horizontally Polarized Long Slot Array, *Stanford Res. Inst. Tech. Rept. 47*, Stanford, Calif., August, 1954.

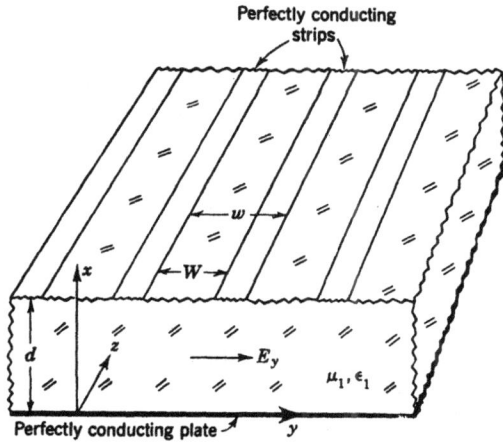

Figure 6-17.   TE capacitive grid structure.

in Sec. 5-2A. As a result of the orientation of the electric field, the analysis of the structure may be based on the parallel plate configuration shown in Fig. 6-18. The equivalent transverse network for the structure in Fig. 6-18 is shown in Fig. 6-19. From transverse resonance the normalized admittance relationship at $x = d$ is

$$j \cot \kappa_1 d = \frac{\kappa_0 \mu_1}{\kappa_1 \mu_0} + j\mathcal{B} \tag{6-52}$$

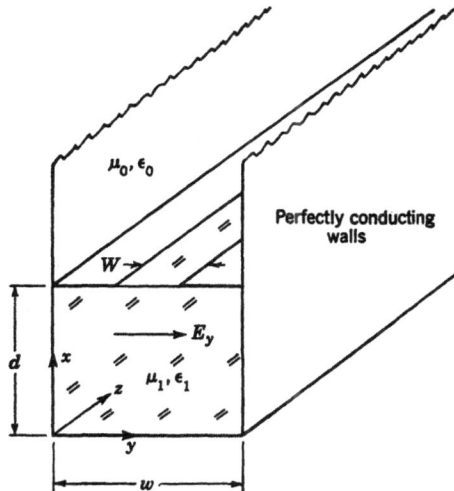

Figure 6-18.   Parallel plate structure with capacitive iris.

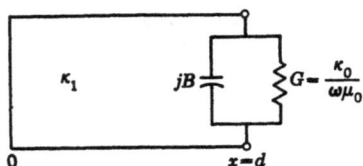

Figure 6-19. Equivalent network for the structure in Fig. 6-18.

If the wall in which the slot is cut is very thin, the susceptance $\mathfrak{B}$ is given by[1]

$$\mathfrak{B} = \mathfrak{B}_{int} + \mathfrak{B}_{ext}$$
$$\approx \frac{\kappa_1 w}{\pi} \ln\left(\csc \frac{\pi W}{2w}\right) + \frac{\kappa_0{}^2 w \mu_1}{\kappa_1 \mu_0 \pi} \ln\left(\csc \frac{\pi W}{2w}\right) \tag{6-53}$$

For operation well above cutoff a solution based on a perturbation about the closed guide (see Sec. 4-8) gives

$$\alpha \approx \frac{\lambda_{g0}}{2d^2} \frac{\mathfrak{G}}{\mathfrak{G}^2 + \mathfrak{B}^2} \tag{6-54}$$

$$\frac{c}{v} \approx \frac{\lambda}{\lambda_{g0}} \left(1 - \frac{\lambda_{g0}^2}{4\pi d^2} \frac{\mathfrak{B}}{\mathfrak{G}^2 + \mathfrak{B}^2}\right) \tag{6-55}$$

[1] Marcuvitz, *op. cit.*, p. 218.

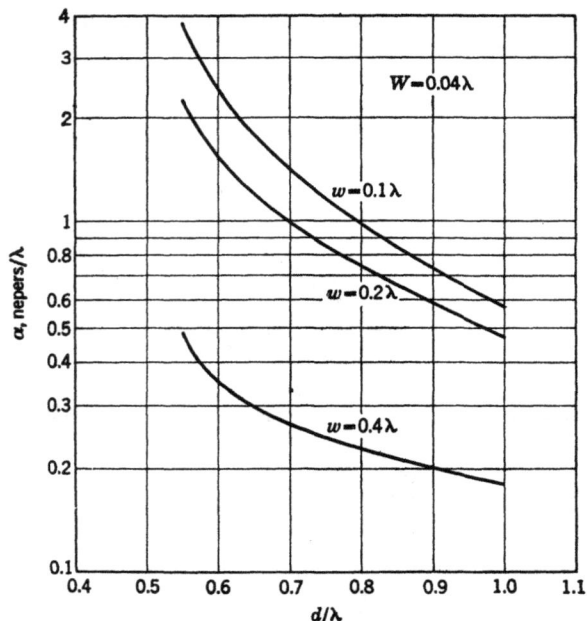

Figure 6-20. Attenuation constant α for the capacitive grid structure in Fig. 6-17. Equation (6-54) is used with $\epsilon_r = \mu_r = 1$.

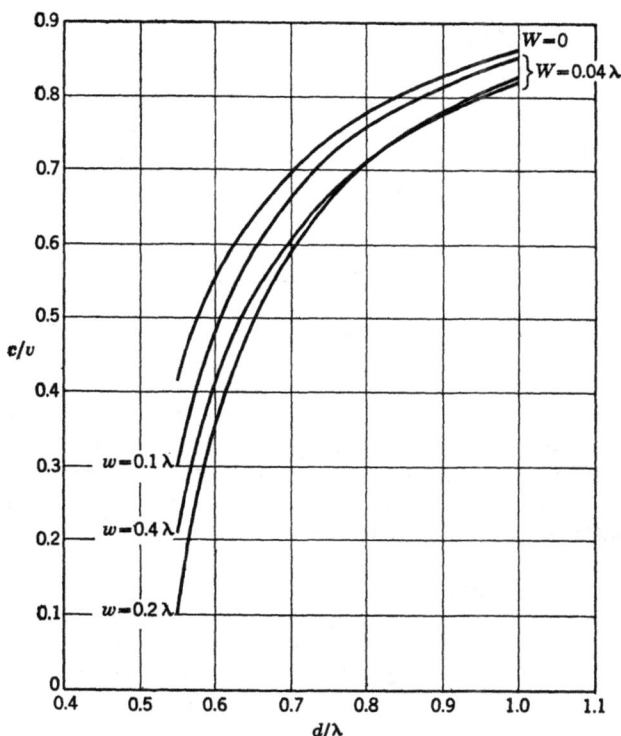

Figure 6-21. Velocity ratio c/v for the capacitive grid structure in Fig. 6-17. Equation (6-55) is used with $\epsilon_r = \mu_r = 1$.

where

$$\lambda_{g0} = \frac{\lambda}{[\epsilon_r - (\lambda/2d)^2]^{\frac{1}{2}}} \tag{6-56}$$

$$\mathcal{G} \approx \frac{\mu_1[\pi^2 - (kd)^2(\epsilon_r - 1)]^{\frac{1}{2}}}{\pi\mu_0} \tag{6-57}$$

$$\mathcal{B} \approx \left[\frac{w}{d} + \frac{[\pi^2 - (kd)^2(\epsilon_r - 1)]w\mu_1}{\pi^2\mu_0 d}\right]\ln\left(\csc\frac{\pi W}{2w}\right) \tag{6-58}$$

Some curves of α and c/v as functions of the geometry of the capacitive grid structure are shown in Figs. 6-20 and 6-21.

6-3. TM PLANAR SOURCES

We have seen in Sec. 6-2 that a number of planar structures support TE waves. Some of these structures also support TM waves. This section describes a number of structures suitable for use as TM planar sources. The

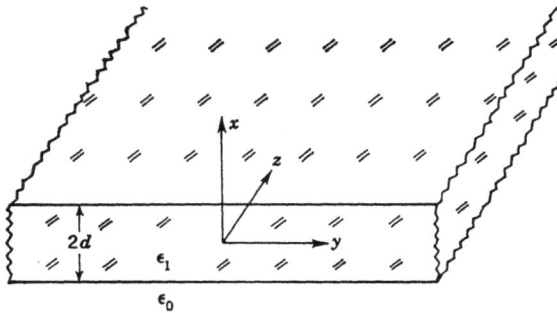

Figure 6-22. Infinite sheet of homogeneous, isotropic dielectric material.

structures will be taken to lie in the yz plane with propagation in the z direction. Hence the TM modes will be characterized by zero magnetic field in the z direction. Some practical TM structures are dielectric and ferrite sheets, dielectric and ferrite sheets on a ground plane, corrugated surfaces, and a bed of metal posts on a ground plane.

A. TM Dielectric Sheet

The natural modes of operation of a dielectric sheet as shown in Fig. 6-22 or a dielectric sheet on a ground plane as shown in Fig. 6-23 include TM and TE modes. A discussion of the TE modes was given in Sec. 6-2A. A TE source gives polarization parallel to the plane of the structure. A TM source, however, gives polarization perpendicular to the plane of the structure. If the dielectric sheet is mounted on a ground plane, it is necessary to use a TM mode if endfire operation is desired.

TM mode propagation in dielectric sheets has been treated by a number of investigators.[1] The structure in Fig. 6-22 may be analyzed in the same manner as for the TE case in Sec. 6-2A. However, rather than go through an analogous analysis we may obtain the desired results by recognizing that

[1] Attwood, *op. cit.*; Zucker, *op. cit.*; Richmond, *op. cit.*

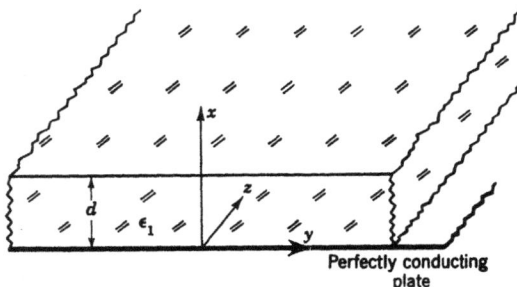

Figure 6-23. Dielectric sheet on a ground plane.

the TM solutions are duals of the TE solutions. The fields for the TM mode may be obtained from the TE case by simply replacing μ by ϵ, ϵ by μ, E by H, and H by $-E$ (see Table 2-1).

Using the results for the TE case in Sec. 6-2A we obtain for the TM field components

$$H_y = B_1 \cos\left(\kappa_1 x - \frac{n\pi}{2}\right) \qquad -d \lessgtr x \lessgtr d \tag{6-59}$$

$$H_y = B_0 e^{-j\kappa_0 x} \qquad\qquad x > d \tag{6-60}$$

$$H_y = (-1)^n B_0 e^{j\kappa_0 x} \qquad x < -d \tag{6-61}$$

$$E_x = \frac{\gamma}{j\omega\epsilon} H_y \tag{6-62}$$

$$E_z = \frac{1}{j\omega\epsilon} \frac{\partial H_y}{\partial x} \tag{6-63}$$

where

$$0 \lessgtr \left(\kappa_1 x - \frac{n\pi}{2}\right) \lessgtr \frac{\pi}{2} \qquad n = 0, 1, 2, \ldots$$

and variation of the form $e^{j\omega t - \gamma z}$ is assumed.

Equating tangential components at $x = d$ or $x = -d$ gives the characteristic equation for TM_n modes as

$$\kappa_1 \epsilon_0 \tan\left(\kappa_1 d - \frac{n\pi}{2}\right) = j\kappa_0 \epsilon_1 \tag{6-64}$$

Equation (6-64) in conjunction with

$$\kappa_1^2 - \gamma^2 = k^2 \epsilon_r \tag{6-65}$$

and

$$\kappa_0^2 - \gamma^2 = k^2 \tag{6-66}$$

may be solved for γ. Solutions for all integral values of n ($n = 0, 1, 2, \ldots$) apply to the dielectric sheet structure in Fig. 6-22. From Eq. (6-63), however, it can be seen that only solutions for n even ($n = 0, 2, 4, \ldots$) apply to the dielectric sheet on a ground plane in Fig. 6-23.

For both the dielectric sheet (Fig. 6-22) and the dielectric sheet on a ground plane (Fig. 6-23) the lowest-order TM mode occurs for $n = 0$ and is denoted the TM_0 mode. Equation (6-64) for this case becomes

$$\kappa_1 \epsilon_0 \tan \kappa_1 d = j\kappa_0 \epsilon_1 \tag{6-67}$$

The TM_0 mode has no cutoff for $\epsilon_r > 1$. As d goes to zero, the TM_0 mode degenerates into a TEM wave. For the nth-order mode there is a critical thickness less than which the mode will not propagate. This is the same for the TM case as for the TE case [see Eq. (6-15)], and the thickness

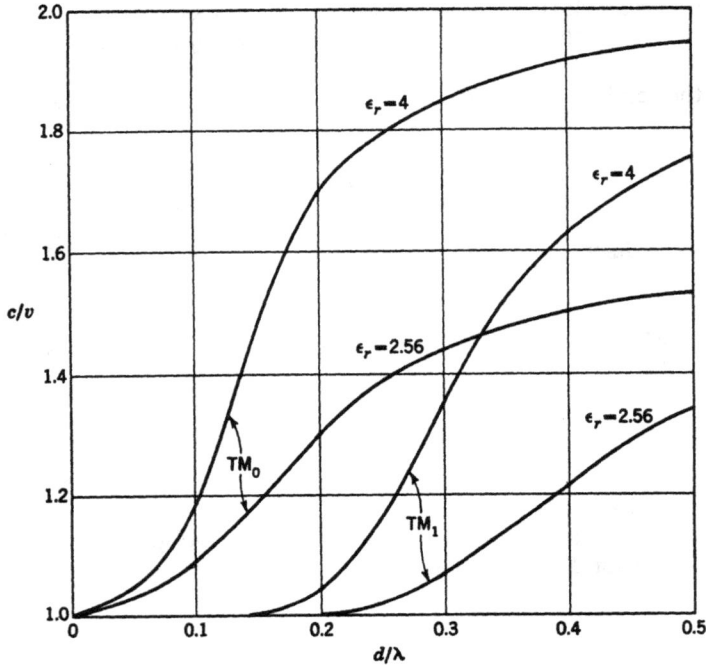

Figure 6-24. Velocity ratio c/v for TM_n modes on an infinite sheet of perfect dielectric. Both even and odd values of n apply for a sheet of thickness $2d$ without ground plane. Even values apply for a sheet of thickness d on a ground plane.

for mode cutoff is given by

$$\frac{2d}{\lambda} = \frac{n}{2\sqrt{\epsilon_r - 1}} \qquad \epsilon_r > 1 \tag{6-68}$$

Some curves of velocity ratio c/v versus dielectric thickness d/λ are given in Fig. 6-24 for TM_0 and TM_1 modes for lossless dielectric and a perfectly conducting ground plane. The solution of Eqs. (6-64) to (6-66) for c/v follows the method described in Sec. 6-2A. For lossy dielectric one may use the above equations for a complex permittivity, or for the low loss case one may use the method of Sec. 5-4C.

Excitation of TM waves on dielectric sheet structures such as those in Figs. 6-22 and 6-23 but of finite extent may be accomplished in practice by butting an open-ended waveguide or sectoral horn against the feed end of the structure. The feed should preferably extend the width of the structure. Polarization of the feed should be perpendicular to the surface of the dielectric sheet, that is, $\mathbf{E} = \mathbf{x}E_z$ in Figs. 6-22 and 6-23. For a sheet of dielectric on a ground plane, TM waves also may be excited through a slot or slots in the ground plane. The excitation problem is considered in more detail in Chap. 7.

Layered Dielectric

The layered dielectric structure of Fig. 6-4 also supports TM modes. Again, duality applies and H_y may be substituted for E_y in Eqs. (6-17) to (6-19). Using Eq. (6-63) for E_z and matching E_z as well as H_y at the boundaries give a set of four equations. Setting the determinant equal to zero gives the transcendental equation

$$\tanh\left(\gamma_2 d_2 - j\frac{n\pi}{2}\right) = -\frac{\epsilon_2\gamma_1}{\epsilon_1\gamma_2}\frac{\epsilon_0\gamma_1 \tanh \gamma_1 d_1 + \epsilon_1\alpha_0}{\epsilon_1\alpha_0 \tanh \gamma_1 d_1 + \epsilon_0\gamma_1}$$

$$n = 0, 1, 2, \ldots \quad (6\text{-}69)$$

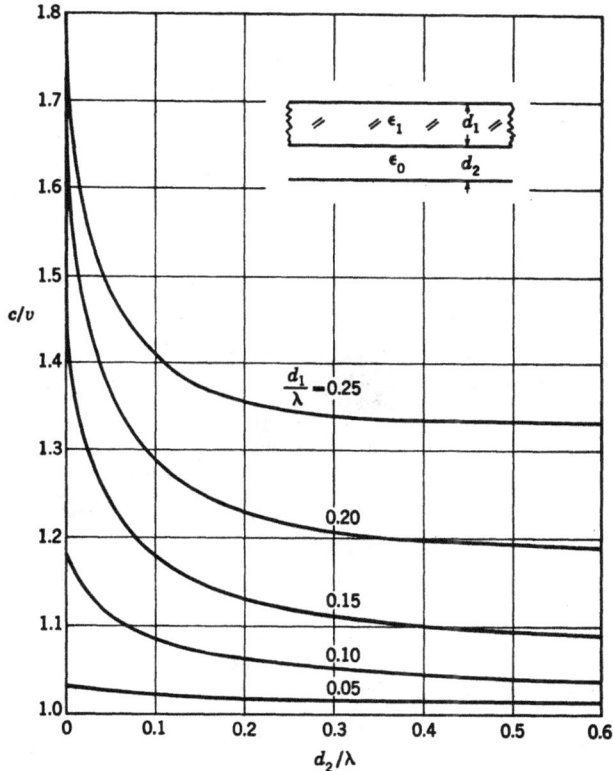

Figure 6-25.   Velocity ratio c/v versus air-layer thickness for dielectric layer ($\epsilon_1/\epsilon_0 = 4$) above a ground plane and with TM_0 surface wave propagation.

which reduces to Eq. (6-64) for $d_1 = 0$, $\alpha_0 = j\kappa_0$, $\gamma_2 = j\kappa_1$, and ϵ_1 substituted for ϵ_2.

Equation (6-69) along with Eqs. (6-20) to (6-22) may be solved for β or β/k $(=c/v)$. The three cases given in Eqs. (6-24) to (6-26) for the TE case also apply for TM modes.

The TM modes may be designated TM_n, and it can be shown that $E_z = 0$ at $x = 0$ for n even; hence a ground plane may be inserted at $x = 0$, and the above results apply for even modes.

The TM_n modes are slow waves ($c/v \gtrless 1$), and cutoff occurs for $\kappa_0 = 0$ for $n = 1, 2, \ldots$. The dominant mode TM_0 has no cutoff.

A case of practical importance occurs in Fig. 6-4 for region 2 equal to free space and for TM_0 operation with a ground plane at $x = 0$. The velocity ratio c/v as a function of layer thickness for this case is given in Fig. 6-25 for a relative dielectric constant of 4 in region 1.

B. TM Corrugated Surface

Cutler[1] appears to have been the first to show that surface waves can exist on a plane, corrugated conductor. The use of corrugated surface structures as antennas has been studied by several groups,[2] and a thorough analysis including the effects of feed and ground plane size has been made by Elliott.[3]

A cross section of a practical corrugated structure is shown in Fig. 6-26. Such a structure will support a longitudinal electric field (E_z) essentially in the form of a TEM wave in the parallel plate (corrugated) region. Assuming propagation in the z direction, the mode of propagation is TM. The geometry and periodicity of the corrugations are not very critical. In practice very good results are obtained for $W \gtrless t$ and $W + t < \lambda/5$. Because

[1] C. C. Cutler, Electromagnetic Waves Guided by Corrugated Conducting Surfaces, *Bell Telephone Lab. Rept.* MM-44-160-218, October, 1944.

[2] D. K. Reynolds and W. S. Lucke, Corrugated Endfire Antennas, *Proc. Natl. Electron. Conf.*, **6**:16–28, September, 1950; W. Rotman, A Study of Single Surface Corrugated Guides, *Proc. IRE*, **39**:952–959, August, 1951; M. J. Ehrlich and L. Newkirk, Corrugated Surface Antennas, *IRE Conv. Record*, Part 2, Antennas and Communications, pp. 18–33, 1953; R. W. Hougardy and R. C. Hansen, Scanning Surface Wave Antennas—Oblique Surface Waves over a Corrugated Conductor, *IRE Trans. Antennas Propagation*, AP-6(4):370–376, October, 1958.

[3] R. S. Elliott, On the Theory of Corrugated Plane Surfaces, *IRE Trans. Antennas Propagation*, AP-2(2):71–81, April, 1954.

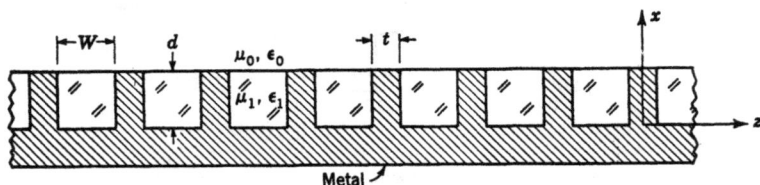

Figure 6-26. Cross section of plane corrugated surface.

of its slow wave characteristic (that is, $c/v > 1$) the TM corrugated surface may be used in transmission-line and endfire antenna applications.

A simple analysis of an infinite TM corrugated surface that gives an approximate expression for the phase velocity of a TM wave that is adequate for many applications may be obtained by matching impedances (transverse resonance) at $x = d$ with the assumptions that only the propagating TM mode exists for $x > d$ and that the field in the region $x \lessapprox d$ is in the form of a TEM wave in a shorted, parallel plate region. This is essentially the method that was employed for the corrugated cylinder in Sec. 5-3E.

Matching impedance at $x = d$ requires that

$$\overleftarrow{Z}_d = -\overrightarrow{Z}_d \tag{6-70}$$

The impedance $\overrightarrow{Z}_d$ is that of a TM wave propagating in the x direction and is given by

$$\overrightarrow{Z}_d = \frac{\kappa_0}{\omega \epsilon_0} \tag{6-71}$$

The impedance $\overleftarrow{Z}_d$ is taken to be zero over the interval t, and that of a shorted line of length d over the interval W. The average impedance is given by

$$\overleftarrow{Z}_d = \frac{jW}{W+t} \sqrt{\frac{\mu_1}{\epsilon_1}} \tan \left(\sqrt{\mu_r \epsilon_r}\, kd\right) \tag{6-72}$$

Substituting (6-71) and (6-72) into (6-70) gives the characteristic equation for the TM corrugated structure,

$$\frac{Wk}{W+t} \sqrt{\frac{\mu_r}{\epsilon_r}} \tan \left(\sqrt{\mu_r \epsilon_r}\, kd\right) = j\kappa_0 \tag{6-73}$$

where μ_r is the relative permeability of the medium and usually may be taken as unity.

Equation (6-73) may be solved for κ_0; then γ may be obtained from Eq. (6-66). For ϵ_r real and greater than unity, κ_0 will be a pure imaginary. This gives an exponential decay to the fields away from the surface in the manner typical of slow wave propagation along the surface.

The corrugated surface is an anisotropic structure. That is, the phase velocity is actually a function of the direction with which a wave travels along the corrugated surface. The geometry for an oblique wave is shown in Fig. 6-27.

Let the variation of the oblique wave be of the form $e^{-j(\kappa x + \tau y + \beta z)}$. It can be readily shown that the angle of propagation is given by

$$\tan \theta = \frac{\tau}{\beta} \tag{6-74}$$

and the velocity ratio c/v is given by

$$\frac{c}{v} = \frac{\sqrt{\tau^2 + \beta^2}}{k} \tag{6-75}$$

The phase constant for the wave at angle θ may be obtained from the analysis for $\theta = 0$ by a proper substitution of phase constants. For $\theta = 0$ the wave is of the form $e^{-j(\kappa x + \beta z)}$ where

$$\kappa^2 + \beta^2 = k^2 \tag{6-76}$$

For oblique propagation the wave has the form $e^{-j(\kappa x + \tau y + \beta z)}$ where

$$\kappa^2 + \beta^2 = k^2 - \tau^2 \tag{6-77}$$

A simple substitution of $\sqrt{k^2 - \tau^2}$ for k in the $\theta = 0$ case will give the solution for oblique propagation. Thus from Eq. (6-73) we obtain for oblique propagation

$$\frac{W\sqrt{k^2 - \tau^2}}{W + t} \sqrt{\frac{\mu_r}{\epsilon_r}} \tan \left(\sqrt{\mu_r \epsilon_r} d \sqrt{k^2 - \tau^2}\right) = j\kappa_0 \tag{6-78}$$

To illustrate the effect of the angle of propagation, consider an air-filled, corrugated structure with $t \ll W$, $W \ll \lambda$, and $d = 0.159 \lambda$. This value of d is convenient, since it gives $kd = 1$. If we pick some value for βd, say $\beta d = \pi/4$, then the value of $d\sqrt{k^2 - \tau^2}$ that satisfies Eqs. (6-77) and (6-78) is found to be 0.202π. From Eq. (6-74), $\tan \theta$ may be expressed as

$$\tan \theta = \frac{\{(kd)^2 - [(kd)^2 - (\tau d)^2]\}^{\frac{1}{2}}}{\beta d} \tag{6-79}$$

which for this case gives $\theta = 44.6$ deg. Solving for τd and utilizing Eq. (6-75) give $c/v = 1.11$. Next consider the same structure with propagation at $\theta = 0$ deg. From Eqs. (6-73) and (6-76), βd is found to be 1.85, which gives

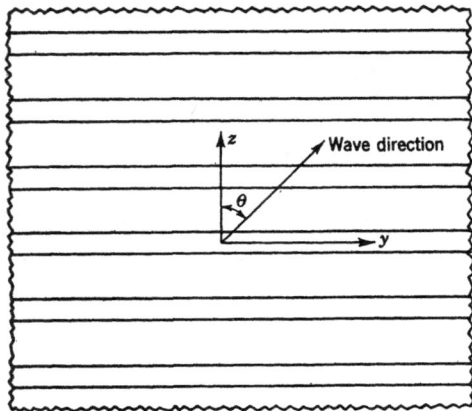

Figure 6-27. Wave traveling obliquely across corrugated surface.

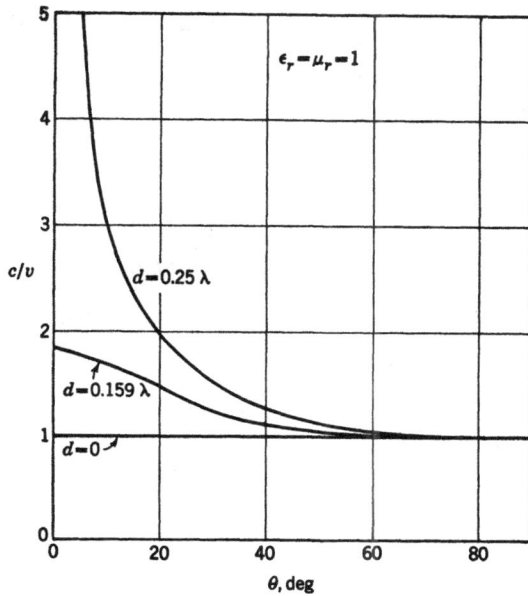

Figure 6-28. Velocity ratio c/v versus angle of propagation for a corrugated surface with $t \ll W$ and $W \ll \lambda$.

a velocity ratio c/v $(=\beta d/kd)$ of 1.85 in this case. Thus it can be seen that for a given corrugated structure the velocity ratio changes significantly with angle of propagation. This effect is further illustrated by the graphs in Fig. 6-28.

The preceding analysis of a corrugated surface is adequate for many applications. More rigorous analyses for normal propagation[1] ($\theta = 0$) and oblique propagation[2] have been obtained by writing expressions for the complete fields, both above the corrugations and in the gaps, and matching tangential fields at the boundaries. The problem in these analyses is idealized to the extent that the corrugations are assumed to be vanishingly thin and, furthermore, that all modes but the dominant mode attenuate to negligible amplitude in traveling from the interface ($x = d$ in Fig. 6-29) to the base of a corrugation ($x = 0$) and back to the interface.

[1] Elliott, *op. cit.*; R. A. Hurd, The Propagation of an Electromagnetic Wave along an Infinite Corrugated Surface, *Can. J. Phys.*, **32**:727–734, December, 1954.

[2] Hougardy and Hansen, *op. cit.*; R. E. Collin, "Field Theory of Guided Waves," pp. 465–468, McGraw-Hill, New York, 1960.

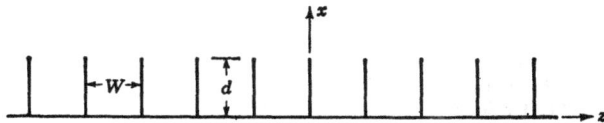

Figure 6-29. Infinite corrugated plane surface.

For TM propagation the field components for the geometry in Fig. 6-29 are E_x, E_z, and H_y. If H_y is specified, the components E_x and E_z can be readily found. In the region above the corrugations ($x > d$) a suitable form for H_y that takes into account the periodic nature of the structure is

$$H_y = \sum_{n=-\infty}^{\infty} A_n e^{-j\left[\kappa_n^o x + \left(\beta + \frac{2n\pi}{W}\right)z\right]} \qquad x > d \qquad (6\text{-}80)$$

where the superscript o denotes the outer region and

$$\left(\beta + \frac{2n\pi}{W}\right)^2 + (\kappa_n^o)^2 = k^2 \qquad (6\text{-}81)$$

The form of the phase constant in the z direction, $\beta + 2n\pi/W$, is a consequence of Floquet's theorem.[1]

A suitable form for H_y in the corrugated region is

$$H_y = \sum_{n=0}^{\infty} B_n \cos \frac{n\pi z}{W} \cos \kappa_n^i x \qquad \begin{matrix} 0 \lessgtr z \lessgtr W \\ 0 \lessgtr x \lessgtr d \end{matrix} \qquad (6\text{-}82)$$

where the superscript i denotes the inner region ($0 \lessgtr x \lessgtr d$) and

$$(\kappa_n^i)^2 + \left(\frac{n\pi}{W}\right)^2 = k^2 \qquad (6\text{-}83)$$

The field in the mth corrugation is related to that in the region $0 \lessgtr z \lessgtr W$ by

$$H_y(x, z + mW) = e^{-j\beta mW} H_y(x,z) \qquad (6\text{-}84)$$

In order to determine the propagation constant β for a given geometry, a calculus of residues technique may be used in evaluating the infinite set of simultaneous equations obtained by matching tangential fields at the boundaries.[2]

The same result may be obtained, however, by taking into account higher-order modes in the transverse resonance method.[3] The transverse equivalent network for the corrugated surface in Fig. 6-29 is illustrated in Fig. 6-30a where the element Z represents the junction discontinuity between the transmission lines which represent the two regions. The transverse network in this case can be reduced to the simplified form in Fig. 6-30b.[4]

[1] J. C. Slater, "Microwave Electronics," pp. 169–177, Van Nostrand, Princeton, N.J., 1950.
[2] Hurd, *op. cit.*
[3] L. O. Goldstone and A. A. Oliner, A Note on Surface Waves along Corrugated Structures, *IRE Trans. Antennas Propagation*, **AP-7**:274–276, July, 1959.
[4] Marcuvitz, *op. cit.*, sec. 5-22.

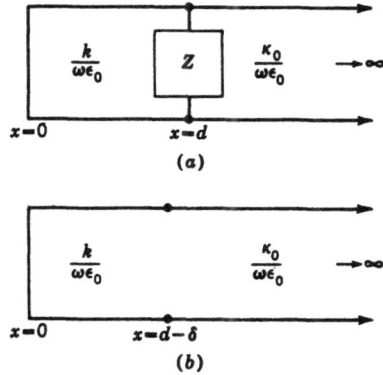

Figure 6-30. Transverse equivalent networks for corrugated plane surface.

(a)

(b)

Application of transverse resonance yields

$$j \frac{k}{\omega \epsilon_0} \tan k(d - \delta) + \frac{\kappa_0}{\omega \epsilon_0} = 0 \qquad (6\text{-}85)$$

which reduces to

$$\tan k(d - \delta) = \frac{j \kappa_0}{k} \qquad (6\text{-}86)$$

where

$$\kappa_0^2 + \beta^2 = k^2 \qquad (6\text{-}87)$$

Equations (6-86) and (6-87) may be combined to yield

$$\cos k(d - \delta) = \frac{k}{\beta} \qquad (6\text{-}88)$$

or

$$kd = \cos^{-1} \frac{k}{\beta} + k\delta \qquad (6\text{-}89)$$

where $k\delta$ is obtained from Marcuvitz[1] as

$$k\delta = \sum_{n=1}^{\infty} \left(\frac{kW}{2n\pi} - \sin^{-1} \frac{k}{\beta_n} \right) + \sum_{n=1}^{\infty} \left(\frac{kW}{2n\pi} - \sin^{-1} \frac{k}{|\beta_{-n}|} \right)$$
$$- \sum_{n=1}^{\infty} \left(\frac{kW}{n\pi} - \sin^{-1} \frac{kW}{n\pi} \right) + \frac{kW \ln 2}{\pi} \qquad (6\text{-}90)$$

Equation (6-90) is valid for $W < \lambda/2$. For $W \ll \lambda$, it is usually safe to neglect δ, in which case Eq. (6-86) reduces to our original characteristic equation (6-73) (for $t \ll W$ and $\mu_r = \epsilon_r = 1$).

Theoretical results showing the effect of W on c/v are given in Fig. 6-31.

[1] *Ibid.*, p. 289, Eq. 3a with $d' = \delta$ and $k \sin \theta = \beta_0$.

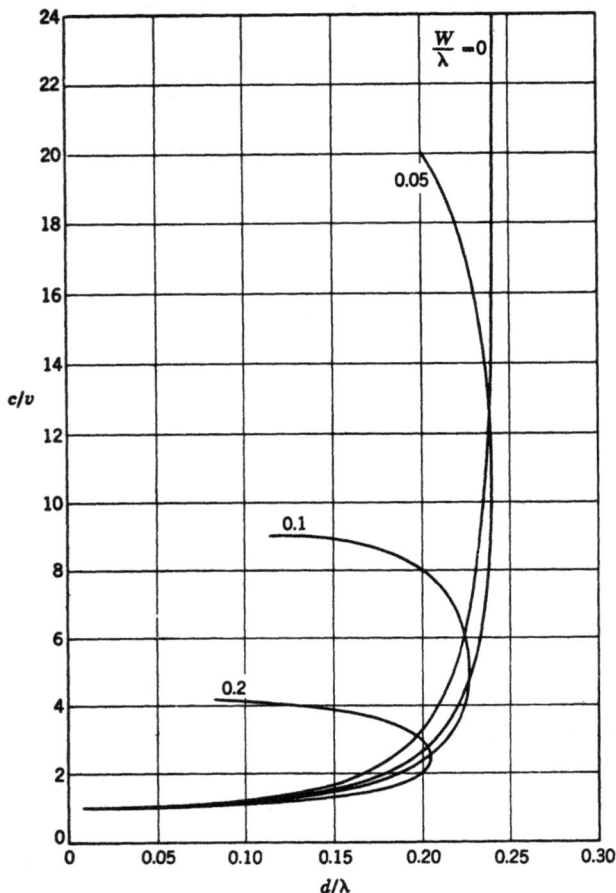

Figure 6-31. Velocity ratio c/v as a function of d/λ for various W/λ. (From Hurd, The Propagation of an Electromagnetic Wave along an Infinite Corrugated Surface, *Can. J. Phys.*, **32**:727–734, 1954.)

Note that, for W nonzero and for a fixed d, there are frequencies for which two values of β occur. For one value of β the slope $d\omega/d\beta$ (which defines group velocity) is positive, and for the other it is negative. The case in which ω/β (phase velocity) is positive and $d\omega/d\beta$ is negative is called *backward wave* operation. This is discussed in more detail in Sec. 8-3.

For TM propagation over a corrugated surface we observe that the impedance in the x direction above the surface is capacitive, that is, $\kappa_0 = -j|\kappa_0|$ in Fig. 6-30. The transverse resonance condition $(\overrightarrow{Z} = -\overleftarrow{Z})$ requires that the impedance of the corrugated surface be inductive. It can be seen from Eq. (6-73) or (6-86) that, as the wavelength is varied, the structure will exhibit passband–stop band characteristics. TM propagation may be realized for $0 < d < \lambda/4$, $\lambda/2 < d < 3\lambda/4$, etc., since these are the regions of inductive surface impedance.

Although corrugated structures apparently are used entirely for TM propagation, the possibility of TE propagation has been pointed out by Collin.[1] For TE propagation the surface impedance must be capacitive, and this would require corrugation depths in the ranges $\lambda/4 < d < \lambda/2$, $3\lambda/4 < d < \lambda$, etc. In this case, however, the corrugation spacing W must be such as to support a TE_{01} mode. For air-filled corrugations this means that $W > \lambda/2$ and the structure may tend to act like a diffraction grating and support a fast wave mode. For TE surface wave propagation the corrugations may be filled with a high-dielectric-constant material in order to support the TE_{01} mode in the corrugations and have perhaps five or more corrugations per wavelength.

C. TM Metal Post Structure

For some applications it is desirable to have a surface wave structure which is isotropic in the plane of the surface. An example is the surface wave Luneberg lens (see Chap. 8). A dielectric sheet on a ground plane has this isotropy; however, the corrugated surface in the previous section does not. The nonisotropy of the corrugated surface can be controlled, however, by forming a doubly corrugated surface[2] as shown in Fig. 6-32. Such a struc-

---

[1] Collin, *op. cit.*, p. 468.

[2] Earliest treatment of the doubly corrugated surface appears to be that of A. F. Kay, Excitation Efficiency of Surface Waves over Corrugated Metal and Doubly Corrugated Metal and in Dielectric Slabs on a Ground Plane, *Tech. Res. Group Sci. Rept.* 5, New York December, 1956.

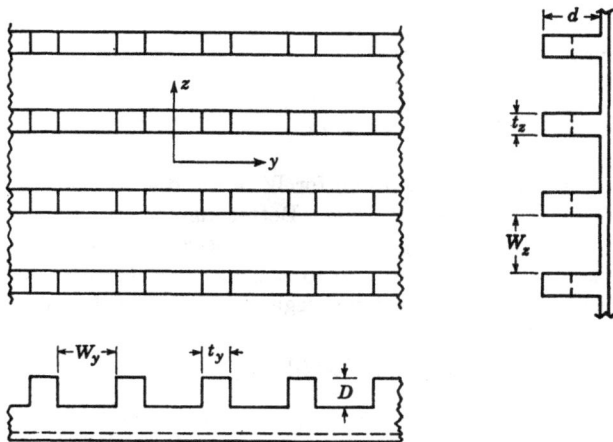

Figure 6-32. Doubly corrugated metal surface.

ture is essentially isotropic when $W_z = W_y$, $t_z = t_y$, and $d = D$. In this case the structure may be thought to be a bed of metal posts on a ground plane.

An approximate analysis of the doubly corrugated surface in Fig. 6-32 can be obtained by applying the analysis of the previous section to each set of corrugations. Assuming that only lowest-order modes exist, we have the impedance relations

$$\frac{E_z}{H_y}\bigg]_{x=0} = j\,\frac{W_z}{W_z + t_z}\,\frac{\kappa_z}{\omega\epsilon_1}\,\tan \kappa_z d \tag{6-91}$$

$$\frac{E_y}{H_z}\bigg]_{x=0} = -j\,\frac{W_y}{W_y + t_y}\,\frac{\kappa_y}{\omega\epsilon_1}\,\tan \kappa_y D \tag{6-92}$$

where κ_y and κ_z are the phase constants for propagation in the x direction in the corrugated region for the two sets of corrugations. For the isotropic case Eqs. (6-91) and (6-92) are identical and reduce to

$$j\,\frac{\kappa_0}{\epsilon_0} = \frac{W}{W + t}\,\frac{\kappa_1}{\epsilon_1}\,\tan \kappa_1 d \tag{6-93}$$

To a first approximation one can assume that $\beta = 0$ in the post region as was done in the corrugated region in the previous section. This gives

$$j\kappa_0 = \frac{W}{W + t}\,\frac{k}{\sqrt{\epsilon_r}}\,\tan (\sqrt{\epsilon_r}\,kd) \tag{6-94}$$

Combining Eq. (6-94) with

$$\kappa_0{}^2 + \beta^2 = k^2 \tag{6-95}$$

gives

$$\frac{\beta}{k} = \frac{c}{v} = \left[1 + \left(\frac{W}{W + t}\right)^2\frac{1}{\epsilon_r}\,\tan^2 (\sqrt{\epsilon_r}\,kd)\right]^{\frac{1}{2}} \tag{6-96}$$

A graph of Eq. (6-96) for $W = t$ and $\epsilon_r = 1$ is given in Fig. 6-33 along with some measured data. Better agreement with measured results is obtained if one recognizes that β is not zero in the post region but is actually imaginary. This gives rise to a value of κ_1 that is somewhat greater than $k\sqrt{\epsilon_r}$. For an air-filled post region Querido[1] has found that $\kappa_1 \approx 1.05k$. Thus

$$\frac{c}{v} = \{1 + [F \tan (1.05kd)]^2\}^{\frac{1}{2}} \tag{6-97}$$

---

[1] H. B. Querido, Surface Wave Fields and Phase Velocity Variations of Grounded Dielectric Sheets and of Periodic Structures of Metal Posts on a Ground Plane, *Ohio State Univ. Res. Found. Antenna Lab. Rept.* 667-46, November, 1958.

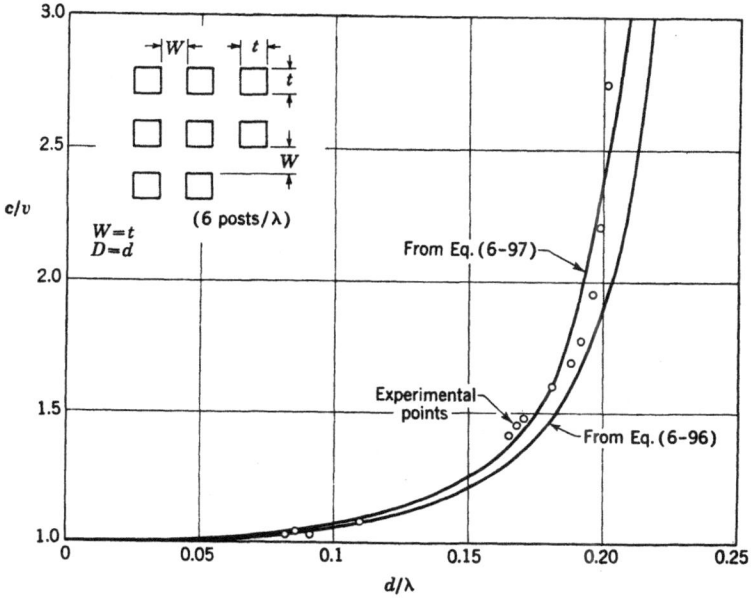

Figure 6-33. Velocity ratio c/v as a function of height for a bed of square metal posts on a ground plane.

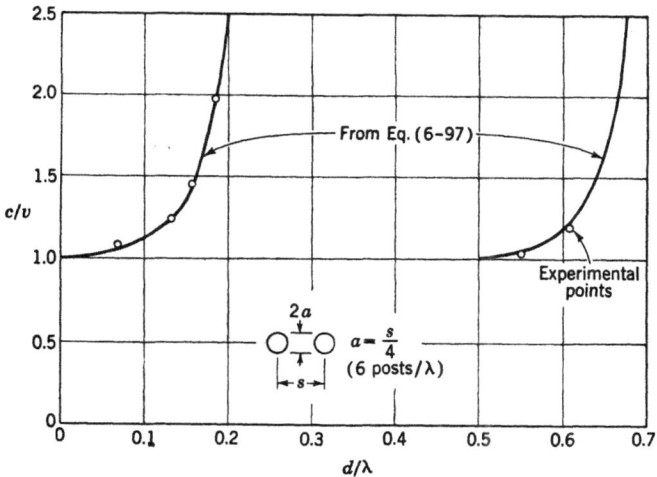

Figure 6-34. Velocity ratio c/v as a function of height for a bed of circular metal posts on a ground plane.

where

$$F = \text{weighting factor}$$

$$= \frac{W}{W + t} \text{ for square posts}$$

$$= 1 - \frac{\pi a}{2s} \text{ for circular posts} \tag{6-98}$$

For the circular posts a is the radius and s is the center-to-center spacing. A comparison of measured and theoretical results for a bed of circular posts is given in Fig. 6-34.

6-4. HYBRID MODE PLANAR SOURCES

Planar structures of infinite extent as illustrated in Fig. 6-35 will support either TE or TM modes but no hybrid modes when regions 0 and 1 are isotropic. Hybrid modes (TE + TM) may be supported, however, when one of the regions is anisotropic, but even then the modes may be predominantly TM or TE. Ferrites, ferroelectrics, and plasmas are examples of substances that may exhibit anisotropic properties and support hybrid modes. The anisotropy is related to an applied external field; hence by varying the magnitude and direction of the applied field, one may change significantly the electrical properties of the medium. The modes of a ferrite sheet on a ground plane will be considered in some detail to illustrate the effects of anisotropy.

A. Ferrite Sheet on Ground Plane[1]

A ferromagnetic material subject to a large constant magnetic field exhibits a tensor permeability toward an additional small alternating field. Ferromagnetic materials known as ferrites have relatively small loss in the microwave region, hence are of practical interest in transmission-line and antenna applications.

[1] A thorough analysis of this problem has been given by R. L. Pease, On the Propagation of Surface Waves over an Infinite Grounded Ferrite Slab, *IRE Trans. Antennas Propagation*, **AP-6**:13–20, January, 1958.

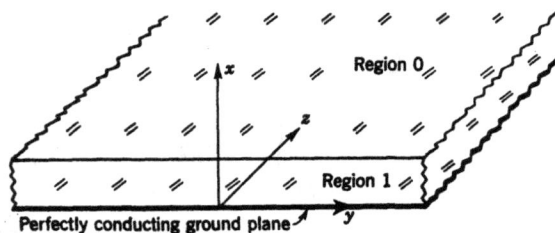

Figure 6-35.   Infinite planar structure.

Let region 0, $x > d$, in Fig. 6-35 be free space and region 1, $0 \lessgtr x \lessgtr d$, be homogeneous, lossless ferrite in which the magnetic flux density **B** is related to magnetic field intensity **H** by

$$
\begin{bmatrix} B_i \\ B_j \\ B_k \end{bmatrix} = \begin{bmatrix} \mu_{ii} & \mu_{ij} & \mu_{ik} \\ \mu_{ji} & \mu_{jj} & \mu_{jk} \\ \mu_{ki} & \mu_{kj} & \mu_{kk} \end{bmatrix} \begin{bmatrix} H_i \\ H_j \\ H_k \end{bmatrix} \qquad i,\, j,\, k = \text{cyclic } 1,\, 2,\, 3 \qquad (6\text{-}99)
$$

If the constant applied magnetic field is along the x_k axis, Eq. (6-99) may be expressed as[1]

$$
\begin{bmatrix} B_i \\ B_j \\ B_k \end{bmatrix} = \begin{bmatrix} \mu_1 & -j\mu_2 & 0 \\ j\mu_2 & \mu_1 & 0 \\ 0 & 0 & \mu \end{bmatrix} \begin{bmatrix} H_i \\ H_j \\ H_k \end{bmatrix} \qquad (6\text{-}100)
$$

Based on the theory of Polder[2] the components of the tensor permeability for saturated, lossless ferrite material are given by

$$
\frac{\mu_1}{\mu_0} = 1 + \frac{4\pi M_s H_0 \Gamma^2}{\Gamma^2 H_0^2 - \omega^2} \qquad (6\text{-}101)
$$

$$
\frac{\mu_2}{\mu_0} = \frac{4\pi M_s \omega \Gamma}{\Gamma^2 H_0^2 - \omega^2} \qquad (6\text{-}102)
$$

$$
\frac{\mu}{\mu_0} = 1 \qquad (6\text{-}103)
$$

where M_s = saturation magnetization
$\quad\quad H_0$ = applied d-c magnetic field
$\quad\quad \Gamma$ = gyromagnetic ratio

Although the mks system of units is used for quantities describing the electromagnetic fields, it is customary to use the guassian system for quantities describing the static magnetization of the ferrite.[3]

For the unsaturated state a generalization of Polder's theory by Rado[4] leads to $\mu_1 = \mu_0$ and

$$
\frac{\mu_2}{\mu_0} = -\frac{4\pi M \Gamma}{\omega} \qquad (6\text{-}104)
$$

where magnetization M is related to H_0 by the magnetization curve of the ferrite.

[1] H. Suhl and L. R. Walker, Topics in Guided Wave Propagation through Gyromagnetic Media, *Bell System Tech. J.*, **33**:579–659, May, 1954.

[2] D. Polder, On the Theory of Ferromagnetic Resonance, *Phil. Mag.*, **40**:99–115, January, 1949.

[3] B. Lax and K. J. Button, "Microwave Ferrites and Ferrimagnetics," p. 150, McGraw-Hill, New York, 1962.

[4] G. T. Rado, Theory of the Microwave Permeability Tensor and Faraday Effect in Nonsaturated Ferromagnetic Materials, *Phys. Rev.*, **89**:529, January, 1953.

Assuming the regions to be charge and current free and with time variation of the form $e^{j\omega t}$, Maxwell's equations may be written

$$\nabla \times \mathbf{E} = -j\omega \mathbf{B} \qquad (6\text{-}105)$$
$$\nabla \times \mathbf{H} = j\omega \mathbf{D} \qquad (6\text{-}106)$$
$$\nabla \cdot \mathbf{B} = 0 \qquad (6\text{-}107)$$
$$\nabla \cdot \mathbf{D} = 0 \qquad (6\text{-}108)$$

If propagation is taken to be in the z direction ($e^{-j\beta z}$) and if there is no variation in the y direction, the fields in region 0 satisfy the transverse wave equation

$$\left[\frac{\partial^2}{\partial x^2} + (\omega^2 \mu_0 \epsilon_0 - \beta^2) \right] \left\{ \begin{matrix} \mathbf{E} \\ \mathbf{H} \end{matrix} \right\} = 0 \qquad (6\text{-}109)$$

For surface wave solutions the fields vary in the transverse (x) direction as $e^{-|\kappa_0|x}$ where

$$\kappa_0{}^2 = \omega^2 \mu_0 \epsilon_0 - \beta^2 \qquad (6\text{-}110)$$

For TM modes the fields in region 0, $x > d$, will be of the form

$$E_x = \frac{-j\beta A}{|\kappa_0|} e^{-(|\kappa_0|x + j\beta z)} \qquad (6\text{-}111)$$
$$E_z = A e^{-(|\kappa_0|x + j\beta z)} \qquad (6\text{-}112)$$
$$H_y = \frac{-j\omega\epsilon_0 A}{|\kappa_0|} e^{-(|\kappa_0|x + j\beta z)} \qquad (6\text{-}113)$$

where A is a proportionality factor which includes $e^{j\omega t}$.

For TE modes in region 0, $x > d$, the fields are

$$H_x = \frac{-j\beta B}{|\kappa_0|} e^{-(|\kappa_0|x + j\beta z)} \qquad (6\text{-}114)$$
$$H_z = B e^{-(|\kappa_0|x + j\beta z)} \qquad (6\text{-}115)$$
$$E_y = \frac{j\omega\mu_0 B}{|\kappa_0|} e^{-(|\kappa_0|x + j\beta z)} \qquad (6\text{-}116)$$

Note that the TE fields are the duals of the TM fields. This will not be true in the anisotropic region 1, $0 \lessgtr x \lessgtr d$.

For the fields in region 1, three cases may be considered corresponding to an applied constant magnetic field in each of the three coordinate directions.

Case I: Applied Magnetic Field in x Direction

From Eq. (6-99) the relations between $\mathbf{B}$ and $\mathbf{H}$ in region 1 for the magnetic field in the x direction are

$$B_x = \mu_0 H_x \qquad (6\text{-}117)$$
$$B_y = \mu_1 H_y - j\mu_2 H_z \qquad (6\text{-}118)$$
$$B_z = j\mu_2 H_y + \mu_1 H_z \qquad (6\text{-}119)$$

Equations (6-117) to (6-119) along with $\mathbf{D} = \epsilon\mathbf{E}$ when substituted into Maxwell's equations give for $0 \lessgtr x \lessgtr d$

$$E_y = -\frac{\omega\mu_0 H_z}{\beta} \tag{6-120}$$

$$j\beta E_x + \frac{\partial E_z}{\partial x} = j\omega\mu_1 H_y + \omega\mu_2 H_z \tag{6-121}$$

$$\frac{\partial E_y}{\partial x} = \omega\mu_2 H_y - j\omega\mu_1 H_z \tag{6-122}$$

$$H_y = \frac{\omega\epsilon_1}{\beta} E_x \tag{6-123}$$

$$j\beta H_x + \frac{\partial H_z}{\partial x} = -j\omega\epsilon_1 E_y \tag{6-124}$$

$$\frac{\partial H_y}{\partial x} = j\omega\epsilon_1 E_z \tag{6-125}$$

$$\mu_0 \frac{\partial H_z}{\partial x} + \mu_2\beta H_y - j\beta\mu_1 H_z = 0 \tag{6-126}$$

$$\frac{\partial E_x}{\partial x} - j\beta E_z = 0 \tag{6-127}$$

The permittivity ϵ_1 is taken to be constant, but it may vary significantly with frequency in practical ferrite materials.

Fields in region 1 that satisfy Eqs. (6-120) to (6-127) as well as the boundary conditions at the perfect conductor ($x = 0$) are

$$E_x = \frac{\beta}{\omega^2\mu_2\epsilon_1}\left[\frac{h^+C^+}{\kappa_1^+}\cos(\kappa_1^+x) + \frac{h^-C^-}{\kappa_1^-}\cos(\kappa_1^-x)\right] \tag{6-128}$$

$$E_y = C^+\sin(\kappa_1^+x) + C^-\sin(\kappa_1^-x) \tag{6-129}$$

$$E_z = \frac{j}{\omega^2\mu_2\epsilon_1}[h^+C^+\sin(\kappa_1^+x) + h^-C^-\sin(\kappa_1^-x)] \tag{6-130}$$

$$H_x = \frac{-\beta}{\omega\mu_0}[C^+\sin(\kappa_1^+x) + C^-\sin(\kappa_1^-x)] \tag{6-131}$$

$$H_y = \frac{1}{\omega\mu_2}\left[\frac{h^+C^+}{\kappa_1^+}\cos(\kappa_1^+x) + \frac{h^-C^-}{\kappa_1^-}\cos(\kappa_1^-x)\right] \tag{6-132}$$

$$H_z = \frac{-j(\beta^2 - \omega^2\mu_0\epsilon_1)}{\omega\mu_0}\left[\frac{C^+}{\kappa_1^+}\cos(\kappa_1^+x) + \frac{C^-}{\kappa_1^-}\cos(\kappa_1^-x)\right] \tag{6-133}$$

where[1]

$$(\kappa_1^\pm)^2 = -\frac{1}{2}\left[\beta^2\left(1 + \frac{\mu_1}{\mu_0}\right) - 2\omega^2\mu_1\epsilon_1\right]$$
$$\pm\frac{1}{2}\left[\beta^4\left(1 - \frac{2\mu_1}{\mu_0} + \frac{\mu_1^2}{\mu_0^2}\right) - \frac{4\beta^2\omega^2\mu_2^2\epsilon_1}{\mu_0} + 4\omega^4\mu_2^2\epsilon_1^2\right]^{1/2} \tag{6-134}$$

and

$$h^\pm = (\kappa_1^\pm)^2 + \frac{\mu_1}{\mu_0}\beta^2 - \omega^2\mu_1\epsilon_1 \tag{6-135}$$

[1] Further details may be found in Pease, *op. cit.*

The phase constant β may be found from a transcendental equation obtained by matching fields at $x = d$ or, what is the same thing, by matching impedances (transverse resonance). Matching impedances gives

$$\frac{-j\omega\mu_0}{|\kappa_0|} = \frac{E_y}{H_z}\bigg]_{\substack{x=d \\ \text{region 1}}} \tag{6-136}$$

and

$$\frac{|\kappa_0|}{j\omega\epsilon_0} = \frac{E_z}{H_y}\bigg]_{\substack{x=d \\ \text{region 1}}} \tag{6-137}$$

Substituting into (6-136) and (6-137) from (6-129), (6-133), (6-130), and (6-132) and eliminating C^+ and C^- give

$$\frac{\epsilon_0(h^+ - h^-)}{\epsilon_1(\beta^2 - \omega^2\mu_0\epsilon_1)} \sin(\kappa_1^+ d) \sin(\kappa_1^- d) + \frac{h^+ - h^-}{\kappa_1^+ \kappa_1^-} \cos(\kappa_1^+ d) \cos(\kappa_1^- d)$$

$$- \left[\frac{\epsilon_0 h^+}{\kappa_1^- \epsilon_1 |\kappa_0|} - \frac{h^- |\kappa_0|}{\kappa_1^-(\beta^2 - \omega^2\mu_0\epsilon_1)} \right] \sin(\kappa_1^+ d) \cos(\kappa_1^- d)$$

$$+ \left[\frac{\epsilon_0 h^-}{\kappa_1^+ \epsilon_1 |\kappa_0|} - \frac{h^+ |\kappa_0|}{\kappa_1^+(\beta^2 - \omega^2\mu_0\epsilon_1)} \right] \sin(\kappa_1^- d) \cos(\kappa_1^+ d) = 0 \tag{6-138}$$

All propagation constants may be found from Eqs. (6-138), (6-134), (6-135), and

$$\kappa_0{}^2 + \beta^2 = \omega^2\mu_0\epsilon_0 \tag{6-139}$$

Solution of these equations is quite involved. However, if the sheet of ferrite is thin enough so that $\sin(\kappa_1 d) \approx \kappa_1 d$ and $\cos(\kappa_1 d) \approx 1$, then $|\kappa_0|$ may be found to be

$$|\kappa_0| = \frac{-(1/d) \pm \{(1/d^2) - 4\omega^2\mu_0\epsilon_0[(\epsilon_0/\epsilon_1)^2 - (\mu_1/\mu_0)^2]\}^{\frac{1}{2}}}{2[(\epsilon_0/\epsilon_1) + (\mu_1/\mu_0)]} \qquad \kappa_1 d \ll 1 \tag{6-140}$$

For

$$\frac{\epsilon_0}{\epsilon_1} + \frac{\mu_1}{\mu_0} > 0 \tag{6-141}$$

and

$$\left(\frac{\mu_1}{\mu_0}\right)^2 > \left(\frac{\epsilon_0}{\epsilon_1}\right)^2 \tag{6-142}$$

the positive root in Eq. (6-140) is chosen and the radical expanded to give

$$|\kappa_0| \approx \frac{\epsilon_0}{\epsilon_1} \omega^2 d(\mu_1\epsilon_1 - \mu_0\epsilon_0) \qquad \kappa_1 d \ll 1 \tag{6-143}$$

The conditions imposed in arriving at (6-143) have reduced the problem to that of a thin, isotropic sheet of ferrite on a ground plane. The same result (6-143) can be obtained for TM propagation over a thin ferrite sheet of isotropic material of permeability μ_1 and dielectric ϵ_1.

Equations (6-143) and (6-139) may be combined to give

$$\frac{\beta}{k} = \frac{c}{v} \approx \left[1 + \left(\frac{\epsilon_0}{\epsilon_1}\right)^2 \left(\frac{\mu_1\epsilon_1}{\mu_0\epsilon_0} - 1\right)^2 (kd)^2 \right]^{\frac{1}{2}} \qquad \kappa_1 d \ll 1 \tag{6-144}$$

Case II: Applied Magnetic Field in the y Direction

If the applied magnetic field is in the y direction, the relations between **B** and **H** in region 1 are, from Eq. (6-99),

$$B_x = j\mu_2 H_y + \mu_1 H_z \tag{6-145}$$
$$B_y = \mu_0 H_y \tag{6-146}$$
$$B_z = \mu_1 H_y - j\mu_2 H_z \tag{6-147}$$

Equations (6-145) to (6-147) along with $\mathbf{D} = \epsilon\mathbf{E}$ when substituted into Maxwell's equations (6-105) to (6-108) give for region 1 in Fig. 6-35

$$E_y = \frac{-j\omega\mu_2}{\beta} H_y - \frac{\omega\mu_1}{\beta} H_z \tag{6-148}$$

$$j\beta E_x + \frac{\partial E_z}{\partial x} = j\omega\mu_0 H_y \tag{6-149}$$

$$\frac{\partial E_y}{\partial x} = -j\omega\mu_1 H_y - \omega\mu_2 H_z \tag{6-150}$$

$$H_y = \frac{\omega\epsilon_1}{\beta} E_z \tag{6-151}$$

$$j\beta H_x + \frac{\partial H_z}{\partial x} = -j\omega\epsilon_1 E_y \tag{6-152}$$

$$\frac{\partial H_y}{\partial x} = j\omega\epsilon_1 E_z \tag{6-153}$$

$$j\mu_2 \frac{\partial H_y}{\partial x} + \mu_1 \frac{\partial H_z}{\partial x} - j\beta\mu_1 H_y - \beta\mu_2 H_z = 0 \tag{6-154}$$

$$\frac{\partial E_x}{\partial x} - j\beta E_z = 0 \tag{6-155}$$

Fields in region 1 that satisfy Eqs. (6-148) to (6-155) and the boundary condition at $x = 0$ are given by

$$\left. \begin{array}{l} E_x = \dfrac{-j\beta}{\kappa_1^e} C \cos \kappa_1^e x \\[2mm] E_z = C \sin \kappa_1^e x \\[2mm] H_y = \dfrac{-j\omega\epsilon_1}{\kappa_1^e} C \cos \kappa_1^e x \end{array} \right\} \text{TM}$$

$$\tag{6-156}$$
$$\tag{6-157}$$
$$\tag{6-158}$$

$$\left. \begin{array}{l} H_x = \dfrac{j\beta}{\kappa_1^h} \dfrac{\mu_1^2}{\mu_1^2 - \mu_2^2} D \sin \kappa_1^h x - \dfrac{j\mu_1\mu_2}{\mu_1^2 - \mu_2^2} D \cos \kappa_1^h x \\[3mm] H_z = \dfrac{\mu_1^2}{\mu_1^2 - \mu_2^2} D \cos \kappa_1^h x - \dfrac{\beta}{\kappa_1^h} \dfrac{\mu_1\mu_2}{\mu_1^2 - \mu_2^2} D \sin \kappa_1^h x \\[3mm] E_y = \dfrac{-j\omega\mu_1}{\kappa_1^h} D \sin \kappa_1^h x \end{array} \right\} \text{TE}$$

$$\tag{6-159}$$
$$\tag{6-160}$$
$$\tag{6-161}$$

where

$$\overset{e}{(\kappa_1)^2} = \omega^2 \mu_0 \epsilon_1 - \beta^2 \tag{6-162}$$

$$\overset{h}{(\kappa_1)^2} = \omega^2 \mu_1 \epsilon_1 \left(1 - \frac{\mu_2^2}{\mu_1^2}\right) - \beta^2 \tag{6-163}$$

and e and h denote TM and TE waves, respectively.

In this case TM and TE fields exist independently. There are no hybrid modes. Two characteristic equations may be obtained by applying transverse resonance, one for TM propagation and the other for TE propagation. These are

$$\overset{e}{\kappa_1} \tan \overset{e}{\kappa_1} d = |\kappa_0| \frac{\epsilon_1}{\epsilon_0} \qquad\qquad \text{TM} \tag{6-164}$$

and

$$\overset{h}{\kappa_1} \cot \overset{h}{\kappa_1} d = \frac{\beta \mu_2}{\mu_1} - |\kappa_0| \frac{\mu_1}{\mu_0}\left(1 - \frac{\mu_2^2}{\mu_1^2}\right) \qquad \text{TE} \tag{6-165}$$

Propagation constants may be found from Eqs. (6-164), (6-165), (6-139), (6-162), and (6-163). For TM propagation, for example, Eqs. (6-164), (6-139), and (6-162) give

$$\frac{c}{v} \approx \left[1 + \left(\frac{\epsilon_0}{\epsilon_1}\right)^2 \left(\frac{\epsilon_1}{\epsilon_0} - 1\right)^2 (kd)^2\right]^{1/2} \qquad \overset{e}{\kappa_1} d \ll 1 \tag{6-166}$$

where the thin-sheet approximation has been used. Equation (6-166) turns out to be the same as Eq. (6-144) with $\mu_1 = \mu_0$.

Propagation constants for TE waves may be found from Eqs. (6-165), (6-139), and (6-163). The presence of the off-diagonal term μ_2, however, adds considerable complexity to the TE case.

Case III: Applied Magnetic Field in the Direction of Propagation (z Direction)

If the applied magnetic field is in the z direction (i.e., in the direction of propagation), the relations between $\mathbf{B}$ and $\mathbf{H}$ in region 1 are

$$B_x = \mu_1 H_x - j\mu_2 H_y \tag{6-167}$$
$$B_y = j\mu_2 H_x + \mu_1 H_y \tag{6-168}$$
$$B_z = \mu_0 H_z \tag{6-169}$$

The fields in region 0 in Fig. 6-34 are the same as in the previous two cases. In region 1, TM and TE waves are not independent. Equations (6-167) to (6-169) along with $\mathbf{D} = \epsilon\mathbf{E}$ when substituted into Maxwell's equations (6-105) to (6-108) give for region 1, $0 \lessgtr x \lessgtr d$,

$$j\beta E_y = -j\omega\mu_1 H_z - \omega\mu_2 H_y \tag{6-170}$$

$$j\beta E_x + \frac{\partial E_z}{\partial x} = -\omega\mu_2 H_z + j\omega\mu_1 H_y \tag{6-171}$$

$$\frac{\partial E_y}{\partial x} = -j\omega\mu_0 H_z \tag{6-172}$$

$$H_y = \frac{\omega\epsilon_1}{\beta} E_x \tag{6-173}$$

$$j\beta H_x + \frac{\partial H_z}{\partial x} = -j\omega\epsilon_1 E_y \tag{6-174}$$

$$\frac{\partial H_y}{\partial x} = j\omega\epsilon_1 E_z \tag{6-175}$$

$$\mu_1 \frac{\partial H_x}{\partial x} - j\mu_2 \frac{\partial H_y}{\partial x} - j\beta\mu_0 H_z = 0 \tag{6-176}$$

$$\frac{\partial E_x}{\partial x} - j\beta E_z = 0 \tag{6-177}$$

Fields in region 1 that satisfy Eqs. (6-170) through (6-177) and the boundary condition at $x = 0$ are found to be

$$E_x = \frac{-j\mu_1}{\omega^2\epsilon_1\mu_0\mu_2}\left[C_1\left(\frac{f^+}{\kappa_1^+}\sin\kappa_1^+ x - \frac{f^+}{\kappa_1^-}\sin\kappa_1^- x\right)\right.$$
$$\left. + D_1\left(\frac{f^+}{\kappa_1^+}\cos\kappa_1^+ x - \frac{f^-}{\kappa_1^-}\cos\kappa_1^- x\right)\right] \tag{6-178}$$

$$E_y = C_1\left(\sin\kappa_1^+ x - \frac{f^+}{f^-}\sin\kappa_1^- x\right) + D_1(\cos\kappa_1^+ x - \cos\kappa_1^- x) \tag{6-179}$$

$$E_z = \frac{-\mu_1}{\omega^2\epsilon_1\mu_0\mu_2\beta}[C_1 f^+(\cos\kappa_1^+ x - \cos\kappa_1^- x)$$
$$+ D_1(-f^+\sin\kappa_1^+ x + f^-\sin\kappa_1^- x)] \tag{6-180}$$

$$H_x = \frac{\beta}{\omega\mu_0}\left\{ C_1\left[\left(\frac{f^+}{\beta^2\kappa_1^+} - \frac{\mu_0}{\mu_1}\right)\sin\kappa_1^+ x + \left(\frac{-f^+}{\beta^2\kappa_1^-} + \frac{\mu_0 f^+}{\mu_1 f^-}\right)\sin\kappa_1^- x\right]\right.$$
$$\left. + D_1\left[\left(\frac{f^+}{\beta^2\kappa_1^+} - \frac{\mu_0}{\mu_1}\right)\cos\kappa_1^+ x + \left(\frac{-f^-}{\beta^2\kappa_1^-} + \frac{\mu_0}{\mu_1}\right)\cos\kappa_1^- x\right]\right\} \tag{6-181}$$

$$H_y = \frac{-j\mu_1}{\beta\omega\mu_0\mu_2}\left[C_1\left(\frac{f^+}{\kappa_1^+}\sin\kappa_1^+ x - \frac{f^+}{\kappa_1^-}\sin\kappa_1^- x\right)\right.$$
$$\left. + D_1\left(\frac{f^+}{\kappa_1^+}\cos\kappa_1^+ x - \frac{f^-}{\kappa_1^-}\cos\kappa_1^- x\right)\right] \tag{6-182}$$

$$H_z = \frac{j}{\omega\mu_0}\left[C_1\left(\kappa_1^+\cos\kappa_1^+ x - \frac{f^+}{f^-}\kappa_1^-\cos\kappa_1^- x\right)\right.$$
$$\left. + D_1(-\kappa_1^+\sin\kappa_1^+ x + \kappa_1^-\sin\kappa_1^- x)\right] \tag{6-183}$$

where

$$(\kappa_1^{\pm})^2 = -\frac{1}{2}\left[(\beta^2 - \omega^2\mu_1\epsilon_1)\left(1 + \frac{\mu_0}{\mu_1}\right) + \frac{\omega^2\mu_2{}^2\epsilon_1}{\mu_1}\right]$$

$$\pm \frac{1}{2}\left[(\beta^4 + \omega^4\mu_1{}^2\epsilon_1{}^2)\left(1 - \frac{2\mu_0}{\mu_1} + \frac{\mu_0{}^2}{\mu_1{}^2}\right) - 2\beta^2\omega^2\mu_1\epsilon_1\left(1 - \frac{2\mu_0}{\mu_1} + \frac{\mu_0{}^2}{\mu_1{}^2}\right)\right.$$

$$\left. + \frac{2\beta^2\omega^2\mu_2{}^2\epsilon_1}{\mu_1}\left(1 + \frac{\mu_0}{\mu_1}\right) - 2\omega^4\mu_2{}^2\epsilon_1{}^2\left(1 - \frac{\mu_0}{\mu_1}\right) + \frac{\omega^4\mu_2{}^4\epsilon_1{}^2}{\mu_1{}^2}\right]^{\frac{1}{2}} \quad (6\text{-}184)$$

and

$$f^{\pm} = \kappa_1^{\pm}\left[\frac{\mu_0}{\mu_1}(\beta^2 - \omega^2\mu_1\epsilon_1) + (\kappa_1^{\pm})^2\right] \quad (6\text{-}185)$$

Matching fields at $x = d$ yields four independent equations which, after eliminating amplitude factors, give

$$\left[\frac{\epsilon_1}{\epsilon_0}\left(\frac{\kappa_1^-}{\kappa_1^+} + \frac{\kappa_1^+}{\kappa_1^-}\right) - \left(\frac{f^+}{f^-} + \frac{f^-}{f^+}\right)\right]\sin \kappa_1^+d \sin \kappa_1^-d$$

$$+ \left[-2 + \frac{\epsilon_1}{\epsilon_0}\left(\frac{f^+}{f^-}\frac{\kappa_1^-}{\kappa_1^+} + \frac{f^-}{f^+}\frac{\kappa_1^+}{\kappa_1^-}\right)\right]\cos \kappa_1^+d \cos \kappa_1^-d$$

$$+ \left[\frac{\kappa_1^+}{|\kappa_0|}\left(1 - \frac{f^+}{f^-}\frac{\kappa_1^-}{\kappa_1^+}\right) - \frac{|\kappa_0|\epsilon_1}{\kappa_1^+\epsilon_0}\left(1 - \frac{f^-}{f^+}\frac{\kappa_1^+}{\kappa_1^-}\right)\right]\sin \kappa_1^+d \cos \kappa_1^-d$$

$$+ \left[\frac{\kappa_1^-}{|\kappa_0|}\left(1 - \frac{f^-}{f^+}\frac{\kappa_1^+}{\kappa_1^-}\right) - \frac{|\kappa_0|\epsilon_1}{\kappa_1^-\epsilon_0}\left(1 - \frac{f^+}{f^-}\frac{\kappa_1^-}{\kappa_1^+}\right)\right]\cos \kappa_1^+d \sin \kappa_1^-d$$

$$+ 2 - \frac{2\epsilon_1}{\epsilon_0} = 0 \quad (6\text{-}186)$$

The propagation constants may be determined from Eqs. (6-139) and (6-184) to (6-186) but only with a great deal of labor. Solution for a thin sheet, however, is relatively simple. With the assumptions that $\kappa_1^+d \ll 1$ and $\kappa_1^-d \ll 1$, $|\kappa_0|$ is found to be

$$|\kappa_0| = \frac{-\dfrac{1}{d} \pm \left\{\dfrac{1}{d^2} - 4\left(1 + \dfrac{\epsilon_0}{\epsilon_1}\right)\dfrac{\omega^2\epsilon_0}{\epsilon_1}\left[\mu_0\epsilon_0 - \mu_1\epsilon_1\left(1 - \dfrac{\mu_2{}^2}{\mu_1{}^2}\right)\right]\right\}^{\frac{1}{2}}}{2(1 + \epsilon_0/\epsilon_1)} \qquad \kappa_1d \ll 1 \quad (6\text{-}187)$$

The denominator of (6-187) is always positive. Therefore, to ensure an exponential decay for the fields in region 0 the + sign is used and it is necessary that $\mu_1\epsilon_1[1 - (\mu_2{}^2/\mu_1{}^2)] - \mu_0\epsilon_0 > 0$. Using a binomial expansion on the radical gives

$$|\kappa_0| \approx \frac{\epsilon_0}{\epsilon_1}\omega^2d\left[\mu_1\epsilon_1\left(1 - \frac{\mu_2{}^2}{\mu_1{}^2}\right) - \mu_0\epsilon_0\right] \qquad \kappa_1d \ll 1 \quad (6\text{-}188)$$

Combining Eqs. (6-188) and (6-139) gives

$$\frac{\beta}{k} = \frac{c}{v} \approx \left(1 + \left(\frac{\epsilon_0}{\epsilon_1}\right)^2 \left\{\frac{\mu_1\epsilon_1[1 - (\mu_2{}^2/\mu_1{}^2)] - \mu_0\epsilon_0}{\mu_0\epsilon_0}\right\}^2 (kd)^2\right)^{\frac{1}{2}} \qquad \kappa_1 d \ll 1$$

(6-189)

Equation (6-189) differs from Eq. (6-144) in Case I only by the term with $\mu_2{}^2$.

Although the thin-sheet approximation is good only for phase velocities a few per cent below the free space velocity, the results should be useful in practice where the thin sheet would be desirable from a loss standpoint. In all cases the fields are predominantly TM for thin sheets.

When it is desirable to change propagation constant by varying the applied magnetic field, it is best to use Cases I and III, since in Case II the TM mode is independent of the applied field and no simple solution for the TE mode exists.

The above analysis has been confined to a sheet of ferrite on a ground plane, since this is the form that a practical traveling wave antenna would usually take. The analysis may be extended to a sheet without ground plane and to a sheet of finite width set between parallel conducting plates in an H-guide configuration (see Sec. 5-4C).

B. Plasma Sheet Structures

The term *plasma*[1] describes an ionized gas which is electrically neutral; that is, the gas consists of an equal number of electrons and ions. In addition the gas may contain a number of un-ionized, or neutral, particles. In practice many ionized gases of interest to antenna engineers are less than 1 per cent ionized, so that electron-ion collisions may be neglected in comparison to electron–neutral-particle collisions. Also ion current may be neglected in comparison with electron current.

In a ferromagnetic material the application of a d-c magnetic field causes magnetic dipoles to precess around it. Similarly, in a plasma ions will circle about an applied magnetic field. Therefore, an RF magnetic field in a ferromagnetic material or an RF electric field in a plasma, normal to the d-c magnetic field, will produce a component of motion at right angles to the RF field and in time quadrature with it. This results in the form of the tensor permeability given in Eq. (6-100). The same sort of relation exists between **D** and **E** in a plasma region with an applied d-c magnetic field. The relationship may be expressed as[2]

$$\begin{bmatrix} D_i \\ D_j \\ D_k \end{bmatrix} = \begin{bmatrix} \epsilon_1 & -j\epsilon_2 & 0 \\ j\epsilon_2 & \epsilon_1 & 0 \\ 0 & 0 & \epsilon_1' \end{bmatrix} \begin{bmatrix} E_i \\ E_j \\ E_k \end{bmatrix} \qquad i, j, k = \text{cyclic 1, 2, 3} \qquad (6\text{-}190)$$

[1] This term appears to have been first introduced by I. Langmuir, *Phys. Rev.*, **33**:954, 1929.

[2] See, for example, R. S. Elliott, The Tensor Permittivity of a Weakly-ionized Plasma, *Microwave J.*, **4**(8), 67–73, August, 1961.

If collisions are neglected, the elements of the dielectric tensor are[1]

$$\frac{\epsilon_1}{\epsilon_0} = 1 - \frac{\omega_p^2}{\omega^2 - \omega_c^2} \tag{6-191}$$

$$\frac{\epsilon_2}{\epsilon_0} = \frac{\omega_p^2 \omega_c}{\omega(\omega^2 - \omega_c^2)} \tag{6-192}$$

$$\frac{\epsilon_1'}{\epsilon_0} = 1 - \frac{\omega_p^2}{\omega^2} \tag{6-193}$$

where ω_p is the plasma frequency (2π times frequency at which the electrons oscillate about their rest positions) and ω_c is the cyclotron frequency (angular frequency of electron motion around magnetic field lines). The plasma frequency is given by

$$\omega_p^2 = \frac{ne^2}{m\epsilon_0} \tag{6-194}$$

where n = electron density
  e = charge of an electron
  m = mass of an electron
The cyclotron frequency is given by

$$\omega_c = \frac{-e\mu_0}{m} H_0 \tag{6-195}$$

where H_0 is the applied d-c magnetic field. Note that, when there is no applied d-c field, $\epsilon_2 = 0$ and $\epsilon_1 = \epsilon_1'$ where ϵ_1'/ϵ_0 is the usual effective dielectric constant for a plasma with no applied magnetic field.

The analysis of a plasma sheet on a ground plane parallels that of the ferrite sheet in the previous section with the tensor dielectric producing hybrid modes (see Probs. 6-6 to 6-8). If the applied field H_0 is not present, the analysis is that of an isotropic sheet on a ground plane. In this case TM and TE modes exist independently and the analysis follows that of Secs. 6-2A and 6-3A with $\epsilon_r < 1$.

In general the natural waves over a plasma layer are slow waves ($c/v > 1$), and under certain conditions both forward and backward waves can exist (see Sec. 8-3). Fast waves also can exist, but they are forced waves and occur only when a source is present.[2]

ADDITIONAL REFERENCES

1. H. M. Barlow and J. Brown, "Radio Surface Waves," Oxford, Fair Lawn, N.J., 1962.
2. H. Jasik (ed), "Antenna Engineering Handbook," chap. 16, McGraw-Hill, New York, 1961.

[1] *Ibid.*; R. F. Whitmer, Principles of Microwave Interactions with Ionized Media, *Microwave J.*, vol. 2, February and March, 1959.
[2] T. Tamir and A. A. Oliner, The Spectrum of Electromagnetic Waves Guided by a Plasma Layer, *Polytech. Inst. Brooklyn Res. Rept.* PIB-MRI-970-61, December, 1961.

3. R. S. Elliott, Azimuthal Surface Waves on Circular Cylinders, *J. Appl. Phys.*, **26**:368, 1955.

4. L. Hatkin, Analysis of Propagating Modes in Dielectric Sheets, *Proc. IRE*, **42**:1565, 1954.

5. R. E. Plummer and R. C. Hansen, Single Slab Arbitrary Polarization Surface Wave Structure, *Proc. IEE*, Part C, Monograph 238R, 1957.

6. R. C. Hansen, Single Slab Arbitrary Polarization Surface Wave Structure, *IRE Trans. Microwave Theory Tech.*, **MTT-5**:115, 1957.

7. H. M. Barlow and A. L. Cullen, Surface Waves, *Proc. IEE*, **100**(III):329, 1953.

8. K. Horiuchi, Surface Wave Propagation over a Coated Conductor with Small Cylindrical Curvature in Direction of Travel, *J. Appl. Phys.*, **24**:961, 1953.

9. R. W. Hougardy, Periodically Slotted Surface Wave Structure, *Systems Develop. Lab. Sci. Rept.* 3508/4, Hughes Aircraft Company, Culver City, Calif., August, 1959.

10. J. R. Wait, Guiding of Electromagnetic Waves by Uniformly Rough Surfaces (Parts I and II), *IRE Trans. Antennas Propagation*, **AP-7**:S154–S168, December, 1959 (special supplement).

11. A. D. Bresler, TE_{no} Surface Waves at Ferrite-Air Interfaces, *Mem.* 48, Electrophysics Group, Microwave Research Institute, Polytechnic Institute of Brooklyn, February, 1959.

12. S. Dmitrevsky, The Character of Modes Propagating in a Grounded Ferrite Slab, *Univ. Toronto Dept. Elec. Eng. Res. Rept.* 20, September, 1959.

PROBLEMS

6-1. Solve the TE dielectric sheet problem (Sec. 6-2) for cylindrical surface waves, i.e., waves traveling radially outward from the z axis in a cylindrical coordinate system (ρ, ϕ, z) for a dielectric sheet lying on a conducting plane at $z = 0$.

6-2. Derive Eq. (6-23).

6-3. Repeat Prob. 6-1 for TM waves.

6-4. Derive Eq. (6-69).

6-5. Verify Eqs. (6-74) and (6-75).

6-6. Following the analysis for the ferrite sheet on a ground plane in Sec. 6-4, show that for a thin plasma sheet on the ground plane and with applied d-c magnetic field in the x direction the transverse propagation constant of Eq. (6-140) becomes

$$|\kappa_0| = \frac{-\dfrac{1}{d} \pm \left\{ \dfrac{1}{d^2} + 4\omega^2\mu_0\epsilon_0 \left[\left(\dfrac{\mu_1}{\mu_0}\right)^2 - \left(\dfrac{\epsilon_0}{\epsilon_1'}\right)^2 \right] \right\}^{1/2}}{2[(\mu_1/\mu_0) + (\epsilon_0/\epsilon_1')]}$$

6-7. Following the analysis for the ferrite sheet on a ground plane in Sec. 6-4, show that for a plasma sheet on a ground plane with applied magnetic field in the y direction the transcendental equations comparable to Eqs. (6-164) and (6-165) are

$$\overset{e}{\kappa_1} \cot \overset{e}{\kappa_1} d = \frac{\epsilon_0}{|\overset{e}{\kappa_0}|\epsilon_1} (\omega^2\mu_1\epsilon_1 - \beta^2) - \frac{\beta\epsilon_2}{\epsilon_1} \qquad \text{TM}$$

and

$$\overset{h}{\kappa_1} \cot \overset{h}{\kappa_1} d = \frac{-\mu_1}{\mu_0} |\kappa_0| \qquad \text{TE}$$

6-8. Repeat Prob. 6-6 for an applied field in the z direction, and show that the equation comparable to Eq. (6-187) is

$$|\kappa_0| = \frac{-\dfrac{1}{d} \pm \left[\dfrac{1}{d^2} + 4 \left(\dfrac{\mu_1}{\mu_0} + \dfrac{\epsilon_0}{\epsilon_1} \right) \dfrac{\omega^2 \epsilon_0}{\epsilon_1} (\mu_1 \epsilon_1 - \mu_0 \epsilon_0) \right]^{\frac{1}{2}}}{2(\mu_1/\mu_0 + \epsilon_0/\epsilon_1)}$$

6-9. The parallel plate region at $x = d$ in Fig. 6-9P is a reactive sheet loaded with tunnel diodes. A possible configuration is illustrated in the figure. Assume that the sheet acts as a lumped susceptance B, and that the tunnel diodes can be

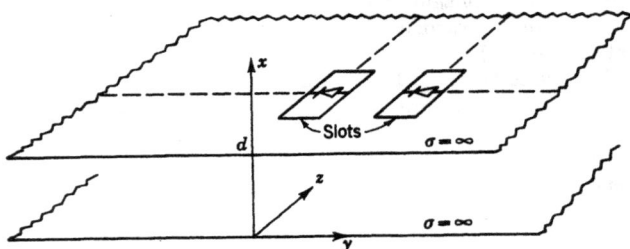

Figure 6-9P

represented simply as a negative conductance $-G_D$. Let the wave transmitted through the surface be dissipated in conductance $G = \kappa/\omega\mu$ where κ is the transverse wave number. Assume that there is no variation in the y direction and z variation is of the form $e^{-\gamma z}$.

a. Find γ for TE$_{01}$ excitation of the parallel plate region. Use a perturbation solution of a transverse resonance equation to get α and β explicitly.

b. If a uniform structure of length L and width W and without diodes is operated as an antenna with gain g_0 in the usual sense of antenna gain, what is the new gain g after the addition of $-G_D$? What is the insertion gain in the usual amplifier sense?

c. What is the significance of $|G_D| = G$?

Clearly state all assumptions and approximations that are employed.

CHAPTER **7**

TRAVELING

WAVE

EXCITATION

7-1. INTRODUCTION

For a given traveling wave antenna structure, it generally is not difficult to excite traveling waves that correspond to the natural modes of the structure. Almost any source will do this. Considerable difficulty may be experienced, however, in obtaining efficient excitation of one or more desired modes.

Exciting, or feeding, a traveling wave antenna or an "open" transmission line differs from the excitation of a conventional closed waveguide in that some of the power radiated by the feed goes into the radiation field of the traveling wave structure. The efficiency of traveling wave excitation may be defined as the ratio of the power delivered to the traveling wave to the total power delivered by the feed. Thus for efficient excitation the power available from the feed should be transferred as completely as possible to the desired traveling wave. The usual approach is to design a feed with, as nearly as possible, the same field distribution in amplitude and phase as that of the traveling wave mode.

Most practical feeds for traveling wave antennas are sections of conventional waveguides perhaps with special transition sections to improve the excitation efficiency. These will be discussed in some detail in this chapter. Much of the discussion will be qualitative, since there are only a few special cases for which rigorous analyses of the combined feed and traveling wave structure have been obtained.

7-2. FEED STRUCTURES

A. Feeds for Perturbed Waveguide Structures

The simplest traveling wave structures to feed are perturbed waveguides such as the slotted and holey waveguides in Chap. 5. Usually a section of unperturbed, standard waveguide is used as part of the feed system. The excitation of standard waveguides by posts, loops, and apertures is well covered in the literature.[1] The only problem in perturbed waveguide structures is that of providing an adequate transition from the unperturbed to the perturbed guide. This may be accomplished by tapering slot width or hole size over a wavelength or so as shown in Fig. 7-1 or matching by means of discontinuity minimizers as described in Sec. 7-3.

For a low sidelobe design the amplitude taper is such that a gradual transition between unperturbed and perturbed guide almost always exists and the excitation efficiency may approach 100 per cent.

An abrupt beginning of the radiating portion of the waveguide, however, will in general create a sizable discontinuity, and serious pattern degradation may occur owing to higher mode radiation in the vicinity of the discontinuity. There may be cases, though, in which the radiation due to the discontinuity is desired in order to obtain a particular beam shape.

B. Horn and Open-ended-waveguide Feeds

A common form of feed for many traveling wave antennas is conventional waveguide either used as open-ended guide or flared to form a horn. Some examples are illustrated in Fig. 7-2.

Open-ended-waveguide and horn feeds work quite well for line source endfire radiators and also for "open" transmission lines such as the dielectric-

---

[1] See, for example, R. E. Collin, "Field Theory of Guided Waves," chap. 7, McGraw-Hill, New York, 1960.

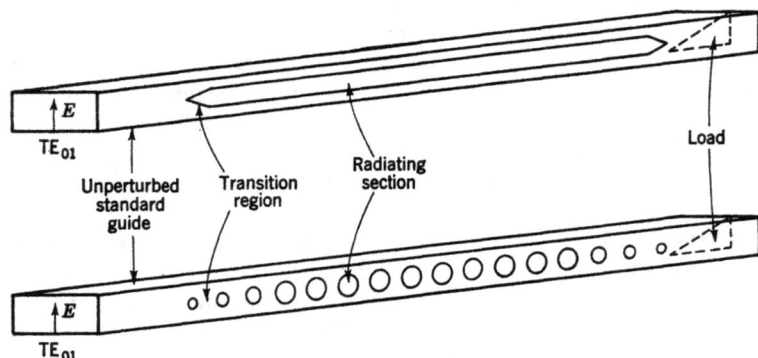

Figure 7-1. Examples of traveling wave antennas fed from conventional waveguide.

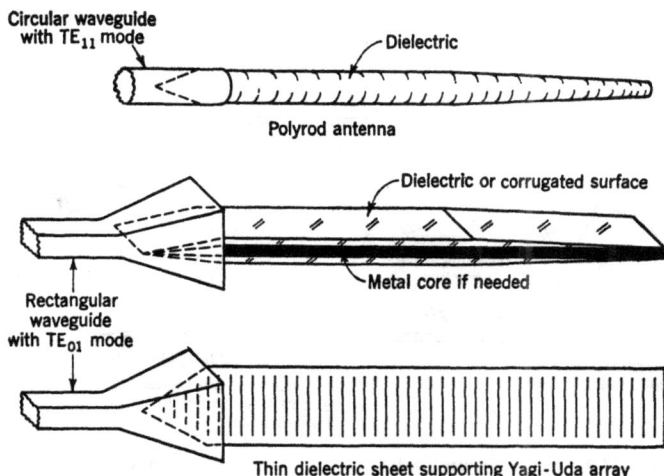

Figure 7-2. Horn and open-ended-waveguide feeds.

coated conductor (Goubau line). This is because such devices incorporate slow wave structures in which the traveling wave is bound to the surface and has a transverse field distribution which in many practical cases is very nearly that of a conventional mode in a standard waveguide. A good transition from closed to open structure may be obtained by tapering the open structure where it fits into the conventional guide. In many practical structures of the type shown in Fig. 7-2 it is possible to get a VSWR of less than 1.05 in the feed waveguide over a relatively wide range of frequencies.

C. Line Source Feeds

In some applications line source feeds may be necessary. This is generally true when the traveling wave structure is a planar structure many wavelengths wide. In this case a feed in the form of a transverse line source may be necessary to utilize fully the width of the traveling wave structure.

Some examples of transverse line feeds are illustrated in Fig. 7-3. A horn aperture may be necessary to ensure good excitation efficiency. The line source feed may be a broadside radiator or, for scanning purposes, may be a waveguide source with adjustable phase velocity as indicated in Fig. 7-3c. Transverse line source feeds are considered in more detail in Sec. 7-4.

The feeds illustrated so far in Figs. 7-2 and 7-3 generally will not be flush with the surface of the traveling wave structure. This is definitely the case when a horn aperture is utilized. For applications where flush mounting is desired, efficient excitation may be obtained by utilizing a source that extends longitudinally along the traveling wave structure and

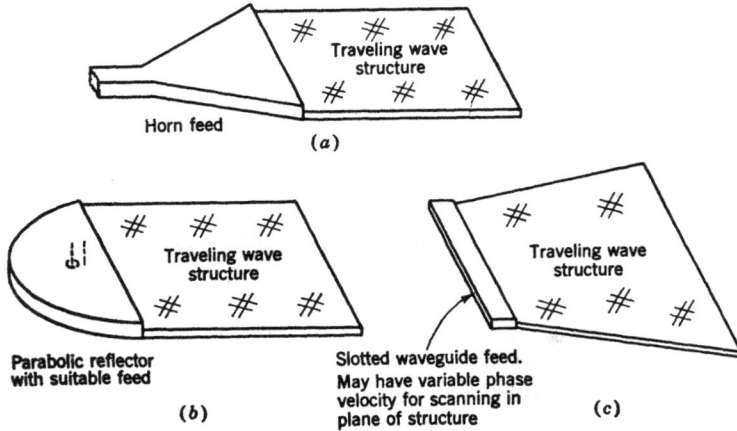

Figure 7-3.   Traveling wave antennas with transverse line source feeds.

in or below it. Some examples of this type of feed are illustrated by the side views in Fig. 7-4. Figure 7-4a, however, depicts one way in which a parallel plate line feed may be used in a flush structure.

Some planar structures such as the grid structures in Chap. 6 may be thought of as perturbed parallel plate waveguides. These may be classified with the structures shown in Fig. 7-1, and in practice a grid structure may be tapered (by varying hole size or wire separation, for example) to obtain

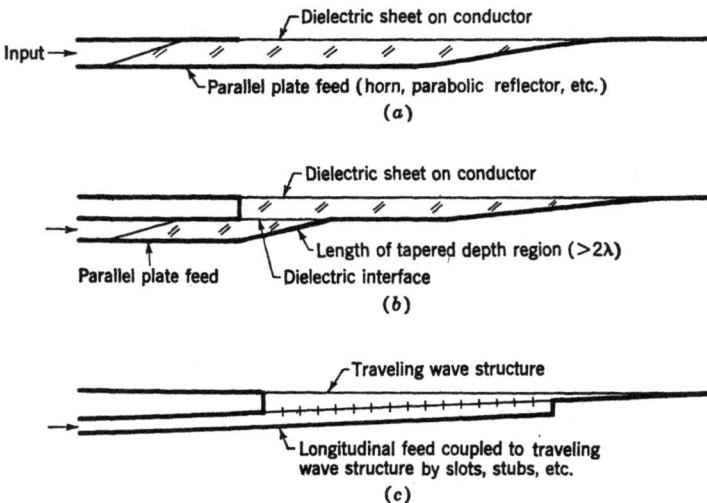

Figure 7-4.   Feeds for flush-mounted antennas.

a good transition with no change in plate separation. Thus grid structures are ideal for flush mounting.

In the longitudinal feeds illustrated in Fig. 7-4b and c both the feed and traveling wave antenna structure may be thought of as transmission lines. The problem then is one of coupling between two transmission lines. The usual procedure is to couple between principal field components in the two lines and adjust phase velocities and coupling coefficient so that complete energy transfer takes place over some specified length that is less than the total length of the antenna structure. Further discussion on coupled transmission lines is given in Sec. 7-6.

The structure in Fig. 7-4b indicates that the energy is forced from the feed line to the traveling wave structure without regard for matching of phase velocities. Actually phase velocities for such feeds may be properly matched over a part of the feed length, and in practice such feeds work quite well, although the feed may tend to produce some unwanted radiation. It is quite difficult to achieve better than about 90 per cent excitation efficiency for the structure in Fig. 7-4b, whereas the properly designed structure in Fig. 7-4c may approach 100 per cent efficiency.

7-3. DISCONTINUITY MINIMIZERS

In principle the feed section of a traveling wave antenna should produce the same field configuration that exists in the aperture, or radiating, section. In practice it is desirable to feed the aperture section from conventional closed waveguide. This usually necessitates a transition section between the closed guide and the open structure, which forms the antenna aperture, in order to obtain a gradual change from the field configuration of the closed guide to that of the aperture section.

When the aperture section consists of a traveling wave structure that is not more than a moderate perturbation of the closed waveguide feed section, a suitable transition usually may be obtained by tapering the perturbation (slot width, for example) over a wavelength or so. Such a taper may be called a "discontinuity minimizer," since it reduces the discontinuity that would exist if the feed and aperture sections were simply butted together. Such a discontinuity by setting up higher-order modes may substantially reduce the excitation efficiency of the desired mode. If the aperture section is designed to support only the dominant mode, the higher modes will be rapidly attenuated but nevertheless may contribute significantly to the radiation pattern of the structure.

When the aperture and feed sections are quite dissimilar as in the case of a corrugated structure fed from a conventional waveguide or parallel plate line, the corrugated structure may be tapered so as to reduce its thickness to a negligible amount in the feed section and at the same time a horn aperture may be added to the feed section. These measures constitute

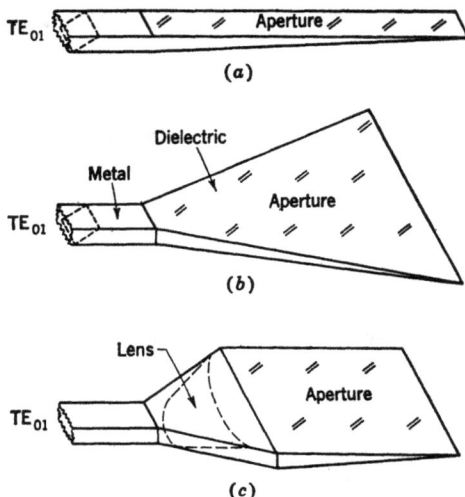

TE$_{01}$

(a)

Dielectric

Metal

TE$_{01}$

Aperture

(b)

Figure 7-5. Practical endfire antennas. The aperture usually would be in
a ground plane.

Lens

TE$_{01}$

Aperture

(c)

a discontinuity minimizer. The horn aperture may be objectionable, however, because it does not permit flush mounting of the antenna.

In some cases it is possible to incorporate a discontinuity minimizer
into the transition from closed to open structure without detracting from
the flush profile of the structure. Figure 7-5 shows several versions of a
practical flush-mounted endfire antenna that represent significant perturbations of the TE$_{01}$ waveguide feed section. The antennas in Fig. 7-5
make excellent endfire antennas, although the pattern and impedance bandwidth are limited by the discontinuity at the beginning of the aperture
section. The discontinuity at the end of the structure may be made negligible by gradually tapering the depth of the structure to a very small value
as indicated in Fig. 7-5.

The aperture sections of the antennas in Fig. 7-5 may be considered
as hybrid-mode dielectric-slab structures (see Sec. 5-4C). As such the discontinuity at the beginning of the aperture may be reduced by making the
dielectric relatively thick in this region so that as a transmission line prac-
tically all the energy is contained in the dielectric region. This technique
may jeopardize the performance as an endfire antenna, however, by causing
an excessive change in phase velocity along the structure as the depth is
tapered to provide a good transition to free space.

For the antennas in Fig. 7-5 the beginning of the aperture is essentially
a TE$_{01}$ dielectric-filled waveguide with a wide wall removed. The resulting
wide slot amounts to a significant perturbation of the closed guide. In many
instances, air-filled slotted waveguides in particular, a V taper at the
beginning of the slot makes an effective discontinuity minimizer. However,
the addition of a V taper to an endfire dielectric-filled antenna such as

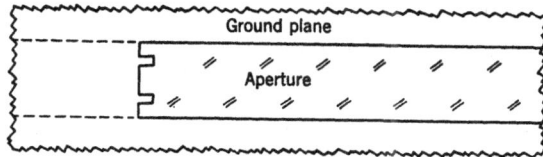

(a) Fingers discontinuity minimizer

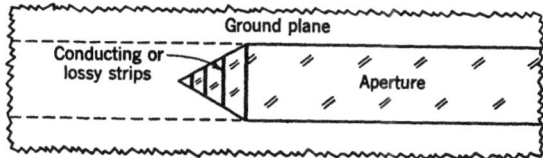

(b) Tapered-ladder discontinuity minimizer

Figure 7-6. Practical discontinuity minimizers for flush endfire antennas.

shown in Fig. 7-5, while improving the pattern in the principal plane normal to the aperture, may seriously distort the pattern in the plane of the aperture.

Distortion in the aperture plane may be attributed to a transverse electric field set up by the V taper. This suggests that a transition section should be designed to modify the longitudinal component of electric field without introducing a transverse component or, if it is not possible to avoid the transverse component, to provide a means for restricting its radiation. Two practical discontinuity minimizers[1] that were developed from these arguments are shown in Fig. 7-6.

The fingers in Fig. 7-6a are merely short stubs usually less than $\lambda/4$ long. Their purpose is to short out the tangential fields of undesired modes present at the discontinuity. The tapered-ladder discontinuity minimizer in Fig. 7-6b consists of a V taper with transverse strips to restrict radiation from any transverse electric field. Best results are obtained with the transverse strips composed of lossy material. The lossy strips reduce antenna efficiency somewhat, but the improvement in antenna bandwidth may more than offset the loss in efficiency.[2]

In general, the lossy tapered ladder is more effective than the fingers. A tapered-ladder discontinuity minimizer on an antenna such as shown in Fig. 7-5b has operated satisfactorily over a 4:1 band.[3]

7-4. EXCITATION EFFICIENCY

The previous sections of this chapter presented a qualitative discussion of the excitation of traveling wave antenna structures. The present section

[1] B. T. Stephenson and C. H. Walter, Endfire Slot Antennas, *IRE Trans. Antennas Propagation*, **AP-3**:81–86, April, 1955.

[2] *Ibid.*

[3] J. W. Eberle, C. A. Levis, and D. McCoy, The Flared Slot: A Moderately Directive Flush-mounted Broad-band Antenna, *IRE Trans. Antennas Propagation*, **AP-8**:461–468, September, 1960.

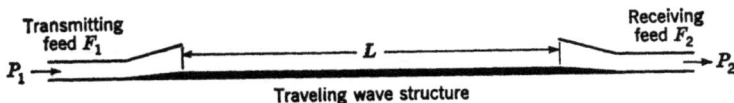

Figure 7-7. System for measuring excitation efficiency. The transmission lines should be matched to the feeds for VSWR < 1.05.

will consider the efficiency of traveling wave excitation in more detail. The experimental determination of excitation efficiency will be discussed, and methods for analytically determining the excitation efficiency will be described and illustrated by examples.

A. Experimental Determination of Excitation Efficiency

The excitation efficiency has been defined as the ratio of the power delivered to the traveling wave to the total power delivered by the source. The excitation efficiency may be determined experimentally by the system illustrated in Fig. 7-7. Feed F_1 is the feed whose excitation efficiency is to be determined. If P_1 is the power delivered to F_1, then the power received at F_2 is given by

$$P_2 = P_1 T_1 T_2 e^{-2\alpha L} \tag{7-1}$$

where T_1 is the power transmission coefficient for the junction between F_1 and the traveling wave line and T_2 is the power transmission coefficient for F_2 and the traveling wave line. It is convenient to use identical feeds having the same locations with respect to the line so that $T_1 = T_2$. Then

$$P_2 = P_1 T_1^2 e^{-2\alpha L} \tag{7-2}$$

By definition T_1 is the excitation efficiency; hence from Eq. (7-2)

$$\text{Excitation efficiency} = \sqrt{\frac{P_2}{P_1}} \, e^{\alpha L} \tag{7-3}$$

The attenuation constant α would include radiation, dielectric, and conductor losses along the traveling wave structure. This quantity can be determined analytically (see Chaps. 5 and 6) or experimentally (see Chap. 4). With α and L known, measurements of P_2 and P_1 are sufficient to determine the excitation efficiency. Note that for planar structures the feeds should be the width of the structure to eliminate errors due to spreading of the fields.

Equation (7-3) neglects interaction between the feeds F_1 and F_2. If reflections from F_1 and F_2 cause a significant standing wave on the traveling wave structure, its effect usually may be eliminated by finding an average value for P_2 when F_2 is moved a distance of one-half of the line wavelength.

For low loss slow wave structures, $\alpha \approx 0$ and the excitation efficiency is simply $\sqrt{P_2/P_1}$.

B. Analytical Determination of Excitation Efficiency

Aperture Method

For slow wave structures with feeds such as horns which have well-defined physical apertures, an approximate excitation efficiency may be determined from the fraction of traveling wave energy intercepted by the feed aperture. This is illustrated in Fig. 7-8. Here we are considering the structure as a receiving antenna. By reciprocity the excitation efficiency for the transmitting case will be the same as for the receiving case.

It is assumed that over the surface denoted as aperture plane in Fig. 7-8 the fields of the traveling wave are the same as would exist on this surface for a traveling wave structure extending indefinitely with no feed. For the mode that is of interest, let the fields tangential to the aperture plane be E_t and H_t. These fields are given in Chaps. 5 and 6 for many practical structures. The power intercepted by the aperture A is

$$P_A = \frac{1}{2}\mathrm{Re}\int_A \mathbf{E}_t \times \mathbf{H}_t^* \cdot \mathbf{ds} \tag{7-4}$$

whereas the total power in the traveling wave is

$$P_T = \frac{1}{2}\mathrm{Re}\int_{\substack{\text{Aperture}\\\text{plane}}} \mathbf{E}_t \times \mathbf{H}_t^* \cdot \mathbf{ds} \tag{7-5}$$

The excitation efficiency is simply the ratio P_A/P_T.

As an example of the aperture method let us consider a planar slow wave structure excited by an aperture source such as that illustrated in Fig. 7-8a. For simplicity let us assume that most of the energy of the surface wave is contained in the air region above the structure. Thus the fields

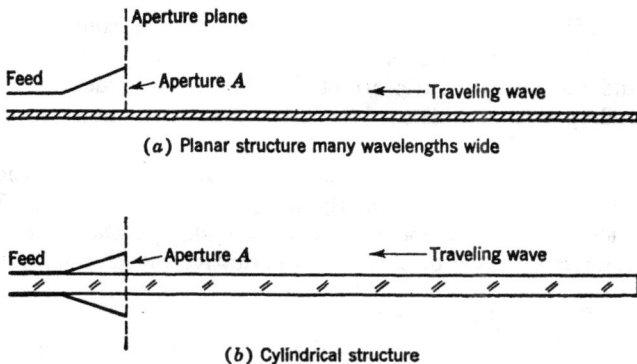

(a) Planar structure many wavelengths wide

(b) Cylindrical structure

Figure 7-8. Wave intercepted by feed aperture.

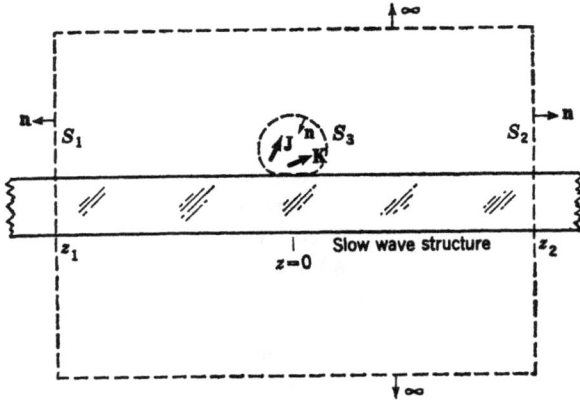

Figure 7-9. Slow wave structure excited by electric- or magnetic-current elements.

may be expressed as (see Sec. 6-3A, for example)

$$H_t = B_0 e^{-|\kappa_0|x} e^{-\gamma z} \tag{7-6}$$

$$E_t = \frac{\gamma B_0}{j\omega\epsilon_0} e^{-|\kappa_0|x} e^{-\gamma z} \tag{7-7}$$

Performing the integration in Eqs. (7-4) and (7-5) where the aperture is of height h and taking the ratio P_A/P_T give

$$\text{Excitation efficiency} = 1 - e^{-2|\kappa_0|h} \tag{7-8}$$

where

$$|\kappa_0| = k \left[\left(\frac{c}{v} \right)^2 - 1 \right]^{\frac{1}{2}} \tag{7-9}$$

This is a very simple result from which excitation efficiency may be found for either TE or TM modes in terms of the height of the aperture and the velocity ratio c/v of the slow wave structure. The simplifications and approximations used in deriving Eq. (7-8) must be kept in mind when using it.

The aperture method for determining excitation efficiency is based on a highly idealized model of the actual antenna structure. Efficiencies determined by this method are generally high, and the method is recommended more as a means of obtaining qualitative rather than quantitative results.

Mode Orthogonality Method

It is well known that the TM and TE modes of closed waveguides are orthogonal to each other.[1] The various modes of open, slow wave

---

[1] Collin, *op. cit.*, pp. 174–179.

structures also have orthogonal properties.[1] For both open and closed waveguides, mode orthogonality may be proved by use of the Lorentz reciprocity theorem. In this section we shall derive the orthogonality relation for slow waves and utilize it in obtaining excitation efficiency.

Consider a lossless slow wave structure as shown in Fig. 7-9. A closed surface S_3 contains the source, which may consist of either electric- or magnetic-current elements or both. A closed surface S consists of the surfaces S_1 and S_2 at the transverse planes z_1 and z_2, respectively, and an infinite cylinder surrounding the traveling wave structure.

Assume that the source produces one or more slow wave (surface wave) modes along the structure as well as a radiation field $\mathbf{E}_R, \mathbf{H}_R$. Let the nth slow wave be denoted as $\mathbf{E}_{sn}, \mathbf{H}_{sn}$. The total field may be represented as

$$\mathbf{E}_T = \sum_n a_n \mathbf{E}_{sn}^{+} + \mathbf{E}_R^{+} \tag{7-10}$$

$$z > 0$$

$$\mathbf{H}_T = \sum_n a_n \mathbf{H}_{sn}^{+} + \mathbf{H}_R^{+} \tag{7-11}$$

$$\mathbf{E}_T = \sum_n b_n \mathbf{E}_{sn}^{-} + \mathbf{E}_R^{-} \tag{7-12}$$

$$z < 0$$

$$\mathbf{H}_T = \sum_n b_n \mathbf{H}_{sn}^{-} + \mathbf{H}_R^{-} \tag{7-13}$$

The traveling waves for a uniform structure have z variation of the form $e^{-\gamma_n z}$; thus the transverse fields $\mathbf{E}^t$ may be represented by

$$\mathbf{E}_{sn}^{t+} = \mathbf{e}_{sn} e^{-\gamma_n z} \tag{7-14}$$
$$\mathbf{E}_{sn}^{t-} = \mathbf{e}_{sn} e^{\gamma_n z} \tag{7-15}$$
$$\mathbf{H}_{sn}^{t+} = \mathbf{h}_{sn} e^{-\gamma_n z} \tag{7-16}$$
$$\mathbf{H}_{sn}^{t-} = -\mathbf{h}_{sn} e^{\gamma_n z} \tag{7-17}$$

where $\mathbf{e}_{sn}$, $\mathbf{h}_{sn}$ are transverse vector functions of x and y.

Maxwell's curl equations with electric- and magnetic-current elements present are

$$\nabla \times \mathbf{H} = j\omega\epsilon \mathbf{E} + \mathbf{J} \tag{7-18}$$
$$\nabla \times \mathbf{E} = -j\omega\mu \mathbf{H} - \mathbf{K} \tag{7-19}$$

The total field $\mathbf{E}_T, \mathbf{H}_T$ will be a solution of Eqs. (7-18) and (7-19). Each slow wave field $\mathbf{E}_{sn}, \mathbf{H}_{sn}$ of the traveling wave structure will be a solution of Eqs. (7-18) and (7-19) with $\mathbf{J}$ and $\mathbf{K}$ equal to zero (source-free field equations).

The Lorentz reciprocity theorem[2] gives

$$\int_S (\mathbf{E} \times \mathbf{H}_0 - \mathbf{E}_0 \times \mathbf{H}) \cdot \mathbf{n} \, ds = \int_V (\mathbf{J} \cdot \mathbf{E}_0 - \mathbf{K} \cdot \mathbf{H}_0) \, dv \tag{7-20}$$

[1] G. Goubau, On the Excitation of Surface Waves, *Proc. IRE*, **40**:865–868, July, 1952; Collin, *op. cit.*, pp. 483–485.
[2] See Eq. (1-19).

where E_0, H_0 are solutions to the source-free field equations and n is a unit vector directed out of S. The surface S encloses volume V.

Consider first the source-free case ($J = K = 0$). The fields in Eq. (7-20) for this case consist only of slow waves along the traveling wave structure which approach zero on the cylindrical surface as it is made infinitely large. Let E, H be the mth slow wave mode and E_0, H_0 the nth slow wave mode. Then Eq. (7-20) reduces to

$$\int e_{sm} \times h_{sn} \cdot z \, ds = \int e_{sn} \times h_{sm} \cdot z \, ds = 0 \qquad m \neq n \qquad (7\text{-}21)$$

where integration is over a transverse plane and z is a unit vector in the z direction. Equation (7-21) is the orthogonality relation for slow wave (surface wave) modes.

Next consider Eq. (7-20) with a source present. The integral over the cylinder at infinity remains zero, since the radiation fields as well as the slow wave fields approach zero as the cylinder becomes infinitely large. Let E, H be the total field and E_0, H_0 the nth surface wave mode propagating in the positive z direction. Integrating Eq. (7-20) over the transverse planes S_1 and S_2 at z_1 and z_2, respectively, and utilizing the orthogonality relation (7-21) give

$$-e^{-\gamma_n z_1} \int_{S_1} (E_R{}^- \times h_{sn} - e_{sn} \times H_R{}^-) \cdot z \, ds$$
$$+ e^{-\gamma_n z_2} \int_{S_2} (E_R{}^+ \times h_{sn} - e_{sn} \times H_R{}^+) \cdot z \, ds$$
$$- 2b_n \int_{S_1} e_{sn} \times h_{sn} \cdot z \, ds = \int_V (E_{sn}{}^+ \cdot J - H_{sn}{}^+ \cdot K) \, dv \qquad (7\text{-}22)$$

The first two terms of Eq. (7-22) must each be equal to a constant, since z_1 and z_2 are arbitrary. Furthermore, the constants must be zero, since E_R and H_R approach zero as z_1 and z_2 go to infinity. Thus we obtain from (7-22) the relation

$$2b_n = \frac{-\int (E_{sn}{}^+ \cdot J - H_{sn}{}^+ \cdot K) \, dv}{\int e_{sn} \times h_{sn} \cdot z \, ds} \qquad (7\text{-}23)$$

By similar arguments for waves propagating in the $-z$ direction we obtain

$$2a_n = \frac{-\int (E_{sn}{}^- \cdot J - H_{sn}{}^- \cdot K) \, dv}{\int e_{sn} \times h_{sn} \cdot z \, ds} \qquad (7\text{-}24)$$

where the volume integral is taken over a volume containing the source and the surface integral is over any transverse plane on either side of the source.

If the source is an electric dipole of length dl and moment P, then $J \, dl = j\omega P$. Similarly if the source is a magnetic dipole (electric loop) of

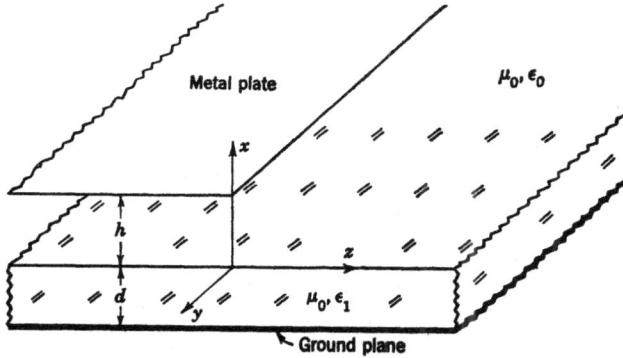

Figure 7-10.   Parallel plate waveguide feeding dielectric slab on ground plane.

moment $\mathbf{M}$, then $\mathbf{K}\,dl = j\omega\mathbf{M}$. For these sources Eqs. (7-23) and (7-24) reduce to

$$2b_n = \frac{-j\omega\mathbf{E}_{sn}{}^+ \cdot \mathbf{P} + j\omega\mathbf{H}_{sn}{}^+ \cdot \mathbf{M}}{\int\mathbf{e}_{sn} \times \mathbf{h}_{sn} \cdot \mathbf{z}\,ds} \tag{7-25}$$

and

$$2a_n = \frac{-j\omega\mathbf{E}_{sn}{}^- \cdot \mathbf{P} + j\omega\mathbf{H}_{sn}{}^- \cdot \mathbf{M}}{\int\mathbf{e}_{sn} \times \mathbf{h}_{sn} \cdot \mathbf{z}\,ds} \tag{7-26}$$

If the source is the aperture of a waveguide structure such as shown in Fig. 7-10, the mode orthogonality method must be modified somewhat. Since all source currents occur in the region $z \lessgtr 0$, the left half plane can be replaced by equivalent currents on the plane $z = 0$ (Huygens' principle). This is illustrated in Fig. 7-11. The currents are given by

$$\mathbf{J}_s = \mathbf{z} \times \mathbf{H}_s \tag{7-27}$$
$$\mathbf{K}_s = \mathbf{E}_s \times \mathbf{z} \tag{7-28}$$

where $\mathbf{E}_s$ and $\mathbf{H}_s$ are the fields existing on the plane $z = 0$ in Fig. 7-10.

Figure 7-11.   Replacing waveguide feed by an equivalent current sheet.

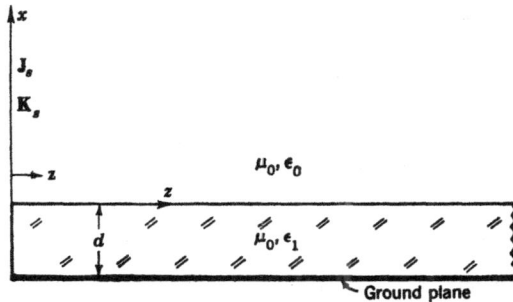

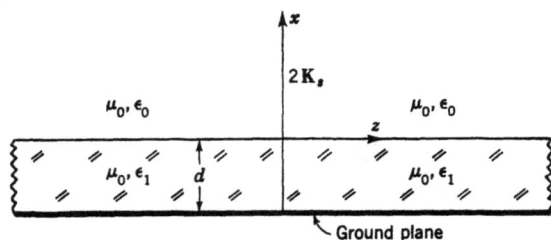

Figure 7-12. Symmetrical structure representation of Fig. 7-11.

In many cases a good approximation would be to use the waveguide fields for $-d \lesssim x \lesssim h$ and take the fields to be zero for $x > h$.

If a sheet that is a perfect conductor of electricity is moved infinitesimally close to $z = 0$ on the left, the fields for $z > 0$ will be unchanged but the current $\mathbf{J}_s$ will be shorted out; the fields are given in terms of $\mathbf{K}_s$ radiating in the presence of the conducting plane at $z = 0$. By image theory the electric conductor can be removed if the magnetic current $\mathbf{K}_s$ is doubled and the surface wave structure imaged about $z = 0$. This results in the system shown in Fig. 7-12, which gives the fields of the original structure in the region $z > 0$.

The original problem of an asymmetrical structure has now been reduced to the problem of a symmetrical structure, and the previous analysis may be applied. In the present problem the volume integral in Eq. (7-24) is replaced by a surface integral over the plane $z = 0$ and Eq. (7-24) becomes

$$a_n = \frac{\int_{z=0} \mathbf{E}_s \times \mathbf{h}_{sn} \cdot \mathbf{z}\, ds}{\int_{z=\text{const}} \mathbf{e}_{sn} \times \mathbf{h}_{sn} \cdot \mathbf{z}\, ds} \qquad (7\text{-}29)$$

Equations (7-23), (7-24), and (7-29) give the amplitudes of the slow wave (surface wave) modes in terms of the source and the fields of the natural modes of the traveling wave structure. The equations show at once whether or not a given source will excite a particular mode.

The power P_{sn} in the nth mode is given by

$$P_{sn} = \frac{a_n^2}{2} \operatorname{Re} \int_{z=\text{const}} \mathbf{e}_{sn} \times \mathbf{h}_{sn} \cdot \mathbf{z}\, ds \qquad z > 0 \qquad (7\text{-}30)$$

$$P_{sn} = \frac{b_n^2}{2} \operatorname{Re} \int_{z=\text{const}} \mathbf{e}_{sn} \times \mathbf{h}_{sn} \cdot \mathbf{z}\, ds \qquad z < 0 \qquad (7\text{-}31)$$

where b_n and a_n are given by Eqs. (7-23), (7-24), and (7-29). The power P_R radiated by the source in Fig. 7-9 is given by

$$P_R = -\tfrac{1}{2} \operatorname{Re} \int_{S_3} \mathbf{E}_T \times \mathbf{H}_T^* \cdot \mathbf{n}\, ds \qquad (7\text{-}32)$$

where $\mathbf{E}_T$ and $\mathbf{H}_T$ are the total fields on the surface S_3. The power P_R

radiated by the source in Fig. 7-11 is

$$P_R = \tfrac{1}{2}\mathrm{Re} \int_{z=0} \mathbf{E}_s \times \mathbf{H}_s^* \cdot \mathbf{z} \, ds \tag{7-33}$$

The excitation efficiency is simply the ratio P_{sn}/P_R.

Fourier Transform Method

For the parallel plate waveguide feed in Fig. 7-10 both excitation efficiency and input impedance can be determined by a Fourier transform method as used by Angulo and Chang.[1] The structure is viewed as a transverse transmission line, and the complete fields are determined.

The structure in Fig. 7-10 may be considered as a transverse transmission line in the form of two parallel plate waveguides in tandem as illustrated in Fig. 7-13. One is an air-filled parallel plate guide with walls at $z = \pm \infty$ extending from $x = 0$ to $x = \infty$. The other is a dielectric-filled parallel plate guide, also with walls at $z = \pm \infty$, extending from $x = 0$ to $x = -d$. The latter guide is terminated by a perfect conductor of electric current at $x = -d$. The first waveguide has an obstacle at $x = h$ in the form of a semi-infinite perfectly conducting plate.

A solution for the structure in Fig. 7-13 has been worked out in detail by Angulo and Chang[1] for excitation of the lowest-order TM surface wave mode. The fields transverse to the x direction (that is, H_y, E_z) may be represented everywhere as

$$H_y = H_{0y} + H_{sy} \tag{7-34}$$
$$E_z = E_{0z} + E_{sz} \tag{7-35}$$

[1] C. M. Angulo and W. S. C. Chang, The Launching of Surface Waves by Parallel Plate Waveguide, *IRE Trans. Antennas Propagation*, **AP-7**(4):359–368, October, 1959.

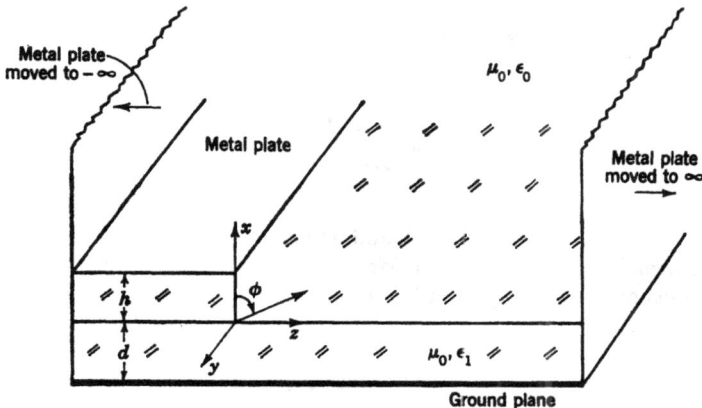

Figure 7-13.   Dielectric sheet on ground plane fed by parallel plate waveguide.

where H_{0y}, E_{0z} are the components of the dominant TM mode in the feed structure ($z < 0$, $-d \lesssim x \lesssim h$) and H_{sy}, E_{sz} are the scattered fields everywhere. The fields H_{0y} and E_{0z} are given by

$$H_{0y} = \left[u(x) \frac{\cosh k(x-h)s'}{\cosh khs'} + u(-x) \frac{\cos k(x+d)r'}{\cos kdr'} \right] e^{jkz(1+s'^2)^{1/2}}$$

(7-36)

$$E_{0z} = \frac{-jks' \tanh khs'}{\omega \epsilon_1} \left[u(x) \frac{\sinh k(x-h)s'}{\sinh khs'} \right.$$
$$\left. - u(-x) \frac{\sin k(x+d)r'}{\sin kdr'} \right] e^{jkz(1+s'^2)^{1/2}}$$ (7-37)

where ks' and kr' are the wave numbers in the x direction in the air and dielectric regions, respectively, and $u(x)$ is the unit step function which is zero for negative arguments and unity for positive arguments. The quantities s' and r' are solutions of

$$s'^2 + r'^2 = \frac{\epsilon_1}{\epsilon_0} - 1$$

(7-38)

$$\frac{\epsilon_1 s'}{\epsilon_0 r'} = \frac{\tan kdr'}{\tanh khs'}$$

(7-39)

For $x > h$, H_{0y} and E_{0z} are zero.

The scattered fields may be represented by the Fourier integrals

$$H_{sy}(x,z) = \frac{1}{\sqrt{2\pi}} \int_{-\infty}^{\infty} I(\beta,x) e^{-j\beta z} \, d\beta$$

(7-40)

$$E_{sz}(x,z) = -\frac{1}{\sqrt{2\pi}} \int_{-\infty}^{\infty} V(\beta,x) e^{-j\beta z} \, d\beta$$

(7-41)

with inverse transforms

$$I(\beta,x) = \frac{1}{\sqrt{2\pi}} \int_{-\infty}^{\infty} H_{sy}(x,z) e^{j\beta z} \, dz$$

(7-42)

$$V(\beta,x) = -\frac{1}{\sqrt{2\pi}} \int_{-\infty}^{\infty} E_{sz}(x,z) e^{j\beta z} \, dz$$

(7-43)

The quantities $V(\beta,x)$ and $I(\beta,x)$ may be interpreted as voltage and current, respectively, and Maxwell's equations require that they be solutions of the transmission-line equations

$$\frac{dV(\beta,x)}{dx} = -j\kappa Z I(\beta,x)$$

(7-44)

$$\frac{dI(\beta,x)}{dx} = -j\kappa Y V(\beta,x)$$ $x > -d$

(7-45)

For the air region $(x > 0)$

$$\kappa = \kappa_0 = (k^2 - \beta^2)^{1/2} \tag{7-46}$$

$$Y = Y_0 = \frac{1}{Z_0} = \frac{\omega\epsilon_0}{\kappa_0} \tag{7-47}$$

In the dielectric $(-d < x < 0)$

$$\kappa_1 = \left(\frac{k^2\epsilon_1}{\epsilon_0} - \beta^2\right)^{1/2} \tag{7-48}$$

$$Y_1 = \frac{1}{Z_1} = \frac{\omega\epsilon_1}{\kappa_1} \tag{7-49}$$

From transmission-line theory V and I may be written immediately as

$$V(\beta,x) = V(\beta,h^+)e^{-j\kappa_0(x-h)} \qquad \tag{7-50}$$
$$I(\beta,x) = Y_0V(\beta,x) \qquad x > h \tag{7-51}$$
$$V(\beta,x) = V(\beta,h^-)\cos\kappa_0(x-h) - jZ_0I(\beta,h^-)\sin\kappa_0(x-h) \tag{7-52}$$
$$0 < x < h$$
$$I(\beta,x) = I(\beta,h^-)\cos\kappa_0(x-h) - jY_0V(\beta,h^-)\sin\kappa_0(x-h) \tag{7-53}$$

$$V(\beta,x) = [V(\beta,h^-)\cos\kappa_0h + jZ_0I(\beta,h^-)\sin\kappa_0h]\frac{\sin\kappa_1(x+d)}{\sin\kappa_1d} \tag{7-54}$$
$$-d < x < 0$$

$$I(\beta,x) = [I(\beta,h^-)\cos\kappa_0h + jY_0V(\beta,h^-)\sin\kappa_0h]\frac{\cos\kappa_1(x+d)}{\cos\kappa_1d} \tag{7-55}$$

where

$$I(\beta,h^-) = -jY_0V(\beta,h^-)\frac{(\kappa_1\epsilon_0/\epsilon_1)\tan(\kappa_0h)\tan(\kappa_1d) - \kappa_0}{(\kappa_1\epsilon_0/\epsilon_1)\tan\kappa_1d + \kappa_1\tan\kappa_0h} \tag{7-56}$$

A knowledge of V and I at the boundary $x = h$ gives V and I everywhere, and the inversion of V and I by Eqs. (7-40) and (7-41) gives the scattered fields everywhere. The equation for the transform may be solved by the Wiener-Hopf technique.[1] It is sufficient to carry out the inversion only for far fields (i.e., fields far away from the discontinuity at $z = 0$), and the integrals may be evaluated by the method of steepest descents. For $z \gg 0$ the field at the surface of the dielectric is essentially that of the dominant TM surface wave mode. For $z \ll 0$ the field is that of the dominant mode in the partially dielectric-filled parallel plate waveguide. After the complete fields have been determined, the reflected power in the parallel plate feed, the surface wave power, and the radiated power can be found by Poynting's theorem.

[1] Details are given in *ibid.* For a general discussion of Wiener-Hopf methods, see for example, B. Noble, "The Wiener-Hopf Technique," Pergamon Press, New York, 1958.

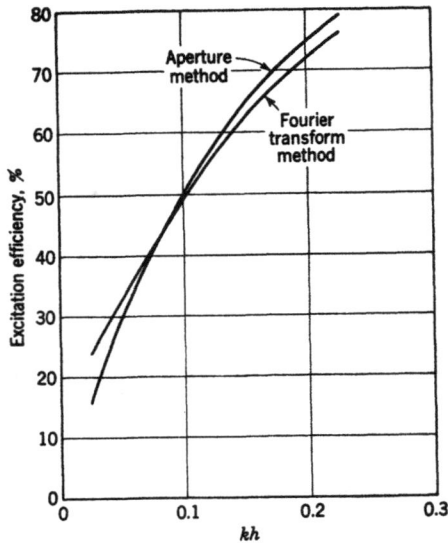

Figure 7-14. Excitation efficiency of the dominant TM surface wave mode on a grounded dielectric sheet ($\epsilon_1/\epsilon_0 = 2.49$ and $kd = 0.5$) with parallel plate feed (see Fig. 7-10 or 7-13).

The percentages of reflected power, surface wave power, and radiated power as found by Angulo and Chang are, respectively,

$$P_{ref} = \frac{(1 + a_2)^2(a_2 - a_1)^2}{(a_2 - 1)^2(a_2 + a_1)^2} \exp\left[\frac{-4a_2}{\pi} \int_0^1 \frac{\Omega(y)}{1 + s'^2 - y^2} \, dy\right] \qquad (7\text{-}57)$$

$$P_{sw} = \frac{4(a_1 - 1)a_2(a_2 + 1)a_1 \operatorname{sech}^2(khs')}{s(a_2 - 1)(a_2 + a_1)^2[1 + \tanh(khs)]^2} \exp\left[2khs - \frac{a_2}{2a_1} \ln \frac{a_1 + 1}{a_1 - 1}\right.$$
$$\left. - \frac{2}{\pi} \int_0^1 \left(\frac{a_2}{a_2^2 - y^2} - \frac{a_1}{a_1^2 - y^2}\right)\Omega(y) \, dy\right] \qquad (7\text{-}58)$$

and in the xz plane

$$P_{rad}(\phi) = \frac{2a_2(a_2 + 1)(s'^2 - s^2)(1 + \sin\phi)(s^2 + \cos^2\phi)}{\pi(a_1 + \sin\phi)^2(a_2 + a_1)^2(a_2 - 1)}$$

$$\times \frac{\left|\tan(kh\cos\phi) + \dfrac{(\epsilon_r - \sin^2\phi)^{\frac{1}{2}}}{\epsilon_r \cos\phi} \tan[kd(\epsilon_r - \sin^2\phi)^{\frac{1}{2}}]\right|}{\left|\operatorname{sech}(kh\cos\phi)\right| \left(s'^2 + \cos^2\phi\right)\left\{1 + \dfrac{\epsilon_r - \sin^2\phi}{\epsilon_r^2 \cos^2\phi} \tan^2[kd(\epsilon_r - \sin^2\phi)^{\frac{1}{2}}]\right\}^{\frac{1}{2}}}$$

$$\times \exp\left[\frac{-2a_2}{\pi} \int_0^1 \frac{\Omega(y) \, dy}{a_2^2 - y^2} + \frac{2\sin\phi}{\pi} p \int_0^1 \frac{(\pi/2) - \Omega(y)}{\sin^2\phi - y^2} \, dy\right] \qquad (7\text{-}59)$$

where p stands for Cauchy's principal value and

$$a_1 = (1 + s^2)^{\frac{1}{2}} \qquad a_1 > 0 \qquad\qquad\qquad (7\text{-}60)$$
$$a_2 = (1 + s'^2)^{\frac{1}{2}} \qquad a_2 > 0 \qquad\qquad\qquad (7\text{-}61)$$
$$\Omega(y) = kh(1 - y^2)^{\frac{1}{2}} + \tan^{-1}\left\{\frac{(\epsilon_r - y^2)^{\frac{1}{2}}}{\epsilon_r(1 - y^2)^{\frac{1}{2}}} \tan[kd(\epsilon_r - y^2)^{\frac{1}{2}}]\right\} \qquad (7\text{-}62)$$

The quantity s' is defined above (see Eq. 7-38), and s is the value of s' for the dominant TM surface wave mode.

Excitation efficiency given by

$$\text{Excitation efficiency} = \frac{P_{sw}}{P_{sw} + P_{rad}} \tag{7-63}$$

is plotted in Fig. 7-14 from data published by Angulo and Chang[1] for $\epsilon_r = 2.49$. Excitation efficiency by the aperture method [Eq. (7-8)] is presented for comparison. The two methods show good agreement except for small h. The Fourier transform method is the more accurate of the two, since the method, itself, is exact and involves only minor approximations in arriving at Eqs. (7-53) to (7-55). The aperture method should be used very carefully for structures with $c/v > 1.1$. On the other hand the mode orthogonality method with only minor approximations should give essentially the same results as the Fourier transform method.

7-5. TOTAL FIELD OF A TRAVELING WAVE STRUCTURE WITH SOURCE

In the discussion that follows we shall consider a traveling wave antenna structure to be a section of "open" transmission line (i.e., structures such as described in Chaps. 5 and 6). This excludes such antennas as horns and reflectors, which may be considered as broadside traveling wave antennas. The radiation field of a section of open transmission line consists of contributions from a space wave set up directly by the feed as well as contributions from a wave or waves guided by the structure. In addition there may be contributions from discontinuities along or at the end of the structure. These will be neglected in the discussion that follows by considering an infinitely long, flat structure with no discontinuities along it other than the source itself. A particular example that will be considered is shown in Fig. 7-15.

The complete spectrum of modes for the semi-infinite region of Fig. 7-15 may consist of a discrete spectrum of proper modes plus a continuous spectrum of improper modes. The discrete modes are proper in the sense that they individually satisfy Maxwell's equations and the appropriate boundary conditions including the radiation condition. They are the natural modes of slow wave structures such as a dielectric sheet on a ground plane or a corrugated surface (see Chaps. 5 and 6), and they possess orthogonality properties (see Sec. 7-4). Furthermore, the proper modes are nonradiating in the absence of any discontinuities along the traveling wave structure.

The continuous modes also individually satisfy Maxwell's equations and the boundary conditions at the structure but are improper in the sense that they do not individually satisfy the radiation condition. The improper

[1] Angulo and Chang, *op. cit.*

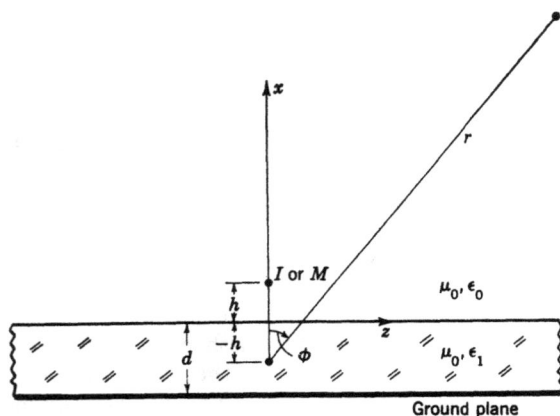

Figure 7-15. Electric- or magnetic-current line source over a grounded dielectric slab. The current is in the y direction.

modes can exist on any open structure, and they superimpose to make up the radiated fields of source-excited open structures.

In this continuous spectrum of improper modes there may be some modes corresponding to fast waves that attenuate along the traveling wave structure owing to radiation and hence increase in amplitude normal to the structure. These are termed leaky waves. These are natural modes for such structures as slotted and holey waveguides (see Chap. 5) but are forced modes for slow wave structures, where they can exist only in the presence of a source or discontinuity (or a modulation of the structure as will be pointed out in Chap. 8).

The leaky wave modes are not proper modes, since for an infinitely long structure they would individually diverge at infinity and furthermore they do not possess orthogonality properties.

To gain more insight into the mode spectrum of a source-excited traveling wave structure let us consider as an example the structure in Fig. 7-15. This has been analyzed in detail by Barone and Hessel,[1] and an excellent treatment for the electric current source is given by Collin.[2]

Because of the invariance of the system in the y direction the fields will be independent of y. For electric-current excitation I, the total electric field will have polarization parallel to the source (that is, $\mathbf{E} = \mathbf{y}E_y$), whereas for magnetic-current excitation M, the total magnetic field will be H_y. A

[1] S. Barone, Leaky Wave Contributions to the Field of a Line Source above a Dielectric Slab, *Polytech. Inst. Brooklyn Rept.* R-532-56, November, 1956; S. Barone and A. Hessel, Leaky Wave Contributions to the Field of a Line Source above a Dielectric Slab, Part II, *Polytech. Inst. Brooklyn Rept.* R-698-58, December, 1958.

[2] Collin, *op. cit.*, p. 485.

transverse network representation of Fig. 7-15 for electric-current excitation is shown in Fig. 7-16.

The superscript h in Fig. 7-16 denotes magnetic-type, or transverse electric, waves. The transverse wave number κ is related to the free-space wave number k and the propagation constant in the z direction γ by

$$\overset{h}{\kappa_1} = (k^2 \epsilon_r + \overset{h}{\gamma^2})^{1/2} \tag{7-64}$$

$$\overset{h}{\kappa_0} = (k^2 + \overset{h}{\gamma^2})^{1/2} \tag{7-65}$$

The admittance of the transverse line is

$$\overset{h}{Y_1} = \frac{\overset{h}{\kappa_1}}{\omega\mu_0} \tag{7-66}$$

$$\overset{h}{Y_0} = \frac{\overset{h}{\kappa_0}}{\omega\mu_0} \tag{7-67}$$

If the mode potentials are chosen to be $e^{-\gamma z}$ $(-\infty < \gamma < \infty)$, then the modal voltage for $x > 0$ in the transverse network is[1]

$$\overset{h}{V}(x) = -\frac{i_g}{2\overset{h}{Y_0}} [e^{-j\overset{h}{\kappa_0}|x-h|} + \overset{h}{\Gamma} e^{-j\overset{h}{\kappa_0}(x+h)}] \tag{7-68}$$

where $\overset{h}{\Gamma}$ is the reflection coefficient looking to the left at $x = 0$. For the ground plane at $x = -d$ the reflection coefficient is

$$\overset{h}{\Gamma} = \frac{\overset{h}{\kappa_0} + j\overset{h}{\kappa_1} \cot \overset{h}{\kappa_1}d}{\overset{h}{\kappa_0} - j\overset{h}{\kappa_1} \cot \overset{h}{\kappa_1}d} \tag{7-69}$$

The total electric field is obtained by integrating the mode functions over the appropriate contour C on the γ plane giving, for a unit current source,

$$\overset{h}{E_y}(x,z) = \frac{-j\omega\mu_0}{4\pi} \int_C \left[\frac{e^{-j\overset{h}{\kappa_0}|x-h|}}{\overset{h}{\kappa_0}} + \overset{h}{\Gamma} \frac{e^{-j\overset{h}{\kappa_0}(x+h)}}{\overset{h}{\kappa_0}} \right] e^{-\gamma z} d\gamma \qquad x > 0 \tag{7-70}$$

[1] Barone, *op. cit.*

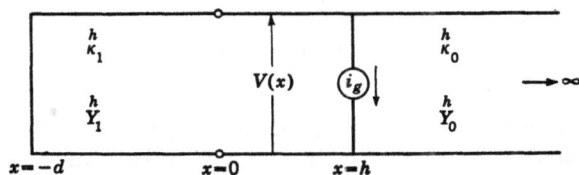

Figure 7-16. Transverse network representation of Fig. 7-15 for electric-current excitation.

For the magnetic source the corresponding equation for H_y is

$$\overset{e}{H}_y = \frac{-j\omega\epsilon_0}{4\pi} \int_C \left[\frac{e^{-j\overset{e}{\kappa}_0|x-h|}}{\overset{e}{\kappa}_0} + \overset{e}{\Gamma}\, \frac{e^{-j\overset{e}{\kappa}_0(x+h)}}{\overset{e}{\kappa}_0} \right] e^{-\gamma z}\, d\overset{e}{\gamma} \qquad x > 0 \qquad (7\text{-}71)$$

where

$$\overset{e}{\Gamma} = \frac{\epsilon_r \overset{e}{\kappa}_0 - j\overset{e}{\kappa}_1 \tan \overset{e}{\kappa}_1 d}{\epsilon_r \overset{e}{\kappa}_0 + j\overset{e}{\kappa}_1 \tan \overset{e}{\kappa}_1 d} \qquad (7\text{-}72)$$

The first term in Eqs. (7-70) and (7-71) is simply the integral representation of the Hankel function of zero order and the second kind. It represents the direct radiation of the source. The second term represents the modification of the field due to the nonzero reflection coefficient introduced by the grounded dielectric slab.

The path of integration for Eq. (7-70) or (7-71) on the complex γ plane must be such that E_y (or H_y) has proper behavior at infinity. The transverse wave number κ_0 must correspond to outward propagating or attenuated waves. This requires that

$$\begin{aligned}\text{Re } \kappa_0 &> 0 \\ \text{Im } \kappa_0 &< 0\end{aligned} \qquad (7\text{-}73)$$

The integrand in Eq. (7-70) or (7-71) is a two-valued function of γ because of the two branches of the function κ_0. The branch points of κ_0 occur at $\pm jk$, and the complex γ plane must be cut by two branch lines running from the branch points to infinity if the integrand is to be single valued. The branch cuts may be chosen arbitrarily as long as they do not intersect the contour of integration in Eq. (7-70) or (7-71).[1]

The integrand of Eq. (7-70) or (7-71) may be considered to be on a two-sheeted Riemann surface. The upper (proper) sheet corresponds to $\text{Im } \kappa_0 < 0$, and the lower (improper) sheet corresponds to $\text{Im } \kappa_0 > 0$. Only fields that go to zero at infinity have poles on the upper sheet, and these will be the surface wave modes (κ_0 = negative imaginary quantity). For κ_0 a positive imaginary quantity or a complex quantity with positive imaginary part, the poles will be on the improper sheet. The complex poles correspond to fast wave (leaky wave) modes. The poles of the integrand in Eq. (7-70) or (7-71) are just the poles of the reflection coefficient $\overset{h}{\Gamma}$ (or $\overset{e}{\Gamma}$). These occur when

$$\overset{h}{\kappa}_0 = j\overset{h}{\kappa}_1 \cot \overset{h}{\kappa}_1 d \qquad \text{TE waves} \qquad (7\text{-}74)$$

$$\epsilon_r \overset{e}{\kappa}_0 = -j\overset{e}{\kappa}_1 \tan \overset{e}{\kappa}_1 d \qquad \text{TM waves} \qquad (7\text{-}75)$$

Note that, for κ_0 a negative imaginary quantity, Eqs. (7-74) and (7-75) are the equations for TE and TM surface wave modes obtained by transverse resonance in Secs. 6-2A and 6-3A, respectively.

[1] For more details see Collin, *op. cit.*, pp. 485–506.

It is convenient to evaluate the integrals of (7-70) or (7-71) by the saddle-point method where the contour of integration is deformed into a contour of steepest descent. Part of this contour may lie on the improper Riemann sheet, and when this occurs, leaky wave modes will contribute to the total field.

In order to carry out the saddle-point integration method it is convenient to use the change of variables

$$\gamma = jk \sin \nu \tag{7-76}$$
$$\nu = \xi + j\eta \tag{7-77}$$
$$\kappa_0 = -k \cos \nu \tag{7-78}$$
$$z = r \sin \phi \tag{7-79}$$
$$x + h = r \cos \phi \tag{7-80}$$

where r, ϕ are cylindrical coordinates of the point of observation. The angle ϕ is measured from the xy plane, and r is measured from $x = -h$, the location of the image of the actual source (see Fig. 7-15). Equations (7-70) and (7-71) may be expressed as

$$\overset{h}{E}_y = \frac{-\omega\mu_0}{4}\left[H_0^{(2)}(k\rho) + \frac{1}{\pi} \int_P \overset{h}{\Gamma}(\nu)e^{-jkr \cos (\nu-\phi)} \, d\nu \right] \tag{7-81}$$

$$\overset{e}{H}_y = \frac{-\omega\epsilon_0}{4}\left[H_0^{(2)}(k\rho) + \frac{1}{\pi} \int_P \overset{e}{\Gamma}(\nu)e^{-jkr \cos (\nu-\phi)} \, d\nu \right] \tag{7-82}$$

where

$$\rho = [(x - h)^2 + z^2]^{\frac{1}{2}} \tag{7-83}$$

The transformation given by Eqs. (7-76) to (7-80) represents a mapping of the complex γ plane onto a strip of the complex ν plane. From Eq. (7-76) we have

$$\gamma = \alpha + j\beta = jk \sin (\xi + j\eta) \tag{7-84}$$

thus

$$\alpha = -k \cos \xi \sinh \eta \tag{7-85}$$
$$\beta = k \sin \xi \cosh \eta \tag{7-86}$$

The two planes are illustrated in Fig. 7-17. To ensure continuity of the integrand on the two sheets of the γ surface, the branch cuts must be taken along the curves Im $\kappa_0 = 0$. For the proper (upper) Riemann sheet, the region between the two branches (shaded region in Fig. 7-17a) corresponds to Re $\kappa_0 > 0$ and the unshaded region corresponds to Re $\kappa_0 < 0$. On the improper (bottom) Riemann sheet these statements are reversed. Since the integrand in Eq. (7-70) or (7-71) should correspond to outgoing waves, Re $\kappa_0 > 0$ and Im $\kappa_0 < 0$; hence the path of integration would be along the imaginary axis of the upper sheet.

The eight quadrants of the two-sheeted Riemann surface map into

(a) γ plane

(b) ν plane

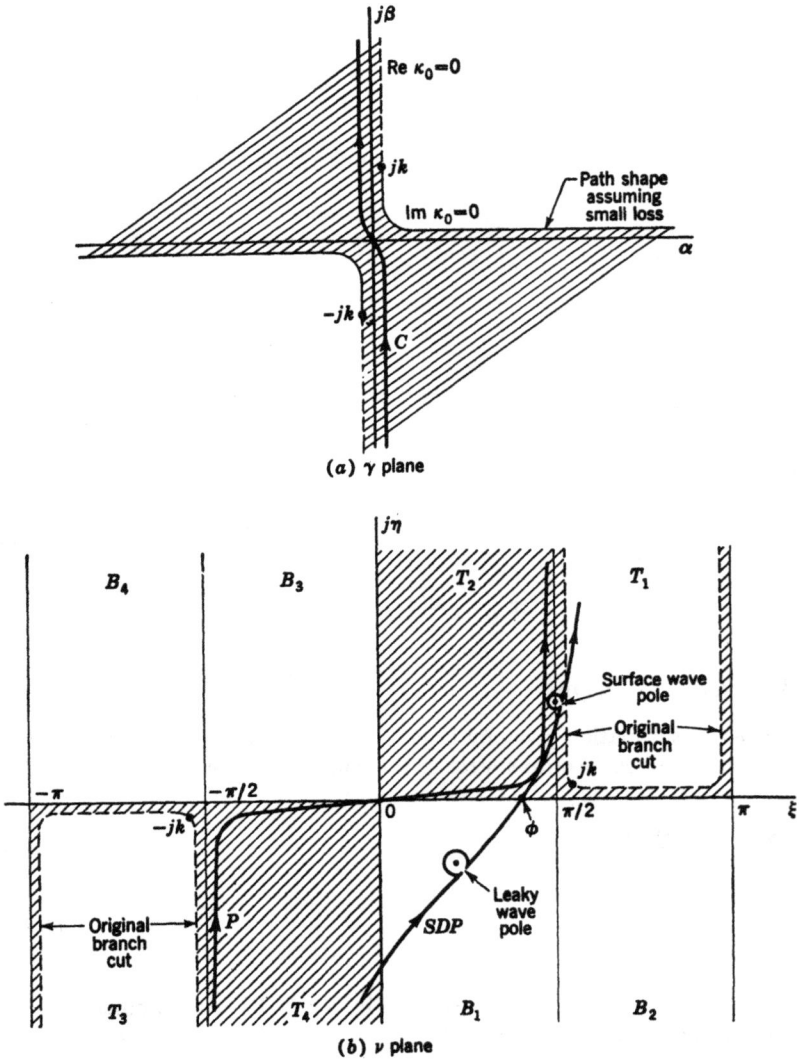

Figure 7-17.   Mapping the γ plane onto the ν plane.

the eight sections that make up the strip of ν plane in Fig. 7-17b. The four quadrants associated with the upper Riemann sheet are denoted by T_n, and the quadrants associated with the bottom sheet are denoted B_n. The negative sign in Eq. (7-78) permits the quadrants T_1, T_2 and T_3, T_4 to map into adjacent strips. The path of integration C in Fig. 7-17a maps into P in Fig. 7-17b, and the branch cuts of Fig. 7-17a map into the dashed lines of Fig. 7-17b. The crosshatched regions are corresponding regions on the two planes.

The integrands in Eqs. (7-81) and (7-82) can be evaluated by deforming the path P into the steepest descent path SDP, which is given by

$$\cos (\xi - \phi) \cosh \eta = 1 \tag{7-87}$$

and which passes through the saddle point $\nu = \phi$. When the path P is deformed to get to SDP, some poles of $\Gamma(\nu)$ may be encountered. When this happens, the path should be warped around the poles as shown in Fig. 7-17b so as not to pass over the poles.

The first term in Eq. (7-81) or (7-82) plus the integral along the path SDP gives the continuous spectrum of the total field. Any imaginary pole residues are surface wave contributions, whereas residues from complex poles are leaky wave contributions. The steepest descent evaluation of the integral in Eq. (7-81) or (7-82) along SDP gives the reflected contribution to the radiated field

$$E_y{}^R \bigg]_{SDP} \approx \frac{-\omega\mu_0}{4\pi} \sqrt{\frac{2\pi}{kr}} \, \overset{h}{\Gamma}(\phi) e^{-j(kr-\pi/4)} \tag{7-88}$$

or

$$H_y{}^R \bigg]_{SDP} \approx \frac{-\omega\epsilon_0}{4\pi} \sqrt{\frac{2\pi}{kr}} \, \overset{e}{\Gamma}(\phi) e^{-j(kr-\pi/4)} \tag{7-89}$$

where

$$\overset{h}{\Gamma}(\phi) = \frac{j \cos \phi - (\epsilon_r - \sin^2 \phi)^{\frac{1}{2}} \cot [kd(\epsilon_r - \sin^2 \phi)^{\frac{1}{2}}]}{j \cos \phi + (\epsilon_r - \sin^2 \phi)^{\frac{1}{2}} \cot [kd(\epsilon_r - \sin^2 \phi)^{\frac{1}{2}}]} \tag{7-90}$$

$$\overset{e}{\Gamma}(\phi) = \frac{j \cos \phi - (\epsilon_r - \sin^2 \phi)^{\frac{1}{2}} \tan [kd(\epsilon_r - \sin^2 \phi)^{\frac{1}{2}}]}{j \cos \phi + (\epsilon_r - \sin^2 \phi)^{\frac{1}{2}} \tan [kd(\epsilon_r - \sin^2 \phi)^{\frac{1}{2}}]} \tag{7-91}$$

Using the first term in the asymptotic expansion of $H_0{}^{(2)}(k\rho)$ in Eq. (7-81) or (7-82) for the direct contribution from the source gives the radiated field $(k\rho \gg 1)$

$$E_y{}^R \approx \frac{-\omega\mu_0}{4\pi} \sqrt{\frac{2\pi}{k\rho}} \, e^{-j(k\rho-\pi/4)} - \frac{\omega\mu_0}{4\pi} \sqrt{\frac{2\pi}{kr}} \, \overset{h}{\Gamma}(\phi) e^{-j(kr-\pi/4)} \tag{7-92}$$

or

$$H_y{}^R \approx \frac{-\omega\epsilon_0}{4\pi} \sqrt{\frac{2\pi}{k\rho}} \, e^{-j(k\rho-\pi/4)} - \frac{\omega\epsilon_0}{4\pi} \sqrt{\frac{2\pi}{kr}} \, \overset{e}{\Gamma}(\phi) e^{-j(kr-\pi/4)} \tag{7-93}$$

where ρ is defined by Eq. (7-83) and r is defined by Eqs. (7-79) and (7-80).

The surface wave and leaky wave contributions, when they exist, are of the form $(j\omega\mu_0/2) \sum_S F(\nu_S) e^{-jkr\cos(\nu_S-\phi)}$ and $(j\omega\mu_0/2) \sum_L F(\nu_L) e^{-jkr\cos(\nu_L-\phi)}$, where S and L denote surface waves and leaky waves, respectively. This would be for an electric-current source; for a magnetic source replace μ_0 by ϵ_0. The function $F(\nu)$ is the residue of the appropriate surface wave or fast wave pole of $\Gamma(\nu)$, and ν_S and ν_L are the pole locations on the ν plane. The series arises from integration around the circles surrounding the poles where the path SDP has been warped to avoid crossing the poles.

The surface wave fields will be of the form[1]

$$\left.\begin{array}{c} E_y{}^S \\ H_y{}^S \end{array}\right\} = A^S e^{-|\kappa_0|x} e^{-j\beta z} \tag{7-94}$$

and Barone[2] has shown that the leaky wave fields are of the form

$$\left.\begin{array}{c} E_y{}^L \\ H_y{}^L \end{array}\right\} = A^L e^{\alpha_L k_\phi{}^r} e^{-j\beta_L k_r r} \tag{7-95}$$

where A = constant specified by h, d, and ϵ_r

$\alpha_L = \sinh \eta_L$

$\beta_L = \cosh \eta_L$

$k_r = k \cos (\phi - \xi_L)$

$k_\phi = -k \sin (\phi - \xi_L)$

$\nu_L = \xi_L + j\eta_L$ = pole location

and r is defined by Eqs. (7-79) and (7-80). The total field of the traveling wave structure with source is the superposition of the radiation, surface wave, and leaky wave contributions; i.e.,

$$E_y = \frac{-\omega\mu_0}{\sqrt{8\pi}} e^{j(\pi/4)} \left(\frac{e^{-jk\rho}}{\sqrt{k\rho}} + \frac{\overset{h}{\Gamma}(\phi)e^{-jkr}}{\sqrt{kr}} \right)$$
$$+ \frac{j\omega\mu_0}{2} \left[\sum_S F(\nu_S)e^{-jkr\cos(\nu_S-\phi)} + \sum_L F(\nu_L)e^{-jkr\cos(\nu_L-\phi)} \right] \tag{7-96}$$

for an electric-current source and

$$H_y = \frac{-\omega\epsilon_0}{\sqrt{8\pi}} e^{j(\pi/4)} \left(\frac{e^{-jk\rho}}{\sqrt{k\rho}} + \frac{\overset{e}{\Gamma}(\phi)e^{-jkr}}{\sqrt{kr}} \right)$$
$$+ \frac{j\omega\epsilon_0}{2} \left[\sum_S F(\nu_S)e^{-jkr\cos(\nu_S-\phi)} + \sum_L F(\nu_L)e^{-jkr\cos(\nu_L-\phi)} \right] \tag{7-97}$$

for a magnetic-current source.

Equations (7-96) and (7-97) are adequate for $k\rho$ and $kr \gg 1$ and when $\Gamma(\nu)$ has no poles near the saddle point ϕ. A more valid representation for the saddle-point contribution for a simple pole near the saddle point is developed by Collin.[3]

For any angle of observation ϕ there will be at most a finite number

---

[1] See Chap. 6 and Sec. 7-4, for example.

[2] Barone, *op. cit.*

[3] Collin, *op. cit.*, pp. 503–506.

of poles between the original path P and the steepest descent path SDP. The residue contributions at these poles indicate either surface wave or leaky wave fields. Consider a pole of either type. For sufficiently small angles of observation there is no residue contribution. As the angle of observation increases, the path SDP deforms until finally it will lie on the pole if the path is not warped. The angle of observation for which this occurs is called the critical angle ϕ_c. The slight warping of the path to avoid crossing over the pole gives a residue contribution. Thus the surface wave or leaky wave field of this pole will begin to appear at $\phi = \phi_c$. For $\phi > \phi_c$ the full residue contribution will be present. Thus surface and leaky wave contributions occur only in a wedge-shaped region of space for $\phi > \phi_c$. Each pole has its particular ϕ_c which is determined by the frequency and the dielectric constant for the grounded dielectric slab case. Whether the residue contributions are significant or not depends on their amplitudes relative to the radiated fields of the direct and reflected waves of the source-structure combination.

The strength of excitation of surface wave fields was considered in Sec. 7-4. The analysis in the present section (7-5) enables one to determine the strength of leaky wave excitation as well as surface wave excitation. For the surface wave (slow wave) structure considered here, one might question whether leaky wave (fast wave) contributions would really be significant. Experimental results of Cassedy and Cohn[1] have demonstrated that leaky wave fields can alter appreciably the field distribution along a grounded dielectric slab excited as in Fig. 7-15.

The fact that leaky wave fields can contribute significantly to the field distribution along a structure (the near field) means that the leaky waves influence the far field. This is true because by Huygens' principle (see Chap. 2) the far field can be obtained by integrating over an equivalent current distribution close to and enclosing the structure. If a leaky wave is strongly excited, a peak in the far field is to be expected at an angle given approximately by the critical angle ϕ_c. This corresponds to the angle of maximum radiation for a fast wave source (see Chap. 2). For additional discussion of waves along an interface and their relation to the radiated field, the reader is referred to the work of Tamir and Oliner.[2]

7-6. COUPLED LINE EXCITATION

In some cases, especially where it is desired to flush mount a traveling wave antenna, coupled line excitation as illustrated in Fig. 7-4 may be used.

[1] E. S. Cassedy and M. Cohn, On the Existence of Leaky Waves Due to a Line Source above a Grounded Dielectric Slab, *IRE Trans. Microwave Theory Tech.*, MTT-9:243–247, May, 1961.

[2] T. Tamir and A. A. Oliner, Guided Complex Waves, Part I: Fields at an Interface, Part II: Relation to Radiation Pattern, *Proc. Inst. Elec. Engrs.*, 110(2):310–334, February, 1963.

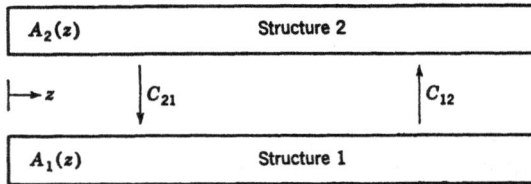

Figure 7-18. Two coupled traveling wave structures.

The amplitude and phase along the antenna aperture will depend on the coupling between the feed line and the traveling wave antenna structure. This fact can be utilized in controlling the aperture distribution of the antenna. Line source excitation has been successfully applied to flush-mounted, slotted waveguide antennas[1] and to Yagi-Uda arrays.[2]

An exact theory, i.e., satisfying Maxwell's equations and the boundary conditions, would be extremely difficult in general. Maxwell's equations, however, can be replaced by the ordinary differential equations of coupled transmission lines which are adequate for determining the relative amplitudes and phases along the coupled lines.

For the coupled traveling wave structures in Fig. 7-18 let $A_1(z)$ be the distribution along structure 1 and $A_2(z)$ the distribution along structure 2. The quantities A_1 and A_2 may represent any field component or suitable scalar function from which the fields can be derived. In general both A_1 and A_2 will be complex. The coupled line equations that apply to Fig. 7-18 can be written[3]

$$\frac{dA_1}{dz} = -(\gamma_1 + C_{11})A_1 + C_{21}A_2 \qquad (7\text{-}98)$$

$$\frac{dA_2}{dz} = C_{12}A_1 - (\gamma_2 + C_{22})A_2 \qquad (7\text{-}99)$$

where only waves traveling in the positive z direction are considered and

γ_1 and γ_2 = uncoupled propagation constants of structures 1 and 2, respectively

C_{12} and C_{21} = coupling coefficients (propagation constants associated with energy transfer from line 1 to 2 and vice versa)

C_{11} and C_{22} = perturbations of propagation constants due to coupling

[1] D. E. Royal, Axially Excited Surface Wave Antennas, *Univ. Illinois Elec. Eng. Res. Lab. Tech. Rept.* 7, October, 1955; W. L. Weeks, Coupled Waveguide Excitation of Traveling Wave Slot Antennas, *Univ. Illinois Elec. Eng. Res. Lab. Tech. Rept.* 27, December, 1957; R. H. MacPhie, Use of Coupled Waveguides in a Traveling Wave Scanning Antenna, *Univ. Illinois Elec. Eng. Res. Lab. Tech. Rept.* 34, April, 1959.

[2] A. J. Giarola, Continuously Excited Traveling Wave Antennas, *Seattle Univ. Elec. Eng. Dept. Sci. Rept.* 1, June, 1959.

[3] S. E. Miller, Coupled Wave Theory and Waveguide Applications, *Bell System Tech. J.*, May, 1954, pp. 661–719.

If we restrict ourselves to reciprocal structures, then

$$C_{12} = C_{21} = C \tag{7-100}$$

and Eqs. (7-98) and (7-99) can be written

$$\frac{dA_1}{dz} = -(\Gamma_1 + C)A_1 + CA_2 \tag{7-101}$$

$$\frac{dA_2}{dz} = CA_1 - (\Gamma_2 + C)A_2 \tag{7-102}$$

where

$$\Gamma_1 = \gamma_1 + C_{11} - C \tag{7-103}$$
$$\Gamma_2 = \gamma_2 + C_{22} - C \tag{7-104}$$

In most applications C is small and C_{11} and C_{22} are very nearly equal to C; therefore

$$\Gamma_1 \approx \gamma_1 \tag{7-105}$$
$$\Gamma_2 \approx \gamma_2 \tag{7-106}$$

Equations (7-101) and (7-102) can be combined to give

$$\frac{d^2A_1}{dz^2} + (2C + \Gamma_1 + \Gamma_2)\frac{dA_1}{dz} + [\Gamma_1\Gamma_2 + C(\Gamma_1 + \Gamma_2)]A_1 = 0 \tag{7-107}$$

$$\frac{d^2A_2}{dz^2} + (2C + \Gamma_1 + \Gamma_2)\frac{dA_2}{dz} + [\Gamma_1\Gamma_2 + C(\Gamma_1 + \Gamma_2)]A_2 = 0 \tag{7-108}$$

Solutions for A_1 and A_2 are of the form

$$A_1 = A_1'e^{r_1 z} + A_1''e^{r_2 z} \tag{7-109}$$
$$A_2 = A_2'e^{r_1 z} + A_2''e^{r_2 z} \tag{7-110}$$

where

$$r_1 = -\tfrac{1}{2}(2C + \Gamma_1 + \Gamma_2) + \tfrac{1}{2}[(\Gamma_1 - \Gamma_2)^2 + 4C^2]^{1/2} \tag{7-111}$$
$$r_2 = -\tfrac{1}{2}(2C + \Gamma_1 + \Gamma_2) - \tfrac{1}{2}[(\Gamma_1 - \Gamma_2)^2 + 4C^2]^{1/2} \tag{7-112}$$

The coefficients A_1', A_1'', A_2', and A_2'' will depend on the initial conditions on A_1 and A_2 [and the corresponding initial conditions on the derivatives from Eqs. (7-101) and (7-102)]. If line 1 is the driven line such that $A_1 = 1$ and $A_2 = 0$ at $z = 0$, solutions for A_1 and A_2 are

$$A_1 = \left\{\frac{1}{2} - \frac{(\Gamma_1 - \Gamma_2)}{2[(\Gamma_1 - \Gamma_2)^2 + 4C^2]^{1/2}}\right\} e^{r_1 z}$$
$$+ \left\{\frac{1}{2} + \frac{(\Gamma_1 - \Gamma_2)}{2[(\Gamma_1 - \Gamma_2)^2 + 4C^2]^{1/2}}\right\} e^{r_2 z} \tag{7-113}$$

$$A_2 = \frac{C}{[(\Gamma_1 - \Gamma_2)^2 + 4C^2]^{1/2}} (e^{r_1 z} - e^{r_2 z}) \tag{7-114}$$

If there is no loss in either line or in the coupling mechanism, then conservation of energy requires that

$$|A_1|^2 + |A_2|^2 = \text{const} \tag{7-115}$$

and the coupling coefficient C is pure imaginary. Even when there is some loss, it is still satisfactory to assume that C is imaginary.[1]

For the special case where the lines are identical ($\Gamma_1 = \Gamma_2 = \Gamma$), Eqs. (7-113) and (7-114) reduce to

$$A_1 = \cos(|C|z)e^{-(j|C|+\Gamma)z} \tag{7-116}$$
$$A_2 = j\sin(|C|z)e^{-(j|C|+\Gamma)z} \tag{7-117}$$

In this case there is periodic and complete power transfer between the lines, with period $|C|z = \pi$.

Either line 1 or line 2 could be the antenna structure, although usually the antenna is the coupled line rather than the driven line. The distribution along the antenna is determined by the propagation const. at Γ, the coupling coefficient C, the position of the radiating section relative to the coupled section, and the relative excitations of the lines.

In the above case $A_1 = 1$ and $A_2 = 0$ at $z = 0$. Solution of Eqs. (7-109) and (7-110) for $|A_1| = |A_2|$ at $z = 0$ but variable phasing illustrates a method of beam scanning[2] where over a limited range the beam position is controlled by the relative phase of the two lines. Neglecting attenuation and assuming that the lines have identical propagation constants (that is, $\Gamma_1 = \Gamma_2 = \Gamma$), the distributions A_1 and A_2 are given by

$$A_1 = e^{j(\phi/2)}\left(\cos\frac{\phi}{2}e^{-j\beta_f z} + j\sin\frac{\phi}{2}e^{-j\beta_s z}\right) \tag{7-118}$$

$$A_2 = e^{j(\phi/2)}\left(\cos\frac{\phi}{2}e^{-j\beta_f z} - j\sin\frac{\phi}{2}e^{-j\beta_s z}\right) \tag{7-119}$$

where ϕ is the phase angle of A_1 relative to A_2 at $z = 0$ and β_f and β_s are the phase constants of the two waves that are present in the coupled system. The phase constants are given by

$$\beta_f = |\Gamma| \tag{7-120}$$
$$\beta_s = |\Gamma| + 2|C| \tag{7-121}$$

where attenuation is taken to be zero. One wave, characterized by β_f, is faster than that characterized by the average phase constant β_{av} ($= |\Gamma| + |C|$), whereas the other wave β_s is somewhat slower than the average wave.

As the phase ϕ is varied, it can be observed in Eqs. (7-118) and (7-119) that the resultant wave along either structure has a phase constant β that varies from β_f to β_s. The angle of radiation is $\theta = \cos^{-1}(\beta/k)$, where θ is measured from the z axis. Thus the sector of scan lies between $\cos^{-1}(\beta_s/k)$

---

[1] *Ibid.*
[2] MacPhie, *op. cit.*

and $\cos^{-1}$ (β_f/k). Actually the term "phase constant" is used loosely here, since only for β_f and β_s will the wave along the structure have a constant phase velocity. For in-between values the resultant wave will have a nonuniform phase velocity which will tend to widen the radiation pattern. In extreme cases where β_s is quite a bit larger than β_f, the beam may become double lobed in the center of the scan sector. Thus this scanning technique is limited to rather small scan sectors, perhaps on the order of 20 deg for an aperture 10 λ long. As aperture length is increased, the beams from the individual waves become narrower and beam splitting will be more pronounced. Although the method is limited with regard to beam scanning, it might also be used to obtain beam switching electronically.

ADDITIONAL REFERENCES

1. C. T. Tai, The Effect of a Grounded Slab on the Radiation from a Line Source, *J. Appl. Phys.*, **22**:405, 1951.
2. N. Marcuvitz, On Field Representations in Terms of Leaky Modes or Eigenmodes, Symposium on Electromagnetic Wave Theory, *IRE Trans. Antennas Propagation*, **AP-4**:192–194, July, 1956.
3. A. L. Cullen, "The Excitation of Plane Surface Waves," Monograph 93, The Institution of Electrical Engineers, Radio Section, February, 1954.
4. A. F. Kay, Excitation Efficiency of Surface Waves over Corrugated Metal and Doubly Corrugated Metal and in Dielectric Slabs on a Ground Plane, *Tech. Res. Group Sci. Rept.* 5, New York, December, 1956.
5. A. F. Kay, Yagi Antenna Study, *Tech. Res. Group Rept.* 121-SR-1, Syosset, N.Y., January, 1960.
6. G. J. Rich, The Launching of a Plane Surface Wave, Paper 1783R, *Proc. Inst. Elec. Engrs.*, March, 1955.
7. J. R. Wait, Excitation of Surface Waves on Conducting, Stratified Dielectric Clad and Corrugated Surfaces, *Natl. Bur. Std. Rept.* 5061, Boulder, Colo., April, 1957.
8. J. R. Wait and A. M. Conda, "The Resonance Excitation of a Corrugated-cylinder Antenna," Monograph 386E, The Institution of Electrical Engineers, June, 1960.
9. D. C. Stickler, The Field of an Electric Line Source above a Dielectric Sheet, *Ohio State Univ. Res. Found. Antenna Lab. Rept.* 786-2, April, 1958.
10. B. Friedman and W. E. Williams, Excitation of Surface Waves, *Proc. IEE.*, **105** (pt. C):252–258, March, 1958.
11. J. Brown and K. P. Sharma, The Launching of Radial Cylindrical Surface Waves by a Circumferential Slot, Paper 2738E, *Proc. Inst. Elec. Engrs.*, March, 1959.
12. K. P. Sharma, An Investigation of the Excitation of Radiation by Surface Waves, Paper 2737E, *Proc. Inst. Elec. Engrs.*, March, 1959.
13. J. Kane, Surface Waves on a Reactive Half Plane, *NY Univ. Inst. Math. Sci. Res. Rept.* EM-159, May, 1960.
14. J. Kane, The Efficiency of Launching Surface Waves on a Reactive Half Plane

by an Arbitrary Antenna, *IRE Trans. Antennas Propagation,* **AP-8**(5):500–507, September, 1960.

15. J. Kane, Optimum Parameters for a Class of Surface Wave Antennas, *Univ. Rhode Island Dept. Elec. Eng. Rept.* 7983/2, March, 1962.

16. J. Kane, The Surface Wave Antenna as a Boundary Value Problem, *Univ. Rhode Island Dept. Elec. Eng. Rept.* 7983/3, September, 1962.

17. J. B. Keller and F. C. Karal, Excitation and Propagation of Surface Waves, *NY Univ. Inst. Math. Sci. Res. Rept.* EM-128, February, 1959.

18. J. B. Keller and B. R. Levy, Decay Exponents and Diffraction Coefficients for Surface Waves on Surfaces of Non-constant Curvature, *NY Univ. Inst. Math. Sci. Res. Rept.* EM-147, October, 1959.

19. S. R. Seshadri and K. Iizuka, A Dipole Antenna Coupled Electromagnetically to a Two-wire Transmission Line, *IRE Trans. Antennas Propagation,* **AP-7**(4): 386–392, October, 1959.

20. D. K. Reynolds and R. A. Sigelmann, Research on Traveling Wave Antennas, *Seattle Univ. Elec. Eng. Dept. Rept.* AFCRC-TR-59-160, June, 1959.

21. C. M. Angulo and W. S. C. Chang, "The Launching of Surface Waves by a Magnetic Line Source," Monograph 411E, The Institution of Electrical Engineers, October, 1960.

22. M. G. Andreasen, Flush-mounted Surface-wave Launcher, *Stanford Res. Inst. Sci. Rept.* 11, March, 1961.

23. W. R. Jones, An Approximate Approach to Problems Involving Surface Waves, *Rept.* FL 61–60, Ground Systems, Hughes Aircraft Company, Fullerton, Calif., October, 1961.

24. M. S. Bobrovnikov and R. P. Starovoitova, Concentrated Excitation of a Metal Cylinder with a Dielectric Coating, *Izv. Vysshikh Uchebn. Zavedenii Radiotekhn.,* **4**(2):140–147, 1961.

25. H. H. Kuehl, Radiation from a Gap-excited Cylinder Surrounded by a Uniform Plasma Sheath, *Univ. Southern Calif. Elec. Eng. Dept. Rept.* 71-202, June, 1960.

26. T. Tamir, Leaky Wave Contributions to the Field of an Electric Line Source above a Plasma Slab, *Polytech. Inst. Brooklyn Res. Rept.* PIB MRI-845-60, November, 1960.

27. T. Tamir and A. A. Oliner, The Influence of Complex Waves on the Radiation Field of a Slot Excited Plasma Layer, *IRE Trans. Antennas Propagation,* **AP-10**(1):55–65, January, 1962.

28. T. B. A. Senior, Loop Excitation of Traveling Waves, *Can. J. Appl. Phys.,* **40**:1736–1748, 1962.

29. K. M. Chen and R. W. P. King, Dipole Antennas Coupled Electromagnetically to a Two-wire Transmission Line, *IRE Trans. Antennas Propagation,* **AP-9**(5): 425–432, September, 1961.

30. H. M. Barlow and J. Brown, "Radio Surface Waves," chaps. 9–11, Oxford, Fair Lawn, N.J., 1962.

31. J. R. Wait, An Approach to the Theory of an Antenna over an Inhomogeneous Ground Plane, *Natl. Bur. Std. Rept.* 6739, Boulder, Colo., January, 1961.

32. R. E. Collin, Analytical Solution for a Leaky Wave Antenna, *IRE Trans. Antennas Propagation,* **AP-10**:561–565, September, 1962.

33. C. M. Angulo and W. S. C. Chang, The Excitation of a Dielectric Rod by a Cylindrical Waveguide, *IRE Trans.,* **MTT-6**:389–393, October, 1958.

34. D. B. Brick, Excitation of Surface Waves by a Vertical Antenna, *Proc. IRE*, **43**:721–727, June, 1955.

35. J. Brown, Some Theoretical Results for Surface Wave Launchers, *IRE Trans. Antennas Propagation*, **AP-7**:S169–S174, Special Supplement, December, 1959.

36. R. H. Clarke, A Method of Estimating the Power Radiated Directly at the Feed of a Dielectric Rod Aerial, *Proc. IEE*, **104** (pt. B):511–514, September, 1957.

37. R. H. DuHamel and J. W. Duncan, Launching Efficiency of Wires and Slots for a Dielectric Rod Waveguide, *IRE Trans.*, **MTT-6**:277–284, July, 1958.

38. J. W. Duncan, The Efficiency of Excitation of a Surface Wave on a Dielectric Cylinder, *IRE Trans.*, **MTT-7**:257–267, April, 1959.

PROBLEMS

7-1. Verify Eq. (7-8).

7-2. Verify Eqs. (7-21) and (7-22).

7-3. Derive Eqs. (7-44) and (7-45).

7-4. Verify Eq. (7-68).

7-5. Verify Eqs. (7-107) and (7-108).

7-6. Using the analysis in Sec. 7-6 show that two coupled lines with propagation constants $\Gamma_1 = \alpha_1 + j\beta_1$ and $\Gamma_2 = \alpha_2 + j\beta_2$ will not have appreciable energy exchange if either

$$\frac{|\alpha_1 - \alpha_2|}{|C|} \gg 1 \qquad \text{or} \qquad \frac{|\beta_1 - \beta_2|}{|C|} \gg 1$$

CHAPTER 8

PRACTICAL
TRAVELING
WAVE
ANTENNAS

8-1. INTRODUCTION

In this chapter a number of practical traveling wave antennas are described. Although this is not a complete listing, it does cover most of the traveling wave antennas that have been studied and used as of this writing. With but few exceptions the analysis, synthesis, and design procedures of the previous chapters apply to the antennas in the following sections.

All traveling wave antennas can be classified as either slow wave $(c/v \gtrless 1)$ or fast wave $(c/v < 1)$ antennas. These are the two main categories used here, but in addition a separate section is included on long wire antennas which have the distinction of being the first traveling wave antennas. Furthermore, arrays of traveling wave antennas, curved traveling wave antennas, modulated traveling wave antennas, and surface wave lenses are distinct antenna types and are covered in separate sections.

8-2. LONG WIRE ANTENNAS

The first antennas to be classed as traveling wave antennas were wires one or more wavelengths long used either singly as simple linear radiators or in arrays to form V and rhombic antennas.

A. Single Wire Antennas

The radiation characteristics of a single wire linear radiator have been developed by a number of authors[1] for both the unterminated line (standing wave distribution) and the terminated line (Beverage or wave antenna). Both terminated and unterminated lines are illustrated in Fig. 8-1.

If attenuation along a thin wire due to radiation and ohmic losses is small, the current distribution I_t for the terminated case can be assumed to be of constant amplitude with phase velocity equal to that of free space; hence

$$I_t(z) = I_0 e^{-jkz} \tag{8-1}$$

where time variation is understood to be of the form $e^{j\omega t}$. The electric far field for this case [see Eq. (2-91)] is

$$E_\theta(\theta) = j30kLI_0 \sin \theta e^{jX} \frac{e^{-jkr}}{r} \frac{\sin X}{X} \tag{8-2}$$

[1] P. S. Carter, C. W. Hansell, and N. E. Lindenblad, Development of Directive Transmitting Antennas by RCA Communications, *Proc. IRE*, **19**:1733, October, 1931; F. M. Colebrook, Electric and Magnetic Fields for Linear Radiator Carrying a Progressive Wave, *J. Inst. Elec. Engrs.* (*London*), **89**:169, February, 1940; A. Alford, A Discussion of Methods Employed in Calculations of Electromagnetic Fields of Radiating Conductors, *Elec. Commun.*, **15**:70–88, July, 1936; J. A. Stratton, "Electromagnetic Theory," pp. 439–448, McGraw-Hill, New York, 1941; E. M. Wells, Radiation Resistance of Horizontal and Vertical Aerials Carrying a Progressive Wave, *Marconi Rev.*, No. 83, October-December, 1946; J. D. Kraus, "Antennas," pp. 148–153, McGraw-Hill, New York, 1950.

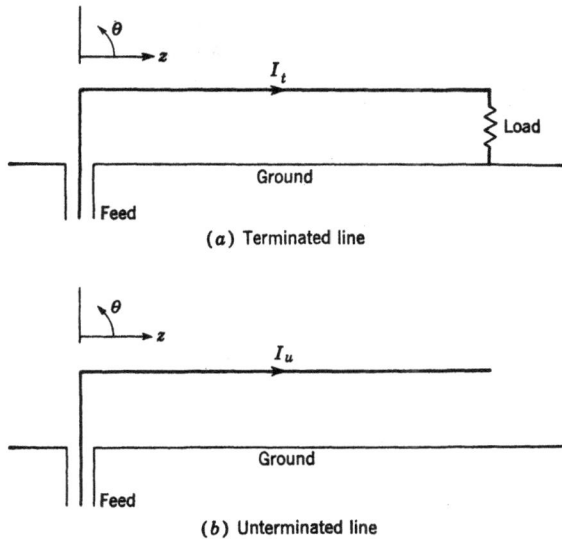

(a) Terminated line

(b) Unterminated line

Figure 8-1   Single wire antennas.

where

$$X = \frac{kL}{2}(1 - \cos\theta) \tag{8-3}$$

Usually only the magnitude of the field is of interest, and Eq. (8-2) is expressed as

$$|E_\theta(\theta)| = \frac{30kLI_0 \sin\theta}{r} \left|\frac{\sin X}{X}\right| \tag{8-4}$$

Graphs of $\sin\theta|(\sin X)/X|$ as a function of source length are given in the appendix.

The radiated power for the terminated line in free space is

$$P_t = \frac{1}{2}\int_0^\pi\int_0^{2\pi}\frac{|E_\theta|^2}{120\pi}r^2\sin\theta\,d\theta\,d\phi$$

$$= 30I_0^2\left[1.415 + \ln\frac{2L}{\lambda} + \frac{\sin(4\pi L/\lambda)}{4\pi L/\lambda} - \text{Ci}\,\frac{4\pi L}{\lambda}\right] \tag{8-5}$$

and the radiation resistance is

$$R = 84.9 + 60\ln\frac{2L}{\lambda} + 60\frac{\sin(4\pi L/\lambda)}{4\pi L/\lambda} - 60\,\text{Ci}\,\frac{4\pi L}{\lambda} \tag{8-6}$$

where Ci is the usual cosine integral given by

$$\text{Ci}(x) = -\int_x^\infty \frac{\cos v}{v}\,dv \tag{8-7}$$

For the unterminated line with a pure sinusoidal current distribution (standing wave) the source I_u can be expressed as a superposition of two uniform waves traveling in opposite directions as

$$I_u(z) = I_0\sin\left[m\pi\left(\frac{z}{L} + \frac{1}{2}\right)\right]$$

$$= \frac{I_0}{2j}\left[e^{jm\pi[(z/L)+\frac{1}{2}]} - e^{-jm\pi[(z/L)+\frac{1}{2}]}\right] \tag{8-8}$$

where m is an integer equal to Lk/π (the number of half wavelengths in L). The resulting far field for the resonant length case is

$$E_\theta(\theta) = \frac{j60I_0}{r\sin\theta}\cos\left(\frac{m\pi}{2}\cos\theta\right)e^{-jkr} \qquad m \text{ odd} \tag{8-9}$$

and

$$E_\theta(\theta) = \frac{60I_0}{r\sin\theta}\sin\left(\frac{m\pi}{2}\cos\theta\right)e^{-jkr} \qquad m \text{ even} \tag{8-10}$$

The radiation resistance for m even or odd is

$$R = 72.45 + 30\ln\frac{2L}{\lambda} - 30\,\text{Ci}\,\frac{4\pi L}{\lambda} \tag{8-11}$$

Equation (8-11) applies only at frequencies for which the antenna length is an integral number of half wavelengths (resonant case). At intermediate frequencies the radiation resistance follows a curve which oscillates above and below the resonance values.[1]

The radiation resistance is the component of input resistance due to power lost by radiation. In the above cases it is with respect to a point on the line where the current is I_0 (peak value). It is interesting to note that the terminated line appears to have a larger radiation resistance than the unterminated line.

The above equations are based on an idealized source that is lossless, thin enough to be represented as a line of current, and located in free space. Typical patterns for the terminated and unterminated cases are shown in Fig. 8-2. The effect of source length can be observed by referring to the $\sin \theta \left| (\sin X)/X \right|$ curves in the appendix. Wire and radiation losses and the presence of ground and obstacles also will affect the pattern of the ideal source. Wire and radiation losses can be easily taken into account [see, for example, Eq. (2-89)]. The presence of the ground, however, considerably complicates the problem. The usual approach is to represent the ground as a complex dielectric and consider the total pattern of the antenna and its properly weighted image. The height of the antenna can be adjusted

[1] Stratton, *op. cit.*, p. 444.

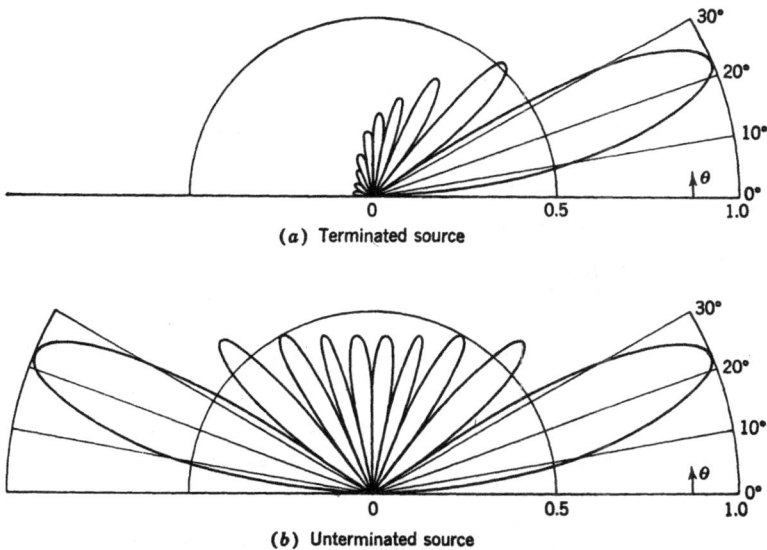

(*a*) Terminated source

(*b*) Unterminated source

Figure 8-2. Patterns of terminated and unterminated long wire antennas. For antennas in free space the lobes are cones about the antenna axis ($\theta = 0$).

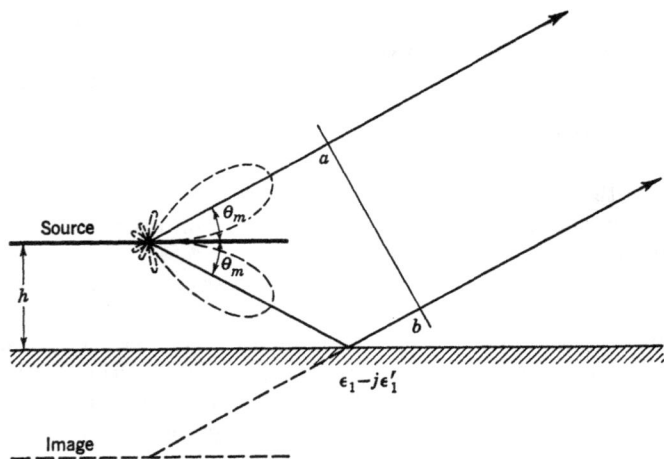

Figure 8-3. Geometry from which h can be adjusted for a given θ_m and complex dielectric so that points a and b have the same phase.

as suggested in Fig. 8-3 so that the reflected wave (or wave from the image) combines in phase with the direct wave from the antenna in the direction of maximum radiation. The important consideration here is the phase shift at the point of reflection.

B. Modified Long Wire Antennas

The radiation properties of an unterminated single wire antenna can be greatly modified by suppressing alternate modes in the standing wave distribution.[1] Antennas in which this is accomplished are generally known as Franklin antennas. Several examples are illustrated in Fig. 8-4. The antenna becomes essentially an end-to-end array of in-phase dipole radiators. The field pattern for such an array is

$$E_\theta(\theta) = \frac{60I_0}{r} \frac{\cos\left[(\pi/2)\cos\theta\right]}{\sin\theta} \frac{\sin\left[(n\pi/2)\cos\theta\right]}{\sin\left[(\pi/2)\cos\theta\right]} \tag{8-12}$$

which has maximum radiation at $\theta = 90$ deg (i.e., broadside). When such an antenna is placed vertically over a perfectly conducting ground, the antenna and its image give the pattern

$$E_\theta(\theta) = \frac{120I_0}{r} \frac{\cos\left[(\pi/2)\cos\theta\right]}{\sin\theta} \frac{\sin\left[(n\pi/2)\cos\theta\right]}{\sin\left[(\pi/2)\cos\theta\right]} \cos\left(kh\cos\theta\right) \tag{8-13}$$

where h is the height above ground of the center of the antenna.

[1] *Ibid.*, pp. 446–448; H. Jasik (ed.), "Antenna Engineering Handbook," chap. 4, McGraw-Hill, New York, 1961.

C. Arrays of Long Wire Antennas

A single long wire antenna has rather high sidelobes as indicated by the patterns in Fig. 8-2. Furthermore the main beam is at an angle with respect to the axis of the antenna, and this must be taken into account when "sighting in" such an antenna or when using it in an array. Because of the beam tilt, "V" and "rhombic" arrays are the most practical arrangements for long wire antennas.

Vertical and horizontal V arrays are illustrated in Fig. 8-5. The inverted V in Fig. 8-5a has principal polarization perpendicular to the ground, whereas the horizontal V has principal polarization parallel to the ground (i.e., in both cases the principal polarization is parallel to the plane of the V). The angles θ_m are the angles of beam maxima with respect to the wires, and these angles are determined by the lengths of wires [see Eq. (8-4) or the graphs in the appendix]. Usually V antennas are symmetrical ($\theta_{m1} = \theta_{m2}$) and the total angle between the wires in Fig. 8-5b is $2\theta_m$. If the total angle is greater than $2\theta_m$, the beam splits into two distinct beams. If the total angle is less than $2\theta_m$, the resultant beam is tilted from the plane of the V.

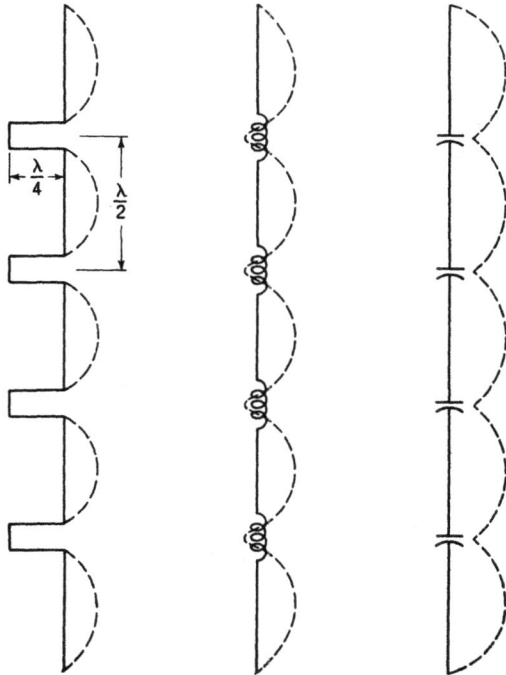

Figure 8-4.   Franklin antennas.

(a) Inverted V

(b) Horizontal V

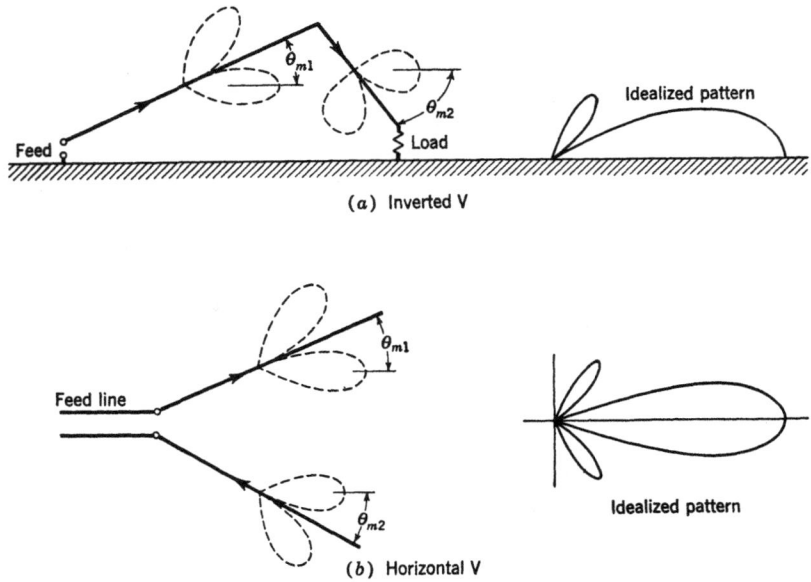

Figure 8-5. V arrays.

The conventional horizontal rhombic antenna is illustrated in Fig. 8-6. This is the most common long wire array. It is essentially two V antennas back to back. Usually the rhomboid is symmetrical (rhombic) and the total angle between the sides at either the feed or load end is equal to or

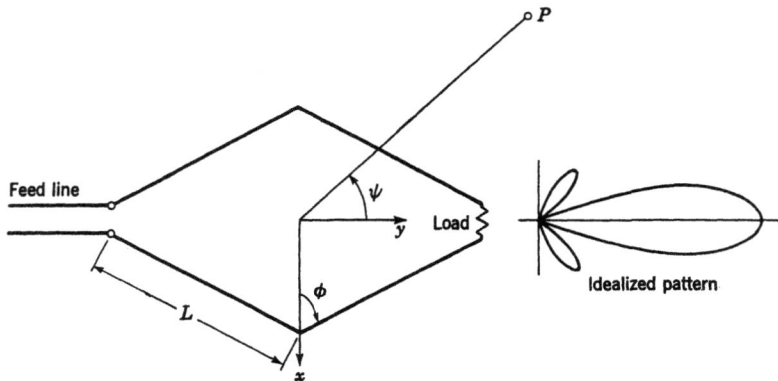

Figure 8-6. Top view of a horizontal rhombic antenna.

less than twice the angle of maximum radiation θ_m of the elements that make up the rhombic. As in the V antenna if the total angle is less than $2\theta_m$, the resultant beam is tilted at an angle with respect to the plane of the rhombic.

The horizontal rhombic (see Fig. 8-6) has principal polarization parallel to the plane of the rhombic. The optimum height above ground for the rhombic can be determined using Fig. 8-3. In practice good results are obtained by assuming the ground to be a perfect reflector, in which case the rhombic and its image form an array with array pattern $E_A(\theta)$ given by

$$E_A(\theta) = \sin (kh \sin \theta) \tag{8-14}$$

The resultant pattern is the product of $E_A(\theta)$ and the pattern of the free-space rhombic.

Assuming a perfectly terminated, symmetrical, horizontal rhombic (see Fig. 8-6) made of perfectly conducting elements with current I_0 and placed over a perfectly conducting ground, Fernandes[1] gives the total far-field pattern (considering both vertical and horizontal components) as

$$E = \frac{480\pi L I_0 \cos \phi}{r\lambda} \left| \frac{\sin X_1}{\sqrt{X_1}} \frac{\sin X_2}{\sqrt{X_2}} \sin \left(\frac{2\pi h \sin \Delta}{\lambda} \right) \right| \tag{8-15}$$

where

$$X_1 = \frac{\pi L}{\lambda} [1 - \cos \Delta \sin (\phi - \psi)] \tag{8-16}$$

$$X_2 = \frac{\pi L}{\lambda} [1 - \cos \Delta \sin (\phi + \psi)] \tag{8-17}$$

The angle Δ is the angle between the horizontal plane and the ray from the origin to the point of observation P, and ψ is the angle between the y axis and the projection of the ray onto the horizontal (xy) plane (see Fig. 8-6).

Considerable control of the main beam of the rhombic can be obtained by varying element length and angle between elements and tilting the plane of the rhombic. Arrays of rhombics are also employed to control the radiation pattern. Arrays are principally used to increase gain and decrease sidelobes. A great deal of design data on rhombic antennas is available in the literature.[2]

[1] A. A. De Carvalho Fernandes, On the Design of Some Rhombic Antenna Arrays, *IRE Trans. Antennas Propagation*, AP-7:39–46, January, 1959.
[2] Jasik, *loc. cit.*; A. E. Harper, "Rhombic Antenna Design," Van Nostrand, Princeton, N.J., 1941; D. Foster, Radiation from Rhombic Antennas, *Proc. IRE*, 25:1327, October, 1937; E. Bruce, A. C. Beck, and L. R. Lowry, Horizontal Rhombic Antennas, *Proc. IRE*, 23:4, January, 1935; W. N. Christiansen, Directional Patterns for Rhombic Antennae, *AWA (Australia) Tech. Rev.*, 7:1, 1946; Fernandes, *op. cit.*

8-3. SLOW WAVE ANTENNAS

Slow wave antennas are characterized by a traveling wave distribution with phase velocity v equal to or less than the velocity of light c in free space (that is, $c/v \gtrless 1$). These are usually endfire or near endfire radiators, and this class of antennas includes the long wire antennas of Sec. 8-2. Because a wave with $c/v \gtrless 1$ in the absence of curvature, discontinuities, and loss may be guided without attenuation along the slow wave structure and in a sense is bound to the surface of the structure, slow wave structures are frequently called surface wave structures.

The present section will be devoted to a discussion of practical slow wave antennas in the form of either line radiators or planar surface radiators. Curved and modulated structures will be considered in later sections.

A. Gain and Radiation Characteristics of Slow Wave Antennas

It has been pointed out that slow wave antennas are usually endfire or near endfire radiators. An endfire line source or quasi line source of transverse (to the direction of propagation) dipole elements radiates essentially a pencil beam. Line sources of longitudinal dipole elements, on the other hand, radiate a conical beam (a good example of this is the long wire antenna). For planar structures with either transverse or longitudinal current elements the beam is usually fan shaped and independent control of the patterns in the two principal planes generally can be obtained. Radiation patterns for a given source distribution can be computed by the methods of Chap. 2, or the source distribution for a given pattern can be determined by the synthesis methods of Chap. 3.

Antenna directivity (or power gain over an isotropic source for a lossless antenna) can be found by applying the usual equation

$$D = \frac{4\pi |F(\theta,\phi)_{\max}|^2}{\int_0^{2\pi} \int_0^{\pi} |F(\theta,\phi)|^2 \sin \theta \, d\theta \, d\phi} \tag{8-18}$$

where $F(\theta,\phi)$ is the far-field pattern of the source. Unless otherwise stated, the directivity is always taken to be in the direction of maximum radiation.

An approximate but useful result can be obtained from Eq. (8-18) by assuming $F(\theta,\phi)$ to be constant between the half-power points and zero elsewhere. That is, the actual pattern is approximated by a wedge-shaped beam. With this simplification Eq. (8-18) reduces to[1]

$$D \approx \frac{4\pi}{\theta_E \theta_H} \tag{8-19}$$

where θ_E and θ_H are the beamwidths in radians between the half-power

[1] Kraus, *op. cit.*, sec. 2-14.

points in the **E** and **H** planes, respectively. Equation (8-19) can be written

$$D \approx \frac{41,253}{\theta_E \theta_H} \qquad (8\text{-}20)$$

where the half-power beamwidths are in degrees. For practical, low loss antennas the gain (efficiency times directivity) is usually somewhat less than Eq. (8-20) would indicate. An approximate gain expression that is quite useful for practical, high gain antennas is[1]

$$G \approx \frac{27,000}{\theta_E \theta_H} \qquad (8\text{-}21)$$

where θ_E and θ_H are half-power beamwidths in degrees in the **E** and **H** planes, respectively.

For a uniform line source of isotropic elements located in free space the directivity is

$$D \approx 2 \frac{L}{\lambda} \qquad (8\text{-}22)$$

where L is the length of the source. Equation (8-22) is applicable to a line source with any angle of radiation from broadside to near endfire. As endfire operation is approached, the directivity rapidly increases to about $4L/\lambda$ (see Fig. 3-27). Usually for endfire sources the phase velocity of the traveling wave is adjusted to give a phase delay of π rad more than the delay of an ordinary endfire source $(c/v = 1)$ over the total length of the source; i.e.,

$$\beta L = \pi + kL \qquad (8\text{-}23)$$

or

$$\frac{c}{v} = 1 + \frac{1}{2L/\lambda} \qquad (8\text{-}24)$$

This is known as the Hansen-Woodyard condition for increased directivity, although the results of Hansen and Woodyard give 2.94 rather than π rad (see Sec. 3-2G). In this case the directivity of the line source with respect to an isotropic source is

$$D \approx \frac{7L}{\lambda} \qquad (8\text{-}25)$$

Equation (8-25) is derived for a uniformly illuminated endfire source where the length of the source is much greater than the wavelength $(L \gg \lambda)$. However for traveling wave endfire structures such as the polyrod and the Yagi antenna, the length is quite important and influences the

---

[1] Jasik, *op. cit.*, chap. 2.

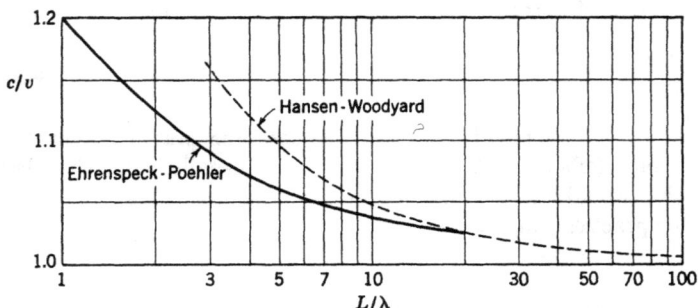

Figure 8-7. Velocity ratio c/v as a function of source length for Hansen-Woodyard and Ehrenspeck-Poehler optimum gain conditions. (After Jasik (ed.), "Antenna Engineering Handbook," Fig. 16-10, McGraw-Hill, New York, 1961.)

design for optimum gain. Ehrenspeck and Poehler[1] in a study of Yagi antennas found that for lengths less than 20 λ the additional phase delay of Hansen and Woodyard is excessive. For lengths 4 to 8 λ about 120 deg is optimum, reducing to about 60 deg for a source 1 λ long. In all cases the phase for optimum gain is somewhat dependent on the amplitude distribution along the source.[1]

Curves of c/v as a function of source length are given in Fig. 8-7 comparing the Hansen-Woodyard and Ehrenspeck-Poehler conditions with the ordinary endfire condition ($c/v = 1$).

For optimum gain endfire antennas Eq. (8-25) is good for moderate antenna lengths, say 10 to 50 λ. For extremely long sources ($L > 50$ λ) a better approximation is

$$D \approx 4 \frac{L}{\lambda} \tag{8-26}$$

and for short sources (3 $\lambda < L < 8$ λ),

$$D \approx 10 \frac{L}{\lambda} \tag{8-27}$$

The effect of source length on gain and beamwidth of an endfire line source is illustrated in Fig. 8-8.

For long sources Eq. (8-24) gives $c/v \approx 1$, and as c/v approaches unity, the wave is no longer tightly bound to a uniform surface wave structure. There is a length beyond which little or no guiding action is achieved. In practice this is probably in the order of 50 λ. For greater lengths one

---

[1] H. W. Ehrenspeck, and H. Poehler, A New Method for Obtaining Maximum Gain from Yagi Antennas, *IRE Trans. Antennas Propagation*, **AP-7**(4):379–386, October, 1959.

must find a way to maintain the guiding action of the structure. One method is to modulate the structure (see Sec. 8-7).

For short sources a sizable fraction of the length may be used in getting the slow wave launched. The desired wave is usually well established at a distance from the feed where the radiated wave from the feed, propagating at velocity of light c, leads the slow wave by about 60 deg.[1] The launch distance l is given by

$$l = \frac{\lambda}{6[(c/v) - 1]} \qquad (8\text{-}28)$$

where v is the phase velocity of the slow wave. The distance l is about half the total length of the structure for an optimum gain design. The amplitude distribution along the source may be quite nonuniform over the distance l. Usually the amplitude has a pronounced hump in this region. Because of the nonuniform amplitude distribution and the fact that the Hansen-Woodyard condition was derived for source length $L \gg \lambda$, Eqs. (8-24) and (8-25) are not adequate for maximum gain designs of short endfire antennas ($L < 5\ \lambda$). The additional phase delay, as mentioned above, should be about 120 deg for $L = 5\ \lambda$ and reduces to about 60 deg for $L = \lambda$. The gain is somewhat higher than for the Hansen-Woodyard case (Eq. 8-25) and is given by Eq. (8-27).

[1] Jasik, *op. cit.*, chap. 16.

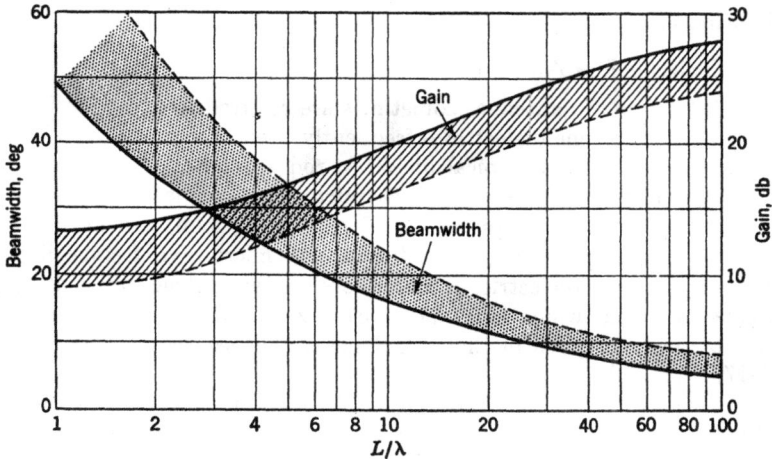

Figure 8-8. Gain (over an isotropic source) and beamwidth of an endfire line source as a function of source length. Solid lines are optimum value and dashed lines are for low sidelobe and broadband design. (After Jasik (ed.), "Antenna Engineering Handbook," Fig. 16-11, McGraw-Hill, New York, 1961.)

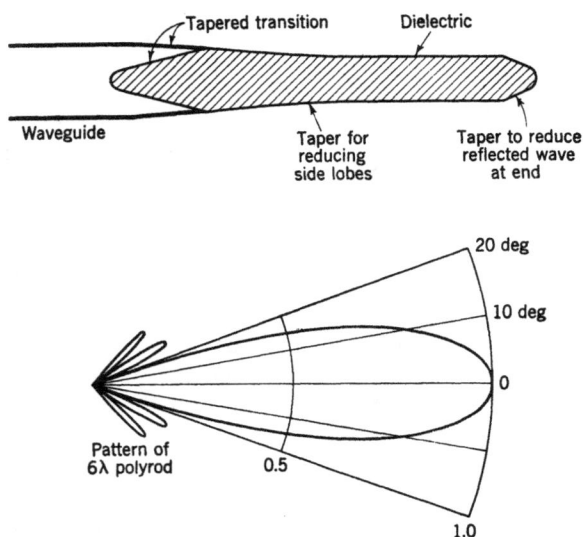

Figure 8-9. Dielectric rod antenna.

B. Slow Wave Line Sources

Some of the more common slow wave line sources are dielectric rods, corrugated rods, slotted waveguides (loaded so $c/v \gtrless 1$), helices, and Yagi-Uda arrays. A brief description of each of these is given along with design information.

Dielectric Rod Antennas

Dielectric rod antennas sometimes are referred to as polyrod antennas because polystyrene dielectric is frequently used. A sketch of the longitudinal cross section of a typical dielectric rod antenna is shown in Fig. 8-9. This antenna is generally used at microwave frequencies and fed from circular or rectangular waveguide. Dipole, stub, loop, and helix feeds are quite practical, however, at uhf and low microwave frequencies. Solid dielectric, hollow dielectric, or metal-core dielectric structures may be used. Equations from which the phase velocity and the field components can be obtained are given in Chap. 5. Some c/v curves are given in Figs. 5-16, 5-17, 5-22, and 5-24.

The usual mode of operation for dielectric rod antennas is the hybrid "dipole" mode in which the source looks like an array of transverse dipoles. TE and TM modes are also possible. The dielectric rod antenna is usually linearly polarized, but circular and elliptical polarization can be obtained easily by using the appropriate feed.

A precise design of dielectric rod antennas is quite difficult because of direct radiation from the feed end and terminating end of the rod. If direct radiation from the ends is kept low by careful design, then the design methods of Chap. 4 can be used to realize a desired amplitude and phase of the surface fields along the rod. Design data are available in Chap. 5 for circular cross sections. For rectangular cross sections a wavelength or more in the wide dimension, the results for infinite dielectric sheets in Chap. 6 give approximate but useful results.

Corrugated Rod Antennas

A corrugated rod antenna is illustrated in Fig. 8-10. It is usually of rectangular or circular cross section and can be thought of as an artificial-dielectric polyrod antenna. Design data for circular rods are available in Chap. 5, and the data in Chap. 6 on planar corrugated surfaces can be used for rectangular cross sections as narrow as a wavelength or so with good results.

Although the corrugated surface is analyzed as a TM structure, the predominant field at the surface is a longitudinal electric field (between corrugations). Hence the structure has the radiation properties of an array of transverse magnetic current elements such that a horizontal surface would have vertical polarization. It is not so versatile as the dielectric rod which supports TE, TM, and hybrid (TE + TM) fields and can be made to give any polarization.

Slotted Waveguide Antennas

All the perturbed circular and rectangular waveguide structures of Chap. 5 can be loaded with dielectric to give $c/v \gtrless 1$ for slow wave operation. Other structures as well as other methods of loading such as ridges, posts, and corrugations are also possible. The chief problem in using modified waveguide structures such as these for slow wave antennas is obtaining an adequate transfer of power from the guiding structure to free space without introducing unwanted discontinuities.

One of the more practical waveguide structures is a dielectric-filled rectangular guide (TE_{01} excitation) with a wide wall removed and with tapered depth.[1] A sketch of this antenna is shown in Fig. 8-11. The propagation constant along such a structure with gradual taper can be obtained

<hr />

[1] B. T. Stephenson and C. H. Walter, Endfire Slot Antennas, *IRE Trans. Antennas Propagation,* **AP-3**(2): 81–86, April, 1955.

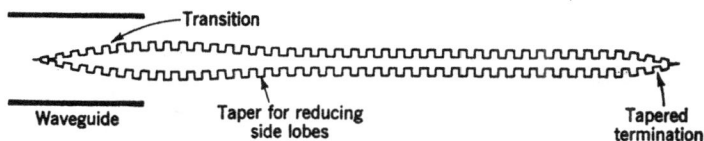

Figure 8-10.   Longitudinal cross section of corrugated rod antenna.

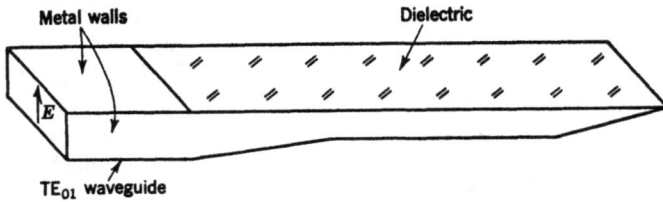

Figure 8-11. Tapered depth slotted waveguide antenna.

from the H-guide analysis as suggested in Chap. 5. Some c/v curves are given in Fig. 5-28.

The tapered depth slot antenna in Fig. 8-11 can be readily flush mounted as shown in Fig. 8-12. An element of length along the slot is equivalent to a transverse electric dipole normal to the slot. Thus a horizontal slot gives vertical polarization. When an endfire antenna is flush mounted in a conducting surface, the shape and extent of the ground plane have considerable effect on the E-plane pattern (see Chap. 2). The principal effect is the tilt of the main beam, but the sidelobes are also affected. An approximate expression for the tilt angle is (see Sec. 2-20)

$$\theta_t \approx \frac{49}{[(L + L_g)/\lambda]^{\frac{1}{2}}} \tag{8-29}$$

This antenna lends itself very nicely to optimum gain designs (see Sec. 8-3A), and it also has good bandwidth characteristics. Bandwidths of better than 50 per cent without modifications and better than 100 per cent with modifications are possible.[1]

[1] Stephenson and Walter, *op. cit.*

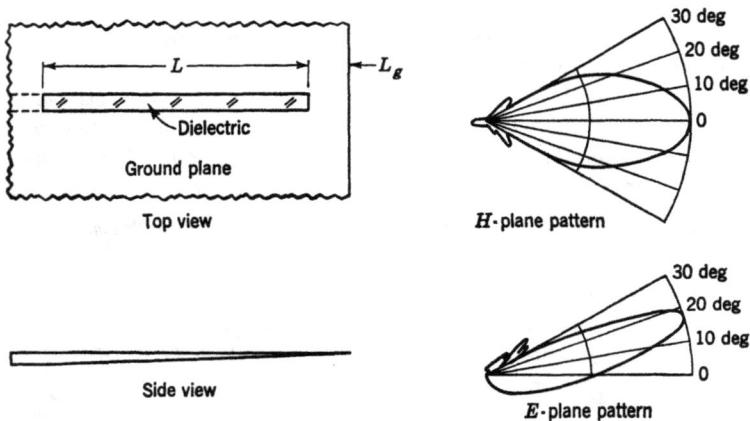

Figure 8-12. Tapered depth slot in ground plane with typical patterns.

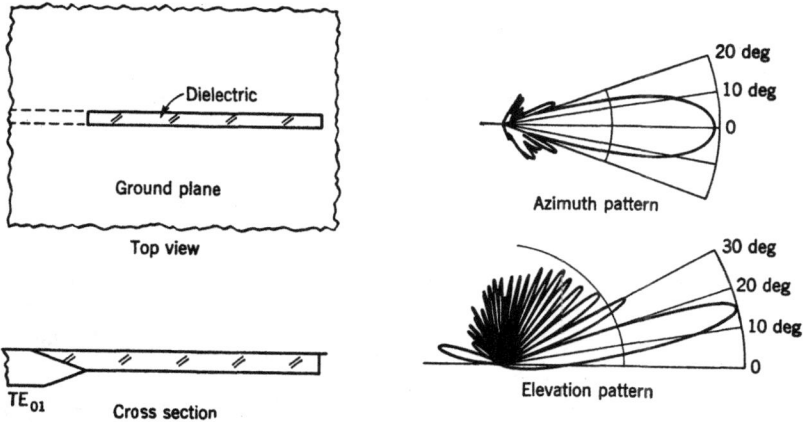

Figure 8-13.   Horizontally polarized endfire slot. Patterns are for a 10 λ source.

The dielectric-filled rectangular guide with TE_{01} excitation can also be opened at the narrow wall as illustrated in Fig. 8-13. An element of length along this slot is equivalent to a longitudinal magnetic dipole (sin θ pattern); thus the slot has the tilted beam of a line current source. The beam is essentially a half cone due to the ground plane. The conical shape is illustrated in Fig. 8-14.

For a horizontal slot the field is horizontally polarized in the elevation pattern of Fig. 8-13, but in the plane of the ground plane vertically polarized lobes appear owing to the conical shape of the beam (see Fig. 8-14).

Design data for this structure can be obtained from Chap. 5. Although it is not a good endfire antenna, since it has a null in the plane of the ground plane, it has been used quite successfully as a flush-mounted runway antenna in an aircraft landing system.[1]

[1] J. R. Baechle and R. H. McFarland, A Flush-mounted Runway Antenna for Use with the FAA Directional Glide-path System, *IRE Trans. Aeron. Navigational Electron.*, **ANE-7**(2):32–39, June, 1960.

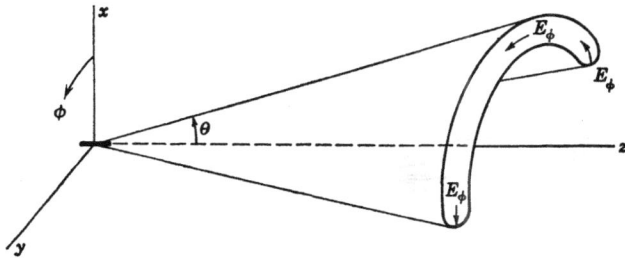

Figure 8-14.   Cross section of the beam of a horizontally polarized slot along z axis.

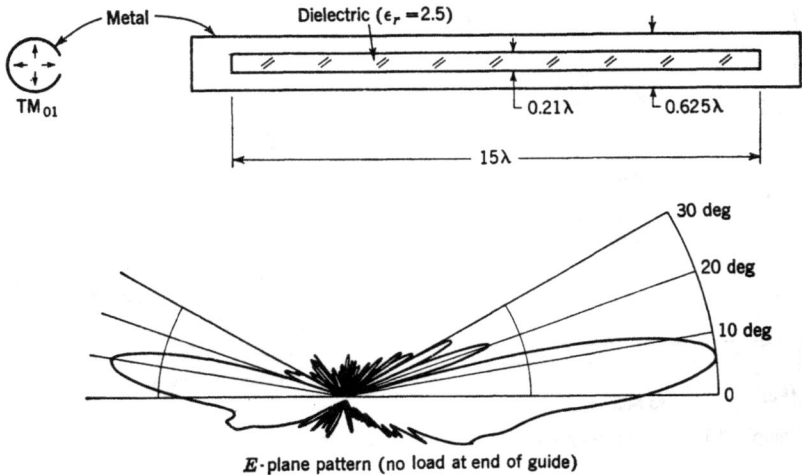

Metal — Dielectric ($\epsilon_r = 2.5$)

TM_{01}

0.21λ 0.625λ

15λ

30 deg
20 deg
10 deg
0

E-plane pattern (no load at end of guide)

Figure 8-15. TM_{01} slotted cylinder endfire antenna.

Slotted circular waveguides also can be used as practical slow wave antennas by dielectric loading. A TM_{01}-excited structure is shown in Fig. 8-15. Design data for this antenna can be obtained from Chap. 5. Figure 8-16 shows the same antenna with a tapered slot designed to give something approaching a gaussian taper to the amplitude of the fields along the aperture. Significant reduction of sidelobes is observed, thus demonstrating that the amplitude distribution along an endfire source can be controlled. Furthermore, tapers of this type tend to reduce pattern deterioration due to end effects.

One can eliminate the beam tilt from the slotted guide by removing

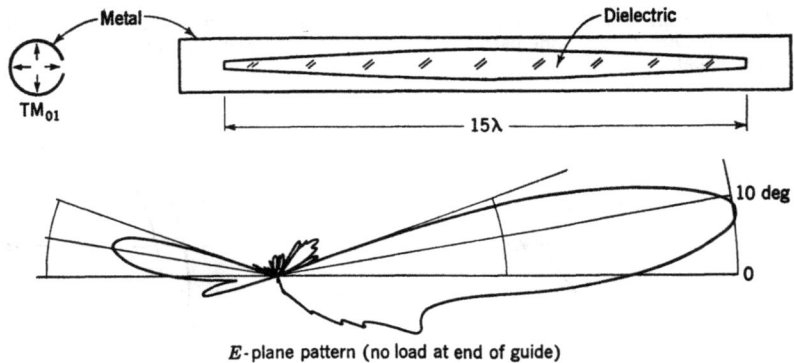

Metal — Dielectric

TM_{01}

15λ

10 deg
0

E-plane pattern (no load at end of guide)

Figure 8-16. TM_{01} slotted cylinder endfire antenna with tapered aperture.

any ground plane and placing another slot diametrically opposite the first as shown in Fig. 8-17. To get the fields of the slots in phase in the endfire direction TM_{11} excitation is used for a circular cross section.

Helical Antennas

The helical antenna can radiate in many modes; however, the principal modes are the axial (endfire) and normal (broadside) modes.[1] Both modes are capable of circular polarization, but the wide bandwidth (almost 2:1) of the axial mode makes it more useful than the normal mode. The normal mode operates over a relatively narrow band, and also it is less efficient than the axial mode.

A sketch of a typical helical antenna is shown in Fig. 8-18. The helix is very easily fed by a coaxial line as indicated in the figure, and to a good approximation the input impedance of the antenna in the axial mode is nearly a pure resistance given by

$$R \approx \frac{140C}{\lambda} \tag{8-30}$$

where C/λ is the circumference in wavelengths. Other types of feeds are possible, particularly at microwave frequencies where waveguide and dielectric rod feeds may be more practical than coaxial line feeds.

A good deal of design information is available on helical antennas.[2,3] The axial mode of operation is most commonly used, and it is defined by[2] $\frac{3}{4} < C/\lambda < \frac{4}{3}$. For the axial mode the velocity ratio of the wave along the structure is[2]

$$\frac{c}{v} \approx \sin \alpha + \frac{\cos \alpha}{C/\lambda} \tag{8-31}$$

[1] Kraus, *op. cit.*, chap. 7.
[2] *Ibid.*
[3] Jasik, *op. cit.*, chap. 7.

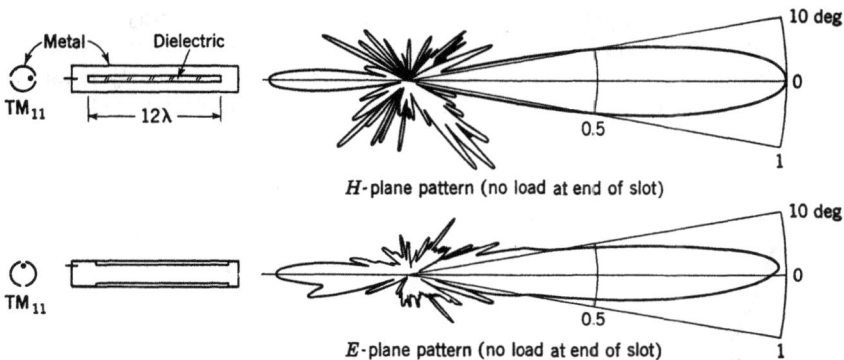

Figure 8-17. TM_{11} slotted cylinder endfire antenna.

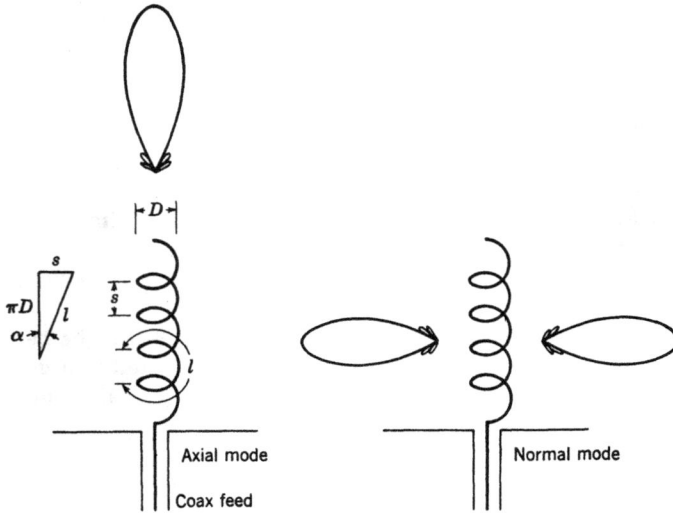

Figure 8-18.   Helical antennas.

where α is the pitch angle (see Fig. 8-18). The ellipticity, or axial ratio AR, of the polarization on axis is given by

$$AR = \frac{2n + 1}{2n} \qquad (8\text{-}32)$$

where n is the number of turns.

It is quite easy to design the axial mode helix for increased directivity (see Sec. 8-3A). A single, long helix is capable of high gain operation (especially with modulation; see Sec. 8-7). For high gain applications, however, it is difficult to support a very long helix mechanically, so arrays of shorter elements are generally used. In addition to being used as an individual antenna or in arrays, the helix is also quite useful as a circularly polarized feed for other antennas such as horns, reflectors, and dielectric rods.

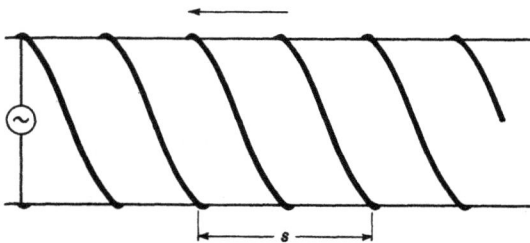

Figure 8-19.   Bifilar helical antenna.

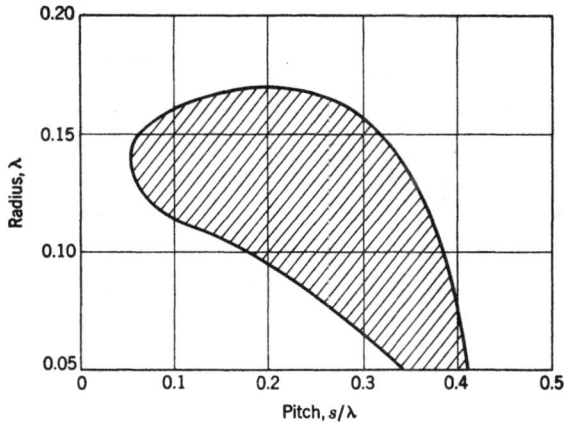

Figure 8-20. Radius and pitch of the bifilar helix in Fig. 8-19. The shaded area gives the region of best backfire operation. (From Patton, The Backfire Bifilar Helical Antenna, Fig. 16, *Univ. Illinois Antenna Lab. Tech. Rept.* 61, September, 1962.)

Although the endfire mode of operation is the most widely used, recently an interesting mode of operation has been studied that is termed the *backfire* mode.[1] In this mode a backward wave exists in the sense that, while the group velocity (which is associated with energy velocity) is away from the feed, the phase velocity is toward the feed and results in a beam in the backward endfire direction; hence the term backfire.

Although this mode of operation is not restricted to a two-winding or bifilar helix, this type is perhaps the simplest to feed and is illustrated in Fig. 8-19. In the backfire mode it has a circularly polarized beam in the direction of the arrow in Fig. 8-19 and has the advantage over the conventional helix that a ground screen is not needed.

A graph of bifilar helix radius versus pitch is given in Fig. 8-20, in which the shaded area shows the region of best backfire operation.

Yagi-Uda Antennas

A sketch of a Yagi-Uda antenna is shown in Fig. 8-21. It is essentially an endfire array of parasitic dipoles, and it is linearly polarized with the polarization of a dipole element of the array. Perhaps the most common feed is the dipole with reflector as shown in Fig. 8-21. This is quite practical in the hf, vhf, and uhf regions. The Yagi-Uda array also may be used in the microwave region, in which case waveguide feeds become practical.

The Yagi-Uda array is inherently a slow wave structure. Design data

[1] W. T. Patton, The Backfire Bifilar Helical Antenna, *Univ. Illinois Antenna Lab. Tech. Rept.* 61, September, 1962.

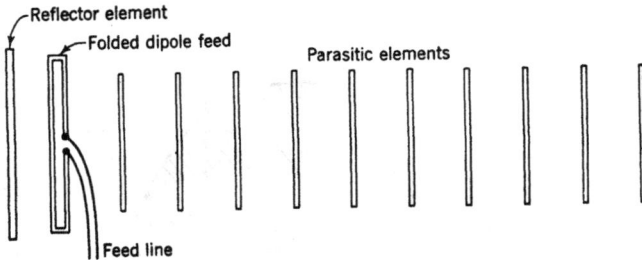

Figure 8-21. Yagi-Uda array. The elements may be shorted dipoles in free space or stubs on a ground plane.

giving phase velocity as a function of geometry are given in Sec. 5-2D. More design data are available in the literature.[1]

The Yagi-Uda array lends itself to optimum gain designs[2] (see Sec. 8-3A) and modulated design (see Sec. 8-7). A single, long Yagi is capable of high gain operation (25 db), but mechanical considerations usually dictate the use of an array for high gain rather than a single Yagi.

The above slow wave line sources by no means exhaust the list of practical antennas. Various modifications of the above structure are possible, such as using half of a dielectric rod in the dipole mode on a ground plane (image line)[3] or introducing discontinuities (posts for example) along a dielectric rod to control radiation.[4] More is said on radiating discontinuities at the end of this section. Another structure that is quite useful as a slow wave or a fast wave source is the trough guide, which is described in Sec. 8-4.

C. Slow Wave Planar Sources

Perhaps the most common slow wave planar, or area, sources are the grounded dielectric sheet and the singly and doubly corrugated surfaces. Grounded ferrite and plasma sheets are also useful as well as arrays of line sources. Arrays, however, are discussed in Sec. 8-6.

Dielectric Sheet Antennas

A dielectric sheet antenna may take the form of a dielectric sheet in free space or a dielectric sheet on a ground plane. Considering a rectangular sheet to be horizontal, the distribution across the width primarily determines the azimuth beam, whereas the distribution along the length primarily determines the elevation beam. The analysis and synthesis methods of

[1] S. Uda and Y. Mushiake, "Yagi-Uda Antenna," Maruzen Co., Ltd., Tokyo, 1954; Jasik, *op. cit.*, chap. 5.

[2] Ehrenspeck and Poehler, *op. cit.*; A. Kay, Yagi Antenna Study, *Tech. Res. Group Sci. Rept.* 1, Syosset, N.Y., January, 1960.

[3] D. D. King, Dielectric Image Line, *J. Appl. Phys.*, **23**:699, 1952.

[4] J. W. Duncan and R. H. DuHamel, A Technique for Controlling the Radiation from Dielectric Rod Waveguides, *IRE Trans. Antennas Propagation*, **AP-5**:284, 1957.

Chaps. 2 and 3 apply. The feed for the dielectric sheet antenna can be any of a number of line sources, some of which are illustrated in Fig. 7-3.

For a sheet whose width is much greater than a wavelength, the infinite sheet analyses of Chap. 6 can be used to obtain design data. Data are given for both TE and TM propagation with and without ground plane. For a horizontal sheet, TM propagation gives vertical polarization and TE propagation gives horizontal polarization. In principle, circular polarization can be obtained by utilizing both TE and TM propagation on a structure for which both modes have the same phase velocity. This can be accomplished by two dielectric layers on a ground plane (see Secs. 6-2 and

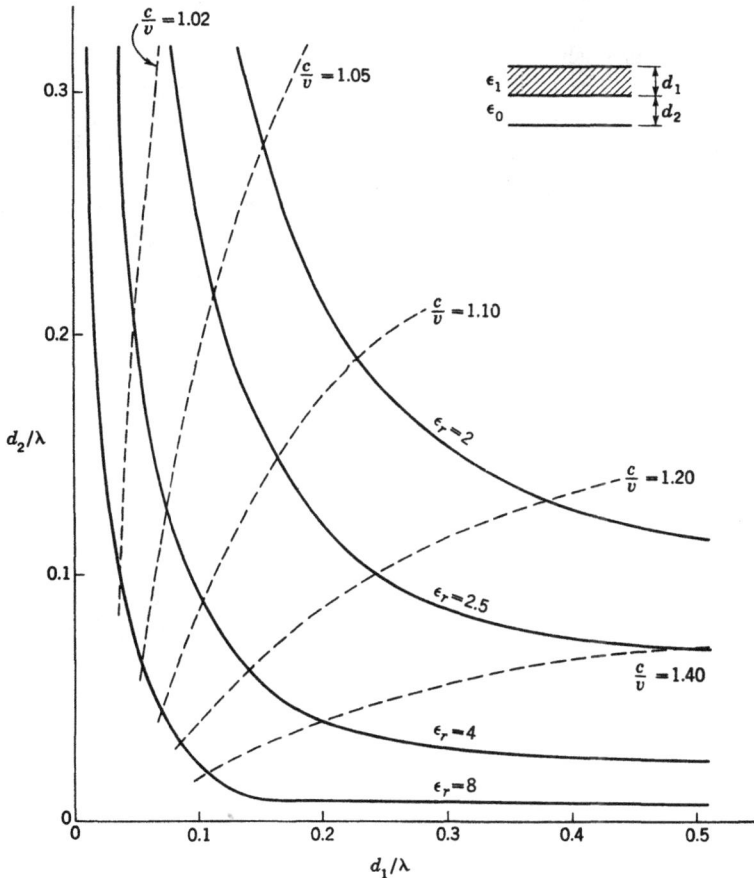

Figure 8-22. Dielectric layer structure for circularly polarized surface wave having c/v between 1.02 and 1.4. (From Plummer and Hansen, Double-slab Arbitrary-polarization Surface-wave Structure, *Proc. Inst. Elec. Engrs.*, Monograph 238 R, 1957.)

6-3)[1] (or three layers in free space) or by a dielectric layer over a corrugated surface.[2] The latter, however, is not isotropic in the plane of the structure. If isotropy is necessary, a post structure (see Sec. 6-3) can be substituted for the corrugated surface. Data for the two dielectric layers on a ground plane are given in Fig. 8-22. Obtaining circular polarization in practice is complicated by the fact that the tilt angles for the TE and TM beams are generally different even though the surface waves have the same phase velocity.

Dielectric sheet antennas can be flush mounted to a conducting surface by recessing the dielectric as illustrated in Fig. 8-23 for a tapered depth design.[3] The beam tilt in the vertical plane is given approximately by Eq. (8-29).

Another flush antenna that is quite practical[4,5] is the flared slot shown in Fig. 8-24. The analysis for the H-guide line in Chap. 5 can be used to

[1] R. E. Plummer and R. C. Hansen, Double-slab Arbitrary-polarization Surface-wave Structure, *Proc. Inst. Elec. Engrs.*, part C, Monograph 238 R, 1957.

[2] R. C. Hansen, Single Slab Arbitrary Polarization Surface Wave Structure, *IRE Trans. Microwave Theory Tech.*, **MTT-5**:115, 1957.

[3] J. N. Hines and J. R. Upson, A Wide Aperture Tapered-depth Scanning Antenna, *Ohio State Univ. Antenna Lab. Rept.* 667-7, Columbus, December, 1957.

[4] Stephenson and Walter, *op. cit.*

[5] J. W. Eberle, C. A. Levis, and D. McCoy, The Flared Slot: A Moderately Directive Flush-mounted Broadband Antenna, *IRE Trans. Antennas Propagation*, **AP-8**(5): 461–468, September, 1960.

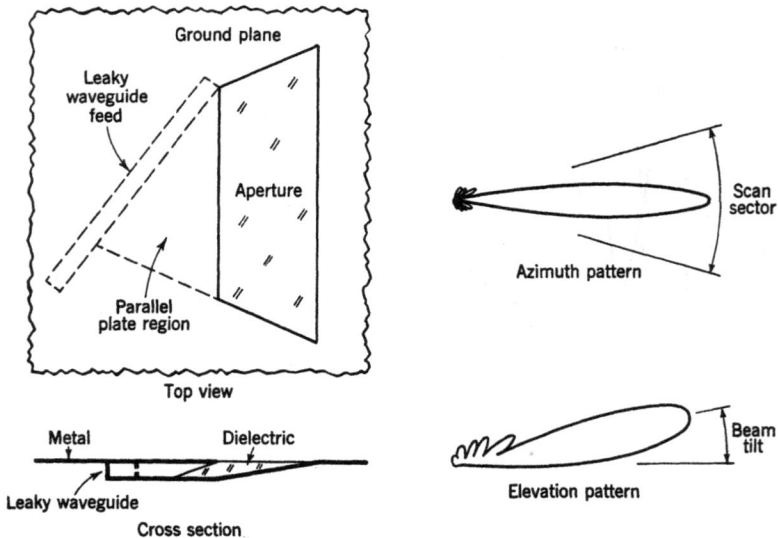

Figure 8-23.   Flush-mounted dielectric sheet antenna.

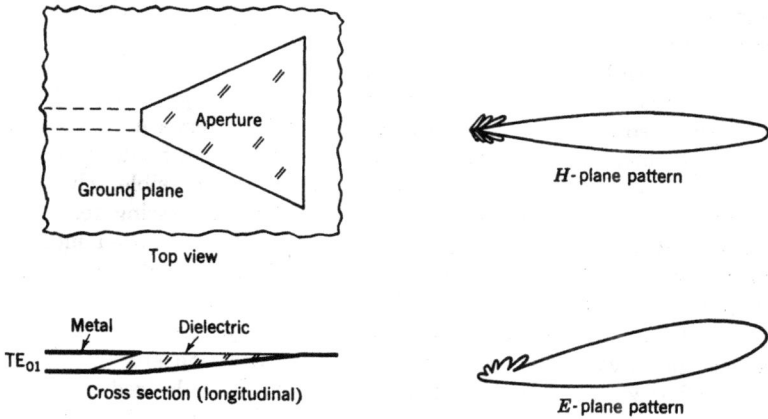

Figure 8-24. Flared slot antenna.

determine the propagation constant along this structure, and patterns can be computed by the methods of Chap. 2. The beam tilt in the vertical plane is given approximately by Eq. (8-29). Flared slot antennas with flare angles of 30 deg or less and uniform transverse cross sections usually give satisfactory H-plane patterns. For larger flare angles it may be necessary to taper the transverse cross section to obtain lens action. This can be treated as a variable index lens where the index is the velocity ratio c/v. The velocity ratio c/v is related to the thickness of the dielectric by the H-guide analysis.

The flared slot with a tapered ladder discontinuity minimizer (see Sec. 7-3) and a ridged-waveguide feed has operated successfully over a 4:1 band.[1]

In addition to rectangular and triangular sheet structures, circular dielectric sheet antennas are also quite practical. Center-fed circular sheets can provide omnidirectional coverage in azimuth with good endfire coverage in elevation.[2] Surface wave lenses are also quite practical, and a circular dielectric disk can be made to operate as a two-dimensional Luneberg lens (see Sec. 8-8).

Corrugated Surface Antennas

A corrugated surface can be viewed as an anisotropic, artificial dielectric. Data are given in Chap. 6 for the phase velocity of TM waves on corrugated and doubly corrugated surfaces. The doubly corrugated structures (metal posts or pins) are very nearly isotropic in the plane of the structure. For the corrugated surface, however, the phase velocity of a

---

[1] *Ibid.*

[2] E. M. T. Jones and R. A. Folsom, Jr., A Note on the Circular Dielectric Dish Antenna, *Proc. IRE*, **41**:798, 1953; R. S. Elliott, Spherical Surface-wave Antennas, *IRE Trans. Antennas Propagation*, **AP-4**:422, 1956.

surface wave very definitely depends on the direction of propagation with respect to the corrugations (see Sec. 6-3B, for example).

A corrugated surface can be used in place of a dielectric sheet in an area antenna (see previous section), but the anisotropy must be taken into account. Also it can be observed (see Sec. 6-3B) that a corrugated surface acts like a filter in that it exhibits passbands and stop bands and also there are ranges over which backward wave operation is possible. These features might be useful in some applications. Another interesting feature of the corrugated surface is that the phase velocity can be varied mechanically by varying the height of the corrugation. For example, posts could be moved up or down through holes in a ground plane.

Although TE surface waves are theoretically possible on a corrugated surface (see Sec. 6-3B), the usual modes of operation are TM. For a TM wave a horizontal surface gives rise to a vertically polarized far field.

As in the dielectric sheet case, corrugated surfaces of circular as well as rectangular shape can be used in practical antennas. Both omnidirectional[1] and high gain[2] corrugated surface antennas have been studied.

Ferrite and Plasma Sheet Antennas

In principle, ferrite or plasma sheets can be used in dielectric sheet antennas to provide a means of electrically varying the phase velocity. In practice, loss may be prohibitive in both types of structures, and with plasma there is also the problem of keeping it confined to the antenna surface. In some cases, however, ferrite losses can be kept to a satisfactory level by using a very thin sheet. Expressions for the phase velocity for a thin ferrite sheet are given in Sec. 6-4A.

An analysis for a plasma sheet would be similar to that for the ferrite sheet (see Probs. 6-6 to 6-8). The plasma sheet, however, does not appear to be as practical as the ferrite sheet because of the confinement problem. There may be situations, however, where a plasma layer is part of the environment of a conducting surface. In such cases one can consider using the plasma sheet as a surface wave antenna.

Additional Structures

Practical antennas also can be obtained by modifying the above structures. One possibility is to modulate the structure (see Sec. 8-7); another is to place small radiating elements such as small dipoles or loops over the surface to form an array of parasitic radiators. To a first approximation the relative amplitude of excitation of the parasitic elements can

---

[1] E. M. T. Jones, An Annular Corrugated Surface Antenna, *Proc. IRE,* **40**:721, 1952; Elliott, *op. cit.*; R. E. Plummer, Surface-wave Beacon Antennas, *IRE Trans. Antennas Propagation,* **AP-6**:105, 1958.

[2] C. H. Walter, Surface-wave Luneberg Lens Antennas, *IRE Trans. Antennas Propagation,* **AP-8**(5): 508, September, 1960.

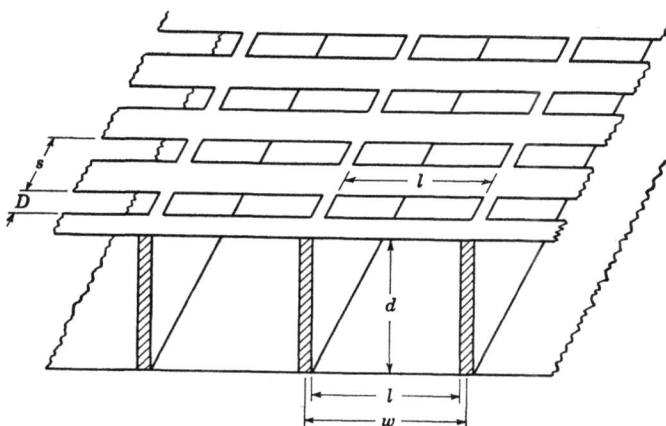

Figure 8-25. Periodically slotted surface wave structure. The slots are nonresonant. Top plate is very thin and all plates have infinite conductivity.

be determined from the exponential decay of the surface wave fields above the unperturbed surface. Similarly the relative phase of excitation can be determined from the phase velocity of the unperturbed surface wave. If the elements are properly spaced, radiation in the endfire direction, in the broadside direction, or even to the rear is possible (see section below on backward wave antennas).

Still other planar, slow wave antennas can be obtained by dielectric loading the holey plate and grid structures of Chap. 6. Metal plate loading

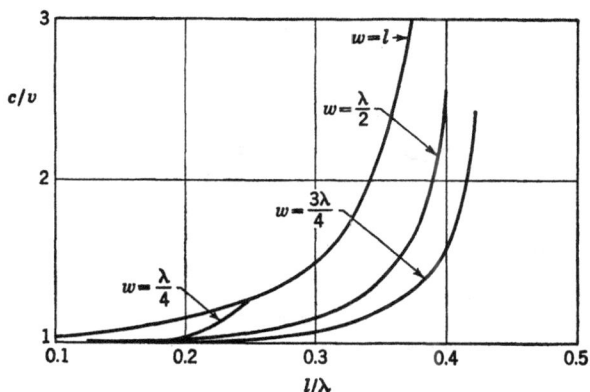

Figure 8-26. Velocity ratio c/v versus slot length l for infinitesimal slot spacing for the structure in Fig. 8-25. (From Hougardy, Periodically Slotted Surface Wave Structure, *Sci. Rept.* 3508/4, Hughes Aircraft Co., Culver City, Calif., August, 1959.)

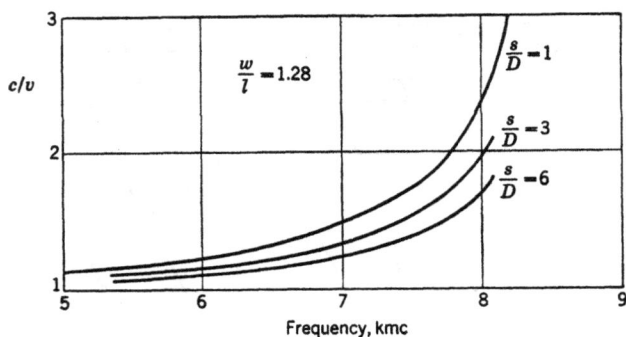

Figure 8-27. Showing the effect of s/D on velocity ratio c/v for the structure in Fig. 8-25. (From Hougardy, Periodically Slotted Surface Wave Structure, *Sci. Rept.* 3508/4, Hughes Aircraft Co., Culver City, Calif., August, 1959.)

also may be used. The slotted plate over the parallel metal plates in Fig. 8-25 makes a practical, vertically polarized slow wave structure when the parallel plates form below-cutoff waveguides ($l < \lambda/2$). The depth d is not critical when greater than l. This structure closely resembles a corrugated surface, and it can be easily fabricated by placing transverse wires across longitudinal metal vanes.

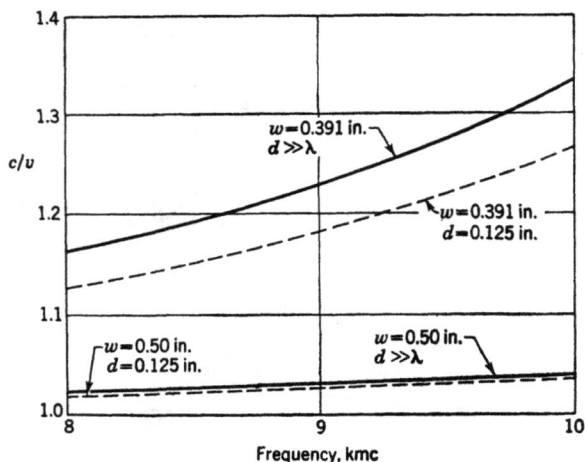

Figure 8-28. Showing the effect of guide depth d on velocity ratio c/v for the structure in Fig. 8-25. (From Hougardy, Periodically Slotted Surface Wave Structure, *Sci. Rept.* 3508/4, Hughes Aircraft Co., Culver City, Calif., August, 1959.)

The structure in Fig. 8-25 can be considered as an array of slotted waveguides of the type studied by Elliott[1] and Hyneman.[2] It also has been analyzed as an infinite plane structure by Hougardy,[3] and some results of his analysis are given in Figs. 8-26 to 8-28.

D. Backward Wave Antennas

Although slow wave antennas are usually endfire, there are a number of slow wave antennas such as the bifilar helix mentioned above which radiate in the back endfire direction. These are sometimes called "backfire" antennas as contrasted to "endfire" antennas, which radiate in the forward direction.

Backward wave structures are characterized by the fact that the phase progression of the radiating wave or the phase progression along the radiating elements is such that a wave travels along the structure toward the feed end often with phase velocity v less than the velocity of light c, i.e., a slow wave ($c/v > 1$). At the same time the group velocity and energy velocity are in the forward direction.

Graphs of phase constant versus frequency are shown in Fig. 8-29 comparing forward and backward wave structures. Forward wave propagation occurs when the phase velocity v ($= \omega/\beta$) is of the same sign as group velocity v_g ($= d\omega/d\beta$). In curve b there is a region between ω_1 and ω_2 in which v and v_g are of opposite sign for $\beta_2 \lesssim \beta < \beta_1$. This is a region of backward wave propagation. Note that for this same range of ω, another value of β

[1] R. S. Elliott, Serrated Waveguide, Part I: Theory, *IRE Trans. Antennas Propagation*, **AP-5**:270, July, 1957.

[2] R. F. Hyneman, Closely Spaced Transverse Slots in Rectangular Waveguide, *IRE Trans. Antennas Propagation*, **AP-7**:335, October, 1959.

[3] R. W. Hougardy, Periodically Slotted Surface Wave Structure, *Sci. Rept.* 3508/4, Hughes Aircraft Co., Culver City, Calif., August, 1959.

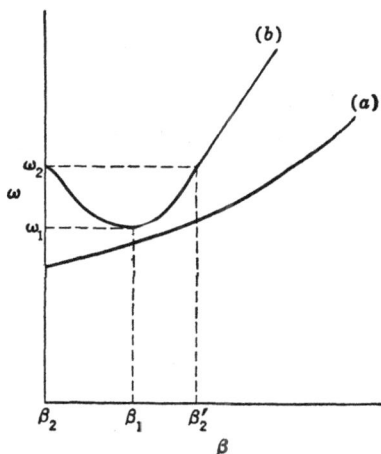

Figure 8-29. Curve (a) shows frequency versus phase constant for a typical forward wave structure. Curve (b) shows frequency versus phase constant for a structure that will support a backward wave.

exists between β_1 and β_2' which corresponds to forward wave propagation. Thus a structure having a β-ω characteristic as shown in Fig. 8-29 can support both a forward and a backward wave simultaneously for $\omega_1 < \omega \lessgtr \omega_2$. This indicates that a test for backward wave propagation is to determine if, for a given value of ω, two values of β can be found which satisfy the characteristic equation of the structure.

Figure 8-29 shows a curve of angular frequency ω versus phase constant β. Since $\omega = kc$, where k is the free-space phase constant and c is the velocity of light then, except for a change in scale, the same curve can be plotted as k versus β. Such a k-β diagram is commonly called a Brillouin diagram and has been used extensively in studying the behavior of periodic structures.[1]

Backward wave propagation appears to be more readily obtainable on periodic structures than on continuous ones. Two continuous structures that support a backward wave are the plasma sheet[2] (see Probs. 6-6 to 6-8) and a circular waveguide with an axial dielectric rod.[3] The latter has an ω-β characteristic like b in Fig. 8-29 for ϵ_r of the rod greater than about 9.4 and for the mode that becomes TE_{11} when the rod completely fills the guide. Note that, in a waveguide structure having this characteristic, the group velocity is zero at $\omega = \omega_1$.

The backward wave in the case of the circular guide is actually a fast wave $(c/v < 1)$, and it is interesting to note that, if a longitudinal slot is cut in the wall of the guide, a beam off axis in the back direction results from the leaky backward wave.

As mentioned above, it appears to be easier to get backward wave propagation from a periodic structure than from a continuous structure. In addition to the corrugated structure in Sec. 6-3 and the bifilar helix of this section, a periodic array of radiating elements fed by a slow wave transmission line can be adjusted to produce a backward wave.[4]

Consider the infinite array in Fig. 8-30. Let θ_n denote the angle at which all elements radiate in phase. The angle θ_n may be imaginary, in which case the antenna beam is in the invisible region (stored energy). A traveling wave producing a beam in the direction θ_n has a phase constant β_n given by

$$\beta_n = k \cos \theta_n \qquad (8\text{-}33)$$

If the separation of elements (or period) of the array is s and the phase

[1] L. N. Brillouin, "Wave Propagation in Periodic Structures," McGraw-Hill, New York, 1946.

[2] A. A. Oliner and T. Tamir, Backward Wave on Isotropic Plasma Slabs, *J. Appl. Phys.*, **33**:231, 1962.

[3] P. J. B. Clarricoats, "Backward Waves in Waveguides Containing Dielectric," Monograph 451E, The Institution of Electrical Engineers, June, 1961.

[4] P. E. Mayes, G. A. Deschamps, and W. T. Patton, Backward-wave Radiation from Periodic Structures and Application to the Design of Frequency Independent Antennas, *Proc. IRE*, **49**:962–963, May, 1961.

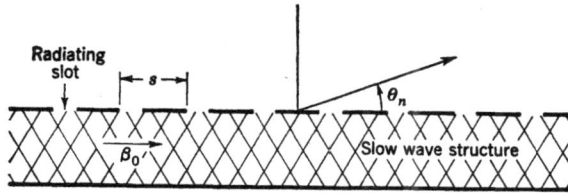

Figure 8-30.   Infinite array of slots fed by a slow wave structure.

constant of the wave feeding the elements is β_0, then from array theory

$$\beta_n = \beta_0 - \frac{2n\pi}{s} \qquad n = 0, \pm 1, \pm 2, \ldots \tag{8-34}$$

where β_n is sometimes referred to as a space harmonic.

Let s be small enough so that β_1 is negative and greater in magnitude than k. This means that $|\cos \theta_1| > 1$, and we say that the beam is in the invisible region. This corresponds to the first triangular region on the k-β diagram of Fig. 8-31, where the broken line represents an idealized k-β_0 curve for the slow wave structure in Fig. 8-30. Actually Fig. 8-31 utilizes the coordinates ks and βs to include the effect of spacing (or period) s. Even though the period of the structure is included, it is customary to refer to a graph such as Fig. 8-31 as a k-β diagram.

If frequency is held constant and s is increased, β_1 becomes equal to $-k$, in which case the structure is an ordinary endfire array at $\theta = 180$ deg or in the backfire direction. This corresponds to point A on the diagram in Fig. 8-31. As s increases still further, β_1 corresponds to a fast wave and the beam swings from 180 deg through broadside (90 deg) and on to endfire (0 deg), which corresponds to point B on the diagram.

For the range 0 to A the structure supports a backward surface wave with phase constant β_1 in addition to the fundamental forward wave β_0. For the range A to B a fast wave occurs and radiates anywhere from 180 to 0 deg depending on s. For the range B to C a forward surface wave exists for β_1 and a backward surface wave for β_2. As s continues to increase, the higher space harmonics β_2, β_3, etc., will produce beams in the visible region.

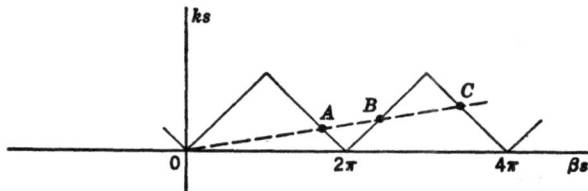

Figure 8-31.   A k-β diagram for a periodically loaded slow wave structure.

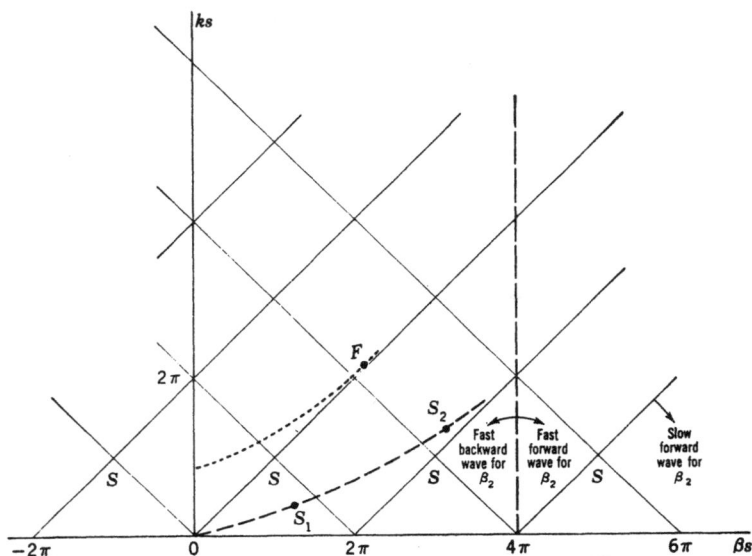

Figure 8-32. k-β diagram showing higher-order space harmonics. Triangular regions labeled S are slow wave regions for all space harmonics.

Figure 8-32 shows a k-β diagram comparing the curves of a slow wave structure (broken line) and a fast wave structure (dotted line). The squares give the fast wave regions of operation, and the triangles labeled S the slow wave regions. For example, a periodically loaded slow wave structure operating at point S_1 has a forward slow wave β_0 and backward slow waves β_1, β_2, etc., whereas operation at point S_2 produces two fast waves with phase constants β_1 and β_2.

For a periodically loaded fast wave structure with k-β_0 curve given by the dotted line, operation at point F would produce visible beams corresponding to phase constants β_0, β_1, and β_2. Whether a beam is backward or forward depends on whether the point of operation is left or right, respectively, of the bisector of the square for the appropriate β (the β_2 case is illustrated in Fig. 8-32).

The k-β_0 curves shown in these examples are idealized. Generally they will not be straight lines, and they may vary rapidly in regions where there is attenuation due to radiation. Each structure will have its own k-β_0 characteristic which, once determined, can be plotted on a k-β diagram to determine what waves will exist on the structure for a particular operating point.

Frequency Independent Antennas

Backward wave radiation appears to be an essential feature of broadband antennas of the type called *frequency independent antennas*. The concept of a frequency independent antenna in which the antenna shape is described entirely by angles was introduced by Rumsey.[1] A successful application of this principle is the log-spiral antenna (frequently called the equiangular spiral antenna) shown in Fig. 8-33.

The general formula for a planar equiangular spiral is

$$\rho = e^{a(\phi+\phi_0)} \tag{8-35}$$

where ρ and ϕ are the usual cylindrical coordinates and a and ϕ_0 are constants. The general form for a three-dimensional frequency independent antenna as given by Rumsey is

$$r = e^{a(\phi+\phi_0)}F(\theta) \tag{8-36}$$

[1] V. H. Rumsey, Frequency Independent Antennas, *IRE Natl. Conv. Record*, part 1, pp. 114–118, 1957.

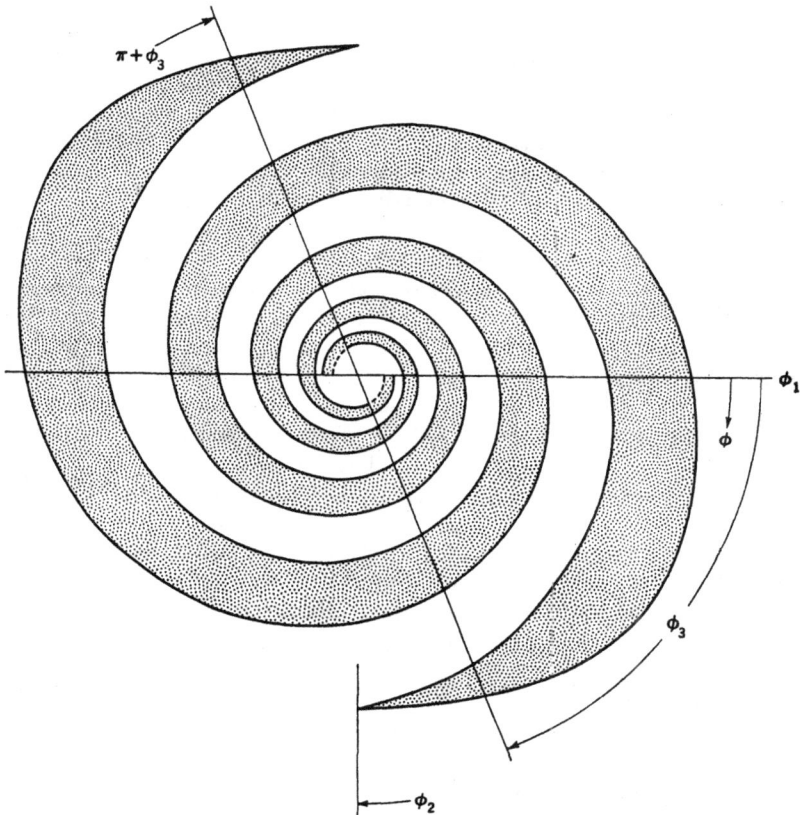

Figure 8-33.   Equiangular spiral antenna.

when r, θ, ϕ are the usual spherical coordinates and $F(\theta)$ is any function of θ.

Assuming a to be positive, ϕ ranges from ϕ_1, which determines the feed point, to some value ϕ_2, which determines the low-frequency limit of the antenna. The antenna in Fig. 8-33 results when ϕ_0 assumes all values from 0 to ϕ_3 and all values from π to $\pi + \phi_3$. For the special case where $\phi_3 = \pi/2$, the spiral is self-complementary (see Sec. 2-21) and its impedance is 60π ohms.

When the two arms in Fig. 8-33 are fed 180 deg out of phase, the planar spiral gives circularly polarized bidirectional radiation along the $\theta = 0$ and $\theta = 180$ deg axes and pattern and impedance bandwidths of greater than 20:1 have been obtained.[1] Unidirectional radiation can be obtained by placing a cavity behind the spiral, but this generally reduces the bandwidth. If it is desired, a null pattern on axis can be obtained by feeding the arms in phase.

Dyson found that unidirectional radiation can be obtained by placing the equiangular spiral on a conical surface.[2] With a two-arm spiral fed at the apex of the cone and with the elements fed 180 deg out of phase, maximum radiation is off the apex. Thus this is a backward wave antenna and the best front-to-back ratio of radiation is found for cone angles of 30 deg or less. As the total cone angle approaches 180 deg, the conical spiral degenerates into a planar spiral with its bidirectional radiation.

To a first approximation one can determine patterns by considering spiral, conical spiral, and helical antennas as arrays of radiating elements. One can look at a generator of the surface containing the spiral or helix and consider a radiating element to exist where the spiral or helix crosses the generator. Assuming a current of the form $I_0 e^{-jkl}$, where k is the usual free-space phase constant and l is distance along the spiral or helix, the relative phases of the elements along the generator can be determined. Better results can be obtained if amplitude is taken into account by introducing attenuation due to radiation. The total structure can be thought of either as an array or as a continuous distribution of generators, whichever is most convenient mathematically. This approach can be used for a spiral on a surface of arbitrary shape. At present the planar spiral is the only case for which a rigorous analysis has been obtained.[3] It was noted that the solution in this case gives an inward traveling wave at infinity when the spiral is excited so as to produce an outward traveling wave at the center.

[1] J. D. Dyson, The Equiangular Spiral Antenna, *IRE Trans. Antennas Propagation*, **AP-7**:181–187, April, 1959.

[2] J. D. Dyson, The Unidirectional Equiangular Spiral Antenna, *IRE Trans. Antennas Propagation*, **AP-7**:329–334, October, 1959.

[3] B. R. S. Cheo, V. H. Rumsey, and W. J. Welch, A Solution to the Frequency-independent Antenna Problem, *IRE Trans. Antennas Propagation*, **AP-9**:527–534, November, 1961.

Log-periodic Antennas

The log periodicity of the radiating elements along a generator of the equiangular spiral antenna has been extended by DuHamel into a whole new class of "log-periodic" antennas.[1] Whereas the frequency independent antenna of Rumsey scales continuously with frequency, i.e., the structures become congruent to themselves except for a possible rotation as frequency is varied, log-periodic structures scale to themselves with logarithmic periodicity.

In symbolic form a log-periodic structure is contrasted to a periodic structure in Fig. 8-34. Each block or cell represents a radiating element or an array of radiating elements. In the periodic structure identical elements are cascaded. In the log-periodic structure each cell scales into the one following it when expanded by a factor τ. The performance of the structure is not uniform with frequency as it is in the continuously scaled frequency independent antennas of Rumsey but rather repeats itself at a log-periodic frequency $f_0\tau^N$, where f_0 is a reference frequency and N is an arbitrary integer. A log-periodic structure can be formed by taking a cell from a successful periodic structure and cascading cells log-periodically.

A cell would be designed to support a backward wave about frequency f_0, and the scaled cells would support the same backward wave about frequencies $f_0\tau^N$. If the frequency ranges of the cells form continuous coverage of the desired band of operation of the antenna, pattern and impedance of the antenna will be essentially uniform with frequency.

In some cases a cell will consist of a single radiating element such as a dipole. In this case the cells should be excited so as to form a backward wave along the cells.

A number of log-periodic antennas have been studied, and these are discussed in the literature.[2] These consist of helices and tapered helices feeding arrays of monopoles, zigzag antennas, and dipole arrays. One of

[1] R. H. DuHamel and D. E. Isbell, Broadband Logarithmically Periodic Antenna Structures, *IRE Natl. Conv. Record*, part 1, pp. 119–128, 1957.

[2] R. S. Elliott, A View of Frequency Independent Antennas, *Microwave J.*, V:61–68, December, 1962; P. E. Mayes, G. A. Deschamps, and W. T. Patton, Backward-wave Radiation from Periodic Structures and Application to the Design of Frequency-independent Antennas, *Univ. Illinois Antenna Lab. Tech. Rept.* 60, Urbana, Ill., December, 1962; R. Mittra and K. E. Jones, On Continuously Scaled and Log-periodic Structures, *Spec. Doc.* FR 63-14-107, Hughes Aircraft Co., Fullerton, Calif., July, 1963.

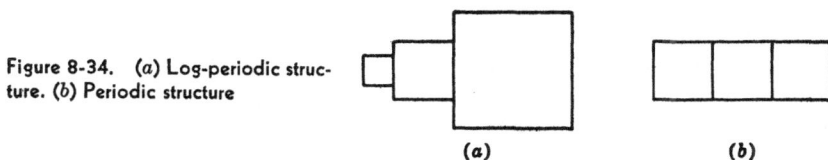

Figure 8-34. (a) Log-periodic structure. (b) Periodic structure

(a)   (b)

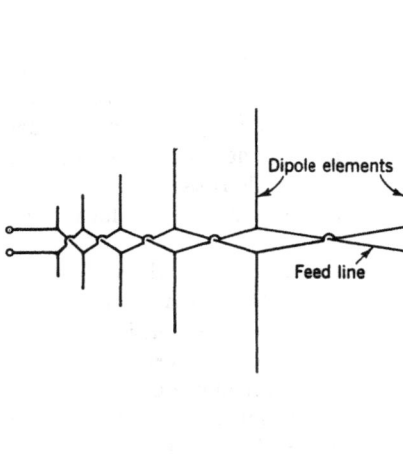

Figure 8-35. Log-periodic dipole array.

the simplest and most useful is the log-periodic dipole array developed by Isbell.[1] This array is illustrated in Fig. 8-35. The element of a cell in this case is a conventional half-wave dipole. Alternating the feed points along the two-wire feed line introduces additional phase delay which has the effect of slow wave excitation. This results in a backward wave on the structure, and radiation is off the apex. The active or radiating region of such a structure is in the region where the dipoles are $\lambda/2$ long. Patterns of this antenna have a very good front-to-back ratio with the usual dipole polarization. Antennas of this type can be readily arrayed.

The success of the broadband antennas described above appears to be due to a great extent to the backward wave operation, which causes radiation toward the smaller end of the structure, thus minimizing scattering effects of the structure. The structures are inefficient in the sense that only a portion of the structure is active, or radiating, at a given frequency. However, it is the ability of this radiating region to move along the structure with frequency and retain essentially the same electrical characteristics that gives the broadband operation.

The active region may consist of essentially a single cell or a number of cells depending on the structure. For the log-periodic dipole array where each cell is a dipole, the active region will consist of a number of cells.

Currents in the active region of the structure decay quite rapidly owing to radiation; hence end effects are usually negligible until the frequency is reduced to the point where the active region approaches the large end of the structure. This constitutes the low-frequency cutoff. High-frequency

[1] D. E. Isbell, Log-periodic Dipole Arrays, *IRE Trans. Antennas Propagation*, **AP-8**:260–267, May, 1960.

cutoff occurs when the separation of the feed terminals becomes an appreciable fraction of a wavelength.

8-4. FAST WAVE ANTENNAS

Fast wave antennas are characterized by a traveling wave distribution with phase velocity v greater than the velocity of light c in free space (that is, $c/v < 1$). For a uniform structure with no dielectric or conductor losses, the fields decay exponentially in the direction of propagation due to radiation losses. There is leakage of power from the guiding structure associated with a power flow normal to the surface of the structure; hence fast wave structures are frequently called leaky wave structures. This section describes a number of practical fast wave antennas. Both line sources and planar sources are considered.

A. Gain and Radiation Characteristics of Fast Wave Antennas

Whereas slow wave antennas $(c/v \gtrless 1)$ are generally endfire, near endfire, or backward wave radiators, fast wave antennas $(c/v < 1)$ have a radiated beam at some angle with respect to the antenna axis. The angle of maximum radiation θ_m is determined by the phase velocity (see Sec. 2-19) and is given by

$$\theta_m \approx \cos^{-1}\frac{c}{v} \qquad (8\text{-}37)$$

A practical upper limit for θ_m is about 85 deg, and the lower limit is about 10 deg. These limits depend on the length of the structure as well as the mode of operation.

Most of the practical fast wave antennas are perturbed waveguide structures such as the slotted and holey guides described in Chaps. 5 and 6. The upper limit on θ_m for such structures is determined by the cutoff of the perturbed guide. That is, at cutoff $v \rightarrow \infty$ and $c/v \rightarrow 0$. Hence broadside radiation from a leaky wave antenna would require the guiding structure to be operating at cutoff. The practical limit appears to be a beam about 5 deg from broadside. This can be obtained with a slotted TM waveguide as in Fig. 5-10, for example. Actually to obtain broadside radiation one must use an in-phase array of elements. This can be accomplished with a waveguide structure by properly spacing discrete elements along the waveguide.[1]

The lower limit for the angle of radiation is not too well defined. For a given traveling wave antenna the beam position θ_m reaches some minimum value as c/v is increased from something less than unity (fast wave operation) to something greater than unity (slow wave operation). For a uniform

---

[1] See discussion on periodic structures in the previous section. See, also, Jasik, *op. cit.*, chap. 9.

structure this minimum value depends on the length of the structure and the mode of operation. For example, a relatively short source ($<5\ \lambda$) may have a significantly larger θ_m for slow wave operation ($c/v > 1$) than a larger source ($>10\ \lambda$) operating as a fast wave structure (say $c/v = 0.95$). Usually, beam angles less than 10 deg correspond to slow wave operation.

For a uniform line source of transverse current elements the directivity is

$$D \approx 4\frac{L}{\lambda} \tag{8-38}$$

where L is the length of the source. Equation (8-38) is essentially independent of the angle of radiation. For longitudinal current elements the gain is approximately $2L/\lambda$ at broadside and decreases monotonically as endfire operation is approached. The gain in the endfire direction will be zero because there will be no radiation on axis with longitudinal current elements.

In terms of the half-power beamwidths, an expression that is useful for a high-gain area source is (see Sec. 8-3A)

$$G \approx \frac{27,000}{\theta_E \theta_H} \tag{8-39}$$

where θ_E and θ_H are half-power beamwidths in degrees in the **E** and **H** planes, respectively.

Fast wave antennas lend themselves very nicely to shaped beam and low sidelobe designs. The fast wave structures of Chaps. 5 and 6 make practical radiators in which relatively independent control of amplitude and phase can be obtained, and the synthesis methods of Chap. 3 are readily applied.

An interesting feature of fast wave sources is that the beam can be scanned by changing phase velocity (see Eq. 8-37). This can be accomplished mechanically by varying the geometry of the structure in some way that changes the phase velocity (moving the back wall in the structure of Fig. 5-1, for example). It can be accomplished electrically by changing the permeability or permittivity of the material filling the structure (changing a d-c magnetic field in a ferrite material, for example). It also can be accomplished electrically by varying frequency.

When a fast wave structure is frequency scanned, the scan angle is limited by cutoff at the low-frequency end and by excitation of higher-order modes at the high-frequency end. A scan sector of 45 deg is not difficult to obtain, and about 60 deg would be the maximum that one could expect to obtain in practice. The dispersion of the structure (change in beam angle/change in frequency) can be controlled by dielectric loading.

If the amplitude distribution along a frequency scanned source does not change significantly over the frequency range, the beamwidth of the

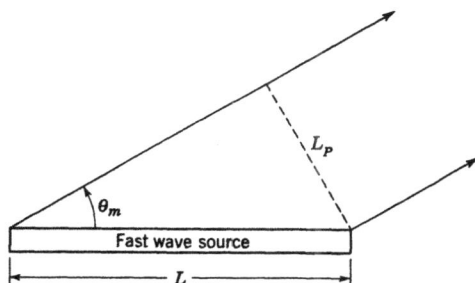

Figure 8-36. Traveling wave source of length L with projected aperture of length L_P.

pattern will be nearly constant over the scan sector. This can be explained by means of Fig. 8-36. The projected aperture is given by

$$L_P = L \sin \theta_m = L(1 - \cos^2 \theta_m)^{1/2} \qquad (8\text{-}40)$$

But $\cos \theta_m = c/v = \lambda/\lambda_g$, where λ_g is the wavelength along the structure; hence Eq. (8-40) can be written

$$\frac{L_P}{\lambda} = \frac{L}{\lambda}\left[1 - \left(\frac{\lambda}{\lambda_g}\right)^2\right]^{1/2} = \frac{L}{\lambda_c} \qquad (8\text{-}41)$$

where λ_c is the cutoff wavelength of the structure. Thus the projected aperture depends only on the antenna geometry and is independent of frequency. In practice the amplitude distribution will vary somewhat with frequency. Also, the projected aperture picture of Fig. 8-36 breaks down as endfire operation is approached. However, the effect can be observed. Practical frequency scanned antennas of this type do exhibit relatively constant beamwidths.[1]

B. Fast Wave Line Sources

Most fast wave antenna structures are perturbations of conventional waveguides. These perturbations may take the form of long slots or closely spaced holes and slots in the walls of closed guides. They also may take the form of some type of radiating discontinuity in an open waveguide such as a trough line. The object is to produce a continuous (or nearly continuous) leakage of energy from the structure. The following sections will describe some practical fast wave antennas.

TE Long Slots

A long, narrow slot in the wall of a TE-excited rectangular or circular guide is quite useful as a fast wave antenna. The slot must be positioned so that wall currents are perpendicular to it. Slotted waveguide cross sec-

---

[1] R. C. Honey, A Flush-mounted Leaky-wave Antenna with Predictable Patterns, *IRE Trans. Antennas Propagation*, **AP-7**:320, 1959.

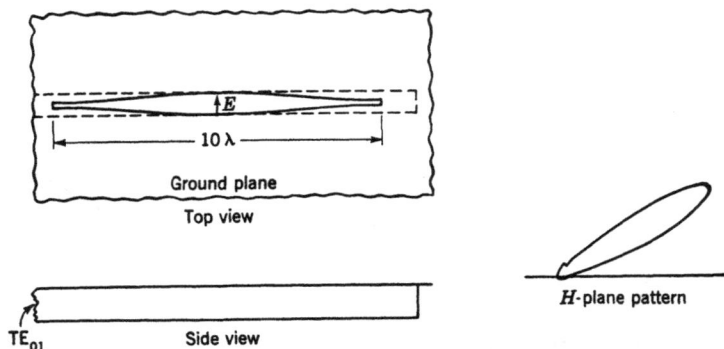

Figure 8-37. TE long slot with tapered aperture.

tions are illustrated in Figs. 5-1 and 5-7. The field in the slot is transverse electric; thus the radiation is that of a longitudinal magnetic-current source.

Design data in the form of equations and graphs are given in Chap. 5. The propagation constant γ ($= \alpha + j\beta$) can be varied over a wide range. The attenuation constant α and the phase constant β are not independent of each other, but within limits one can adjust the cross section of the slotted guide to keep one quantity essentially constant while varying the other. For example, tapering the slot width as illustrated in Fig. 8-37 can give an approximation to a gaussian amplitude taper without seriously altering the phase velocity of the traveling wave.

In practice TE slots greater than 5 λ long give beams from about 20 to about 75 deg from endfire for air-filled guides. Dielectric loading can be used to get c/v near unity to lower the beam still further, but it would be difficult in most practical applications to get the beam lower than about 10 deg. This would be getting into the domain of slow wave operation (see Sec. 8-3).

TM Long Slots

Good radiation characteristics can be obtained from long slots in TM-excited rectangular and circular waveguides. Cross sections of such antennas are shown in Figs. 5-10 and 5-13. In this case a slot is placed in the wall of the guide so that currents are parallel to the slot. For a guide with $\epsilon_r < 2$, the field in the slot is principally transverse magnetic and radiation is that of a longitudinal electric-current source. For $\epsilon_r \gtrsim 2$, the field in the aperture is found to be predominantly longitudinal electric[1] and the radiation is that of transverse magnetic-current elements.

Design data for TM-excited long slot antennas are given in Chap. 5

[1] J. N. Hines, V. H. Rumsey, and C. H. Walter, Traveling Wave Slot Antennas, *Proc. IRE*, **41**:1624, 1953.

in the form of equations and graphs relating propagation constant to the geometry. The propagation constant can be varied over quite a wide range, and considerable pattern control can be exercised by means of the theory in Chaps. 2 and 3.

The air-filled TM slotted guide gives beams from about 20 deg from endfire to within about 5 deg of broadside for $L > 5\lambda$. For dielectric loading ($\epsilon_r \gtrsim 2$) an element of slot has the radiation pattern of a transverse magnetic-current element and a beam at less than 10 deg from endfire is possible. For beam scanning applications the dielectric-loaded TM guide is the most versatile of the long slot structures.

From the standpoint of power-handling capability, the TM slot is superior to the TE slot. The data in Chap. 5 show that, for a given attenuation constant, the TM slot is much wider than the TE slot. For low rates of radiation the slot width in the TE case may be only a few thousandths of an inch at X band. A TM slot with the same attenuation constant would have a slot width approximately an order of magnitude greater. For extremely low rates of radiation it is probably best to go to a holey waveguide structure (see Fig. 5-4, for example).

Hybrid Mode Long Slots

In some cases the fields of perturbed waveguides are not pure TE or TM. For example, the slotted waveguide with TM excitation described above is a hybrid mode structure when filled with dielectric having $\epsilon_r \gtrsim 2$. In this case the predominant field in the aperture is E_z (see Fig. 8-38). This corresponds to radiation from transverse magnetic-current elements which are isotropic in the xz plane. The TM_{11}-excited square waveguide with long slot fits this hybrid category and can be used as a fast wave or a slow wave structure with $\epsilon_r \gtrsim 2$. The polarization is parallel to the xz plane.

TE excitation of dielectric-filled slotted guides also can provide hybrid operation. Modes corresponding to $H_x = 0$ and $E_x = 0$ exist (see Sec. 5-4C), and fast wave operation ($c/v < 1$) appears to be possible for waveguide width w sufficiently small. This hybrid operation, however, is usually associated with slow wave structures (see Sec. 8-3).

Figure 8-38.   Coordinate system for slotted waveguide.

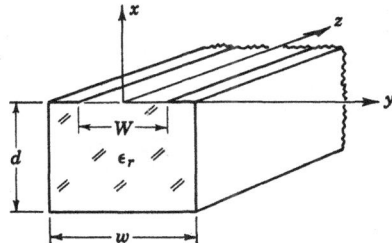

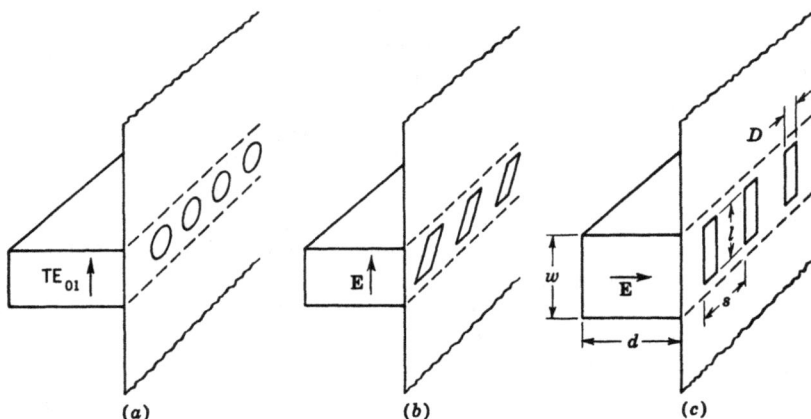

Figure 8-39. Slots and holes in rectangular waveguides.

Holey Waveguides

Closely spaced holes and slots in conventional waveguides provide another means for obtaining fast wave line sources. Rectangular waveguides with TE excitation are chiefly used, and some possible fast wave sources are illustrated in Fig. 8-39. Holes and longitudinal or slanted slots in the narrow wall have the radiation characteristics of longitudinal magnetic-current elements, whereas transverse slots in the wide wall radiate as transverse magnetic-current elements.

These structures are perturbations of the closed, rectangular waveguide with TE_{01} excitation. An analysis based on transverse resonance including design data is given in Sec. 5-2B for a in Fig. 8-39. The same analysis can be applied to the structure of Fig. 8-39b for any hole shape for which an aperture admittance can be determined. Small aperture theory[1] is useful in many cases. Slanted slots as in Fig. 8-39b are generally used as resonant elements in a waveguide-fed array of discrete elements[2] rather than closely spaced in a leaky wave type of structure.

Analyses and design data for the structure in Fig. 8-39c are available in the literature.[3] Design data from Hyneman are given in Fig. 8-40.

[1] H. A. Bethe, Lumped Constants for Small Irises, MIT *Radiation Lab. Rept.* 43-22, March, 1943; N. Marcuvitz, Waveguide Circuit Theory: Coupling of Waveguides by Small Apertures, *Microwave Res. Inst. Rept.* R-157-47, PIB-106, Polytechnic Institute of Brooklyn, 1947; A. A. Oliner, Equivalent Circuits for Small Symmetrical Longitudinal Apertures and Obstacles, *Microwave Res. Inst. Rept.* R-717-59, PIB-645, Polytechnic Institute of Brooklyn, 1959.

[2] Jasik, *op. cit.*, chap. 9.

[3] R. S. Elliott, Serrated Waveguide, Part I: Theory, *IRE Trans. Antennas Propagation*, AP-5:270, 1957; K. C. Kelly and R. S. Elliott, Serrated Waveguide, Part II: Experiment, *IRE Trans. Antennas Propagation*, AP-5:276, 1957; R. F. Hyneman, Closely-spaced Transverse Slots in Rectangular Waveguide, *IRE Trans. Antennas Propagation*, AP-7:335, 1959.

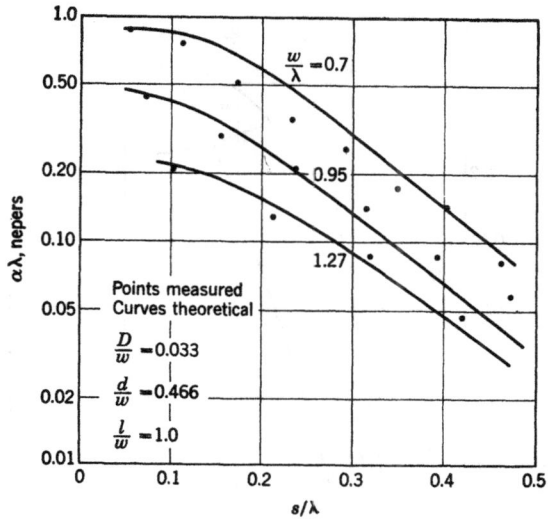

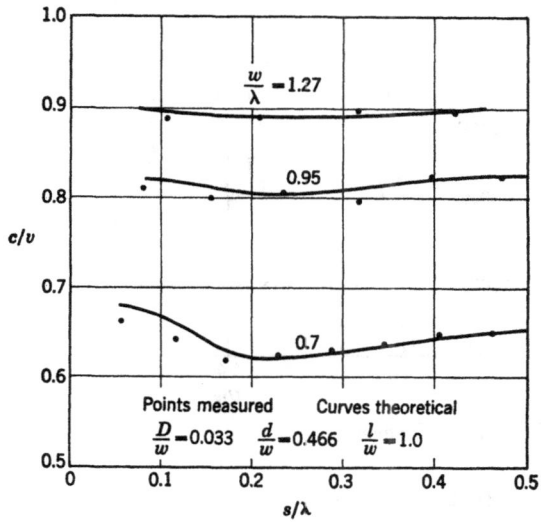

Figure 8-40. Design data for the slotted waveguide of Fig. 8-39c. (From Hyneman, Closely-spaced Transverse Slots in Rectangular Waveguide, IRE Trans. Antennas Propagation, AP-7:335, 1959.)

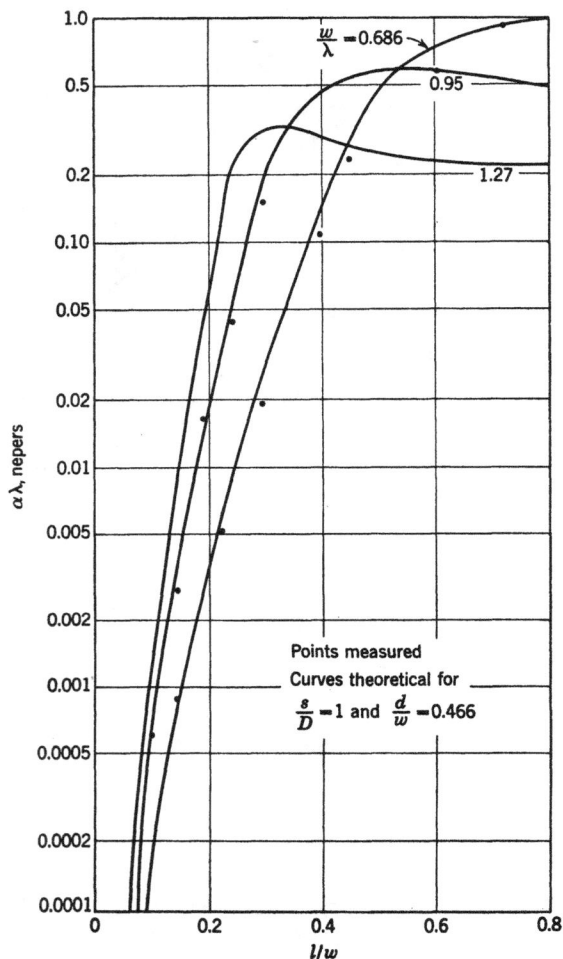

Figure 8-40 (Continued)

In all three cases of Fig. 8-39 the velocity ratio c/v is principally determined by the width of the TE_{01} waveguide and the attenuation constant is principally determined by the size and spacing of the slots and holes. The attenuation and phase constants are not independent, however.

The closely spaced holes and slots are useful in designs where low rates of attenuation are required. Proper design and placement of closely spaced small apertures also can provide any desired polarization.[1] An illustration of an elliptically polarized leaky wave antenna is given in Fig. 8-41.

[1] W. J. Getsinger, Elliptically Polarized Leaky Wave Array, *IRE Trans. Antennas Propagation*, **AP-10**(2): 165–171, March, 1962.

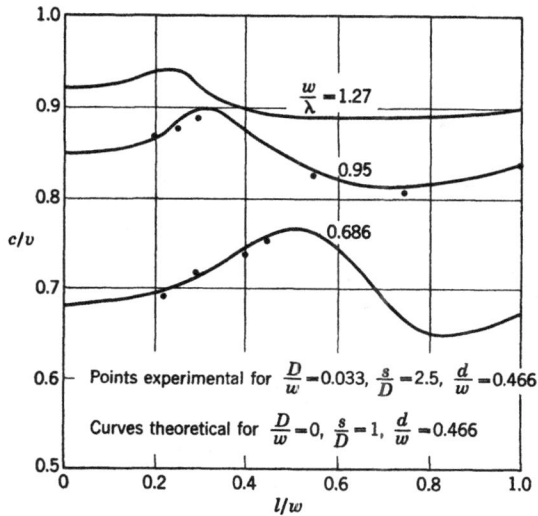

Figure 8-40 (Continued)

In addition to the waveguide sources described above, a slotted coaxial line[1] also has interesting and useful properties. A leaky TEM-mode coaxial line is less dispersive than a leaky rectangular or circular waveguide; hence the beam position can be held reasonably constant over a frequency band. Leakage can be controlled by slot size and displacement of linear conductor from the center.[1]

Trough Guide

The trough guide was mentioned as a slow wave structure in Sec. 8-3B. It is also useful as a fast wave antenna.[2] A sketch of the cross section of

[1] R. J. Stegen and R. H. Reed, Arrays of Closely Spaced Non-resonant Slots, *IRE Trans. Antennas Propagation,* **AP-2** :109, 1954.

[2] W. Rotman and S. J. Naumann, The Design of Trough Wave-guide Antenna Arrays, *Air Force Cambridge Res. Center Rept.* AFCRC-TR-58-154, Bedford, Mass., 1958; W. Rotman and A. A. Oliner, Asymmetrical Trough Waveguide Antennas, *IRE Trans. Antennas Propagation,* **AP-7** :153, 1959; W. Rotman and N. Karas, Trough Wave-guide Radiators with Periodic Posts, *Air Force Cambridge Res. Center Rept.* AFCRC-TR-58-356, Bedford, Mass., 1958; Jasik, *op. cit.,* chap. 16.

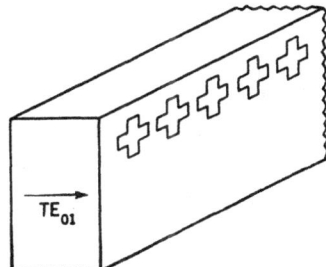

Figure 8-41. Crossed-slot leaky wave antenna for elliptical polarization.

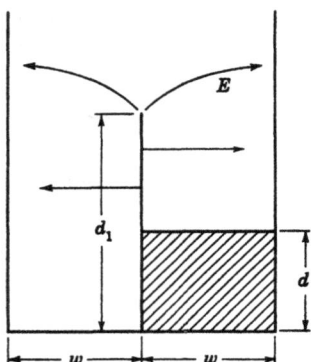

Figure 8-42. Asymmetrical trough guide. Asymmetry is produced by metal insert of height d.

an asymmetrical trough guide is shown in Fig. 8-42. Design data for fast wave operation are given in Figs. 8-43 and 8-44.

A wide range of attenuation and phase constants can be obtained by varying the asymmetry (either continuously or periodically) and by introducing serrations in the center fin.[1] It is possible to obtain a main beam anywhere from endfire through and just beyond broadside. Data for a symmetrical trough guide with serrated fin are given in Fig. 8-45. In this case either fast or slow wave propagation is possible.

[1] A. A. Oliner and W. Rotman, Periodic Structures in Trough Waveguides, *IRE Trans. Microwave Theory Tech.*, **MTT-7**:134, 1959.

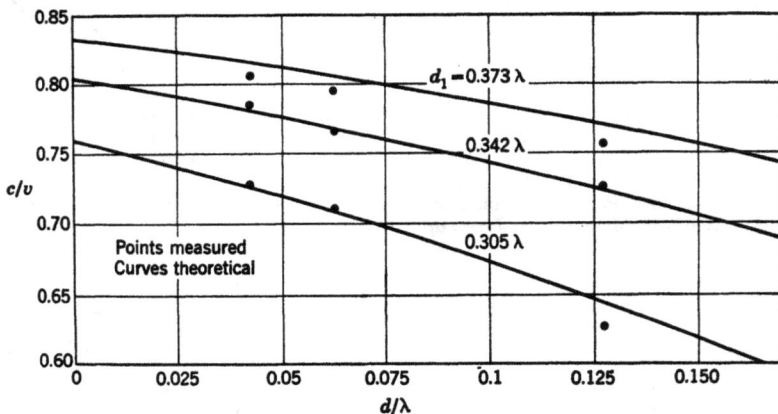

Figure 8-43. Velocity ratio c/v for the trough guide in Fig. 8-42. (From Rotman and Oliner, Asymmetrical Trough Waveguide Antennas, IRE Trans. Antennas Propagation, **AP-7**:153, 1959.)

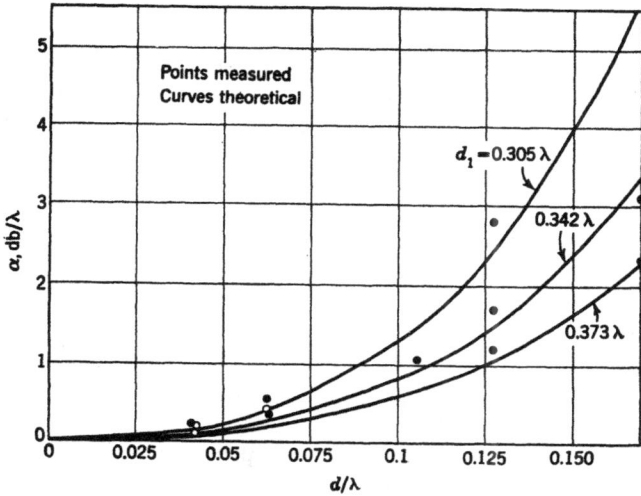

Figure 8-44.   Attenuation constant α for the trough guide of Fig. 8-42. Note that α in db/λ divided by 8.686 gives α in nepers/λ. (From Rotman and Oliner, Asymmetrical Trough Waveguide Antennas, IRE Trans. Antennas Propagation, AP-7:153, 1959.)

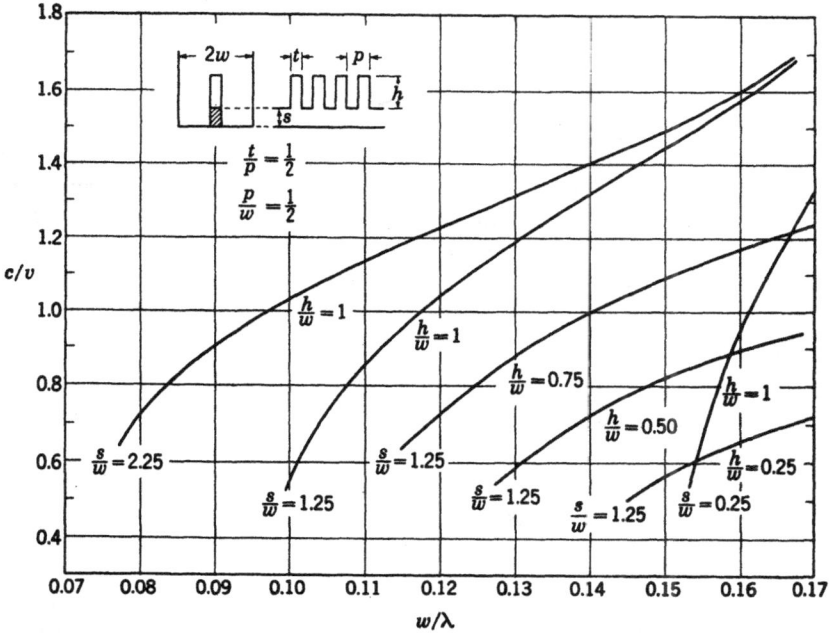

Figure 8-45.   Relative phase velocity c/v for symmetrical trough guide with serrated metal fin. (From Oliner and Rotman, Periodic Structures in Trough Waveguides, IRE Trans. Microwave Theory Tech., MTT-7:134, 1959.)

C. Fast Wave Planar Sources

Planar sources can be obtained by arraying line sources or by using a parallel plate structure in which one plate (or sometimes both plates) has radiating elements such as slots or holes. The arraying of traveling wave sources is discussed in Sec. 8-6. The present section describes some of the parallel plate structures.

Holey Plate Structures

A parallel plate structure with one plate full of holes is described in Chap. 6 (see Fig. 6-6). With the electric field parallel to the plates and transverse to the direction of propagation the analysis becomes that of a rectangular waveguide with holes in a narrow wall and radiating into a parallel plate region. Design data are obtained by transverse resonance and presented in Chap. 6.

The velocity ratio c/v depends primarily on the plate separation (TE_{01} mode), and the attenuation depends primarily on the hole size and spacing. The radiation is that of longitudinal magnetic-current elements. Beam angles can be obtained from about 20 to 80 deg with respect to the plane of the aperture for an air-filled parallel plate region greater than 5λ long. Dielectric loading can be used to bring the beam closer to endfire. Dielectric loading also increases the dispersion of the structure, which may be desirable for frequency scan applications.

Inductive Grid Structure

The holey plate described above is effectively an inductive surface. Another form of inductive surface is the inductive grid in Fig. 6-11. A grid of this type can be formed by round conducting wires or flat conducting strips. The electric field is parallel to the elements of the grid, and radiation is that of longitudinal magnetic-current elements.

Design data from an analysis by transverse resonance are given in Sec. 6-2C. The inductive grid provides a greater range of attenuation constant α than the holey plate structure, but the holey plate structure can be made more nearly isotropic in the plane of the structure.

Capacitive Grid Structure

A TE_{01} parallel plate structure with one plate made of longitudinal conducting strips is quite useful as a fast wave antenna (see Fig. 6-17). An electric field exists across the gaps between the strips; hence the strips form a capacitive grid. The radiation is that of longitudinal magnetic-current elements.

Design data from an analysis based on transverse resonance are given in Sec. 6-2D. This structure is not as good for low rates of radiation as the inductive grid.

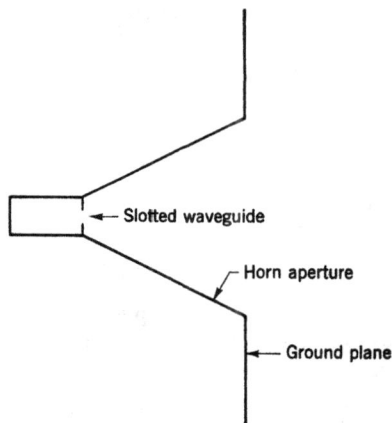

Figure 8-46.   Cross section of slorn antenna.

Slotted waveguide

Horn aperture

Ground plane

D. Line Source with Horn Aperture

In addition to arraying line sources or using leaky parallel plate structures, area sources also can be obtained by placing a horn aperture on a fast wave line source. The cross section of a slot antenna with horn (called a slorn) is shown in Fig. 8-46.

A longitudinal section showing how such an antenna might be flush mounted is shown in Fig. 8-47. When the fields radiated from the line source are viewed as a portion of a plane wave at angle θ_m, the end plates should be slanted to avoid reflecting a portion of this wave.

The propagation constant for the line source with horn depends not only on the geometry of the line source but also on the flare angle and length of the horn (see Fig. 8-46). The attenuation constant is influenced more than the phase constant by the horn aperture; thus varying the horn length provides a practical means of tapering the amplitude distribution at the aperture of the slorn antenna.

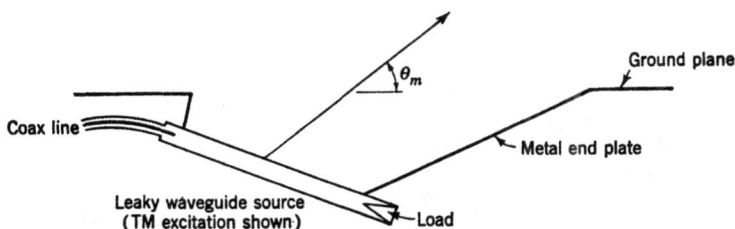

Ground plane

θ_m

Coax line

Metal end plate

Leaky waveguide source
(TM excitation shown)

Load

Figure 8-47.   Longitudinal section of flush-mounted slorn.

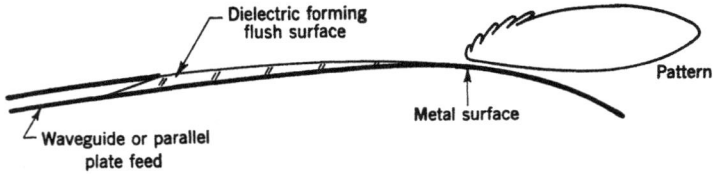

Figure 8-48. Flush-mounted antenna on a curved surface.

8-5. CURVED TRAVELING WAVE ANTENNAS

In many practical applications it is desirable or even necessary to place an antenna on a curved surface. For fast wave structures one can usually design the antenna to compensate for the curvature. In some cases the curvature can be used to advantage in obtaining a specified beam shape for either a slow or fast wave structure.

A. Curved Slow Wave Structures

Lossless, linear, slow wave structures can guide a wave along the surface of the structure without attenuation. Curving a slow wave structure introduces leakage attenuation. For gradual curvature this generally has the effect of filling in nulls and increasing the width of the beam. It provides a method of controlling the source distribution of the antenna and hence the radiation pattern. An example of a practical slow wave antenna on a curved surface is shown in Fig. 8-48.

Beam shaping with curved, slow wave antennas is usually on a cut-and-try basis, since the shape and extent of the conducting surface beyond the end of the antenna may have considerable effect on the pattern. One can, however, obtain approximate results by using an optical synthesis method (see Sec. 3-2E) as suggested by Fig. 8-49. The power from an element of source Δs is associated with a wedge-shaped region of space $\Delta\theta$. One can use as an intermediate step the equivalent linear source illustrated in Fig. 8-49. Approximate relations between the distributions along curved and equivalent linear sources are developed in the next section.

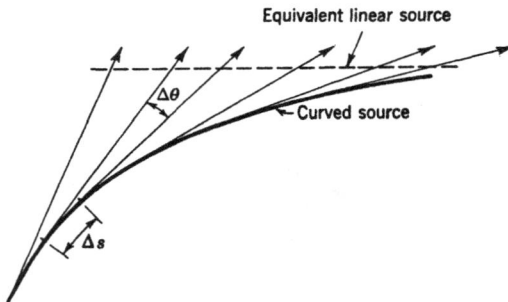

Figure 8-49. Optical representation of a curved source.

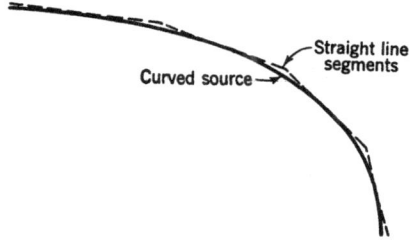

Figure 8-50. Representing curved source by straight-line segments.

If an equivalent linear source can be found, the pattern can be determined by the usual methods (see Chap. 2). The pattern also can be determined by representing the curved source as a series of straight-line segments as shown in Fig. 8-50. The radiation pattern can be obtained by properly superimposing the patterns of the linear elements. If the linear elements are identical, the resultant pattern is obtained by rotating the pattern of one element the correct amount for each element position and superimposing the patterns with proper amplitudes and phases.

The propagation constant for a slow wave structure such as a grounded sheet of dielectric or a corrugated surface depends on the curvature of the surface. The effect of radius of curvature can be determined from an analysis of azimuthal surface waves on a circular cylinder.[1] For the dielectric-coated cylinder shown in Fig. 8-51, the propagation constant may be determined from

$$\frac{J'_{\gamma a}(\sqrt{\epsilon_r}\,kb)}{N'_{\gamma a}(\sqrt{\epsilon_r}\,kb)} = \frac{\sqrt{\epsilon_r}\,\dfrac{J_{\gamma a}(\sqrt{\epsilon_r}\,ka)}{J'_{\gamma a}(\sqrt{\epsilon_r}\,ka)}\dfrac{H_{\gamma a}^{(2)'}(ka)}{H_{\gamma a}^{(2)}(ka)} - 1}{\sqrt{\epsilon_r}\,\dfrac{N_{\gamma a}(\sqrt{\epsilon_r}\,ka)}{N'_{\gamma a}(\sqrt{\epsilon_r}\,ka)}\dfrac{H_{\gamma a}^{(2)'}(ka)}{H_{\gamma a}^{(2)}(ka)} - 1}\,\frac{J'_{\gamma a}(\sqrt{\epsilon_r}\,ka)}{N'_{\gamma a}(\sqrt{\epsilon_r}\,ka)} \qquad (8\text{-}42)$$

[1] R. S. Elliott, Azimuthal Surface Waves on Circular Cylinders, *J. Appl. Phys.*, **26**(4):368–376, April, 1955.

Figure 8-51.   Dielectric-clad cylinder.

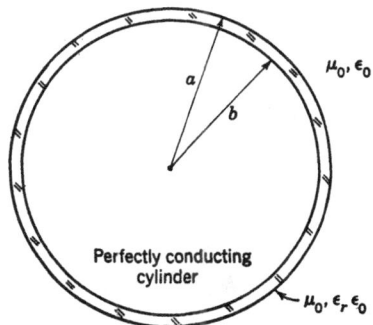

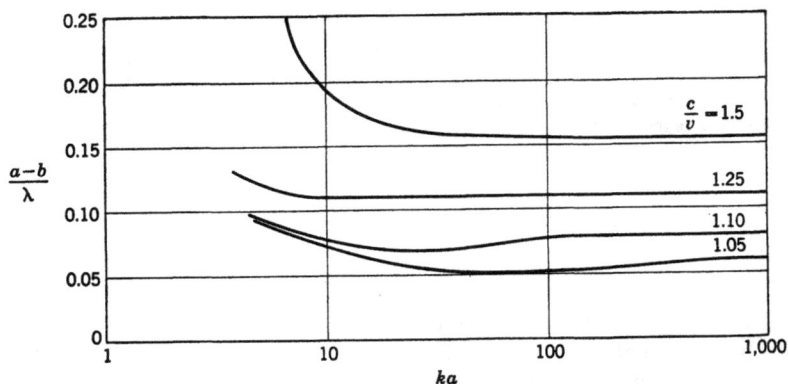

Figure 8-52. Velocity ratio c/v as a function of dielectric thickness and cylinder radius for $\epsilon_r = 4$ and TM wave. (From Elliott, Azimuthal Surface Waves on Circular Cylinders, J. Appl. Phys., **26**(4): 368–376, April, 1955.)

for TM waves and

$$\frac{J_{\gamma a}(\sqrt{\epsilon_r}\,kb)}{N_{\gamma a}(\sqrt{\epsilon_r}\,kb)} = \frac{\dfrac{1}{\sqrt{\epsilon_r}}\dfrac{J_{\gamma a}(\sqrt{\epsilon_r}\,ka)}{J'_{\gamma a}(\sqrt{\epsilon_r}\,ka)}\dfrac{H_{\gamma a}^{(2)\prime}(ka)}{H_{\gamma a}^{(2)}(ka)} - 1}{\dfrac{1}{\sqrt{\epsilon_r}}\dfrac{N_{\gamma a}(\sqrt{\epsilon_r}\,ka)}{N'_{\gamma a}(\sqrt{\epsilon_r}\,ka)}\dfrac{H_{\gamma a}^{(2)\prime}(ka)}{H_{\gamma a}^{(2)}(ka)} - 1}\frac{J'_{\gamma a}(\sqrt{\epsilon_r}\,ka)}{N'_{\gamma a}(\sqrt{\epsilon_r}\,ka)} \qquad (8\text{-}43)$$

for TE waves.

For ka large and $\beta > k$, γ $(=\alpha + j\beta)$ is principally imaginary and a first approximation to the phase constant β can be found from

$$\frac{J'_{\beta a}(\sqrt{\epsilon_r}\,kb)}{N'_{\beta a}(\sqrt{\epsilon_r}\,kb)} = \frac{\sqrt{\epsilon_r}\dfrac{J_{\beta a}(\sqrt{\epsilon_r}\,ka)}{J'_{\beta a}(\sqrt{\epsilon_r}\,ka)}\dfrac{N'_{\beta a}(ka)}{N_{\beta a}(ka)} - 1}{\sqrt{\epsilon_r}\dfrac{N_{\beta a}(\sqrt{\epsilon_r}\,ka)}{N'_{\beta a}(\sqrt{\epsilon_r}\,ka)}\dfrac{N'_{\beta a}(ka)}{N_{\beta a}(ka)} - 1}\frac{J'_{\beta a}(\sqrt{\epsilon_r}\,ka)}{N'_{\beta a}(\sqrt{\epsilon_r}\,ka)} \qquad (8\text{-}44)$$

for TM waves and

$$\frac{J_{\beta a}(\sqrt{\epsilon_r}\,kb)}{N_{\beta a}(\sqrt{\epsilon_r}\,kb)} = \frac{\dfrac{1}{\sqrt{\epsilon_r}}\dfrac{J_{\beta a}(\sqrt{\epsilon_r}\,ka)}{J'_{\beta a}(\sqrt{\epsilon_r}\,ka)}\dfrac{N'_{\beta a}(ka)}{N_{\beta a}(ka)} - 1}{\dfrac{1}{\sqrt{\epsilon_r}}\dfrac{N_{\beta a}(\sqrt{\epsilon_r}\,ka)}{N'_{\beta a}(\sqrt{\epsilon_r}\,ka)}\dfrac{N'_{\beta a}(ka)}{N_{\beta a}(ka)} - 1}\frac{J'_{\beta a}(\sqrt{\epsilon_r}\,ka)}{N'_{\beta a}(\sqrt{\epsilon_r}\,ka)} \qquad (8\text{-}45)$$

for TE waves.

Equations (8-44) and (8-45) reduce to Eqs. (6-67) and (6-9), respectively, as b goes to infinity; i.e., the cylinder approaches the plane sheet case. Results from Eqs. (8-44) and (8-45) are shown in Figs. 8-52 and 8-53 for $\epsilon_r = 4$.

[Recently Baechle[1] has obtained essentially the same numerical results as Elliott[2] by using a coordinate transformation and by assuming series representations for field quantities, propagation constant, and surface impedance. Assuming a separable solution to the wave equation reduces the problem to one of integrating a sequence of easily solved differential equations. Solution of these equations and application of appropriate boundary conditions determine the constants in the series expansions. Using only the first two terms, the resulting expression for the propagation constant is

$$\gamma \approx \gamma_0 \left(1 + \frac{cK}{2}\right)$$

where K is the curvature of the surface, γ_0 is the propagation constant of the plane structure, and the coefficient c depends upon the type of surface wave structure. For a TM corrugated surface with air dielectric,

$$\gamma \approx k(1 + \tan^2 kd)^{1/2}\left[1 + \frac{(1 - kd\tan^3 kd)K}{2k\tan(kd)(1 + \tan^2 kd)}\right]$$

where the corrugations are of depth d and negligible width. For a TM

---

[1] J. R. Baechle, "The Effects of Curvature on Traveling Wave Antennas," Ph.D. Dissertation, The Ohio State University, Columbus, December, 1963.
[2] Elliott, *op. cit.*

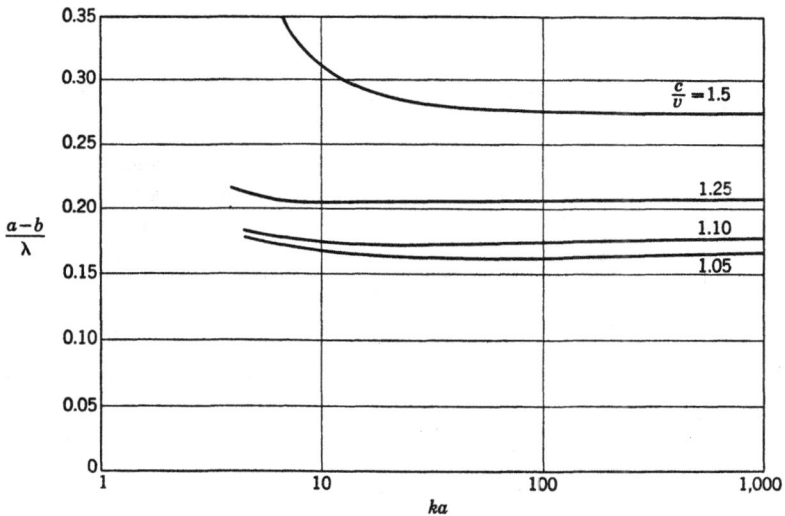

Figure 8-53.   Velocity ratio c/v as a function of dielectric thickness and cylinder radius for $\epsilon_r = 4$ and TE wave. (From Elliott, Azimuthal Surface Waves on Circular Cylinders, J. Appl. Phys., **26**(4): 368–376, April, 1955.)

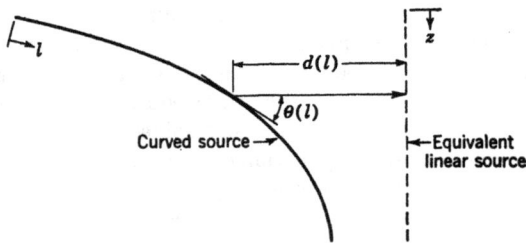

Figure 8-54. Geometry of curved leaky wave antenna.

dielectric sheet on ground plane,

$$\gamma \approx k \left[1 + \left(\frac{X_s}{Z_0} \right)^2 \right]^{\frac{1}{2}} \left\{ 1 + \frac{K}{2k(X_s/Z_0)[1 + (X_s/Z_0)^2]} \right\}$$

where X_s is the surface reactance of the plane surface wave structure, Z_0 is the intrinsic impedance of the ambient medium (usually free space), and k is the usual free-space wave number $(2\pi/\lambda)$.]

B. Curved Fast Wave Structures

Fast wave structures lend themselves very nicely to curved surfaces.[1] Because of the wide range of angle of radiation (near endfire to near broadside) a fast wave source is more versatile for curved surface applications than a slow wave source. Both shaped and pencil beams are quite practical with curved fast wave antennas.

The geometry of a curved leaky wave antenna is shown in Fig. 8-54. The antenna may be a line source or an area source (extending into page). The phase velocity along the source can be found from the relation

$$\frac{c}{v}(l) = \cos\theta(l) \tag{8-46}$$

where $\theta(l)$ is the angle of maximium radiation of a segment of the source centered at l. Geometrical optics can be utilized to relate the curved source amplitude distribution $A_c(l)$ to the amplitude $A_e(z)$ of an equivalent linear source. When the equivalent linear source is perpendicular to rays from the curved source, conservation of energy between two ray surfaces gives

$$A_c{}^2(l) = A_e{}^2(z) \sin\theta(l) \tag{8-47}$$

for an area source where $A_c{}^2(l)$ and $A_e{}^2(z)$ have dimensions of power per unit area and

$$A_c{}^2(l) = A_e{}^2(z)d(l) \sin\theta(l) \tag{8-48}$$

for a line source where $A_e{}^2(z)$ has dimensions of power per unit area and

---

[1] C. H. Walter, Curved Slot Antennas, *Ohio State Univ. Res. Found. Rept.* 667-1, Columbus, 1956; R. C. Honey and J. K. Shimizu, A Leaky Wave Antenna with Curved Aperture, *Stanford Res. Inst. Sci. Rept.* 5, Project 2605, February, 1960; Jasik, *op. cit.,* chap. 16.

$A_c{}^2(l)$ has dimensions of power per unit length. By use of the equivalent linear source, the analysis and synthesis methods of Chaps. 2 and 3 are directly applicable.

Very good results with curved traveling wave antennas have been obtained with slotted waveguides[1] and with inductive grid structures.[2] Beam scanning is also quite practical either by changing frequency[1] or by electrically or mechanically varying c/v along the structure.

For practical fast wave structures (see Chaps. 5 and 6) $\theta(l)$ in Fig. 8-54 can range from approximately 20 to 85 deg. Somewhat more versatility is achieved if the fast wave structure is loaded with dielectric so that $c/v \gtrsim 1$ can be realized. In this way $\theta(l)$ can be extended to less than 10 deg.

8-6. ARRAYS OF TRAVELING WAVE ANTENNAS

The usual method of array design for an array with identical elements employs pattern multiplication (array pattern times element pattern). This is an approximation which is often adequate, but it will never predict exactly the pattern of a practical array. An exact method should be used when the pattern structure at low levels (below 20 db) is important. Both the pattern multiplication method and an exact method are discussed as well as a method in which a large array is analyzed as a traveling wave structure of infinite extent.

A. Pattern Multiplication

Pattern multiplication can be applied to an array of identical elements. In pattern multiplication it is assumed that the array pattern is given by the product of an array factor (the pattern of the same array if the elements were all isotropic) and the pattern of an element of the array.

The usual assumption in pattern multiplication is that the pattern of each element in the presence of the remaining elements is the same as the pattern of the element by itself. This method gives good results for dipole arrays, for example, since the dipole pattern is relatively unaffected by the presence of other dipoles in the array. For arrays of traveling wave antennas, however, coupling effects can alter element patterns significantly and pattern multiplication should be used with care.

For endfire sources the field decays exponentially away from the surface of the source and interaction between elements is found to be negligible if field crossover between elements is at the 12-db contour. This gives a minimum separation of[3]

$$\frac{d}{\lambda} \approx \frac{1}{2}\sqrt{\frac{L}{\lambda}} \tag{8-49}$$

[1] Walter, *op. cit.*
[2] Honey and Shimizu, *op. cit.*
[3] Jasik, *op. cit.*, chap. 16.

where L is the length of the element. A practical upper limit for separation is obtained by requiring that the position of the first principal sidelobe of the array pattern correspond to the position of the first minimum in the element pattern. For endfire elements this corresponds to a maximum separation of[1]

$$\frac{d}{\lambda} \approx \sqrt{\frac{L}{\lambda}} \tag{8-50}$$

For a planar array of in-phase line-source elements (slow wave or fast wave) the half-power beamwidth due to arraying action is

$$BW_a \approx \frac{65\,\lambda}{nd} \quad \text{deg} \tag{8-51}$$

where n is the number of elements and d is the separation. The beamwidth and pattern shape in the orthogonal plane are essentially that of the individual element. For volume arrays, patterns in both principal planes are the appropriate array patterns.

B. Exact Array Procedure

When the element pattern is known to be distorted by the presence of other elements in the array and a precise array pattern is desired, then an exact array procedure should be employed. An exact procedure is as follows.[2] Let P_n be the pattern (amplitude and phase) when a unit current is injected into the terminals of the nth element of the array with the terminals of all other elements open-circuited. We could just as well let Q_n be the pattern when a unit voltage is applied to the terminals of the nth element with all other elements short-circuited. Let $I_0, I_1, \ldots, I_n$ represent the complex currents flowing into the terminal pairs when all elements of the array are energized. By superposition the pattern of the array is given by

$$P = I_0 P_0 + I_1 P_1 + I_2 P_2 + \cdots \tag{8-52}$$

When the patterns P_n are all identical, this method reduces to the pattern multiplication method.

To apply the exact method it is necessary to construct the array and measure each pattern P_n in the presence of the entire array (or be able to calculate this). This is in contrast to the pattern multiplication method in which the pattern of only one element by itself or in the presence of neighboring elements need be determined. The exact method can be used to deter-

[1] *Ibid.*

[2] J. N. Hines, V. H. Rumsey, and T. E. Tice, On the Design of Arrays, *Proc. IRE,* **42**(8):1262–1267, August, 1954.

mine the excitation coefficients I_n which give the best approximation to a desired pattern, but it is not suitable for determining the number of array elements and their spacing. In practice it is convenient to use the pattern multiplication method to obtain a first approximation to the array structure. When the structure has been fixed, the feed system can be designed by the exact procedure and the resulting pattern will be essentially as predicted.

C. Infinite Array Approximation

In most practical array designs the arraying is done to produce a narrow beam with low sidelobes. This usually means a tapered distribution. Consider the planar array of slotted waveguides in Fig. 8-55. If the array is many wavelengths wide and the excitation tapers gradually to zero at the sides of the array, a propagation constant for the composite structure can be obtained from an analysis of the same type of structure but of infinite extent.

The antenna in Fig. 8-55, for example, is analyzed as a capacitive grid structure in Chap. 6. The propagation constant for an infinite array has been used in the design and analysis of a finite array with excellent results.[1]

D. Determining Coupling Effects in Traveling Wave Arrays

Experimental

Perhaps the most straightforward method of determining coupling effects in antenna arrays is by experiment. If the elements of the array have well defined terminals, mutual impedance measurements can be made. However, for arrays of traveling wave elements the effect on the far field of scattering or diffraction by the array structure is usually of more concern than mutual impedance.

As mentioned in Sec. 8-6B, it is necessary to know the amplitude and phase of the far field of the array with each element individually excited while the other elements are, appropriately, either open- or short-circuited.

[1] Honey, *op. cit.*

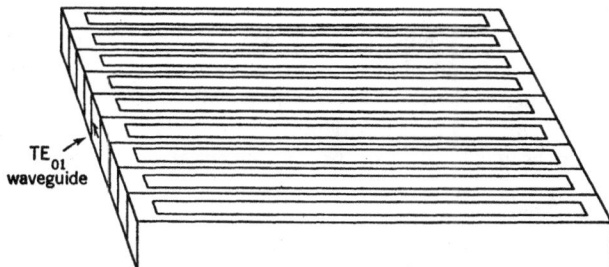

Figure 8-55.   Planar array of slotted waveguides.

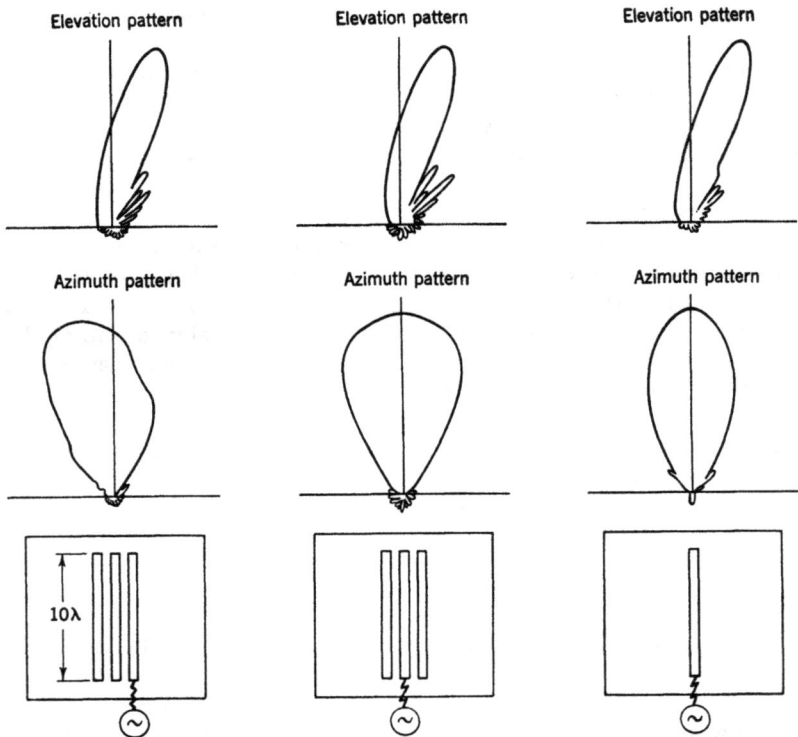

Figure 8-56. Measured patterns of tapered depth slot array. Patterns of a single slot are shown for comparison.

This is generally easier to do experimentally than analytically but still may be quite difficult. There is no problem in determining the amplitude pattern; the difficulty is in determining the phase pattern. It is also somewhat of a problem to determine accurately the relation between the far-field pattern and the input current (or voltage). Nevertheless, measurement is generally the most practical way of determining coupling effects.

Figure 8-56 shows an example of pattern amplitude distortion for a 10 λ tapered depth slot antenna (see Sec. 8-3B). Pattern distortion is rather severe, and pattern multiplication involving the element pattern by itself could be used only as a first approximation in this case.

Analytical

A qualitative picture of coupling effects in arrays of traveling wave antennas can be obtained by extending the analysis of Sec. 7-6 to more than two elements. For example, the equations for three lines, assuming coupling only between adjacent elements, are

$$\frac{dA_1}{dz} = -(\gamma_1 + C_{11})A_1 + C_{21}A_2 \tag{8-53}$$

$$\frac{dA_2}{dz} = -(\gamma_2 + C_{22})A_2 + C_{32}A_3 + C_{12}A_1 \tag{8-54}$$

$$\frac{dA_3}{dz} = -(\gamma_3 + C_{33})A_3 + C_{23}A_2 \tag{8-55}$$

If the unperturbed propagation constants and the coupling coefficients are known, then for a given set of initial conditions (excitation coefficients) the amplitude and phase along each line in the presence of the others can be determined. Calculation of the array pattern is then straightforward. The main difficulty is in determining the coupling coefficients. This may have to be done experimentally in many cases.

Coupling effects in arrays of traveling wave sources also can be determined by a transverse resonance method. This has been done for two parallel, slotted waveguides on a plane[1] and on a cylinder.[2] For more than two sources, however, the analysis gets very involved. If the array is wide in terms of wavelengths, as it usually is in practice, the best approach is to use the infinite array approximation (Sec. 8-6C).

E. Monopulse Arrays of Traveling Wave Antennas

Precise target location is not possible with a single line source because of its conical beam shape. An exception to this is the endfire line source which has a pencil beam. Arraying line sources in the conventional manner improves the resolution by giving a narrow fan- or pencil-type beam. Accurate target location also can be obtained by a monopulse array of traveling wave sources.[3]

Consider the coordinate system in Fig. 8-57. A line source lying along the z axis will have the conical pattern shown in the figure. Only half of a cone is shown to illustrate a practical situation in which a ground surface would tend to restrict radiation to the region $x \gtrless 0$. A TE source will have E_ϕ polarization, whereas a TM source will have E_θ polarization. Accurate θ-direction information can be obtained from a monopulse array of two collinear sources. Similarly, accurate ϕ-direction information can be obtained from an array of two parallel (side-by-side) sources. A sketch of a θ, ϕ monopulse system is shown in Fig. 8-58.

The monopulse pattern can be scanned in the ϕ direction by rotating the array or by having a number of the monopulse arrays disposed in a

[1] S. Nishida, Coupled Leaky Waveguides I: Two Parallel Slits in a Plane, *IRE Trans. Antennas Propagation*, **AP-8**:323–330, May, 1960.

[2] S. Nishida, Coupled Leaky Waveguides II: Two Parallel Slits in a Cylinder, *IRE Trans. Antennas Propagation*, **AP-8**:354–360, July, 1960.

[3] D. E. McDowell and J. N. Hines, A Monopulse Array of Traveling-wave Slot Antennas, *Ohio State Univ. Antenna Lab. Rept.* 667-8, Columbus, September, 1956.

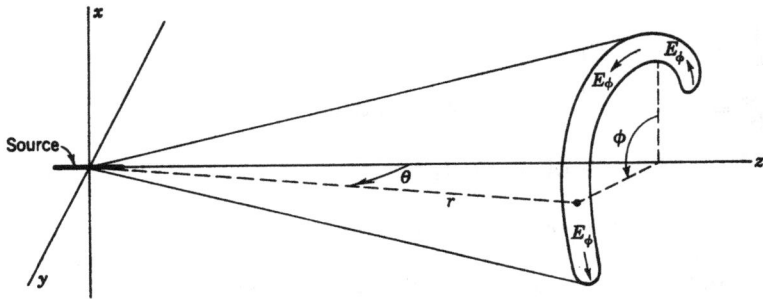

Figure 8-57.   Cross section of the pattern of a TE traveling wave line source.

circle about the z axis and switching from one to another to get the necessary coverage.

Scanning in the θ direction can be accomplished by varying the phase velocity along the individual line sources. However, it is necessary to ensure that the array pattern of the two collinear sources tracks with the element patterns. Let E_1 be the pattern of a line source by itself. The array pattern E_2 of two collinear sources is approximately (by pattern multiplication) the product of the element pattern times the array factor. The result in this case for equal lengths of feed lines is

$$E_2 = E_1 \cos \left[\tfrac{1}{2}(kD \cos \theta + \delta) \right] \tag{8-56}$$

where D is the spacing between centers of the elements and δ is either 0 or π depending on the output terminals of the monopulse feed system (see Fig. 8-58).

From Eq. (8-56) it can be seen that the array factor is independent of c/v; hence it will not track with the element beam as c/v is varied [recall that the beam maximum is at $\cos^{-1}(c/v)$]. If, now, a phase shifter is added

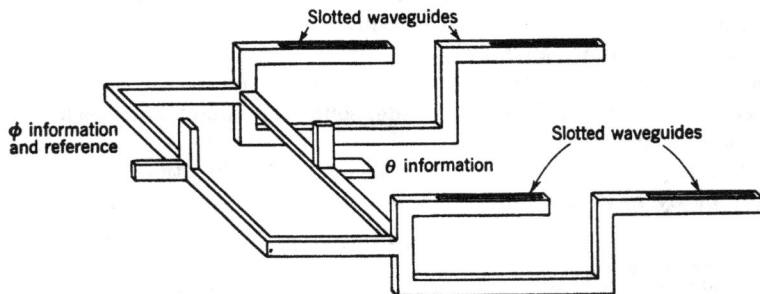

Figure 8-58.   Sketch of a θ, ϕ monopulse system using slotted waveguide line source radiators.

to one of the feed lines as shown in Fig. 8-59, the array pattern becomes

$$E_3 = E_1 \cos [\tfrac{1}{2}(\xi - kD \cos \theta - \delta)] \tag{8-57}$$

where ξ is the additional phase delay of the phase shifter. If ξ is adjusted so that

$$\xi = kD \frac{c}{v} \tag{8-58}$$

then the array factor will track with the element pattern as c/v is varied.

The in-phase ($\delta = 0$) pattern of the θ-monopulse system is

$$E_i = E_1 \cos \left[\frac{kD}{2} \left(\cos \theta - \frac{c}{v} \right) \right] \tag{8-59}$$

and the out-of-phase pattern is

$$E_o = -E_1 \sin \left[\frac{kD}{2} \left(\cos \theta - \frac{c}{v} \right) \right] \tag{8-60}$$

A maximum of the array factor coincides with the maximum of the element pattern for the in-phase case, whereas for the out-of-phase case an array factor null coincides with the element maximum. This is true, now, regardless of beam position.

8-7. MODULATED SLOW WAVE STRUCTURES

Linear, uniform, slow wave sources are inherently endfire, near-endfire, or backfire, and sidelobe levels are relatively high. Experience has shown that tapering polyrods and other slow wave sources will reduce the level of the sidelobes. Introducing a periodicity to the tapering (i.e., modulating the source) also provides control of the sidelobe level as well as some control of angle of radiation from a slow wave source.

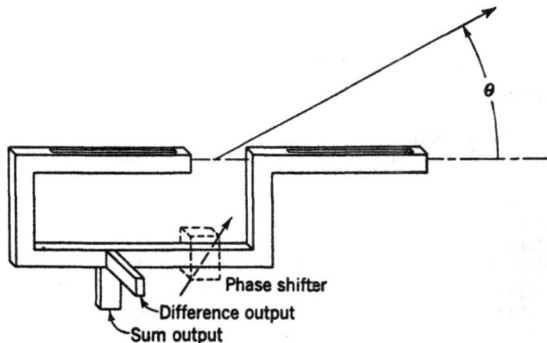

Figure 8-59. θ-monopulse system with phase shifter.

Phase shifter
Difference output
Sum output

The first study of modulated antenna structures appears to be that of Simon and his work on high-gain endfire antennas.[1] Simon's work was primarily experimental and it led to widespread interest in modulated structures.[2]

This section will describe some of the analysis and synthesis methods that are obtained. Although slow wave structures are the prime concern here, modulation techniques are also quite useful for fast wave structures.

A. Analysis of Modulated Structures

Modulation may be in the form of amplitude modulation, phase modulation, or a combination of the two. Time modulation is possible but will not be included in this discussion.

A modulation form can be specified, and the resulting pattern computed. However, this still leaves the problem of determining a surface that will produce the desired modulation. The surface parameters can be brought into the problem through the analyses of Chaps. 5 and 6, where traveling wave phase and amplitude can be related to a physical structure. The results of Chaps. 5 and 6 for infinitely long structures can be applied to modulated structures if the variations due to modulation are sufficiently gradual (say a modulation period greater than the free-space wavelength). In some cases it may be most convenient to modulate the surface reactance of a traveling wave source. This case can be readily analyzed and is described later.

In the analysis and synthesis of modulated structures it is convenient to work with the wave number spectrum of the structure. Let β be the wave number (phase constant) of a uniform wave along a traveling wave structure. The wave number is related to the angle of maximum radiation θ_m by

$$\frac{\beta}{k} = \frac{c}{v} = \cos \theta_m \qquad (8\text{-}61)$$

If $|\beta|$ is equal to or greater than the free-space wave number k, the traveling wave is a slow wave. Fast (leaky) wave propagation occurs for $|\beta|$ less than k.

Varying the amplitude or phase velocity (or both) of a traveling wave gives rise to a spectrum of wave numbers. Each wave number β_n of the

[1] J. C. Simon and V. Biggi, Un nouveau type d'aerien et son application à la transmission de television a grande distance, *L'Onde Elec.*, No. 332, pp. 883–896, November, 1954.

[2] A. Thomas and F. J. Zucker, Radiation from Modulated Surface Wave Structures—I, *IRE Natl. Conv. Record*, part 1, pp. 153–160, 1957; R. L. Pease, Radiation from Modulated Surface Wave Structures—II, *IRE Natl. Conv. Record*, part 1, pp. 161–165, 1957; A. A. Oliner and A. Hessel, Guided Waves on Sinusoidally-modulated Reactance Surfaces, *IRE Trans. Antennas Propagation*, **AP-7**:S201–S218, December, 1959; A. Ishimaru and G. D. Bernard, Closed Form Analysis for the Radiation Pattern of the Modulated Antenna, *IRE Trans. Antennas Propagation*, **AP-10**:151–159, March, 1962.

spectrum corresponds to a uniform traveling wave over the length of the structure. The pattern of each wave is the usual $(\sin X_n)/X_n$ where

$$X_n = \frac{kL}{2}\left(\cos\theta - \frac{\beta_n}{k}\right) \qquad (8\text{-}62)$$

with θ measured from the longitudinal axis of the source. The resultant pattern is a superposition of $(\sin X_n)/X_n$ functions with the appropriate amplitudes and phases.

From Chap. 2 the far-field pattern of a one-dimensional source of length L can be expressed as

$$F(\theta) = f(\theta)\int_{-L/2}^{L/2}|A(z)|e^{j[kz\cos\theta+\psi(z)]}\,dz \qquad (8\text{-}63)$$

where $f(\theta)$ = pattern of element dz
$|A(z)|$ = amplitude of source distribution
$\psi(z)$ = phase of source distribution
For phase modulation $|A(z)|$ is constant and $\psi(z)$ varies in the appropriate manner. For amplitude modulation $\psi(z)$ is a linear function of z [that is, $\psi(z) = -\beta z$] and $|A(z)|$ is varied.

Phase Modulation

Consider a constant amplitude traveling wave $(|A(z)| = 1)$ with phase variation of the form

$$\psi(z) = -\beta_0 z - m\sin\left(\frac{2\pi z}{p}+\xi\right) \qquad (8\text{-}64)$$

where β_0 = average phase constant
m = amplitude of modulation
p = period of modulation
ξ = phase of modulation
The far field (Eq. 8-63) becomes

$$F(\theta) = f(\theta)\int_{-L/2}^{L/2}e^{j\left[kz\cos\theta-\beta_0 z-m\sin\left(\frac{2\pi z}{p}+\xi\right)\right]}\,dz \qquad (8\text{-}65)$$

Using the identity

$$e^{-jm\sin b} = \sum_n e^{-jnb}J_n(m) \qquad (8\text{-}66)$$

with the substitution

$$b = \frac{2\pi z}{p}+\xi \qquad (8\text{-}67)$$

and

$$\beta_n = \beta_0 + \frac{2n\pi}{p} \qquad (8\text{-}68)$$

gives

$$F(\theta) = f(\theta) \sum_n L e^{-jn\xi} J_n(m) \frac{\sin\left[(kL/2)(\cos\theta - \beta_n/k)\right]}{(kL/2)(\cos\theta - \beta_n/k)} \tag{8-69}$$

Thus the far-field pattern of a phase modulated source can be expressed as a sum of $(\sin X)/X$ functions whose amplitudes are given by the Bessel function $J_n(m)$.

Although Eq. (8-69) appears to be quite useful, it has several disadvantages. First, it will give only approximate results in many practical cases because the series would be truncated to aid in computation. Second, the form of Eq. (8-69) makes it difficult to see the effects on the pattern due to changes in m and p.

To observe the effects of m and p a more convenient form for the pattern can be obtained by considering the modulated structure to be an array of elements where each element is one period of the modulated structure.[1] Neglecting coupling effects, the total pattern is a product of the element pattern and the array factor of q identical elements. The array factor is simply

$$\text{Array factor} = \frac{\sin qY}{q \sin Y} \tag{8-70}$$

where

$$Y = \frac{kL}{2q}\left(\cos\theta - \frac{\beta_0}{k}\right) \tag{8-71}$$

For q odd, the center of the element representing one cycle of modulation is at $z = 0$. The pattern of the element for $\xi = 0$ and $f(\theta) = 1$ is

$$\left.\text{Element pattern}\right]_{q\text{ odd}} = \int_{-L/2q}^{L/2q} e^{j\left(kz\cos\theta - \beta_0 z - m\sin\frac{2\pi qz}{L}\right)} dz \tag{8-72}$$

For q even the element pattern is given by

$$\left.\text{Element pattern}\right]_{q\text{ even}} = \int_0^{L/q} e^{j\left(kz\cos\theta - \beta_0 z - m\sin\frac{2\pi qz}{L}\right)} dz \tag{8-73}$$

The two integrals can be expressed in terms of the Anger functions[2] defined by

$$J_\nu(w) = \frac{1}{2\pi}\int_{-\pi}^{\pi} e^{j(\nu\eta - w\sin\eta)} d\eta \tag{8-74}$$

---

[1] Ishimaru and Bernard, *op. cit.*

[2] G. N. Watson, "A Treatise on the Theory of Bessel Functions," 2d ed., pp. 308–319, Cambridge, New York, 1944; E. Jahnke, F. Emde, and F. Losch, "Tables of Higher Functions," 6th ed., p. 251, McGraw-Hill, New York, 1960; G. D. Bernard and A. Ishimaru, "Tables of the Anger and Lommel-Weber Functions," University of Washington Press, Seattle, 1962.

For integer ν the Anger function is identical with the Bessel function of the first kind.

The total pattern $F(\theta)$ is simply the product of the element pattern and the array factor. That is,

$$F(\theta) = \frac{\sin qY}{q \sin Y} J_{Y/\pi}[(-1)^{q-1}m] \tag{8-75}$$

Letting

$$\nu = \frac{Y}{\pi}$$

and

$$m' = (-1)^{q-1}m$$

we can rewrite Eq. (8-75) as

$$F(\theta) = \frac{\sin (\pi q\nu)}{q \sin (\pi\nu)} J_\nu(m') \tag{8-76}$$

The chief advantage of the array method is that the effect of the amplitude and period of the modulation upon the radiation pattern can be seen more easily than by the previous method which resulted in a series of $(\sin X)/X$ functions.

Amplitude Modulation

For an amplitude modulated source of length L and phase constant β_0 the far-field pattern is

$$F(\theta) = \int_{-L/2}^{L/2} A(z)e^{j(kz \cos \theta - \beta_0 z)} dz \tag{8-77}$$

In general the amplitude can be resolved into a series of traveling waves by Fourier analysis (see Chap. 2). As an example consider the sinusoidal modulation

$$A(z) = A_0 + m \sin\left(\frac{2\pi z}{p} + \xi\right) \tag{8-78}$$

which can be rewritten as

$$A(z) = A_0 + \frac{m}{2j} e^{j[(2\pi z/p)+\xi]} - \frac{m}{2j} e^{-j[(2\pi z/p)+\xi]} \tag{8-79}$$

where m = amplitude of modulation
p = period of modulation
ξ = phase of modulation

Substituting Eq. (8-79) into (8-77) gives

$$F(\theta) = \int_{-L/2}^{L/2} A_0 e^{j(kz \cos \theta - \beta_0 z)} \, dz - \frac{jm}{2} \int_{-L/2}^{L/2} e^{j[kz \cos \theta - \beta_0 z + (2\pi z/p) + \xi]} \, dz$$

$$+ \frac{jm}{2} \int_{-L/2}^{L/2} e^{j[kz \cos \theta - \beta_0 z - (2\pi z/p) - \xi]} \, dz \quad (8\text{-}80)$$

which upon integration reduces to

$$F(\theta) = LA_0 \frac{\sin[(L/2)(k \cos \theta - \beta_0)]}{(L/2)(k \cos \theta - \beta_0)}$$

$$- \frac{jmL}{2} e^{j\xi} \frac{\sin [(L/2)(k \cos \theta - \beta_0 + 2\pi/p)]}{(L/2)(k \cos \theta - \beta_0 + 2\pi/p)}$$

$$+ \frac{jmL}{2} e^{-j\xi} \frac{\sin [(L/2)(k \cos \theta - \beta_0 - 2\pi/p)]}{(L/2)(k \cos \theta - \beta_0 - 2\pi/p)} \quad (8\text{-}81)$$

As in the case of phase modulation, the pattern of an amplitude modulated wave can be expressed as a series of $(\sin X)/X$ terms. Alternatively, the amplitude modulated structure also can be considered as an array using pattern multiplication as described in the previous section. The chief difficulty with the array method is the evaluation of the element pattern.

Combined Amplitude and Phase Modulation

In the general case where a source distribution $A(z)e^{j\psi(z)}$ has periodicity in both $A(z)$ and $\psi(z)$, one can always resort to Fourier analysis and arrive at a series of uniform traveling waves for the source distribution and the corresponding series of $(\sin X)/X$ functions for the far field (see Chap. 2). One can also apply the array method of the previous sections.

Consider the special case where both amplitude and phase are sinusoidally modulated. Let the source distribution be of the form

$$A(z)e^{j\psi(z)} = \cos \frac{q\pi z}{L} e^{-j[\beta_0 z + m \sin (2\pi z/p)]} \quad (8\text{-}82)$$

where q is an integer and L is the length of the source. The far-field pattern $F(\theta)$ is

$$F(\theta) = \int_{-L/2}^{L/2} \cos \frac{q\pi z}{L} e^{j[kz \cos \theta - \beta_0 z - m \sin (2\pi z/p)]} \, dz \quad (8\text{-}83)$$

Writing $\cos (q\pi z/L)$ in exponential form and expanding $e^{-jm \sin (2\pi z/p)}$ into a Fourier exponential series, interchanging summation and integration, and performing the integration result in

$$F(\theta) = \sum_{n=-\infty}^{\infty} \frac{LJ_n(m)}{2} \left\{ \frac{\sin [(L/2)(k \cos \theta - \beta_0 + q\pi/L - 2n\pi/p)]}{(L/2)(k \cos \theta - \beta_0 + q\pi/L - 2n\pi/p)} \right.$$

$$+ \left. \frac{\sin [(L/2)(k \cos \theta - \beta_0 - q\pi/L - 2n\pi/p)]}{(L/2)(k \cos \theta - \beta_0 - q\pi/L - 2n\pi/p)} \right\} \quad (8\text{-}84)$$

For the case of constant amplitude ($q = 0$) Eq. (8-84) reduces to Eq. (8-69).

Surface Reactance Modulation

Surface reactance modulation is a special form of combined amplitude and phase modulation. A rigorous treatment of a sinusoidally modulated surface which follows that of Oliner and Hessel[1] is presented here.

If the period of the modulation of a reactive surface is sufficiently small, the propagation is in the form of a slow wave and one or more stop bands may exist. If the spacing is larger than some critical value, the modulated wave gives rise to one or more leaky waves which radiate away from the surface at an angle θ_m [see Eq. (8-61); also see Sec. 8-3D].

A solution can be obtained by viewing the geometry in terms of modes associated with propagation perpendicular to the surface. The nature of the impedance boundary at the surface is such that each of these modes couples only to itself and the next higher and lower modes. This is in contrast to a conventional diffracting surface in which all modes couple to each other. The result is that the problem reduces to the solution of an infinite set of linear equations in which each equation involves only three modes. A simple explicit expression for propagation constants can be obtained by a perturbation procedure that is valid for small modulations.

For propagation in the z direction let the modulated reactance of a surface in the yz plane be

$$X(z) = X_s\left(1 + m \cos \frac{2\pi z}{p}\right) \tag{8-85}$$

where m is the amplitude of modulation, p is the period, and X_s is the average value of the reactance. The modulation amplitude is restricted to $m \leqslant 1$.

Assuming TM operation and no variation in the y direction, the tangential fields are E_z and H_y and they are related by

$$-E_z(0,y,z) = jX(z)H_y(0,y,z) \tag{8-86}$$

The total field can be expressed in terms of modes defined with respect to propagation in the transverse (or x) direction. The corresponding vector mode functions $\mathbf{h}_n$ and $\mathbf{e}_n$ which form an orthonormal set are (see Chap. 4)

$$\mathbf{h}_n(z) = \mathbf{e}_n(z) \times \mathbf{x} = \mathbf{y}h_n(z) = \mathbf{y}\frac{1}{\sqrt{2\pi}}e^{-j(\beta_0 + 2n\pi/p)z} \tag{8-87}$$

where p is the period of the modulation and β_0 is the phase constant at the zeroth modulation (the origin or reference point). At the nth modulation the phase constant β_n is

$$\beta_n = \beta_0 + \frac{2n\pi}{p} \qquad n = 0, \pm 1, \pm 2, \ldots \tag{8-88}$$

Equation (8-88) is actually a statement of Floquet's theorem.

[1] Oliner and Hessel, *op. cit.*

The transverse (or x direction) phase constant is

$$\kappa_n = \sqrt{k^2 - \left(\beta_0 + \frac{2n\pi}{p}\right)^2} \tag{8-89}$$

The total magnetic field can be expressed in terms of the vector mode functions $\mathbf{h}_n$ and the modal currents I_n as

$$H_y(x,z) = \sum_{n=-\infty}^{\infty} I_n(x)h_n(z) \tag{8-90}$$

where

$$I_n(x) = \int_0^p \mathbf{H}(x,z) \cdot \mathbf{h}_n^*(z)\, dz \tag{8-91}$$

owing to the orthogonality properties of the vector mode functions (see Sec. 4-8).

Similarly the modal voltage V_n of the nth transverse mode is

$$V_n(x) = \int_0^p \mathbf{E}(x,z) \cdot \mathbf{e}_n^*(z)\, dz \tag{8-92}$$

With the boundary condition at $x = 0$ given by Eq. (8-86) and using Eq. (8-85), Eq. (8-92) becomes

$$V_n(0) = j\int_0^p \left(1 + m\cos\frac{2\pi z}{p}\right) X_s H_y(0,z)e_n^*(z)\, dz \tag{8-93}$$

With the use of e_n ($= ze_n$) from Eq. (8-87) and of Eq. (8-91), Eq. (8-93) becomes

$$V_n = jX_s I_n + jX_s \frac{m}{2}(I_{n+1} + I_{n-1}) \tag{8-94}$$

Thus it can be seen that the sinusoidal modulation in reactance couples each transverse mode to its nearest neighbors rather than to all other transverse modes as usually occurs for diffracting surfaces.

The propagation constants of waves guided by the modulated surface can be found by transverse resonance (see Sec. 4-8). For the TM operation that is assumed in the present case, transverse resonance gives

$$\frac{V_n}{I_n} = \frac{\kappa_n}{\omega\epsilon} \tag{8-95}$$

Combining Eq. (8-95) with (8-94) results in

$$I_{n+1} + \frac{2I_n}{m}\left(1 - \frac{j\kappa_n}{X_s\omega\epsilon}\right) + I_{n-1} = 0 \qquad n = 0, \pm1, \pm2, \ldots \tag{8-96}$$

Equation (8-96) gives an infinite set of linear, homogeneous equations for the infinite number of modal currents I_n. Writing Eq. (8-96) as

$$I_{n+1} + D_n I_n + I_{n-1} = 0 \tag{8-97}$$

where

$$D_n = \frac{2}{m}\left(1 - \frac{j\kappa_n}{X_s\omega\epsilon}\right) \tag{8-98}$$

the first two equations of the semi-infinite set beginning at n are

$$I_n + D_{n-1}I_{n-1} + I_{n-2} = 0 \tag{8-99}$$
$$I_{n+1} + D_nI_n + I_{n-1} = 0 \tag{8-100}$$

which combine to give

$$\frac{I_n}{I_{n+1}} = -\cfrac{1}{D_n - \cfrac{1}{D_{n-1} + I_{n-2}/I_{n-1}}} \tag{8-101}$$

Continuing in this manner gives the continued fraction solution

$$\frac{I_n}{I_{n+1}} = -\cfrac{1}{D_n - \cfrac{1}{D_{n-1} - 1/(D_{n-2} - \cdots)}} \tag{8-102}$$

Similarly from the remaining semi-infinite set we obtain

$$\frac{I_{n+1}}{I_n} = -\cfrac{1}{D_{n+1} - \cfrac{1}{D_{n+2} - 1/(D_{n+3} - \cdots)}} \tag{8-103}$$

In order that the infinite set of Eqs. (8-96) have a nonvanishing solution, Eq. (8-102) must equal the inverse of Eq. (8-103). This gives

$$D_n - \frac{1}{D_{n-1} - 1/(D_{n-2} - \cdots)} - \frac{1}{D_{n+1} - 1/(D_{n+2} - \cdots)} = 0 \tag{8-104}$$

Substituting Eqs. (8-98) and (8-89) into (8-104) results in

$$1 - \frac{j\sqrt{\mu/\epsilon}}{X_s}\left[1 - \left(\frac{\beta_0}{k} + \frac{2\pi n}{kp}\right)^2\right]^{\frac{1}{2}}$$

$$= \cfrac{m^2/4}{1 - \cfrac{j\sqrt{\mu/\epsilon}}{X_s}\left\{1 - \left[\frac{\beta_0}{k} + \frac{2\pi(n-1)}{kp}\right]^2\right\}^{\frac{1}{2}} - \cfrac{m^2/4}{1 - \frac{j\sqrt{\mu/\epsilon}}{X_s}\left\{1 - \left[\frac{\beta_0}{k} + \frac{2\pi(n-2)}{kp}\right]^2\right\}^{\frac{1}{2}} - \cdots}}$$

$$+ \cfrac{m^2/4}{1 - \cfrac{j\sqrt{\mu/\epsilon}}{X_s}\left\{1 - \left[\frac{\beta_0}{k} + \frac{2\pi(n+1)}{kp}\right]^2\right\}^{\frac{1}{2}} - \cfrac{m^2/4}{1 - \frac{j\sqrt{\mu/\epsilon}}{X_s}\left\{1 - \left[\frac{\beta_0}{k} + \frac{2\pi(n+2)}{kp}\right]^2\right\}^{\frac{1}{2}} - \cdots}} \tag{8-105}$$

Owing to $m^2/4$ in each term, the continued fractions converge rapidly for all values of modulation. For numerical results n is set equal to zero in Eq. (8-105) and the five terms are sufficient to yield the wave number β_0 to satisfactory accuracy in most cases.[1]

For complex β_0 (leaky wave) a perturbation solution of Eq. (8-105) that is good for small values of modulation can be employed. Writing β_0 as

$$\beta_0 = \beta + \Delta\beta - j\alpha \tag{8-106}$$

where β is the unperturbed phase constant corresponding to $m = 0$, Eq. (8-105) yields

$$
\Delta\beta - j\alpha = \frac{-m^2 k X_s{}^2}{(4\mu/\epsilon)(1 + X_s{}^2\epsilon/\mu)^{\frac{1}{2}}}
$$
$$
\times \left(\frac{1}{1 - \dfrac{j\sqrt{\mu/\epsilon}}{X_s}\left\{1 - \left[(1 + X_s{}^2\epsilon/\mu)^{\frac{1}{2}} - \dfrac{2\pi}{kp}\right]^2\right\}^{\frac{1}{2}}} \right.
$$
$$
\left. + \frac{1}{1 - \dfrac{j\sqrt{\mu/\epsilon}}{X_s}\left\{1 - \left[(1 + X_s{}^2\epsilon/\mu)^{\frac{1}{2}} + \dfrac{2\pi}{kp}\right]^2\right\}^{\frac{1}{2}}} \right) \tag{8-107}
$$

Equation (8-107) takes into account only the $n = 0$ and ±1 modes. It yields accurate results except within certain narrow ranges of kp.[1]

B. Synthesis of Modulated Structures

An exact procedure for the design of a surface that will support a prescribed group of traveling waves simultaneously was developed by Bolljahn.[2] The resulting surface is modulated in the direction of propagation and has a surface impedance that varies periodically in the direction of propagation. The technique is limited in that it is not valid for cylindrical endfire structures and it is inherently applicable only to the corrugated-surface type of structure. The former limitation is due to the lack of axial symmetry in the hybrid modes that are required for true endfire operation. The latter limitation arises because the analysis postulates that there is no energy flow across the interface between the guiding structure and the free-space region above the structures. For a corrugated surface there is no energy flow below the interface; hence the method is applicable.

[1] *Ibid.*
[2] J. T. Bolljahn, Synthesis of Modulated Corrugated Surface Wave Structures, *IRE Trans. Antennas Propagation*, **AP-9**:236–241, May, 1961.

Consider the corrugated surface in Fig. 6-26. Slow waves on this structure have the components

$$E_x = \frac{B}{\omega\epsilon}\frac{\beta}{|\kappa|} e^{-|\kappa|x}e^{-j\beta z} \tag{8-108}$$

$$E_z = \frac{jB}{\omega\epsilon} e^{-|\kappa|x}e^{-j\beta z} \tag{8-109}$$

$$H_y = \frac{B}{|\kappa|} e^{-|\kappa|x}e^{-j\beta z} \tag{8-110}$$

where

$$|\kappa|^2 = \beta^2 - k^2 \tag{8-111}$$

In Bolljahn's method a group of waves having different $|\kappa|$'s and β's but the same frequency are postulated and the flow lines represented by the directions of the real part of the complex Poynting vector are found. These flow lines or surfaces are the surfaces across which there is no energy flow; hence they are surfaces which can support the set of waves postulated. The imaginary part of the complex Poynting vector can be used to determine the necessary reactance variation along these surfaces.

The procedure is to find the complex Poynting vector $\mathbf{E} \times \mathbf{H}^*$ and separate it into real and imaginary parts. Since there are no sources of energy in the region under consideration, we note that

$$\nabla \cdot \text{Re}\,(\mathbf{E} \times \mathbf{H}^*) \equiv 0 \tag{8-112}$$

Assuming that the surfaces sought are of the form shown in Fig. 8-60, we can apply Gauss's theorem to the crosshatched region and obtain

$$-\int_0^{x_0} \text{Re}\,(\mathbf{E} \times \mathbf{H}^*)_z \Big]_{z=0} dx - \int_0^z \text{Re}\,(\mathbf{E} \times \mathbf{H}^*)_x \Big]_{x=0} dz$$
$$+ \int_0^x \text{Re}\,(\mathbf{E} \times \mathbf{H}^*)_z \Big]_{z=z} dx = 0 \tag{8-113}$$

There will actually be a family of surfaces in Fig. 8-60 corresponding to different values of x_0. The $\mathbf{E}$ and $\mathbf{H}$ fields for a corrugated surface can be

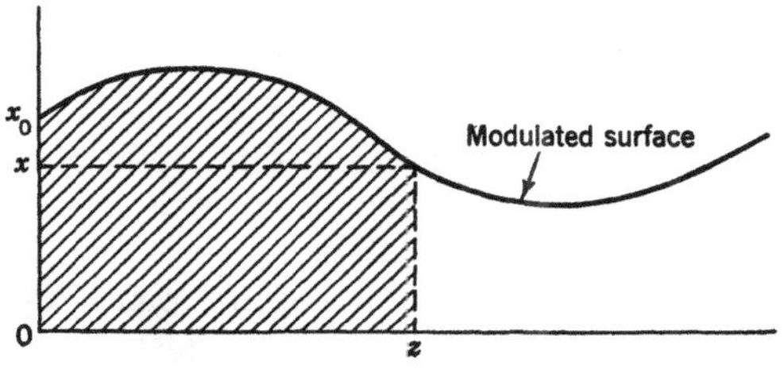

Figure 8-60. Application of Gauss's theorem to surface.

substituted into Eq. (8-113), and the integrations carried out. The surface for a particular x_0 is determined by substituting values of z into the resulting equation and finding the corresponding values of x. The reactance along the surface can be found from tangential **E** and **H** to be

$$X_s = \frac{E_t}{H_t} = \frac{\text{Im } (\mathbf{E} \times \mathbf{H}^*) \times \text{Re } (\mathbf{E} \times \mathbf{H}^*)}{|\mathbf{H}|^2 \, |\text{Re } (\mathbf{E} \times \mathbf{H}^*)|} \qquad (8\text{-}114)$$

Bolljahn's method constitutes a true synthesis method, since it leads directly to the determination of a surface that will support a prescribed set of surface waves. The surface waves, themselves, are determined by the antenna pattern that is desired. Methods for determining the surface waves when the pattern is specified are given in Chap. 3.

The analysis methods of this section (8-7) also can be used to determine a surface that will produce a desired set of traveling waves, but not directly. The analysis methods must be used on a trial-and-error basis. That is, the spectrum or set of waves for a given modulation is first determined. If this is not the desired spectrum or a good enough approximation to it, the modulation is modified and the new spectrum is determined. The process can be repeated until a satisfactory approximation to the desired set of waves is obtained. Excellent results can be obtained in this manner in many cases.

8-8. SURFACE WAVE LENS ANTENNAS

A wave propagates with phase velocity v over a surface wave structure. By definition, the ratio of the velocity of light c in free space to the phase velocity of a wave in some medium is the index of refraction η of that medium. That is,

$$\eta = \frac{c}{v} \qquad (8\text{-}115)$$

Hence a surface wave structure can be thought of as having an index of refraction given by the ratio c/v, and lens action in the plane of the structure can be obtained by properly varying the phase velocity.

Isotropy in the plane of the structure is usually required. Examples of surface wave structures having this isotropy are dielectric sheet in free space or on ground plane, bed of metal posts on ground plane, and parallel plate structure with holey top plate (see Chap. 6). For structures such as these the radiation pattern of a lens can be obtained in the plane of the structure while in the orthogonal plane the pattern is that of a traveling wave endfire antenna. Two examples of surface wave lenses are shown in Fig. 8-61. Rays are drawn to illustrate the lens action.

Surface wave lens designs can be obtained by application of geometrical optics.[1] From Fermat's principle and the calculus of variations the ray paths in a medium where the index varies only with radius is given by

$$\eta r \sin \tau = K \tag{8-116}$$

where K is a constant for each ray and r and τ are defined in Fig. 8-62. Equation (8-116) is a generalization of Snell's law. If the index is known in a radially symmetric region, Eq. (8-116) can be used to determine the ray paths. One can also sketch out some ray paths and use Eq. (8-116) to synthesize an index variation.

[1] C. H. Walter, Surface Wave Luneberg Lens Antennas, *IRE Trans. Antennas Propagation*, **AP-8**:508–515, September, 1960.

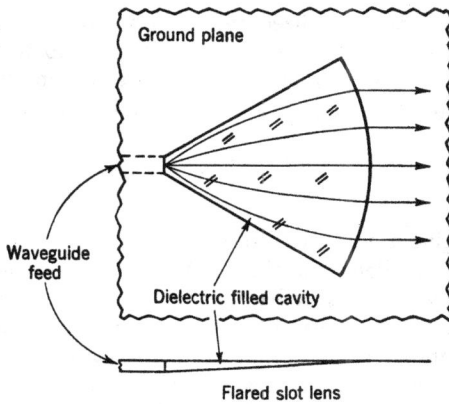

Figure 8-61. Surface-wave lens antennas.

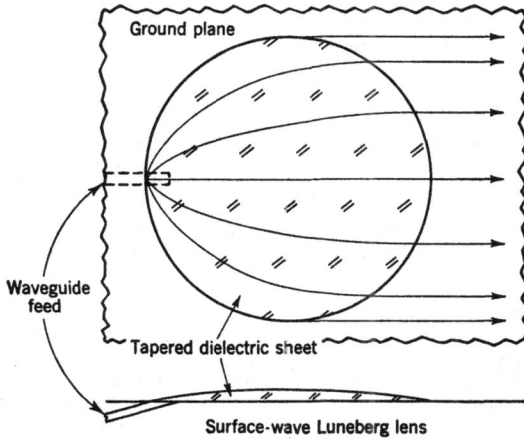

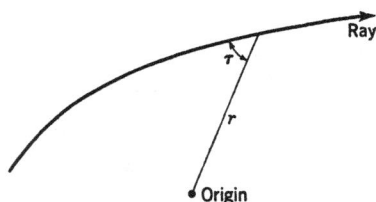

Figure 8-62. Geometry for generalized Snell's law.

A class of radially symmetric lenses called Luneberg lenses[1] is very useful as antennas.[2] The most common index variation and the one usually implied when the term Luneberg lens is used is

$$\eta = \sqrt{2 - r^2} \tag{8-117}$$

where r is normalized radius. This index corresponds to the ray paths in Fig. 8-61. For this index variation the portion of a plane wave incident on the lens will be focused diametrically opposite at the rim of the lens. For an internal feed at radius r_1 the index variation

$$\eta = \frac{(1 + r_1^2 - r^2)^{1/2}}{r_1} \tag{8-118}$$

will give a collimated beam.[3]

For dielectric sheet and metal post structures the required index variation is obtained by simply varying the thickness of the dielectric or the height of the posts. The feed can be an open-ended waveguide butted against the rim or a short tapered depth antenna (see Fig. 8-12) butted against the surface of the lens. If the lens is on a ground plane the tapered depth feed can be flush with the surface of the ground plane and butted against the lens as in Fig. 8-61. A line feed of this type gives some defocusing, but it is usually not serious for lens diameters greater than 15 λ if the feed length is less than 3 λ.

The fields of a surface wave lens can be obtained by application of Huygens' principle. One can find equivalent currents **J** and **K** on the surface enclosing the antenna as described in Sec. 2-2. Usually this surface would be taken as the surface of the lens and the ground plane. The amplitude and phase of the fields over the surface will depend on the thickness and dielectric constant of the material forming the surface wave structure and also on the radiation pattern of the feed.

[1] R. K. Luneberg, "Mathematical Theory of Optics," pp. 189–213, Brown University Press, Providence, R.I., 1944.

[2] S. Adachi, R. C. Rudduck, and C. H. Walter, A General Analysis of Nonplanar, Two-dimensional Luneberg Lenses, *IRE Trans. Antennas Propagation*, **AP-9**:353–357, July, 1961; R. C. Rudduck and C. H. Walter, A General Analysis of Geodesic Luneberg Lenses, *IRE Trans. Antennas Propagation*, **AP-10**:444–450, July, 1962.

[3] A. S. Gutman, Modified Luneberg Lens, *J. Appl. Phys.*, **25**:855–859, July, 1954.

The fields **E** and **H** over the lens and ground plane can be found approximately by optical methods. Referring to Fig. 8-63, one can assume that all the energy in an incremental angle $\Delta\sigma$ at the feed (P_1) travels as a surface wave between two surfaces designated as rays in the figure. The ray paths can be determined by means of Eq. (8-116). If one knows the relative energy density leaving the feed at angle σ, then conservation of energy in the channel ΔW enables one to find the relative amplitude of the surface wave fields at some point P. The relative phase ψ of the fields at P with respect to the feed point P_1 can be obtained from the path length equation

$$\text{Electrical path length} = \int_{P_1}^{P} \eta \, ds \qquad (8\text{-}119)$$

where ds is an element of path length along the ray through P_1 and P.

The equivalent currents on the Huygens' surface S are $\mathbf{J} = \mathbf{n} \times \mathbf{H}$ and $\mathbf{K} = \mathbf{E} \times \mathbf{n}$, where $\mathbf{n}$ is the unit vector normal to the surface and directed toward the source-free side. The electric field on the source-free side of the surface S is given by

$$\mathbf{E} = -\frac{1}{\epsilon} \nabla \times \mathbf{F} - \frac{j}{\omega\mu\epsilon} \nabla \times \nabla \times \mathbf{A} \qquad (8\text{-}120)$$

where $\mathbf{F}$ and $\mathbf{A}$ are the usual potential functions given by

$$\mathbf{F} = \frac{\epsilon}{4\pi} \int_S \frac{\mathbf{K}e^{-jkr}}{r} \, dS \qquad (8\text{-}121)$$

$$\mathbf{A} = \frac{\mu}{4\pi} \int_S \frac{\mathbf{J}e^{-jkr}}{r} \, dS \qquad (8\text{-}122)$$

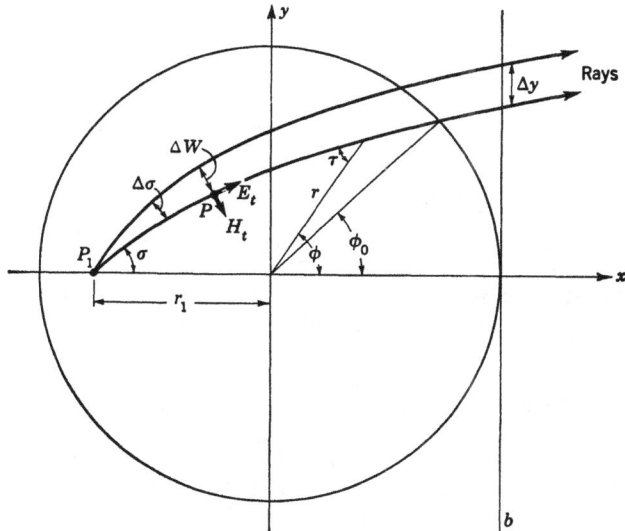

Figure 8-63. Optical analysis of surface wave Luneberg lens antenna.

The far-field vertical pattern of a 15 λ diameter lens with about 15 λ ground plane extending beyond the lens is given in Fig. 2-23. The distribution used in computing the pattern was obtained by the above method.

If only the lens patterns of surface wave lenses as in Fig. 8-61 or of two-dimensional rim-radiating Luneberg lenses are desired, a good approximation is to replace the lens by an equivalent line source such as line b in Fig. 8-63. Assuming that there is no power in the z direction (normal to the surface), conservation of energy gives

$$U(\sigma) \, \Delta\sigma = U_b(y) \, \Delta y \tag{8-123}$$

where $U(\sigma)$ is the energy distribution per unit angle in the pattern of the feed and $U_b(y)$ is the energy distribution per unit length along line b. If the lens is focused at infinity, the rays are parallel when they leave the lens (the phase is constant along b) and Eq. (8-123) gives the field relations

$$E_b{}^2(y) = \frac{E^2(\sigma)}{\eta_1 r_1 \cos \sigma} = \frac{E^2(\sigma)}{(\eta_1{}^2 r_1{}^2 - y^2)^{1/2}} \tag{8-124}$$

where $E(\sigma)$ = field pattern of feed
$\quad E_b(y)$ = amplitude of field along line b
$\quad \eta_1$ = refractive index at feed point

ADDITIONAL REFERENCES

Long-wire and Rhombic Antennas

1. C. W. Harrison, Jr., Radiation from Vee Antennas, *Proc. IRE*, **31**:362, July, 1943.
2. C. W. Harrison, Jr., Radiation Field of Long Wires with Application to Vee Antennas, *J. Appl. Phys.*, **14**:537, October, 1943.
3. E. A. Laport, "Radio Antenna Engineering," pp. 301–339, McGraw-Hill, New York, 1952.
4. K. L. Phillips, Theory of Radiation Patterns and Gains for Horizontal Rhombic Antennas Adapted to Electronic Computation, *Natl. Bur. Std. Rept.* 7200, Boulder, Colo., October, 1961.
5. R. P. Decker, The Influence of Gain and Current Attentuation on the Design of Rhombic Antennas, *IRE Trans. Antennas Propagation*, **AP-7**:188–196, April, 1959.

Modulated Structures

6. C. C. Wang and E. T. Kornhauser, Propagation on Modulated Corrugated Rods, *Brown Univ. Div. Eng. Sci. Rept.* AF 4561/12, August, 1961.
7. M. G. Andreasen, Radiation from Variable Reactance Surface-wave Antennas, *Stanford Res. Inst. Sci. Rept.* 12, March, 1961.
8. A. Ishimaru, Sinusoidally Modulated Wave Guides for Traveling Wave Antennas, *Univ. Washington Dept. Elec. Eng. Tech. Rept.* 46, September, 1960.

9. R. N. Fenton and D. J. Angelakos, "A Modulated Ferrite Traveling Wave Antenna," Institute of Engineering Research, University of California, Berkeley, Calif., July, 1960 (Astia No. AD 243 726).

Traveling Wave Antenna Applications

10. W. Rotman and N. Karas, Printed Circuit Radiators: The Sandwich Wire Antenna, *Microwave J.*, **2**(8):29, 1959.
11. W. F. Croswell, Arrays of Frequency-scanned Antennas, *Ohio State Univ. Antenna Lab. Rept.* 667-44, July, 1958.
12. "An Electromechanically Scannable Trough Waveguide Antenna," Final Report on Contract No. AF 19(604)-4056 prepared by Melpar, Inc., Falls Church, Va., for Air Force Cambridge Research Center, Bedford, Mass., October, 1959.
13. A. Ishimaru and C. Hsieh, Frequency Scanning of Slow Wave Antennas, *Univ. Washington Dept. Elec. Eng. Tech. Rept.* 57, August, 1961.
14. R. A. Sigelmann and D. K. Reynolds, Traveling Wave Antenna with Broad Bandwidth, *Univ. Washington Dept. Elec. Eng. Tech. Rept.* 42, June, 1960.
15. D. J. Sletten, R. B. Mack, W. G. Mavroides, and H. M. Johanson, Corrective Line Sources for Paraboloids, *Air Force Cambridge Res. Center Rept.* AFCRC-TR-55-122, Bedford, Mass., December, 1955.
16. Research Directed toward Theoretical and Experimental Investigation of Corrected Line Source Feeds, Report on Contract AF 19(604)-5718 by Wiley Electronics, Phoenix, Ariz., for Air Force Cambridge Research Center, Bedford, Mass., February, 1960 (Astia No. AD 23 647).
17. A. F. Kay, A Line Source Feed for a Spherical Reflector, *Tech. Res. Group Rept.* AFCRL 529, Somerville, Mass., May, 1961.
18. J. L. Yen, Coupled Surface Waves and Broadside Arrays of Endfire Antennas, *IRE Trans. Antennas Propagation*, **AP-9**:296–304, May, 1961.
19. E. M. T. Jones and J. K. Shimizu, A Wide-band Transverse-slot Flush-mounted Array, *IRE Trans. Antennas Propagation*, **AP-8**:401–407, July, 1960.
20. E. D. Sharp and E. M. T. Jones, An Antenna Array of Longitudinally-slotted Dielectric-loaded Waveguides, *Stanford Res. Inst. Sci. Rept.* 6, Menlo Park, Calif., March, 1961.
21. J. Aasted and R. C. Honey, A Dielectric Loaded Leaky-wave Antenna, *Stanford Res. Inst. Sci. Rept.* 8, Menlo Park, Calif., March, 1961.
22. W. J. Getsinger, Leaky-wave Antennas Using Periodically Spaced Small Apertures, *Stanford Res. Inst. Sci. Rept.* 10, Menlo Park, Calif., March, 1961.
23. H. W. Ehrenspeck, The Backfire Antenna, A New Type of Directional Line Source, *Air Force Cambridge Res. Lab. Rept.* AFCRL 722, Bedford, Mass., August, 1961.

Spiral and Log-periodic Antennas

24. W. L. Curtis, Spiral Antennas, *IRE Trans. Antennas Propagation*, **AP-8**:298–306, 1960.
25. J. R. Copeland, Radiation from the Balanced Conical Equiangular Spiral Antenna, *Ohio State Univ. Antenna Lab. Rept.* 903-12, Columbus, September, 1960.

26. W. C. Y. Lee, "Analysis of Non-planar Equiangular Spiral Antenna," M.S. Thesis, Department of Electrical Engineering, The Ohio State University, Columbus, 1960.

27. R. Sussman, The Equiangular Plane Spiral Antenna, *Univ. Calif. (Berkeley) Electron. Res. Lab. Rept.* AFOSR 2265, September, 1961.

28. R. H. DuHamel, Log-periodic Antennas and Circuits, Formal Document FR 63-14-74, Hughes Aircraft Company, Fullerton, Calif., April, 1963.

29. R. Mittra and K. E. Jones, Theoretical Brillouin (k-β) Diagram for Monopole and Dipole Arrays and Their Application to Log-periodic Antennas, *Univ. Illinois Antenna Lab. Tech. Rept.* 70, Urbana, April, 1963.

PROBLEMS

8-1. Referring to Fig. 8-3, find optimum heights (*a*) for a horizontally polarized line source and (*b*) for a vertically polarized line source. Assume that the ground is a complex dielectric $\epsilon_1 - j\epsilon_1'$.

8-2. A line source of length L has maximum radiation at angle θ_m. Two such identical sources make up a V antenna with angle ϕ between the elements where ϕ is less than $2\theta_m$. Find the angle that the resultant beam is tilted with respect to the plane of the V.

8-3. Show that Eq. (8-44) reduces to Eq. (6-67) as $b \to \infty$.

8-4. Derive Eqs. (8-22) and (8-25).

8-5. Verify Eq. (8-69).

8-6. Verify Eq. (8-84).

8-7. Verify Eq. (8-113).

8-8. Derive Eq. (8-114).

8-9. Show that Eq. (8-116) can be expressed in differential form as

$$\eta r = K[1 + (1/r^2)(dr/d\phi)^2]^{1/2}$$

APPENDIX

The function $(\sin X)/X$ gives the far-field pattern of a line source of length L and having constant amplitude and linear phase (see Sec. 2-7). The quantity X is given by

$$X = \frac{kL}{2}\left(\frac{c}{v} - \cos\theta\right)$$
$$= \frac{\pi L}{\lambda}\left(\frac{c}{v} - \cos\theta\right)$$

where c is the velocity of light in free space and v is the phase velocity of the wave along the source. The linear phase ψ along such a source is related to c/v by

$$\psi = -360\,\frac{c}{v}\,l \qquad \text{deg}$$

where l is distance along the source and the negative sign denotes a phase delay for a forward traveling wave. The angle of maximum radiation θ_m is given by

$$\theta_m = \cos^{-1}\frac{c}{v}$$

Figures A-1 through A-12 illustrate the behavior of the $(\sin X)/X$ pattern as L and c/v are varied. Beam positions θ_m of zero (endfire), 45, and 90 deg (broadside) are shown for source lengths of 5, 10, and 20 λ. Only the far-field region zero to 90 deg is shown. For the broadside case, the pattern is symmetrical about 90 deg. Figures A-1, A-5, and A-9 show the effect of the Hansen-Woodyard increased directivity condition (see Sec. 3-2G).

Many practical sources have a $\sin\theta$ element pattern where θ is measured from the axis of propagation. Figure A-13 gives $\sin\theta$ plotted in decibels so that values from Fig. A-13 can be added to decibel values from the patterns in Figs. A-1 through A-12 to obtain $\sin\theta$ $[(\sin X)/X]$ patterns.

Because of the usefulness of the $(\sin X)/X$ function a universal curve is given in Figs. A-14 to A-28. Values of X to 105 have been used to carry sidelobes out to the 40 db level. Zeros of $(\sin X)/X$ occur for $X = n\pi$, where $n = 1, 2, 3, \ldots$ The envelope of the sidelobes is given by $1/X$. The decibel scale has been adjusted so that 60 db corresponds to 6 in. on the $|(\sin X)/X|$ scale so that one may overlay conventional 10-divisions-per-inch graph paper.

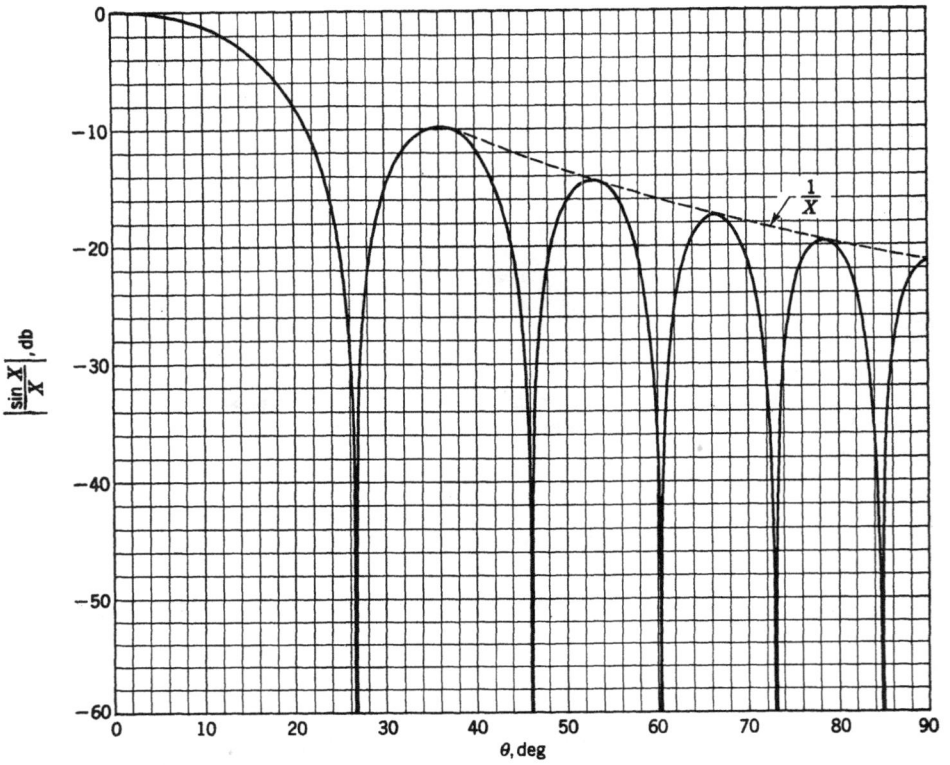

Figure A-1. $|(\sin X)/X|$ pattern for a 5 λ source with Hansen-Woodyard increased directivity ($c/v = 1.094$).

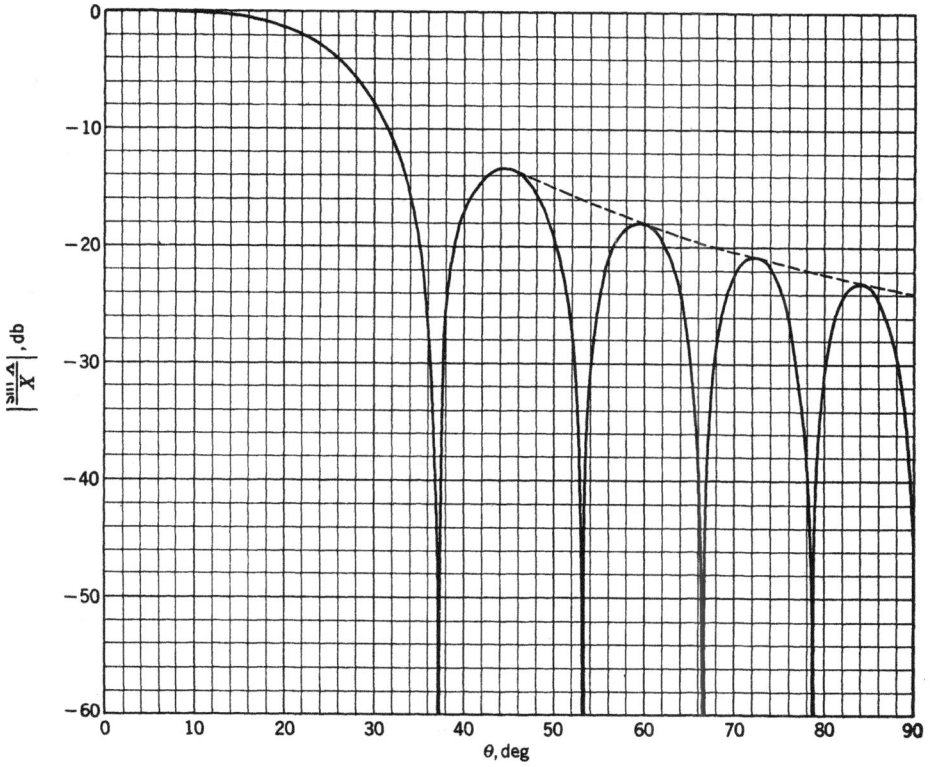

Figure A-2. $|(\sin X)/X|$ pattern for a 5 λ source with ordinary endfire operation ($c/v = 1$).

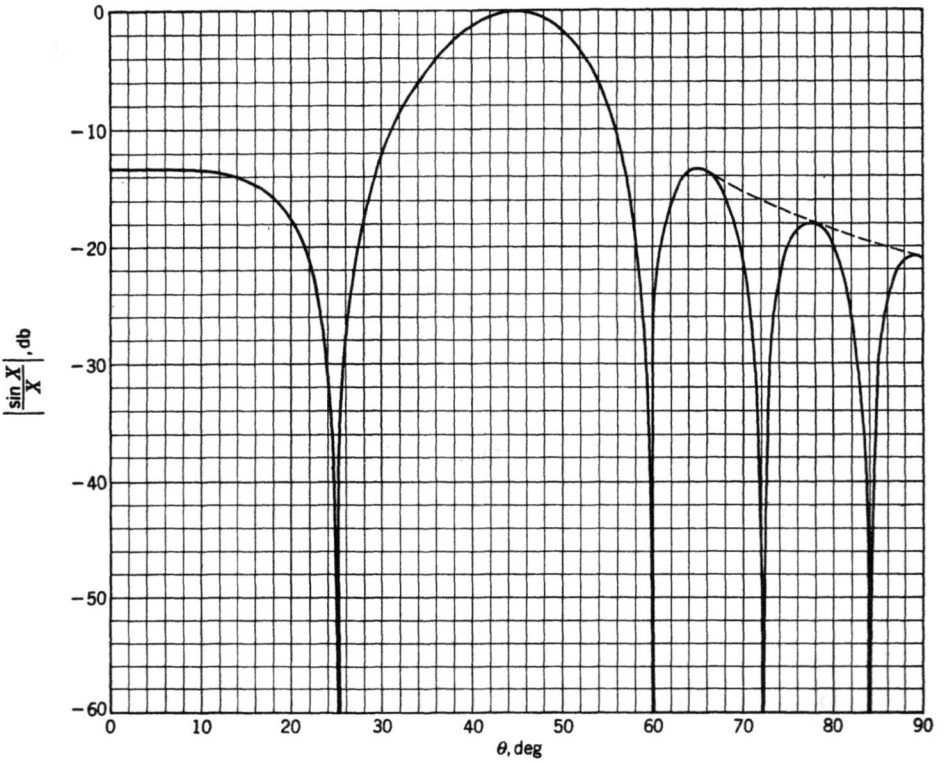

Figure A-3. $|(\sin X)/X|$ pattern for a 5 λ source with beam at 45 deg ($c/v = 0.707$).

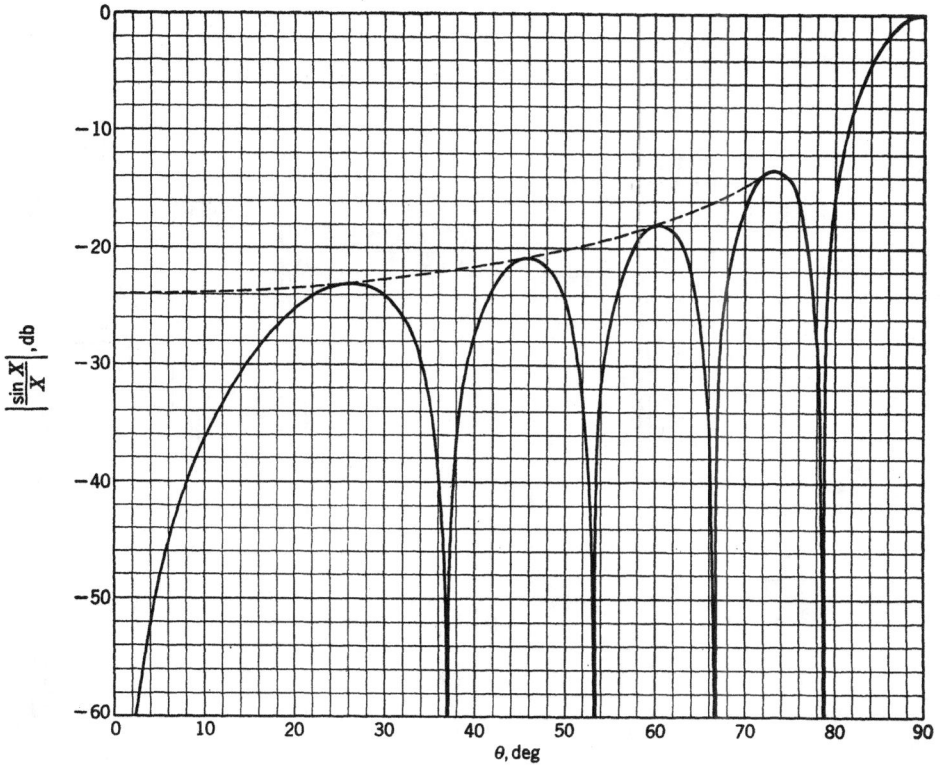

Figure A-4. $|(\sin X)/X|$ pattern for a 5 λ source with broadside operation ($c/v = 0$).

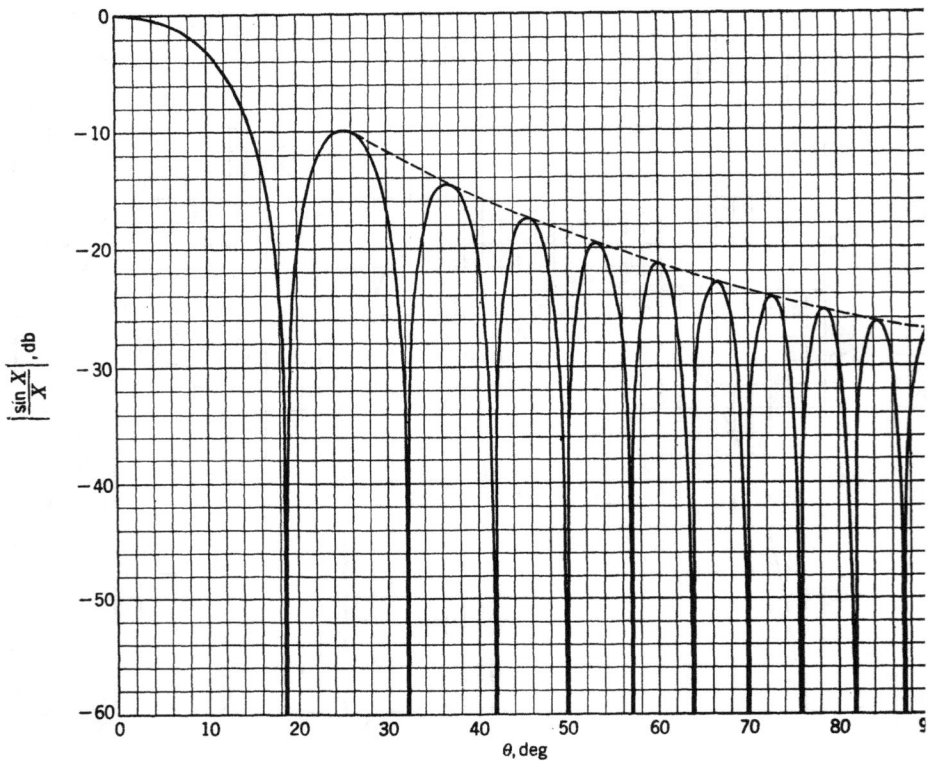

Figure A-5. $|(\sin X)/X|$ pattern for a 10 λ source with Hansen-Woodyard increased directivity $(c/v = 1.047)$.

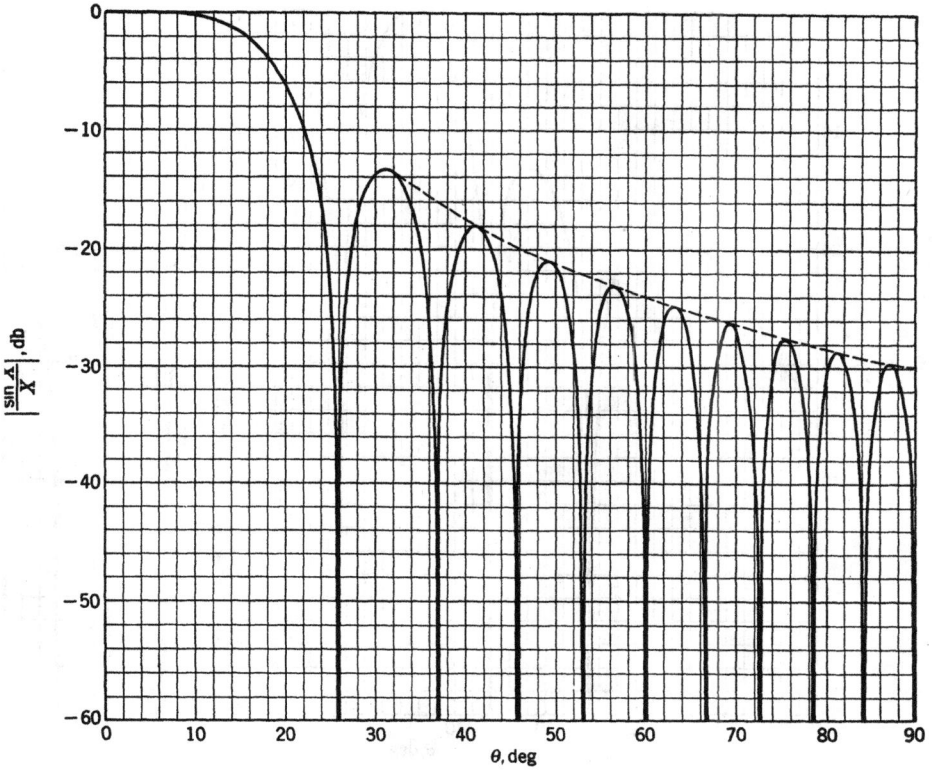

Figure A-6. $|(\sin X)/X|$ pattern for a 10 λ source with ordinary endfire operation ($c/v = 1$).

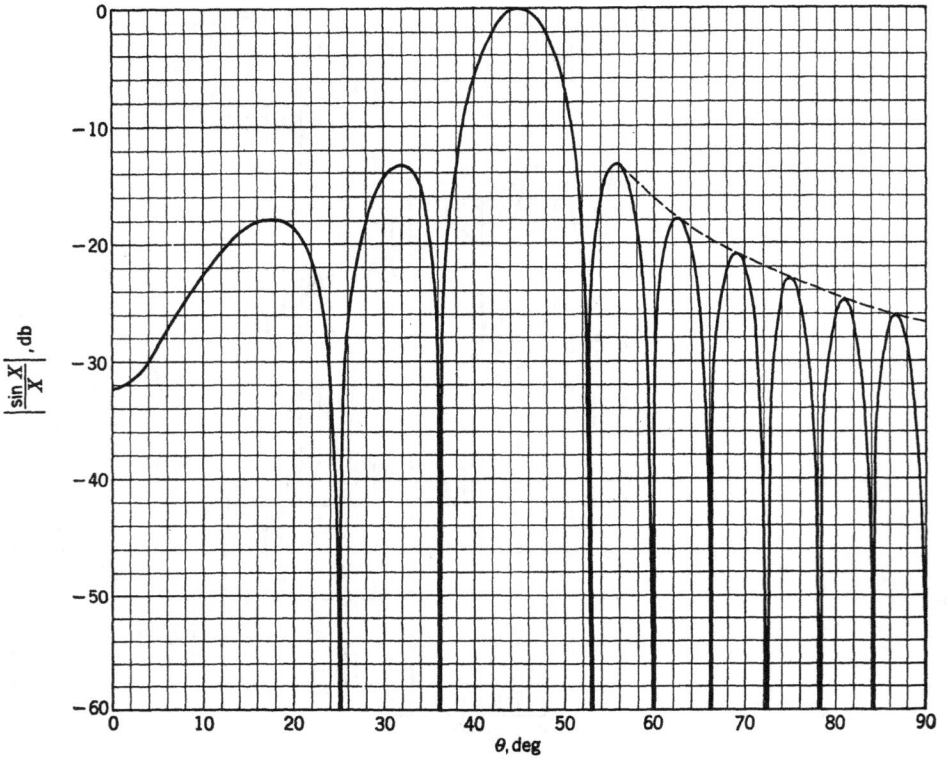

Figure A-7. $|(\sin X)/X|$ pattern for a 10 λ source with beam at 45 deg $(c/v = 0.707)$.

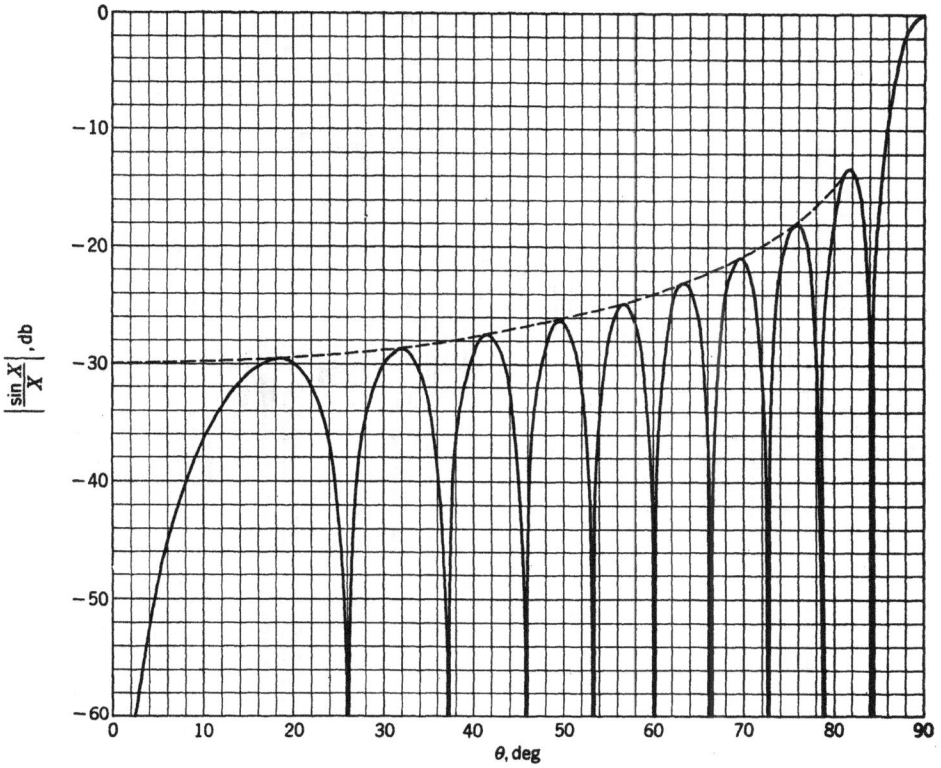

Figure A-8. $|(\sin X)/X|$ pattern for a 10 λ source with broadside operation ($c/v = 0$).

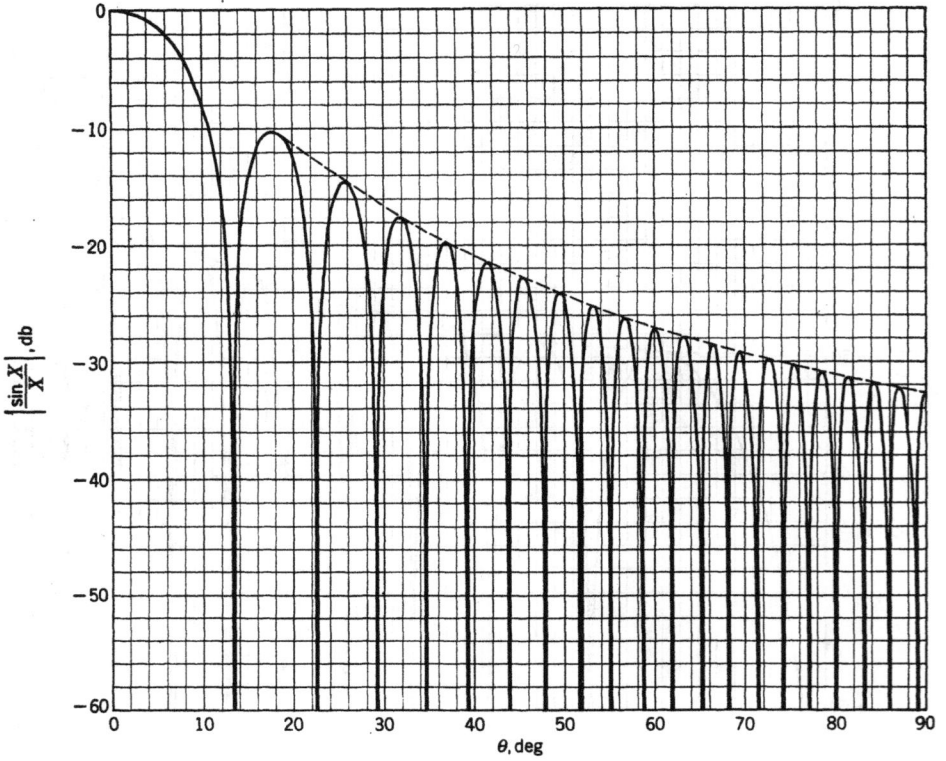

Figure A-9. $|(\sin X)/X|$ pattern for a 20 λ source with Hansen-Woodyard increased directivity ($c/v = 1.023$).

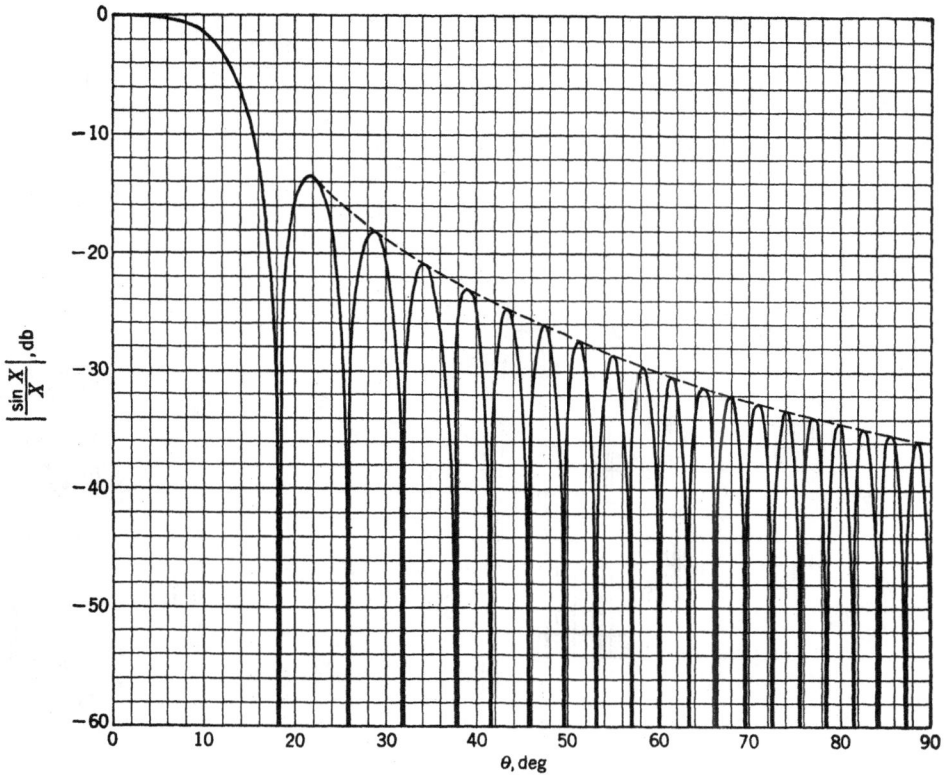

Figure A-10. |(sin X)/X| pattern for a 20 λ source with ordinary endfire operation ($c/v = 1$).

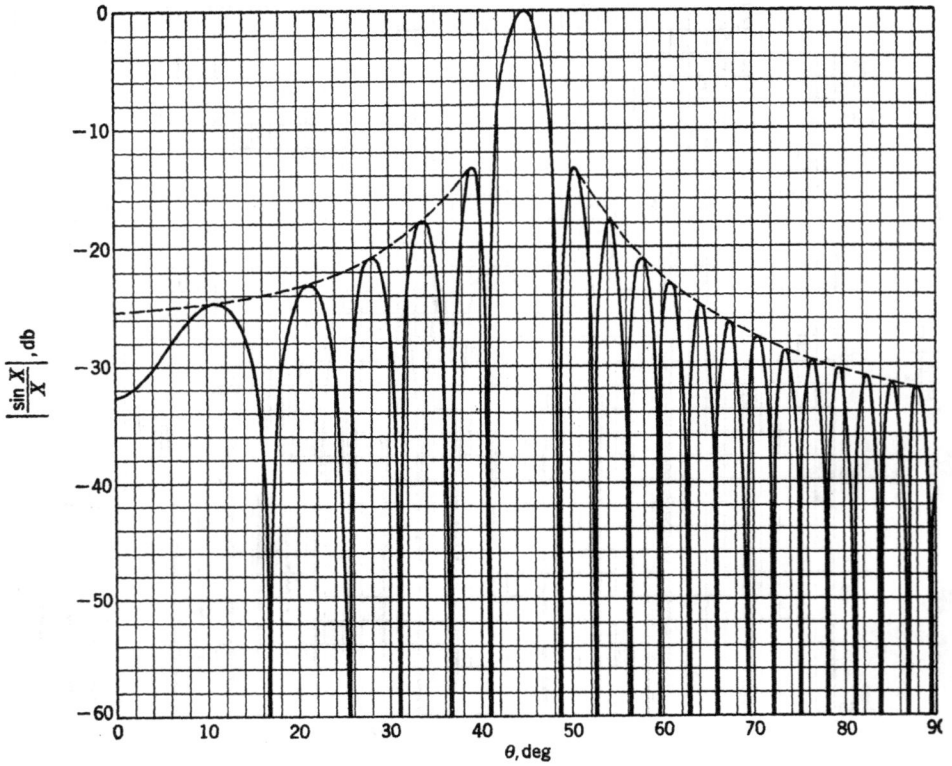

Figure A-11. $|(\sin X)/X|$ pattern for a 20λ source with beam at 45 deg ($c/v = 0.707$).

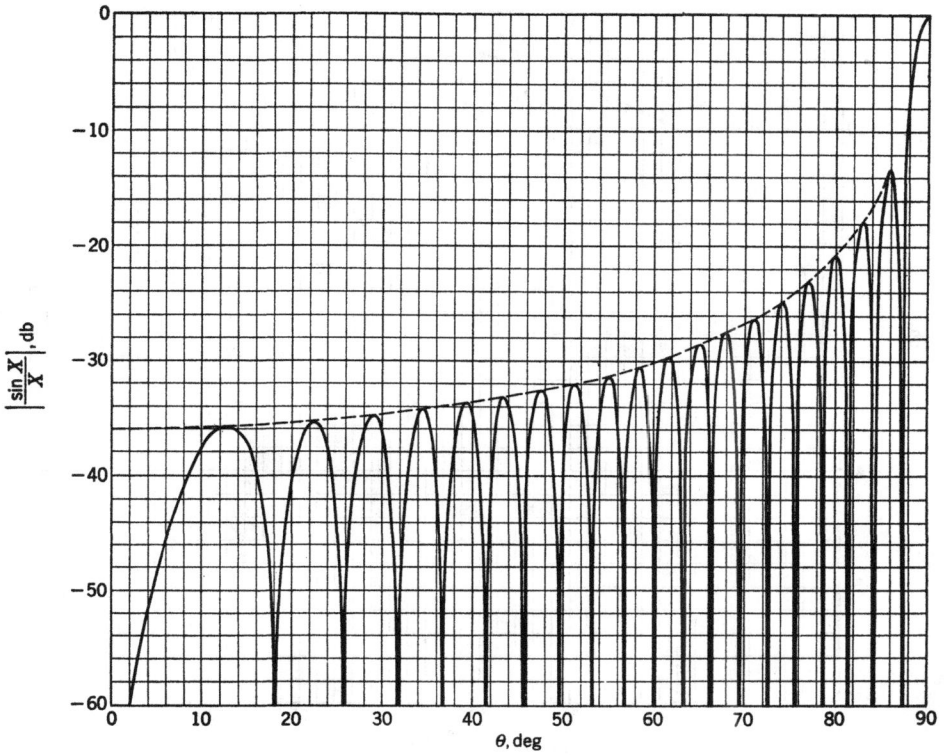

Figure A-12. $|(\sin X)/X|$ pattern for a 20 λ source with broadside operation ($c/v = 0$).

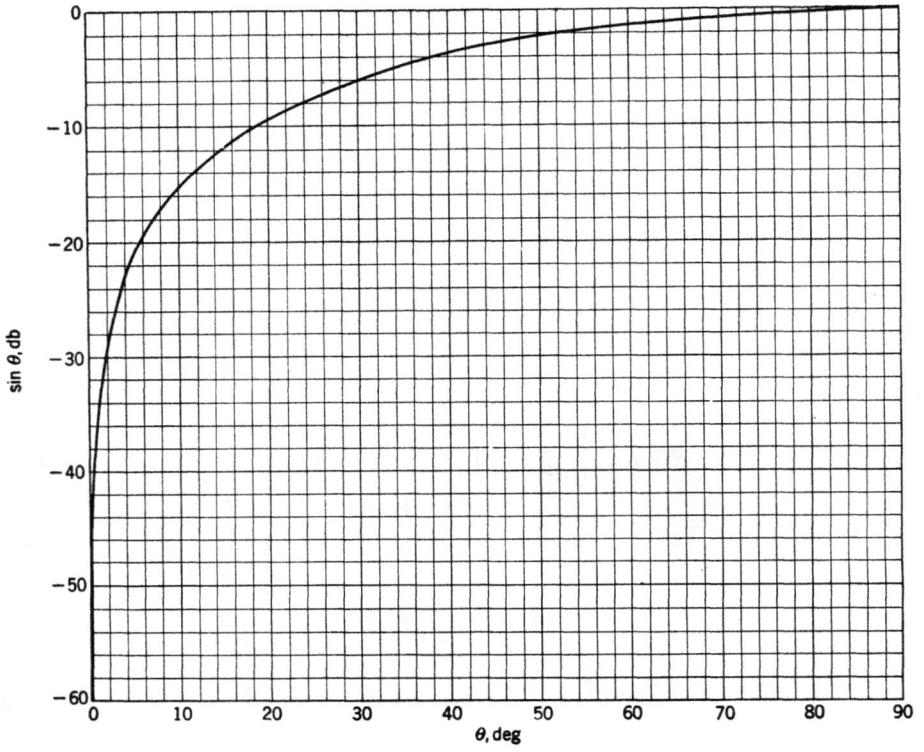

Figure A-13.   sin θ pattern.

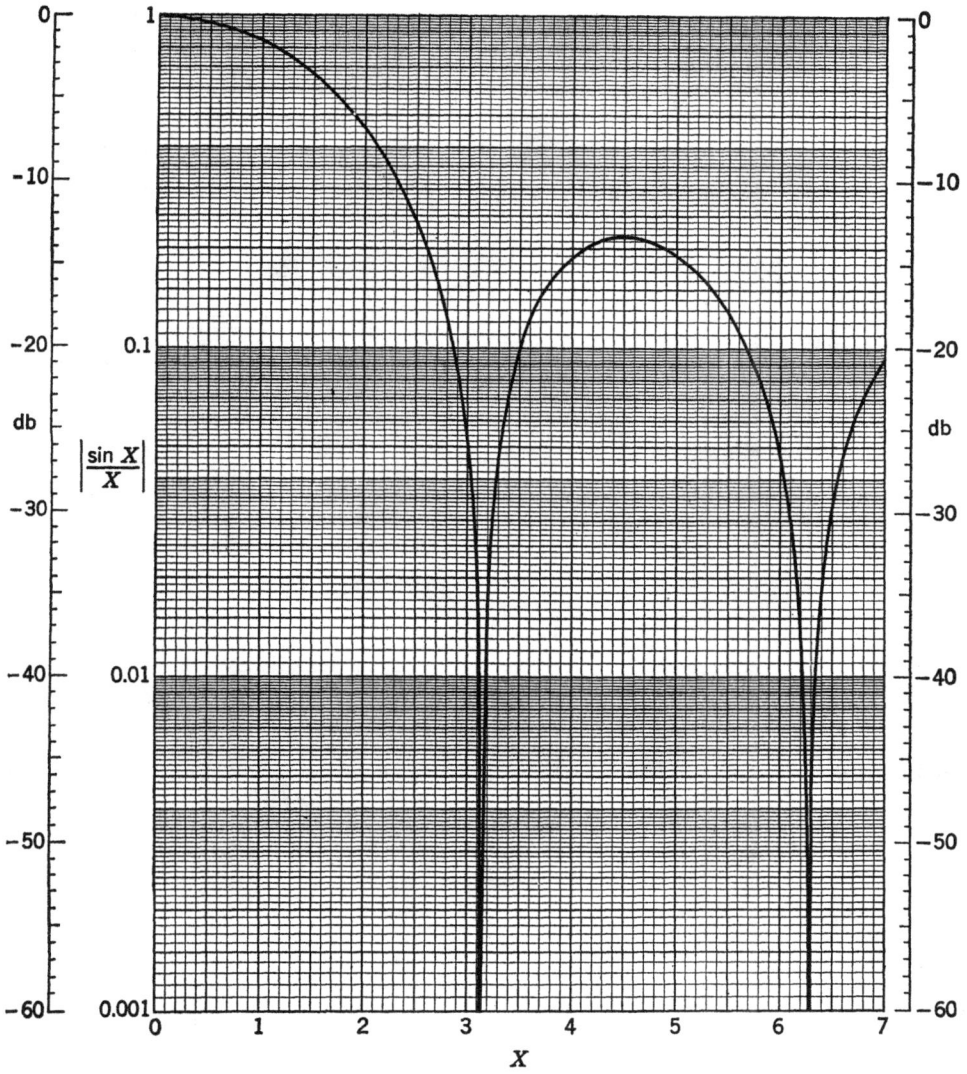

Figure A-14

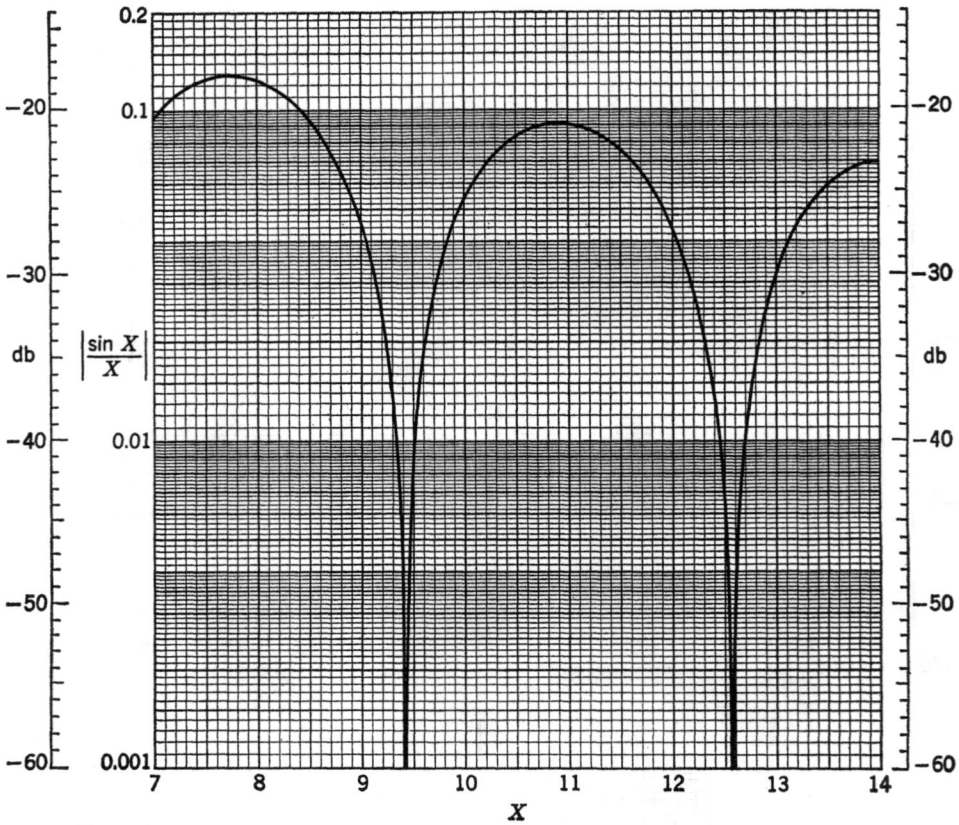

Figure A-15

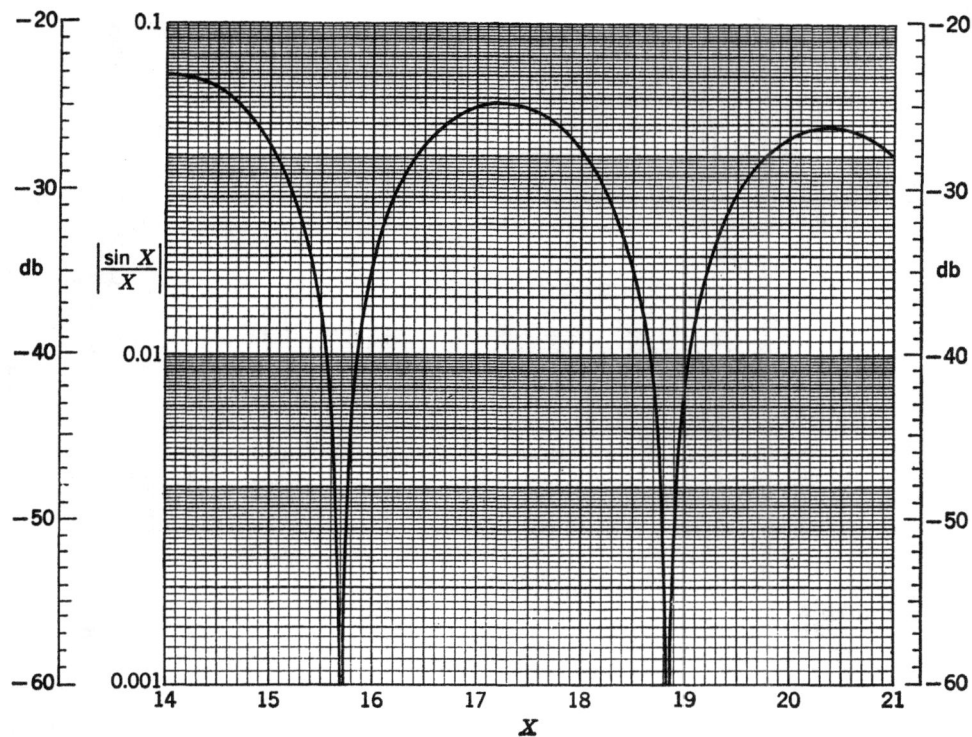

Figure A-16

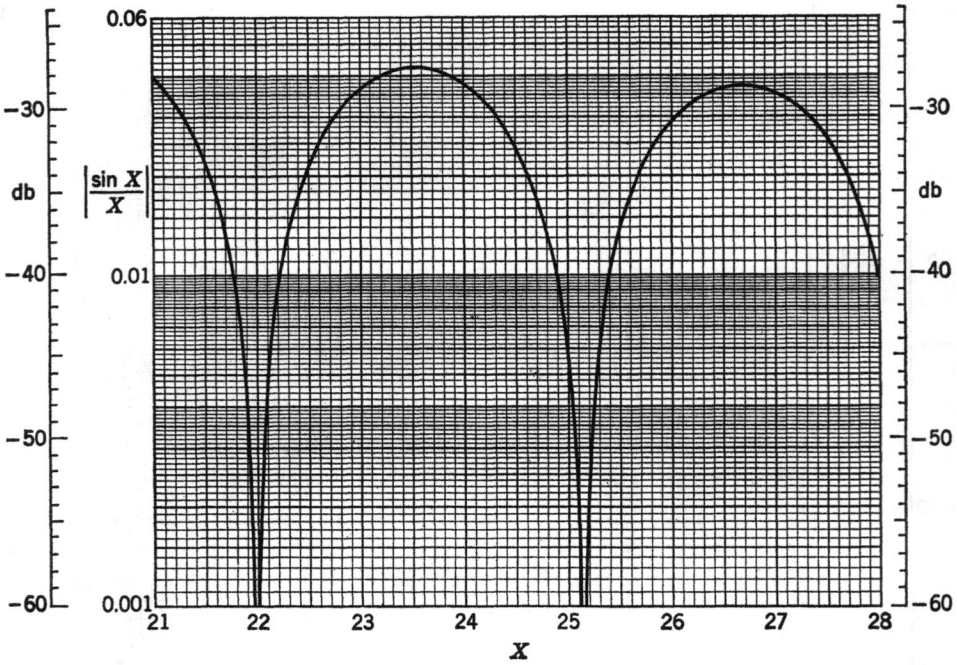

Figure A-17

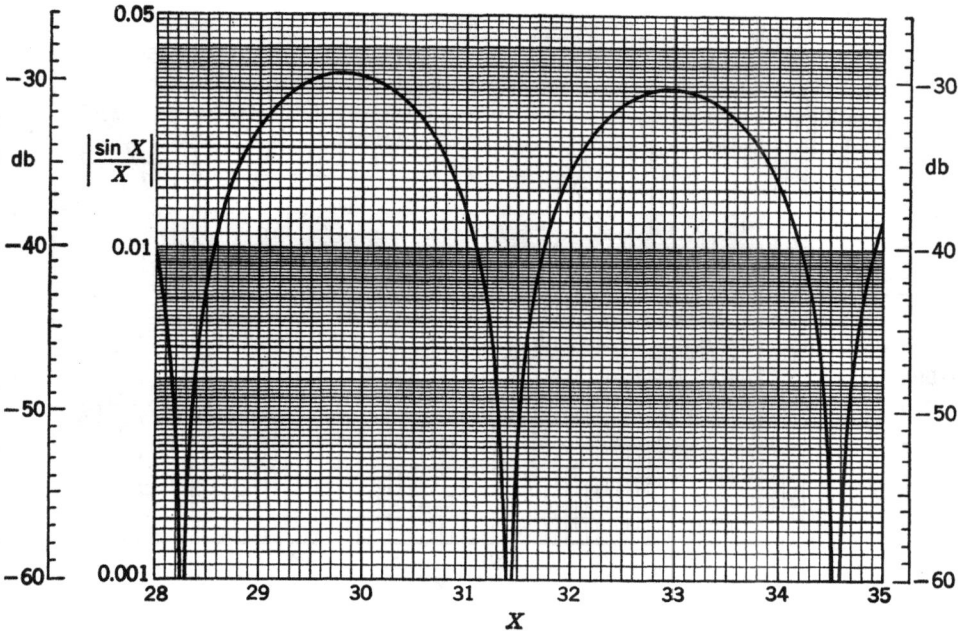

Figure A-18

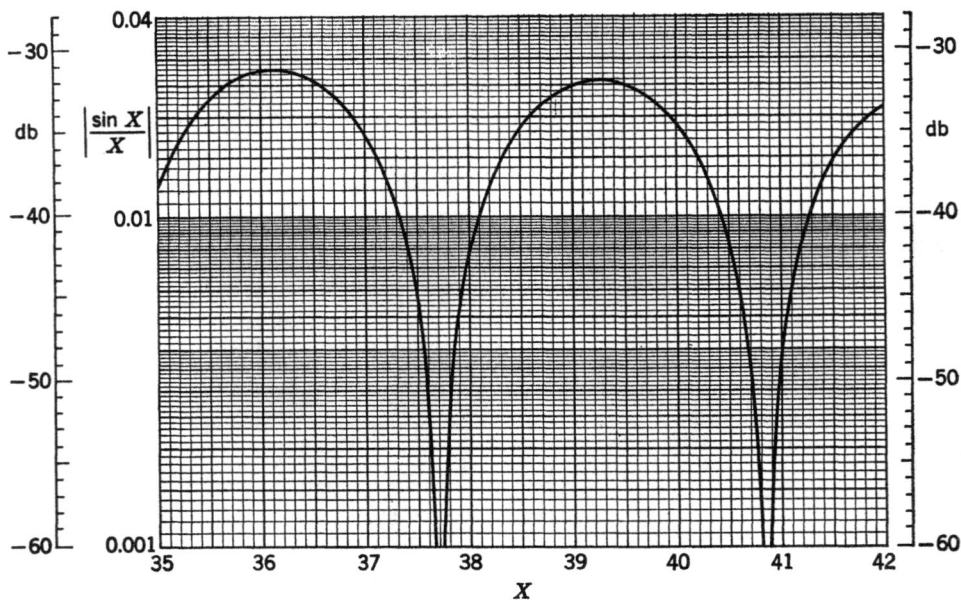

Figure A-19

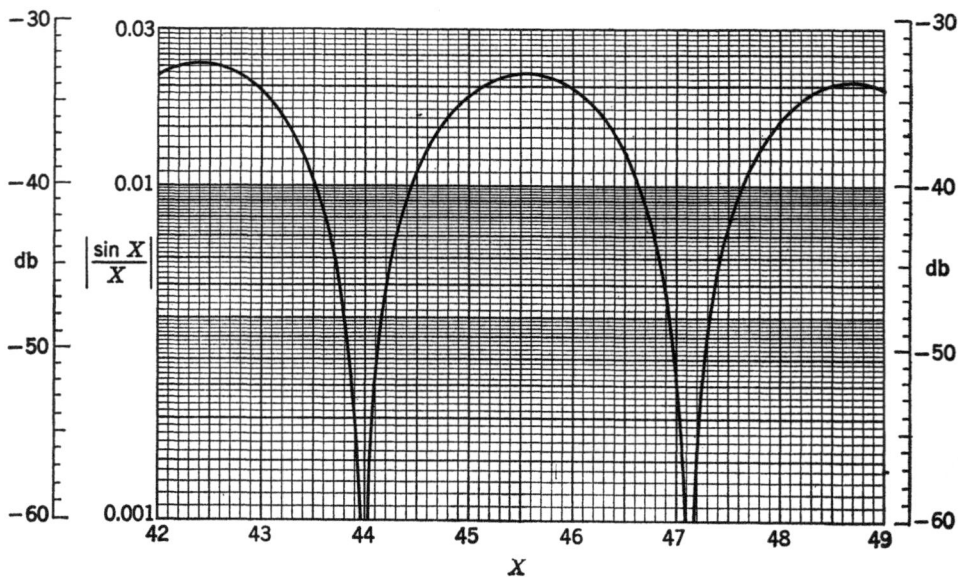

Figure A-20

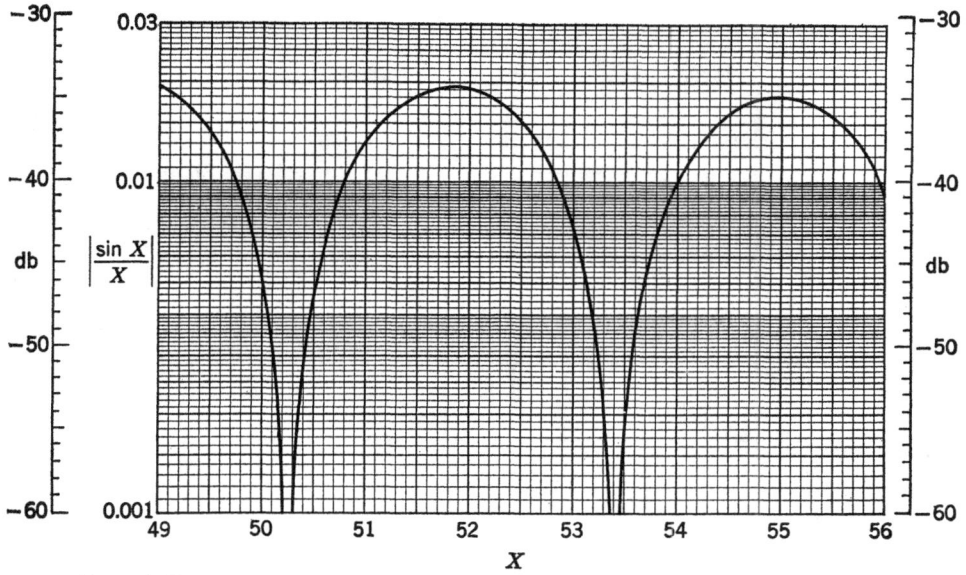

$$\left| \frac{\sin X}{X} \right|$$

Figure A-21

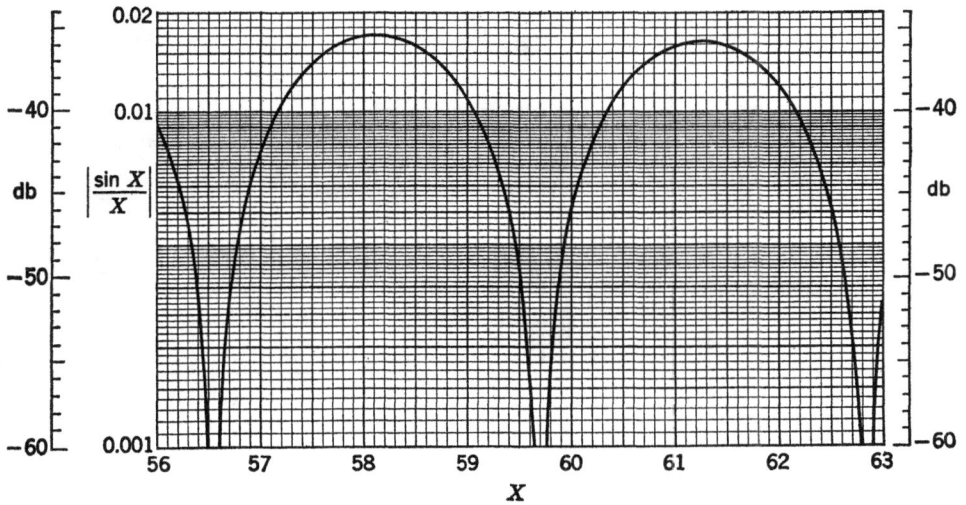

$$\left| \frac{\sin X}{X} \right|$$

Figure A-22

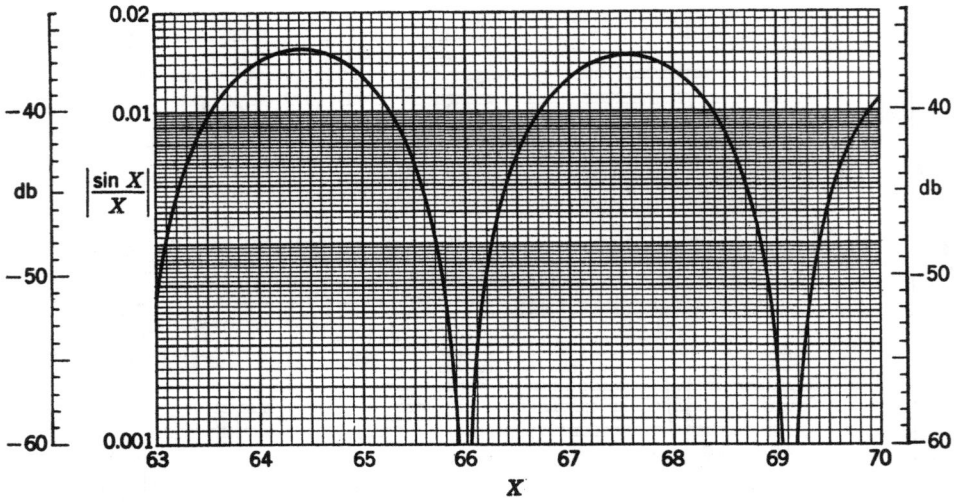

Figure A-23

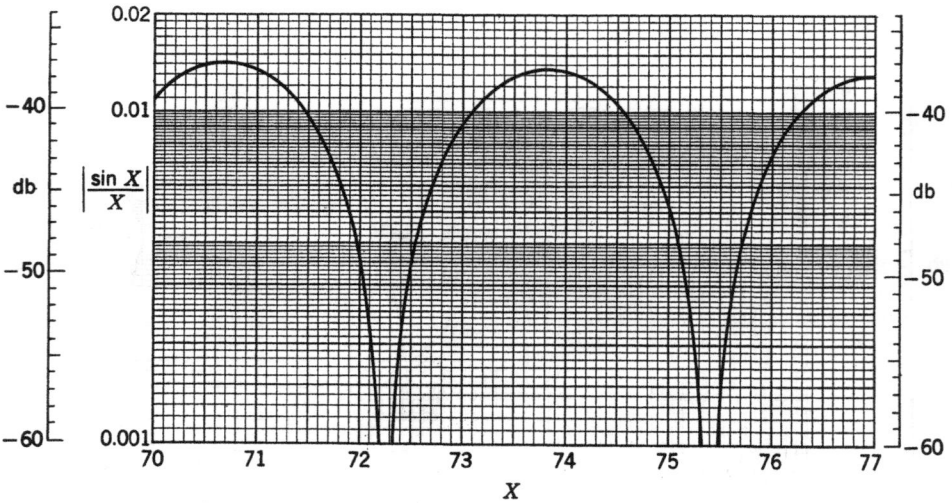

Figure A-24

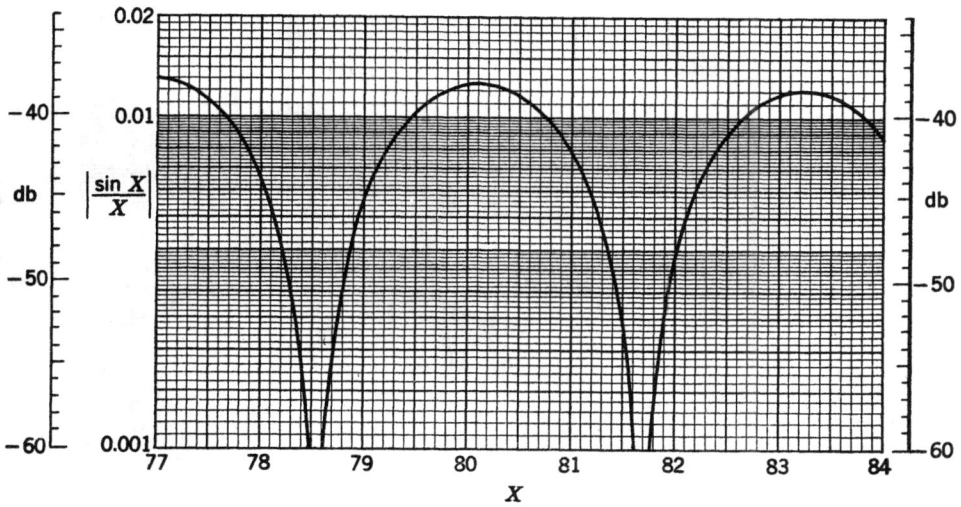

Figure A-25

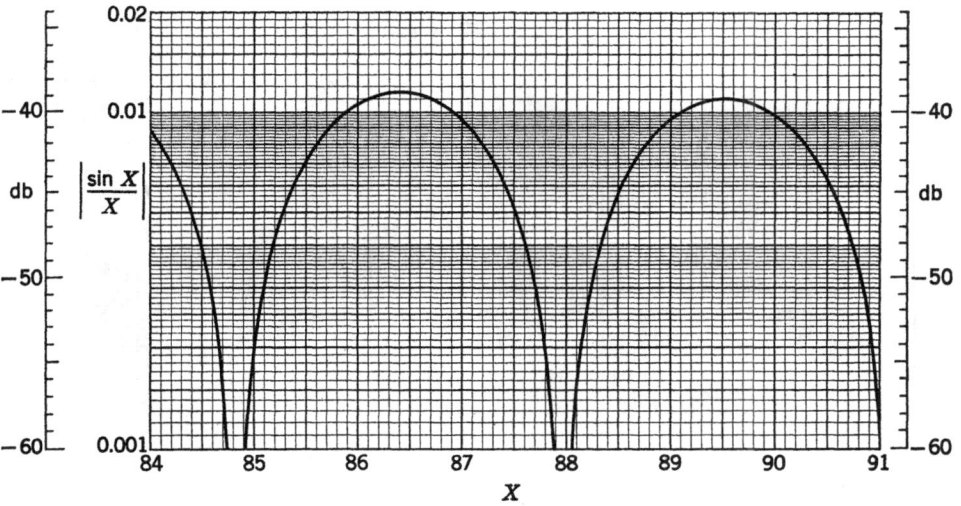

Figure A-26

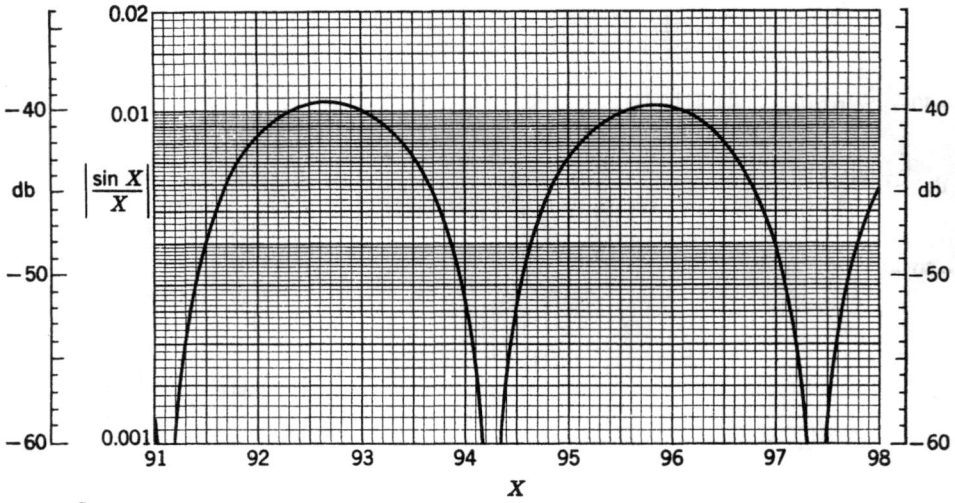

Figure A-27

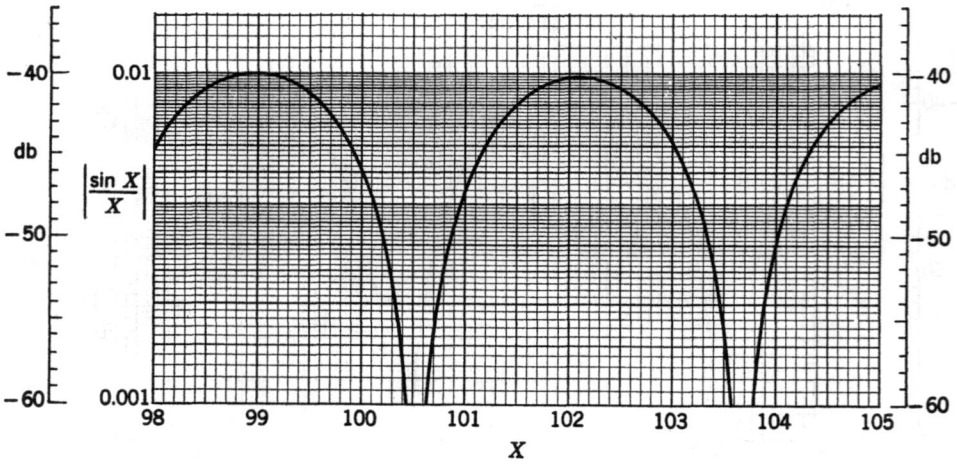

Figure A-28

NAME INDEX

415

SUBJECT INDEX

419

Attenuation constant α, for slotted
    cylinders, 180, 195, 202
  for slotted rectangular guides, 189, 199
  for surface wave structure with
    source, 303
  for trough guide, 359
  due to wall loss, 231
  for waveguide with transverse slots,
    355
Autocorrelation function, 85
Axial mode, helical antenna, 331
Axial ratio, helical antenna, 332
Azimuthal surface wave, 363

Babinet's principle, 74–76
Backfire antenna, 341
Backfire mode, helical antenna, 333
Backward wave, 264, 333
Backward wave antennas, 341–349
Beam scanning, 128, 129, 350
  with curved source, 367
Beam shaping with curved, slow wave
    antenna, 362
Beam tilt, 73
  elimination of, 330
  of flared slot, 337
  of tapered depth slot, 328
Bed of metal posts on ground plane, 266
  circular posts, 267
  square posts, 267
Beverage antenna, 315
Bifilar helix, 332
Brillouin diagram, 343, 344 (see k-β
    diagram)
Broadband antennas (see Frequency
    independent antennas)
Broadside source, 17
Butterworth pattern, 89–92
  by Laplace transform synthesis, 89
  as low-pass filter, 89
  as maximally flat (Butterworth)
    response, 90

c/v (see Velocity ratio c/v)
Capacitive grid (see TE capacitive grid
    structure)

Characteristic equation, for dielectric-
    coated metal rod, dipole mode,
    223
  HE_{nm} mode, 224
  TE mode, 223
  for dielectric cylinder, HE_{nm} mode,
    215
  symmetrical TE mode, 216
  symmetrical TM mode, 216
  for ferrite sheet on ground plane, 274
  for H guide, $E_z = 0$ modes, 229
  $H_z = 0$ modes, 227
  TE modes, 229
  for TE dielectric sheet, 237
  for TM corrugated surface, 259
  for TM dielectric sheet, 255
  layered, 257
Chebyshev pattern, 92–94
Chebyshev polynomials, 108
Chebyshev (equal-ripple) response, 92
Circular dielectric sheet antennas, 337,
    385
Circular polarization, with dielectric
    sheets, 335
  with helix, 331–333
Circular source, 58
  far-field region of, 59–61
  Fresnel and near-field region of, 61–63
  quantized, 63
  ring source, 61, 78
  semicircular source, 78
    angle of maximum radiation, 78
  synthesis of, 130–139
Circular source method of Fresnel region
    synthesis, 142–144
Classical method of analysis of traveling
    wave structure, 162
Coaxial dielectric cylinder (see Dielectric
    coaxial waveguide)
Collecting aperture, 6
Collinear sources, 371
Complementary structures, 74, 75
Complex propagation constant, 88
Conical spiral, 346
Continuity equation, 21, 22
Continuous modes, 299
Continuous source, definition of, 1, 15
Convolution integral, 83

About the Author

Dr. Carlton H. Walter, author of *Traveling Wave Antennas,* was associated with the ElectroScience Laboratory (formerly Antenna Laboratory) and the Electrical Engineering Department of The Ohio State University until his retirement in 1983. He began working at the Antenna Lab while a graduate student and obtained an M.S. in Physics in 1951 and a Ph.D. in Electrical Engineering in 1957. After graduate work, he joined the faculty of the department of Electrical Engineering and presently is Professor Emeritus. At the ElectroScience Lab, Dr. Walter held the position of Director.

After his retirement from Ohio State, Dr. Walter joined the Military Electronics and Avionics Division of TRW in San Diego where he was responsible for antenna and new technology programs.

Dr. Walter's experience includes both teaching and research. He taught numerous electromagnetic fields and circuits courses and originated a graduate course on antennas, the notes for which subsequently were organized into this book. In addition to traveling wave antennas, his research has addressed Luneberg lens antennas, broad band antennas, electrically small antennas, super directive antennas and low radar cross section antennas.

Dr. Walter was active outside The Ohio State University as a consultant and as a member of IEEE. He served as president of the Antenna and Propagation Society and was a Fellow and Life Member of IEEE.

www.ingramcontent.com/pod-product-compliance
Lightning Source LLC
Chambersburg PA
CBHW060745220326
41598CB00022B/2332